Molecular Biology of Kinetoplastid Parasites

https://doi.org/10.21775/9781910190715

Edited by

Hemanta K. Majumder

Infectious Diseases and Immunology Division
CSIR-Indian Institute of Chemical Biology
Kolkata
India

Caister Academic Press

Caister Academic Press
Norfolk, UK

www.caister.com

British Library Cataloguing-in-Publication Data
A catalogue record for this book is available from the British Library

ISBN: 978-1-910190-71-5 (paperback)
ISBN: 978-1-910190-72-2 (ebook)

Cover design adapted from images supplied by Hemanta K. Majumder.

Contents

Preface

Parasitic diseases pose an enormous threat to human health and welfare. Admirable research efforts and promising advancement in the field of research on protozoan parasites have taken place in last few decades. The diseases caused by *Leishmania* and *Trypanosoma* affect many millions of people in both tropical and subtropical regions of the world. An estimated 700,000–1 million new cases of leishmaniasis and 20,000–30,000 deaths occur annually. There are three main forms of leishmaniasis: visceral leishmaniasis (VL), cutaneous leishmaniasis (CL) and mucocutaneous leishmaniasis. *Leishmania* species are found throughout Latin America, Africa and Asia. African trypanosomiasis (sleeping sickness) is fatal if untreated, and occurs in 36 African countries, particularly in East and Central Africa, where some 50 million people are at risk of acquiring infection. *Trypanosoma cruzi*, the causative agent of Chagas' disease, is endemic in Latin America. Emergence of parasites resistant to many of the available drugs is also responsible for the depressing scenario and cause of death. So the disease is not only complex but also cosmopolitan.

Leishmania and *Trypanosoma* share common biological traits and they cause low-priority diseases as they offer few commercial incentives to the pharmaceutical companies. These kinetoplastid protozoan parasites have attracted considerable attention from the scientific community because of their unusual biology. These two organisms have special features. They are characterized by the presence of unusual mitochondrion containing a massive intercatenated network structure of DNA called kinetoplast DNA or kDNA. None of the host organisms of these parasites contain DNA which resembles this unique kDNA. Therefore, these kDNAs can be excellent targets for development of therapeutic agents.

Measures to control these diseases have not been very successful and attempts to develop effective vaccines are still far from success. Therefore, improved and rational measures for drug development are still desirable.

Recent progress in molecular biology with reference to whole genome sequencing has greatly facilitated drug design, drug delivery and immunotherapy to provide newer intervention strategies against these parasites.

When I was contacted by Hugh Griffin of Caister Academic Press to edit a book, I accepted the invitation and I felt that it is the right time to address the important subject on molecular biology of kinetoplastid parasites. The book contains 13 chapters contributed by eminent scientists working in this field.

The articles deal with the biology and biochemistry of different targets, molecular immunology in relation to immune evasion and immunotherapy, host–parasite interaction, cellular defence mechanism adopted by the parasites for survival, membrane architecture as targets, life cycle and epigenetic regulation of the parasites.

I am thankful to the scientists for their contribution in this book. Finally the book was made possible because of continuous help from my PhD students Sourav Saha and Somenath Roy Chowdhury.

Hemanta K. Majumder, PhD
Infectious Diseases and Immunology Division,
CSIR-Indian Institute of Chemical Biology,
Kolkata, India

Current Books of Interest

About the Editor

Hemanta K. Majumder is a Senior Scientist Platinum Jubilee Fellow of National Academy of Sciences (India) at CSIR-Indian Institute of Chemical Biology, Kolkata. He was the former Head of the Infectious Diseases and Immunology Division of the Institute. He served the State Council of Science and Technology, Government of West Bengal as the Working Chairman from 2004 to 2011. His main research interests include biochemistry and molecular biology of DNA topoisomerases of *Leishmania* in relation to development of therapeutics targeted to these enzymes and understanding the mechanism of programmed cell death in this unicellular protozoan parasite.

Dr Majumder is in the editorial board of many national and international journals. He received his PhD degree in biochemistry in 1975 from Calcutta University. He did his post doctoral studies at Albert Einstein College of Medicine, New York, USA (1976–9), and at the University of Zurich, Switzerland (1979–81). He was a Visiting Associate Research Molecular Biologist at University of California at Berkeley (1988–9). He was also a visiting Fellow at University of Rome, Italy (2008), University of Aarhus, Denmark (2010), Robert Koch Institute at Berlin (2013), LMU Germany (2013 and 2015) and Leiden University Medical Center, The Netherlands (2015).

Dr Majumder is a Fellow of all the National Academies in the country, e.g. Indian National Science Academy (INSA), Indian Academy of Sciences (FASc), National Academy of Sciences (FNASc) and also The World Academy of Sciences (FTWAS).

Genome-wide Profiling of Unique Domain Architectures Reveals Novel Epigenetic Regulators of *Leishmania infantum*

1

V. S. Gowri, Nimisha Mittal, Rohini Muthuswami and Rentala Madhubala*

School of Life Sciences, Jawaharlal Nehru University, New Delhi, India.

*Correspondence: madhubala@mail.jnu.ac.in and rentala@outlook.com

https://doi.org/10.21775/9781910190715.01

Abstract

Leishmania, a protozoan parasite, constitutes a major source of human mortality and morbidity. Epigenetic gene regulation has emerged as a major mechanism for gene regulation. Not much is known about the post-translational histone modifications and chromatin-modifying enzymes in *Leishmania*. The resolution of the genome of *Leishmania* has enabled us to perform the first-ever genome-wide survey of the epigenetic regulator proteins. In this chapter, the complete repertoire of epigenetic modulators, comprising 238 proteins [40 writers, 16 readers (excluding 32 ankyrin repeat proteins and 121 WD40 proteins) and 18 ATP-dependent chromatin remodellers belonging to SWI2/SNF2 family and 11 erasers] is reported. Our analysis has shown that the organism contains 30 lysine methyltransferases, of which 18 are specific to kinetoplastids. Three DOT1 methyltransferases were identified, in contrast with an earlier report of only two DOT1 homologues in *T. brucei*. Our analysis also showed that *L. infantum*, unlike humans, contains only the second type of demethylases (Jumonji type) and LSD1-type demethylase is absent. Further, the organism contains three class III (sirtuins) HDACs phylogenetically closer to the Gram-negative bacterial sirtuins. The present study provides new insights into a complete repertoire of histone-modifying enzymes that could help in better understanding of epigenetic regulation in *Leishmania*.

Introduction

The epigenome of a eukaryotic cell is regulated by chromatin remodelling, a process mediated by three distinct sets of proteins known as writers, readers and erasers. The histone-modifying enzymes catalysing modification of the N-terminus of histones are known as writers, and the enzymes catalysing the removal of the group are termed as erasers. The cell uses a plethora of modifications including acetylation, methylation, phosphorylation, ubiquitination and SUMOylation (Bannister and Kouzarides, 2011; Suganuma and Workman, 2011). Of these, histone acetylation, catalysed by histone acetyltransferases (HATs), is strongly associated with transcription activation while histone deacetylation mediated by histone deacetylases (HDACs) is associated with transcription repression (Berger, 2007; Bannister and Kouzarides, 2011). The other well-characterized modification, histone methylation catalysed by histone methyltransferases (HMTs), has been found to mediate both transcription activation as well as repression (Martin and Zhang, 2005; Berger, 2007). Collectively, HATs, HMTs, and other enzymes catalysing modification of histones are known as writers while HDACs and other enzymes that remove the modifications are known as erasers.

The third arm of chromatin remodelling is the readers that recognize the histone modification and bind to the modified histones to elicit the biological response. The ATP-dependent chromatin remodelling enzymes that reposition nucleosomes are components of the reader class of proteins (Hargreaves and Crabtree, 2011). These proteins possess specific domains, for example the bromodomain and chromodomain, that enable them to recognize and bind to modified histones.

Mutations in histone-modifying enzymes, as well as ATP-dependent chromatin remodelling enzymes, have been shown to result in diseases (Cho *et al.*, 2004; Portela and Esteller, 2010; Hargreaves and Crabtree, 2011). In addition, chromatin remodelling enzymes have also been reported to play a critical role during parasitic infections. Chromatin remodelling enzymes have been characterized from many parasites; however, characterization of these key enzymes remains to be explored in eukaryotic pathogens except in *Schistosoma manosoni*, in which an *in silico* analysis and drug target screening has been recently reported (Anderson *et al.*, 2012; Pierce *et al.*, 2012).

Leishmaniasis is a parasitic infection caused by various species of the genus *Leishmania*. There is a growing need for new drugs and novel drug targets due to increasing resistance to the currently available anti-leishmanial drugs. No information is currently available regarding the presence of epigenetic writers, readers, and erasers in this organism. Therefore, a comprehensive *in silico* dissection of all the epigenetic regulator proteins and their subcellular localization is expected to provide preliminary clues about the functioning of these critical enzymes. We believe that identifying the epigenetic regulators in *Leishmania* and understanding how they are different from the ones present in human cells is of great importance as it will enable us to design inhibitors that will specifically target the *Leishmania* enzymes.

In this chapter, we report the first comprehensive analysis of the epigenetic regulating proteins from *Leishmania infantum*. Our analysis shows that acetylation, methylation and biotinylation are the main histone modifications occurring in this organism. The reader proteins are present as stand-alone motifs. We report the presence of 18 ATP-dependent chromatin remodelling proteins that are mainly involved in DNA repair. Through our study, we have been able to show that the chromatin remodellers present in *Leishmania* differ substantially from the human counterpart. Of particular importance is the complete absence of histone demethylase LSD1, the lack of many ATP-dependent chromatin remodelling proteins, and the presence of only Jumonji-type demethylases. This is first comprehensive report on the epigenetic regulators present in *Leishmania*, and the uniqueness of the proteins present in it offer novel drug targets to combat the disease caused by this protozoan parasite.

Results and discussion

Identification of chromatin remodelling proteins in *L. infantum*

A total of 238 chromatin remodelling proteins (40 writers, 169 readers, 18 ATP-dependent chromatin remodellers and 11 erasers) were identified in *L. infantum* using Pfam HMMs as query and classified based on the closest human homologue (Tables 1.1–1.6). The analysis of the writers clearly showed that *L. infantum* contains 30 lysine methyltransferases, of which 18 are specific to kinetoplastids (Fig. 1.1a). The organism also contains three arginine methyltransferases, six lysine acetyltransferases and one lysine biotinase (Table 1.1). The analysis indicated that other modifications such as ubiquitination and SUMOylation are possibly absent in *L. infantum* as enzymes catalysing these reactions were not identified in our searches. A study of readers showed that *L. infantum* contains one chromodomain protein and six bromodomain proteins capable of recognizing lysine methylation and acetylation respectively (Tables 1.2–1.5 and Fig. 1.1b). In addition, the organism also contains 18 readers belonging to the ATP-dependent chromatin remodelling protein family (Table 1.5 and Fig. 1.1b). A striking feature of these proteins present in *L. infantum* is that they often contain only the catalytic domain with a maximum of one additional domain appended N- or C-terminus to it, unlike the multidomain enzymes found in humans. Furthermore, many of the domains present in

Table 1.1 Analysis of writer proteins present in *L. infantum*

Enzyme name	Coding Gene(s)	UniprotKB accession of human homologue	Pfam domains in human homologues	TritrypDB accession of *L. infantum* homologues	Pfam domains in *L. infantum* homologues	WolfPsort subcellular localization prediction
Lysine methyltransferases	SUV39H1	Q43463	Chromo; Pre-SET; SET*	LinJ.26.2590	SET*; Dala_Dala_lig_C	Nuclear
	SMYD5	Q6GMV2	SET	LinJ.23.0990	SET	Extracellular
	SUV420H2	Q86Y97	SET	Homologues only in tryps and not in *C. fasciculata* and *L. infantum*		
	DOT1L	Q8TEK3	DOT1	LinJ.33.1890	DOT1	Mitochondrial
				LinJ.07.0030;		Mitochondrial
				LinJ.20.0030		Mitochondrial
	SMYD2	Q9NRG4	SET; Zf-MYND	LinJ.12.0880	SET; Zf-MYND	Cytoplasmic
	ASH1L	Q9NR48	BAH; Bromodomain; PHD; SET	LinJ.36.0230	SET	Cytoplasmic
	SMYD3	Q9H7B4	SET; Zf-MYND	LinJ.23.0960	SET; Zf-MYND	Mitochondrial
				LinJ.34.1120		Mitochondrial
				LinJ.06.1040*		Mitochondrial
				LinJ.27.0160*		Mitochondrial
	SMYD3	A8MXR1	SET	LinJ.33.0520	SET	Cytoplasmic
Kinetoplastid-specific lysine methyltransferases				LinJ.08.0820	SET; Dala_Dala_lig_C, SET*	Nuclear
				LinJ.09.1480		Mitochondrial
				LinJ.12.0232		Extracellular
				LinJ.15.1280		Mitochondrial
				LinJ.18.1240		Mitochondrial
				LinJ.21.2120		Cytoplasmic
				LinJ.22.0005		Mitochondrial
				LinJ.23.0560		Mitochondrial
				LinJ.24.1760		Mitochondrial
				LinJ.24.1820		Cytoplasmic
				LinJ.25.1860*		Mitochondrial
				LinJ.27.0160		Mitochondrial
				LinJ.31.1060		Cytoplasmic
				LinJ.35.0740		Mitochondrial
				LinJ.35.2420		Nuclear
				LinJ.35.4620		Mitochondrial
				LinJ.36.4520		Mitochondrial
				LinJ.36.7140		Mitochondrial
Histone arginine methyltransferases	PRMT2	P55345	PRMT5	LinJ.12.0850	PRMT5	Cytoplasmic_Nuclear
	PRMT5	O14744	PRMT5	LinJ.21.1690	PRMT5	Mitochondrial
	PRMT7	Q9NVM4	PrmA	LinJ.06.0900	PrmA	Mitochondrial

Table 1.1 Continued

Enzyme name	Coding Gene(s)	UniprotKB accession of human homologue	Pfam domains in human homologues	TritrypDB accession of *L. infantum* homologues	Pfam domains in *L. infantum* homologues	WolfPsort subcellular localization prediction
Histone acetyltransferases	CDYL	Q9Y232	Chromo; ECH	LinJ.29.2420	ECH	Mitochondrial
	ELP3	Q9H9T3	Acetyltransf_1 Radical_SAM	LinJ.16.0250	Acetyltransf_1; Radical_SAM	Cytoplasmic;
			Radical_SAM	LinJ.23.1610	Radical_SAM	Cytoplasmic
	KAT5	Q92993	MOZ_SAS; Tudor-knot	LinJ.14.0140	MOZ_SAS; Tudor-knot	Cytoskeleton
	NAT10	Q9H0A0	DUF1726; GNAT_ acetyltr_2; Helicase_ RecD; tRNA_ bind_2	LinJ.17.1350	Helicase_ RecD; DUF1726	Mitochondrial
	MYST3	Q92794	MOZ_SAS; PHD	LinJ.28.2440	MOZ_SAS	Nuclear
Lysine biotinases	HLCS	P50747	BPL_C; BPL_ LplA_LipB	LinJ.31.1070	BPL_LplA_ LipB	Cytoplasmic
Lysine ribosylases	PARP1	P09874	BRCT; PADR1; PARP; PARP_ reg; WGR; zf-PARP	Homologues only in tryps and *C. fasciculata* and not in *L. infantum*		

L. infantum proteins with a specific pfam domain architecture unlike their *Leishmania* homologs within the same enzyme class are indicated with an asterisk.

Table 1.2 Analysis of reader proteins present in *L. infantum*

Histone modifications	Recognition modules	*L. infantum* genes	CDD domain organization	Subcellular localization
Lysine methylation	PHD	LinJ.27.2580	RING, PHD	Nuclear
		LinJ.25.2130	SNF2_N, PHD	Nuclear
		LinJ.25.2340	PHD	Cytoplasmic
		LinJ.24.1260	PHD, NAT_SF	Mitochondrial
	Chromo	LinJ.14.0160	Chromo	Cytoplasmic
	Tudor	LinJ.32.1000	SNc, SNc, SNc, SNc, Tudor, Snc	Cytoplasmic
	Zf-CW	LinJ.33.1160	Zf-CW	Mitochondrial
		LinJ.14.0360	Bromo, zf-CW	Nuclear
	PWWP	LinJ.33.2950	PWWP, Med26, Zinc_ribbon	Cytoplasmic
		LinJ.33.2960	PWWP, Zf-ribbon TFII	Nuclear
		LinJ.22.0460	PWWP, UPF0066	Mitochondrial
Lysine acetylation	Bromo	LinJ.36.3130	Bromo	Nuclear
		LinJ.36.3520	Bromo	Nuclear
		LinJ.36.7210	Bromo	Nuclear
		LinJ.14.0360	Bromo, Zf-CW	Nuclear
	Tandem Bromo	LinJ.09.1320	Bromo, Bromo	Nuclear

Table 1.3 Analysis of WD40 proteins in *L. infantum*

L. infantum genes	CDD domain organization	WolfPsort subcellular localization prediction	Number of transmembranes
LinJ.33.3360	WD40, Coatomer_WDAD	Cytoplasmic	
LinJ.35.3920	WD40, Sof1	Nuclear	
LinJ.24.1970	WD40, PFU	Cytoplasmic	
LinJ.32.0050	WD40	Cytoplasmic	
LinJ.17.0280	WD40	Mitochondrial	
LinJ.08.0420	WD40	Mitochondrial	
LinJ.29.0770	WD40	Nuclear	
LinJ.31.0210	NLE, WD40	Cytoplasmic/mitochondrial	
LinJ.24.1790	WD40	Nuclear	
LinJ.30.1460	WD40	Mitochondrial	
LinJ.21.0100	WD40	Cytoplasmic/nuclear	
LinJ.14.0580	WD40, WD40, UTP12	Cytoplasmic	
LinJ.18.0830	WD40, WD40, UTP12	Extracellular	
LinJ.28.0690	WD40	Cytoplasmic	
LinJ.35.4740	WD40, Katanin_Con80	Nuclear	
LinJ.22.1070	WD40, WD40	Extracellular	
LinJ.11.0410	BOP1NT, WD40	Mitochondrial	
LinJ.33.0300	WD40	Mitochondrial	
LinJ.26.1380	WWE, WD40	Mitochondrial	
LinJ.34.2510	WD40, WD40	Mitochondrial	
LinJ.36.2630	WD40, WD40	Nuclear	
LinJ.26.1090	TROVE, DUF4062, WD40	Plasma membrane	
LinJ.36.4420	WD40, UTP21	Cytoplasmic	
LinJ.27.1140	WD40	Cytoplasmic	
LinJ.16.1650	WD40, WD40, GrpE, ABC_ATPase, Tropomyosin_1	Mitochondrial	
LinJ.30.1010	wD40, CAF1C_H4-bd	Cytoplasmic	
LinJ.25.1930	WD40	Mitochondrial	
LinJ.19.0980	WD40, WD40	Extracellular	
LinJ.23.0040	WD40, UTP15_C	Cytoplasmic	
LinJ.36.4070	WD40	Extracellular	
LinJ.21.1470	WD40	Mitochondrial	
LinJ.36.4940	WD40	Cytoplasmic	
LinJ.20.0020	WD40	Cytoplasmic	
LinJ.33.1500	Prefoldin, WD40, BBP1_C, EGF_alliinase	Cytoplasmic	
LinJ.34.0320	WD40	Mitochondrial	
LinJ.25.1660	WD40, WD40	Cytoplasmic	
LinJ.35.5240	WD40	Cytoplasmic/nuclear	
LinJ.32.0640	WD40	Mitochondrial	
LinJ.11.1140	WD40	Extracellular	
LinJ.19.0380	WD40, WD40, PDZ	Mitochondrial	

Table 1.3 Continued

L. infantum genes	CDD domain organization	WolfPsort subcellular localization prediction	Number of transmembranes
LinJ.24.1550	WD40	Cytoplasmic	
LinJ.27.1530	WD40	Mitochondrial	
LinJ.36.3810	WD40	Mitochondrial	
LinJ.33.0380	WD40	Nuclear	
LinJ.21.1360	WD40	Mitochondrial	
LinJ.21.1480	WD40	Nuclear	
LinJ.33.0780	WD40	Nuclear	
LinJ.32.1120	WD40	Cytoplasmic	
LinJ.27.2090	WD40, Tropomyosin_1, TMF_TATA_bd	Nuclear	
LinJ.30.2160	WD40	Extracellular	
LinJ.36.0630	WD40	Mitochondrial	
LinJ.19.1580	WD40	Nuclear	
LinJ.36.1060	WD40	Mitochondrial	
LinJ.35.0570	WD40, WD40	Plasma membrane	
LinJ.23.1400	BBC, DUF1899, DUF1900, WD40	Mitochondrial	
LinJ.24.0270	WD40	Mitochondrial	
LinJ.26.2340	WD40	Cytoplasmic	
LinJ.15.0260	WD40	Cytoplasmic/nuclear	
LinJ.22.1030	WD40	Mitochondrial	
LinJ.18.0920	WD40	Extracellular	
LinJ.33.1300	WD40	Cytoplasmic	
LinJ.32.1050	WD40	Plasma membrane	
LinJ.19.0330	WD40	Mitochondrial	
LinJ.24.0230	CAF1C_H4-bd, WD40	Cytoplasmic/nuclear	
LinJ.36.4450	WD40	Nuclear	
LinJ.01.0330	WD40	Cytoplasmic/nuclear	
LinJ.33.1290	WD40	Mitochondrial	
LinJ.34.4210	WD40	Plasma membrane	
LinJ.21.1220	Clathrin, Clathrin, WD40	Mitochondrial	
LinJ.27.2590	WD40, LisH	Mitochondrial	
LinJ.29.0100	WD40	Nuclear	
LinJ.33.1000	WD40, WD40	Mitochondrial	
LinJ.13.0010	WD40	Cytoplasmic	
LinJ.33.2770	WD40, WD40	Mitochondrial	
LinJ.23.0790	BING4CT, WD40	Mitochondrial	
LinJ.33.2270	WD40	Cytoplasmic	
LinJ.31.2450	WD40	Extracellular	
LinJ.33.2020	CAF1C_H4-bd, WD40	Mitochondrial	
LinJ.24.0120	WD40	Nuclear	
LinJ.27.2430	Prp19, WD40, RING	Mitochondrial	

Table 1.3 Continued

L. infantum genes	CDD domain organization	WolfPsort subcellular localization prediction	Number of transmembranes
LinJ.29.1690	WD40	Cytoplasmic	
LinJ.35.3350	WD40	Cytoplasmic	
LinJ.06.1170	WD40	Cytoplasmic	
LinJ.30.3460	WD40	Cytoplasmic	
LinJ.18.1440	WD40, TPR	Nuclear	
LinJ.08.1100	PKc_like	Mitochondrial	
LinJ.28.1880	DUF1619, WD40	Mitochondrial	
LinJ.36.2150	WD40	Nuclear	
LinJ.06.1020	WD40	Mitochondrial	
LinJ.27.0690	WD40	Extracellular	
LinJ.14.1530	WD40	Nuclear	
LinJ.32.2720	WD40, LRR_RI	Mitochondrial	
LinJ.35.2450	DUF3639, WD40	Cytoplasmic	1
LinJ.12.0060	Beach, LamG, WD40, PH-like	Cytoplasmic	
LinJ.36.1800	WD40	Plasma membrane	
LinJ.16.1720	WD40	Extracellular	
LinJ.13.1300	WD40	Cytoplasmic	
LinJ.02.0590	WD40	Cytoplasmic	
LinJ.22.0800	WD40	Cytoplasmic	
LinJ.26.1570	WD40	Mitochondrial	
LinJ.32.3970	WD40	Mitochondrial	
LinJ.33.0600	WD40	Cytoplasmic	
LinJ.28.0290	WD40	Cytoplasmic/mitochondrial	
LinJ.34.3590	WD40	Mitochondrial	
LinJ.14.1400	WD40	Cytoplasmic	
LinJ.07.0090	Rav1p_C	Plasma membrane	
LinJ.21.1910	WD40	Nuclear	
LinJ.10.0830	WD40	Cytoplasmic	
LinJ.24.2350	NLE, WD40	Nuclear	
LinJ.06.0030	WD40	Nuclear	
LinJ.28.2940	WD40	Nuclear	
LinJ.28.2970	WD40	Nuclear	
LinJ.18.0470	CenP_F_leu_Zip, WD40	Mitochondrial	
LinJ.25.0440	WD40, WD40, UTP13	Cytoplasmic	
LinJ.36.5050	WD40	Cytoplasmic	
LinJ.34.4140	WD40, Coatomer WDAD, COPI	Mitochondrial	
LinJ.34.0490	WD40	Nuclear	
LinJ.28.1210	WD40	Cytoplasmic	
LinJ.10.0090	WD40	Cytoplasmic	
LinJ.33.0060	WD40	Cytoplasmic	
LinJ.30.0420	WD40	Mitochondrial	1

WD40 proteins with at least one transmembrane helix are highlighted in yellow colour. WD40 proteins with N-terminus CAF-1 histone binding domains are highlighted in grey colour.

Table 1.4 Analysis of Ankyrin repeat proteins in *L. infantum*

L. infantum genes	CDD domain organization	WolfPsort subcellular localization prediction	Number of transmembranes
LinJ.35.2030	Filamin, zf-DHHC, AFD_class_I, ANK(21), HP, ABC_ATPase, ABC_ATPase	Cytoplasmic	
LinJ.24.0130	Dynein_heavy, C2, ANK(12), Esterase_lipase, TPR	Cytoplasmic	
LinJ.05.0640	CULLIN, LAT, ANK(7)	Nuclear	
LinJ.30.3310	ANK, ANK	Nuclear	
LinJ.03.0460	zf-DHHC, ANK, ANK	Plasma membrane	4
LinJ.34.3340	ANK	Cytoplasmic	
LinJ.35.2710	ANK	Mitochondrial	
LinJ.29.1190	ANK, ANK	Cytoplasmic	
LinJ.11.0700	ANK, ANK	Cytoplasmic	
LinJ.18.1110	ANK	Cytoskeleton	
LinJ.19.1110	ANK, ANK	Plasma membrane	
LinJ.25.0960	ANK	Cytoplasmic	
LinJ.04.0510	zf-DHHC, ANK, ANK	Plasma membrane	6
LinJ.34.4110	ANK, ANK	Cytoplasmic	1
LinJ.04.0590	ANK, ANK, ANK	Plasma membrane	
LinJ.28.2190	AFD_class_I, ANK	Mitochondrial	
LinJ.03.0790	ANK, ABC_ATPase	Nuclear	
LinJ.26.0300	ANK, HP, ABC_ATPase	Nuclear	
LinJ.35.0310	ANK	Mitochondrial	
LinJ.28.0720	ANK	Cytoplasmic/Nuclear	
LinJ.33.0150	ANK	Cytoplasmic	
LinJ.29.0590	ANK	Plasma membrane	4
LinJ.06.0730	C2, ANK	Extracellular	
LinJ.32.1240	ANK	Cytoplasmic	
LinJ.35.5310	Dynein_heavy, ANK	Mitochondrial	
LinJ.05.0260	ANK	Cytoplasmic	
LinJ.09.0340	ANK	Mitochondrial	
LinJ.16.0270	ANK, Esterase_lipase	Plasma membrane	1
LinJ.09.1260	ANK	Extracellular	
LinJ.25.0820	LAT, ANK	Plasma membrane	5
LinJ.29.1570	ANK	Cytoplasmic	
LinJ.36.0350	ANK	Mitochondrial	

Ankyrin repeat proteins with at least one transmembrane helix are highlighted in yellow.

chromatin remodelling proteins in humans are present as stand-alone proteins in *L. infantum*. For example, in humans BRG1, an ATP-dependent chromatin remodelling protein contains bromodomain while, in *L. infantum*, bromodomain is present as an independent protein.

Finally, analysis of the erasers shows that *L. infantum* possesses three lysine demethylases, seven lysine deacetylases, and one arginine demethylase (Table 1.6 and Fig. 1.1c). It is pertinent to note that all the three lysine demethylases identified in our study belonged to the Jumonji-type demethylases,

Table 1.5 ATP-dependent chromatin remodelling protein present in *L. infantum*

L. infantum genes	PFAM/CDD domain assignment	Region of assignment in query	SNF2 subfamily names	WolfPsort subcellular localization prediction
LinJ.36.5420	SNF2_N, Helicase_C	267–556, 903–982	Swr1-like	Nuclear
LinJ.33.1800	SNF2_N, Helicase_C, HAND[a], SLIDE	169–451, 508–588, 850–910[a], 978–1084	ISWI/SNF2-like	Mitochondrial
LinJ.28.1950	SNF2_N, Zf-RING_2, Helicase_C	393–868, 959–999, 1097–1179	RAD 5/16-like	Nuclear
LinJ.28.0810	SNF2_N, zf-C3HC4_2, Helicase_C	8–414, 425–471, 540–619	RAD 5/16-like	Nuclear
LinJ.25.0770	Zf-PARP, SNF2_N, SNF2_N, Helicase_C	33–112, 150–298, 552–775, 789–836, 957–1036	RAD 5/16-like	Nuclear
LinJ.29.0210	SNF2_N, Helicase_C	39–439, 524–603	RAD54-like	Plasma
LinJ.24.0770	SNF2_N, Helicase_C	438–732, 827–907	RAD54-like	Nuclear
LinJ.24.1070	SNF2_N, Helicase_C	253–644, 855–932	RAD54-like	Nuclear
LinJ.14.0040	Tet_JBP, SNF2_N, Helicase_C	256–412, 470–771, 851–930	Bifunctional helicase and Thymine dioxygenase	Cytoplasmic
LinJ.14.0900	SNF2_N, Helicase_C	482–812, 895–974	ERCC6/SS01653-like	Nuclear
LinJ.01.0240	SNF2_N, Helicase_C	1–304, 368–446	ERCC6/SS01653-like	Extracellular
LinJ.24.0620	SNF2_N, Helicase_C	189–441, 506–583	SMARCAL1	Nuclear
LinJ.11.0050	SNF2_N, Helicase_C, HNH	178–421, 483–558, 866–915	ZRANB3/AH2*	Plasma
LinJ.26.1540	SNF2_N, SNF2_N, Zf-RING_2, Helicase_C	704–973, 1204–1478, 1565–1613, 1709–1790	Kinetoplastid-specific SNF2-like	Nuclear
LinJ.25.2130	SNF2_N, Helicase_C, PHD[a]	301–567, 638–718,875–920[a]	Kinetoplastid-specific SNF2-like	Nuclear
LinJ.31.2390	SNF2_N,TMHMM, SNF2_N Helicase_C	84–115, 116–138, 139–429, 709–787	*Leishmania*-specific SNF2-like	Nuclear
LinJ.30.1370	SNF2_N, Helicase_C	271–575, 780–854	*Leishmania*-specific SNF2-like	Mitochondrial
LinJ.23.1100	SNF2_N, Helicase_C	85–419, 474–604	(Pseudogene) *Leishmania*-specific SNF2-like	Mitochondrial

[a]Domain assignment based on Conserved Domains Database at NCBI.
*AH2 refers to Annealing helicase 2 (AH2), a DNA-rewinding motor with an HNH motif.

indicating that LSD1-type demethylases might be absent in *L. infantum*.

Subcellular localization of chromatin remodelling proteins in *L. infantum*

The subcellular localization of the chromatin remodelling proteins was predicted using WolFPsort. The analysis showed that the mitochondria were enriched in histone-modifying enzymes but not in readers (Tables 1.1–1.6 and Fig. 1.2). Of the 40 writers, 23 were predicted to be localized to the mitochondria. Similarly, of the 11 erasers, seven were predicted to be present in the mitochondria (Fig. 1.2). However, this organelle contained only three reader proteins. On the other hand, the nucleus was found to be enriched in ATP-dependent chromatin remodelling factors and other readers (Fig. 1.2).

The mitochondrial DNA has not been reported to have a proper nucleosome structure. Studies have indicated that the DNA is bent; however, the packaging of this DNA into a chromatin-like structure has not been shown until now. It is in this context that the presence of histone-modifying enzymes

Table 1.6 Erasers present in *L. infantum*

Enzyme name	Coding gene/(s)	UniprotKB accession of human homologue	Pfam domains in human homologue	TritrypDB accession of *L. infantum* homologues	Pfam domains in *L. infantum* homologues	WolfPsort subcellular localization prediction
Lysine demethylase	KDM4C	Q9H3R0	JmjC; JmjN	LinJ.27.1030	JmjC	Cytoplasmic_ nuclear
	JMJD5	Q8N371	4GJZ	LinJ.30.1250	JmjC	Mitochondrial
	C14orf169	Q9H6W3	Cupin_4	LinJ.31.0250	Cupin_4	Mitochondrial
Arginine demethylase	JMJD6	Q6NYC1	JmjC	LinJ.27.1240	F-box, JmjC	Mitochondrial
Histone deacetylase	HDAC1	Q13547	Hist_deacetyl	LinJ.24.1410	Hist_deacetyl	Cytoplasmic
	HDAC2	Q92769	Hist_deacetyl	LinJ.21.0740	Hist_deacetyl	Cytoplasmic
	HDAC6	Q9UBN7	Hist_deacetyl; zf-UBP	LinJ.08.1000	Hist_deacetyl	Mitochondrial
	HDAC9	Q9UKV0	HDAC4_Gln; Hist_deacetyl	LinJ.08.1300	Hist_deacetyl	Mitochondrial
	SIRT2	Q8IXJ6	SIR2	LinJ.26.0200	SIR2	Cytoplasmic
	SIRT4	Q9Y6E7	SIR2	LinJ.23.1450	SIR2	Mitochondrial
	SIRT5	Q9NXA8	SIR2	LinJ.34.1900	SIR2	Mitochondrial

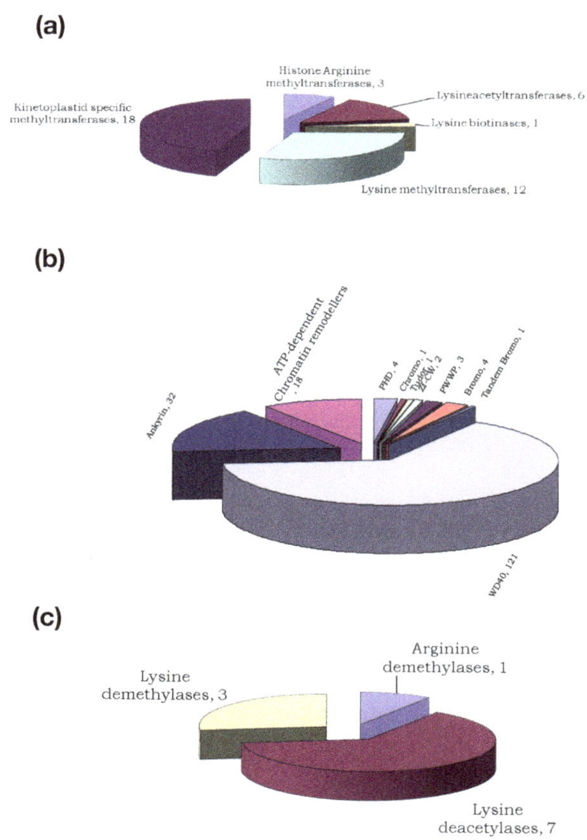

Figure 1.1 Distribution of subclasses of histone-modifying enzymes in *L. infantum*. (a) writers, (b) readers and (c) erasers.

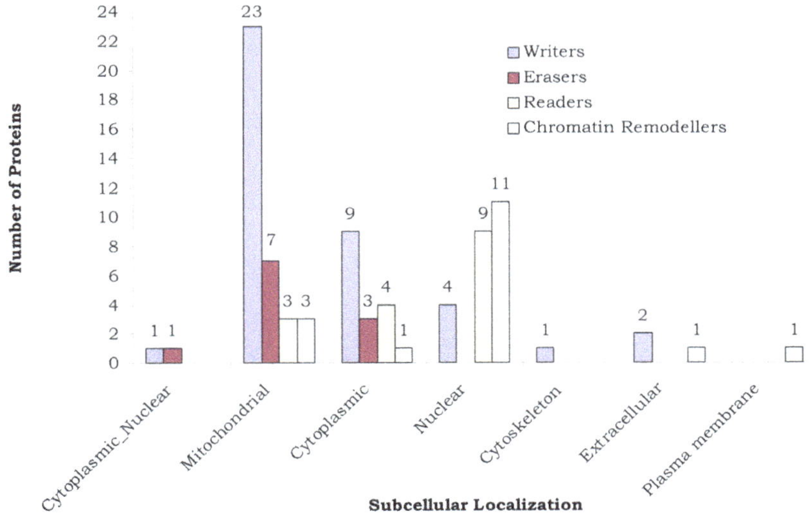

Figure 1.2 Distribution of writers, readers, chromatin remodelling proteins, and erasers in different subcellular compartments. WolFPsort was used to identify the putative location of these proteins present in *L. infantum*.

in the mitochondria is intriguing. It is also possible that these act on substrates other than histones, and that their primary target in mitochondria is not histones but other proteins. However, further studies are needed to understand the role of these proteins in mitochondria and to identify their targets. With this background, we will now discuss in detail the writers, readers, and erasers present in *L. infantum*.

Histone modification writers

As shown in Table 1.1, *L. infantum* contains 40 proteins that catalyse modification of histones, of which 75% are lysine methyltransferases. In the following sections, we discuss in detail the histone modifiers present in *L. infantum*.

Histone methyltransferases

In humans, both lysine and arginine residues have been shown to be methylated by 37 histone methyltransferases (Khare *et al.*, 2012). Lysine methylation is catalysed by class V methyltransferases (KMT) while arginine methylation is mediated by class I methyltransferases (RMT) (Table 1.1).

Our analysis showed that *L. infantum* contains 30 KMTs and three RMTs. Of the 30 *Li*KMTs, 27 were found to possess SET domain. Further, of the 30 *Li*KMTs, 18 *Li*KMTs are specific to kinetoplastids as they do not have a recognizable homologue in other organisms while 12 *Li*KMTs possess a human homologue. There are three *Li*KMTs (LinJ.12.0880, LinJ.23.0960, and LinJ.34.1120) with a Zf-MYND domain tethered and two *Li*KMTs (LinJ.25.1860 and LinJ.26.2590) with Dala_Dala_Lig_C domains appended to the SET domain (Table 1.1). LinJ.06.1040 has only a SET domain in *L. infantum*, while the *L. major* and *T. cruzi* orthologues of (LinJ.06.1040) have a Zf-MYND domain attached at the N-terminus. Finally, while most of the kinetoplastid SET domain proteins have a single orthologue, LinJ.31.1060 has two orthologues in *L. braziliensis* in tandem (LbrM.31.1260 and LbrM.31.1270), two orthologues in *T. cruzi* (Tc00.1047053505007.39 and Tc00.1047053510007.10) and three orthologues of *T. congolense* (TcIL3000.0.01420, TcIL3000.0.15600, and TcIL3000.4.3640).

Recently, DOT1 homologues belonging to the class I methyltransferase have been identified that methylate lysine K79 on histone H3 (Kernstock *et al.*, 2012). Studies have shown that *T. brucei* contains two DOT1 homologues (DOT1A and DOT1B) (Janzen *et al.*, 2006). Experimental studies have shown that DOT1A selectively mono- and di-methylates H3K76 while DOT1B trimethylates H3K76 (Janzen *et al.*, 2006). The reaction mediated by DOTIA is essential for replication regulation while the trimethylation of H3K76 has been shown to be essential in antigenic variation and developmental differentiation (Gassen *et al.*, 2012). Our sequence mining procedure suggests that three DOT1 homologues are present in *L. infantum* (Table 1.1). The sequence comparison of the kinetoplastid homologues with orthologues from other species clearly indicate the conservation of class I methyltransferase signature motifs. However, DOT1A (LinJ.07.0030) and DOT1B (LinJ.20.0030) homologues (characterized earlier in *T. brucei* with the metazoan and yeast homologues) cluster together while the third copy (LinJ.33.1890) shares a close similarity to *C. briggasae* DOT1 and therefore, is remarkably divergent relative to other two DOT1 homologues (Fig. 1.3).

In addition to the lysine methyltransferases, *L. infantum* contains three arginine methyltransferases. Two of these *Li*RMTs (LinJ.06.0900 and LinJ.21.1690) are predicted to be localized in mitochondria while the third copy (LinJ.12.0850) is predicted to have both cytoplasmic and nuclear localization (Table 1.1). All three *Li*RMTs have an authentic PRMT5/PrmA catalytic domain, like their human counterparts (Table 1.1).

Histone acetyltransferases

The lysine acetyltransferases containing acetyl-CoA-binding domain along with a bromodomain are classified into two classes based on the sequence conservation patterns of the histone acetyltransferase domain and their biological function. The MYST family proteins are involved in transcription, DNA replication and repair while the GNAT (Gcn5-related N-acetyltransferases) family mediates transcriptional initiation. *L. infantum* encodes three MYST-type acetyltransferases (LinJ.14.0140, LinJ.17.1350 and LinJ.28.2440) and two GNAT family acetyltransferases (LinJ.16.0250 and LinJ.23.1610) known as elongator complex protein 3 (ELP3) (Table 1.1).

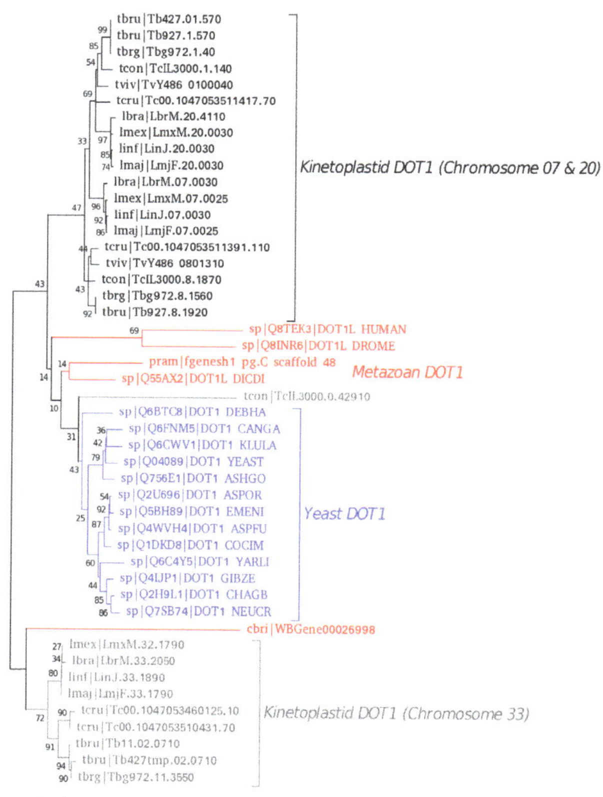

Figure 1.3 Sequence-based phylogeny of DOT1 homologues from kinetoplastids, yeast and other metazoans constructed using MEGA 5.0.

The ELP3 has a specific HAT domain capable of acetylating histones and non-histone substrates and is well-conserved from archaea to human. The expected localization of ELP3 is in the cell cytosol, and a recent study by Alsford and Horn (2011) describes the importance of *Tb*ELP3 in transcription elongation. In addition to HAT domain, *Li*ELP3 proteins contain a Radical SAM (RS) domain. The sequence-based phylogeny clearly indicates the kinetoplastid orthologues are closely related to the Alveolate ELP3 enzymes and are divergent from the eukaryotic and archaeal counterparts (Fig. 1.4).

Histone biotinase

In addition to methylation and acetylation of lysine residues on histones, histone H3 can also be modified by covalent binding of biotin in a reaction catalysed by holo carboxylase synthetase (HLCS) (Kobza *et al.*, 2005, 2008). Although the physiological importance of the modification is unclear, using synthetic peptides, K4, K9 and K18 on histone H3 have been identified as useful targets of biotinylation in a human while K14 and K23 have been shown to be relatively poor targets (Kobza *et al.*, 2005). Based on the experimental results, it is hypothesized that the biotinylation of lysine is likely to affect the transcriptional activity of chromatin.

The functional domain of HLCS proteins is highly conserved across species and is composed of three regions: the N-terminal region, the central catalytic region, and a small C-terminal domain. The catalytic domain shares a sequence identity of ≈30% across human, drosophila and *E. coli* proteins. Our analysis shows that similar to humans, yeast, and bacteria, only one copy of HLCS is present in kinetoplastids (Table 1.1). However, *P. falciparum* encodes two copies of HLCS, *S. mansoni* encodes three copies, and *Entamoeba* encodes multiple copies of HLCS. Furthermore, the N-terminal domain is shorter in kinetoplastids and Alaveolates as compared to the bacterial homologue. Sequence comparisons of kinetoplastid HLCS with other orthologues suggest the conservation or conservative substitution of all the eight biotin ligase residues. The C-terminal domain is thought to interact with ATP and substrates contain two motifs: 'LYY' and 'PDGNSFD' (data not shown). These motifs have high homology among the eukaryotic organisms but are not found in prokaryotes, archaea, kinetoplastids and alveolates. A neighbour-joining bootstrap phylogenetic tree also clearly suggests a close clustering of kinetoplastids and alveolate HLCS with archaeal and bacterial HLCS while all other eukaryotic HLCS form a distinct cluster (Fig. 1.5). The tyrosine in 'LYY' motif is mutated in patients with HLCS deficiency suggesting a critical role for this residue in the enzyme function (Van Hove *et al.*, 2008). This tyrosine is replaced by S/E/L/I/A/G in kinetoplastids and *Plasmodium* species.

Histone ribosylases

Among the kinetoplastids, lysine ribosylases are present only in trypanosomes and *C. fasciculata* and not in *Leishmania* (Table 1.1). Recently, lysine crotonylation enzymes (writers and erasers) were identified in humans (Baumann, 2015). However, these are absent in kinetoplastids, suggesting the absence of this modification in kinetoplastids.

Histone modification readers

Bioinformatic analysis shows that *L. infantum* contains 16 readers (excluding the ankyrin repeats and WD repeat proteins) as well as 18 ATP-dependent chromatin remodelling factors.

Three of the remodeller proteins, however, have been identified to possess mitochondrial targeting sequence by our bioinformatic analysis. The localization of remodeller proteins in mitochondria is interesting especially since the remodellers, unlike histone-modifying enzymes, do not seem to act on targets other than nucleosome. The nuclear remodellers identified in *Leishmania* have been shown in other organisms to play a major role in DNA repair. Many of these proteins have been reported to use naked DNA as substrate. For example, SMARCAL1 is known to remodel naked DNA in the presence of ATP, and this remodelling is essential both for DNA repair as well as for transcription (Sharma *et al.*, 2015; Haokip *et al.*, 2016). Thus, we believe that these proteins might not be regulating transcription by remodelling nucleosomes *per se* but could be essential for mediating DNA repair, possibly in both the nucleus and the mitochondria. The functional role of these proteins needs to be characterized both in the nucleus as well as

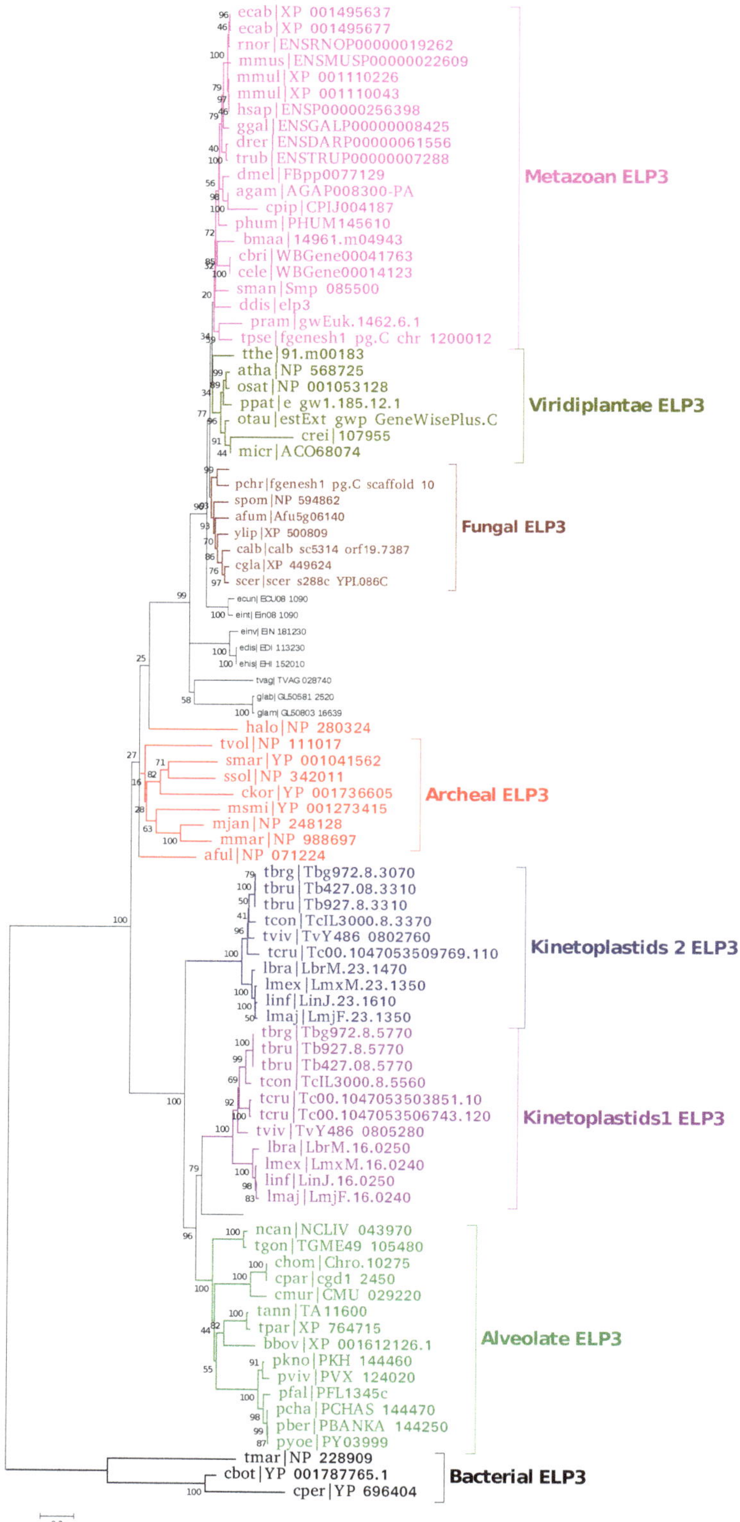

Figure 1.4 Sequence-based phylogeny of ELP3 homologues from kinetoplastids, yeast, metazoa, alveolates, viridiplantae, archea and bacteria.

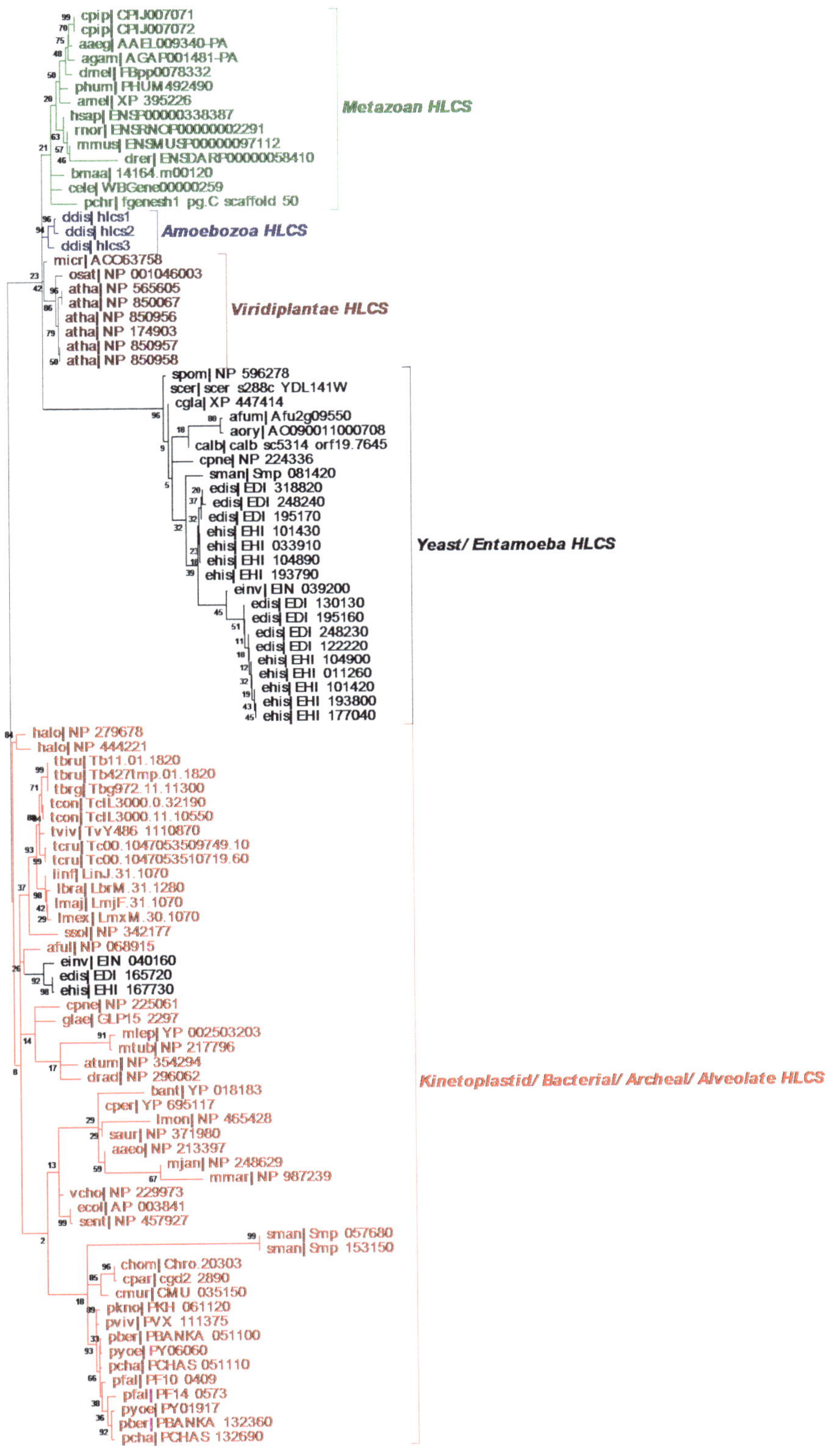

Figure 1.5 A neighbour-joining bootstrap tree of HLCS enzymes from kinetoplastids, yeast, metazoa, plants, archaea and bacteria constructed using MEGA 5.0.

in mitochondria. In the following sections, we will discuss the readers identified in *Leishmania*.

Readers of histone acetylation modification

The bromodomain binds specifically to acetylated lysine and is usually associated with nuclear proteins (Dhalluin *et al.*, 1999). However, *Leishmania* contains only stand-alone bromodomain modules. There are five bromodomain proteins in *L. infantum*, and all of them have been predicted to have nuclear localization (Table 1.2). Of the five bromodomain proteins, one (LinJ.14.0360) has a CW-type zinc finger domain tethered C-terminus to bromodomain while another (LinJ.09.1320) has tandem bromodomains.

Readers of histone methylation modification

The methylation modification on histones can be recognized by several protein motifs such as PHD fingers, Royal superfamily members (chromodomain, Tudor domain, and MBT domain) (Maurer-Stroh *et al.*, 2003), CW-type zinc finger domain (MORC family) and PWWP, WD40, and Ankyrin repeats (Yun *et al.*, 2011).

Chromodomain

Chromodomain is usually 40–50 amino acids in length and present in a wide range of regulatory proteins involved in chromatin remodelling (Koonin *et al.*, 1995). This domain is highly conserved across species, and mammalian chromodomain proteins have been shown to play a significant role in the regulation of gene expression and genome organization (Jones *et al.*, 2000). *L. infantum* has one chromodomain protein (LinJ.14.0160) with a predicted cytoplasmic localization (Table 1.2).

Tudor domain

Tudor domain, a little adaptor belonging to the Royal superfamily, recognizes methylated lysines and arginines on histones H3 and H4 (Yun *et al.*, 2011). Deregulation of Tudor-containing proteins is often associated with human disorders making them novel pharmacological targets for therapeutic intervention (Lu and Wang, 2013). Biochemical characterization of *Pf*TSN shows that it possesses nuclease activity with Tudor domain as the RNA-binding partner (Hossain *et al.*, 2008a). Inhibitor binding studies with a specific micrococcal nuclease inhibitor 3′,5′-deoxythymidine biphosphate (pdTp) is shown to inhibit both the nuclease and RNA-binding activities of *Pf*TSN without any significant effect on the mammalian cell lines even with 4-fold concentrations of the inhibitor (Hossain *et al.*, 2008b; Sunil *et al.*, 2008).

L. infantum contains only one Tudor domain tethered C-terminus to a *Staphylococcus aureus* nuclease (SNc) domain (Table 1.2). The *L. infantum* Tudor-SNc protein (*Li*TSN; LinJ.32.1000) contains four full-length SNc domains followed by a Tudor domain and a partial C-terminus SNc domain with a predicted cytoplasmic localization.

PWWP domain

This is also one of the Royal superfamily members named after a conserved Pro-Trp-Trp-Pro (PWWP) motif wherein the conserved aromatic cage recognizes the methylated lysine and is often involved in crosstalk between various epigenetic marks through their cooperativity with other histone and DNA 'reader' and 'modifier' domains (Qin and Min, 2014). *L. infantum* encodes three PWWP domain-containing proteins. Two of them are tethered at N-terminus to a zinc finger motif of transcription factor II with cytoplasmic (LinJ.33.2950) and nuclear (LinJ.33.2960) localization signals. The third copy of PWWP domain (LinJ.22.0460) has a protein of unknown function (UPF0066 domain) tethered C-terminus to it with a predicted mitochondrial localization signal. The InterPro (EMBL-EB1, Hinxton, UK) (protein sequence analysis and classification tool) search suggests that (LinJ.22.0460) contains a PWWP domain along with a tsaA-like domain, a beta-barrel structural domain found in proteins. The UPF0066 domain that is tethered to the C-terminus of (LinJ.22.0460) is predicted to possess AdoMet-binding ability and methyltransferase activity, suggesting that it could probably be involved in AdoMet recognition to provide this cofactor for methyltransferase activity (Forouhar *et al.*, 2007).

CW-type zinc finger

This is a new reader module belonging to the microrchida (MORC) family that recognizes different methylated states of lysines on histone H3 (Yun *et al.*, 2011). It contains a zinc-binding domain comprising of 60 amino acids. Four invariant cysteines potentially coordinate the zinc ion and three conserved tryptophans (He *et al.*, 2010) found exclusively in vertebrates, vertebrate-infecting parasites and higher plants (Perry and Zhao, 2003). There are two CW-type zinc finger proteins in *L. infantum*. LinJ.14.0360 has a bromodomain tethered N-terminus to the CW-type zinc finger domain with a predicted nuclear localization while LinJ.33.1160 contains only the CW-type zinc finger domain with a predicted mitochondrial localization (Table 1.2).

Plant homeodomain

The plant homeodomain (PHD) zinc fingers that recognize methylated lysines on histone H3 are ≈65 amino acids long characterized by a canonical Cys4-His-Cys3 motif that coordinates two zinc ions. They often occur with other histone-binding modules such as chromodomain, bromodomain, etc. Mutations or deletion of this domain have been associated with a number of human diseases such as cancer, mental retardation and immunodeficiency, making PHD fingers as valuable diagnostic markers (Musselman and Kutateladze, 2009). *L. infantum* encodes four PHD finger-containing proteins (Table 1.2). Apart from LinJ.25.2340 which has only PHD finger domain, the other three PHD finger proteins are associated with a RING (LinJ.27.2580) domain, SNF2_N 9 (LinJ.25.2130), and NAT_SF (LinJ.24.1260).

WD40 repeats

WD40 repeats were first identified in GTP-binding protein (G-protein) (Fong *et al.*, 1986). Since then, several WD40 repeat proteins have been identified in higher eukaryotes. The binding modes of WD40 to histone H3 tail and histone H4 are varied. A very recent investigation on the chromatin assembly factor-1 (CAF-1) protein, which contains WD40 repeats, in *Paramecium tetraurelia* clearly shows that PtCAF1 is essential for RNA-mediated control of DNA elimination (Ignarski *et al.*, 2014). There are nearly 121 WD40 repeat proteins in *L. infantum*. Two of them (LinJ.35.2450 and LinJ.30.0240) have a single transmembrane helix. There are two histone chaperone proteins (LinJ.24.0230 and LinJ.33.2020) in *L. infantum* possessing a histone-binding domain at the N-terminus and a WD40 repeat at the C-terminus (Table 1.3).

Ankyrin repeats

Ankyrin repeats are 33 amino acid motifs tandemly repeated that often bind methylated lysine on the histones. *L. infantum* contains 32 ankyrin repeat proteins (Table 1.4). Several of them exist as tandem repeats while some are tethered to other enzymes. Six of them contain multiple transmembrane helices targeting plasma membrane except LinJ.34.4110, which contains the cytoplasm targeting signal.

ATP-dependent chromatin remodelling proteins

The ATP-dependent chromatin remodelling proteins, also known as SNF2 sucrose non-fermenter or SWI/SNF-related enzymes, hydrolyse ATP and use the energy to reposition the nucleosomes. Based on the sequence homology of the ATPase core, these proteins are subdivided into 24 subfamilies that fall into six subgroups (Flaus *et al.*, 2006).

There are 18 chromatin remodelling enzymes in *L. infantum* falling into five subgroups, of which one consists of kinetoplastid- and *Leishmania*-specific proteins (Table 1.5 and Fig. 1.6a). Thus, the sequence-based phylogeny of kinetoplastid SNF2 enzymes (Fig. 1.6b) resembles the one obtained by Flaus *et al.* (2006) except for the kinetoplastid- and *Leishmania*-specific SNF2 groups which are found as outgroups or distantly related to other SNF2 subfamilies. There are two *Leishmania*-specific SNF2 and two kinetoplastid-specific SNF2 proteins that do not cluster with any other eukaryotic SNF2 homologues (Table 1.5). LinJ.31.2390 (*Leishmania*-specific SNF2) has a transmembrane domain inserted within the SNF2 catalytic domain while LinJ.23.1100 is a pseudogene. A cytoplasmic copy of j-binding protein (LinJ.14.0040) specific to kinetoplastids is also found. The j-binding proteins contain Jmjc catalytic domain and are bifunctional helicase and thymine dioxygenases (Vainio *et al.*, 2009).

Interestingly, of these 18 ATP-dependent chromatin remodeller proteins, only ISWI and SMARCAL1 is known to participate in transcriptional regulation (Corona and Tamkun, 2004; Sharma *et al.*, 2015). All the proteins though are known to function in DNA repair pathways, indicating that these proteins in *L. infantum* might have a critical role only in DNA repair.

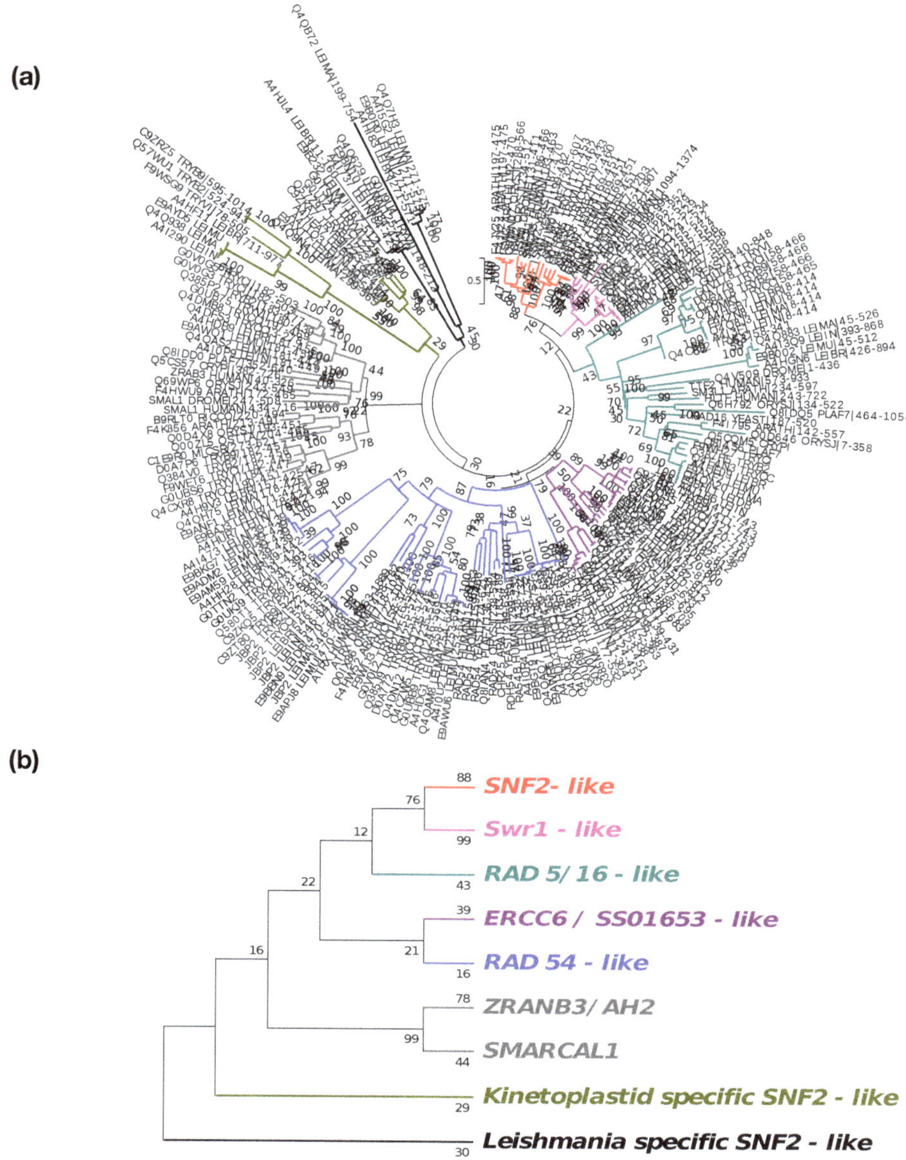

Figure 1.6 Analysis of ATP-dependent chromatin remodelling proteins. (a) A neighbour-joining bootstrap tree of the SNF2 domain sequences of the human, *Drosophila*, yeast and other eukaryotic pathogens. The subfamily assignments for the kinetoplastid enzymes were based on the nearest human or another eukaryotic homologue of known function on the dendrogram. The circular dendrogram is colour coded into different subfamilies. (b) A topology diagram of the sequence-based phylogeny of the kinetoplastid SNF2 enzymes.

There are three RAD5/16-like subfamily SNF2 proteins and three RAD54-like enzymes in *L. infantum*. All the RAD5/16-like enzymes have a zinc finger domain inserted between SNF2_N and helicase domains and have nuclear localization signals. The only exception is LinJ.25.0770 which has a Zf-PARP tethered N-terminus to SNF2_N. Zf-PARPs are DNA-binding modules present in chromatin modifiers (D'Amours *et al.*, 1999). A sequence-based phylogeny of the complete family of zf-PARPs shows the clustering of protozoan Zf-PARP enzymes with fungal Zf-PARP (Petrucco and Percudani, 2008). The RAD54-like enzymes have both nuclear and plasma membrane localization signals (LinJ.29.0210). There are two ERCC6 enzymes in *L. infantum*. A single mitochondrial ISWI (LinJ.33.1800) copy has HAND and SLIDE domains in addition to the SNF2_N and helicase_C domains. While the Swr1-like (LinJ.36.5420) and SMARCAL1 (LinJ.24.0620) have SNF2_N and helicase domains, the ZRANB3/AH2 homologue of *L. infantum* (LinJ.11.0050) has an HNH domain C-terminus to the helicase domain. To date, HARP (Yusufzai and Kadonaga, 2008) and AH2 (Yusufzai and Kadonaga, 2010) are the two annealing helicases known. Although LinJ.24.0620 resembles human SMARCAL1 protein, it cannot have a DNA annealing activity as it lacks the HARP domain. However, *L. infantum* contains an authentic AH2 homologue (LinJ.11.0050) with a C-terminus HNH motif commonly found in bacteria and fungi that are often associated with nuclease activity. Thus, unlike human that have both HARP (SMARCAL1) and AH2 annealing helicases, kinetoplastids have only AH2 type annealing helicase.

Histone modification erasers

The third component of chromatin remodelling comprises the histone modification erasers, enzymes that catalytically mediate the removal of modifications from histones. There are two categories of erasers in *L. infantum*: histone deacetylases (HDACs) and histone demethylases (HDMTs). There are seven HDACs and four HDMTs in *L. infantum* (Table 1.6). In the final section, we, therefore, discuss in detail these enzymes.

Histone deacetylases

In contrast to acetylation, deacetylation of histones by HDACs results in transcription repression. Based on the size and sequence homology, HDACs are classified into three classes: I, II and III (Verdin *et al.*, 2003). Class I HDACs are ubiquitously expressed while class II HDACs have restricted tissue distribution. Class III HDACs are related to yeast silent information regulator 2 (Sir2) and require NAD for deacetylase activity. *L. infantum* contains two class I HDACs (LinJ.21.0740 and LinJ.24.1410) both of which are predicted to be in the cytoplasm, and two class II HDACs (LinJ.08.1000 and LinJ.08.1300) predicted to be located in the mitochondria. There are three class III HDACs (LinJ.26.0200, LinJ.23.1450 and LinJ.34.1900) in *L. infantum* (Table 1.6).

Class I and class II histone deacetylases in *L. infantum*

Two class I HDACs, highly homologous in sequence, are present in *L. infantum* and other *Leishmania* species; in contrast, trypanosomes have only one class I HDAC enzyme. Thus, LinJ.24.1410 is restricted to *Leishmania*, and there is no homologue of this enzyme in trypanosomes. The active site residues and zinc-binding motifs are highly conserved in the kinetoplastid HDACs. However, the sequence-based phylogeny suggests that the kinetoplastid class I HDACs have distinctly diverged from other eukaryotic class I HDACs (data not shown). The class-specific sequence motifs are highly conserved in class I (GGGGY) HDACs and class II (LEGGY) HDACs. The sequence-based phylogeny of class II HDACs suggests that one copy of the kinetoplastid HDACs is closer to *P. chrysosporium* while the other copy has diverged distinctly from the eukaryotic class II HDACs (data not shown).

Class III histone deacetylases in *L. infantum*

The class III HDACs are related to the yeast Sir2. There are three Sir2 proteins in *L. infantum* belonging to class III HDACs. SIR2 homologue (LinJ.26.0200) is predicted to be localized to the cell cytosol while the other two Sir2 copies are predicted to be localized in the mitochondria. The SIR2 homologue of (LinJ.26.0200) has been shown to be localized in the cytosol and proved indispensable for parasite survival under stressful conditions (Vergnes *et al.*, 2002). All Sir2 proteins characterized to date have a conserved core domain with a series of sequence motifs and function as mono-ADP-ribosylases in an NAD-dependent reaction although much weaker than deacetylation of histone lysines (Saunders and Verdin, 2007).

Based on the sequence conservation patterns documented earlier, kinetoplastid sequences could be classified into class Ib, class II and class III sirtuins. A sequence and molecular phylogeny analysis suggest the conservation of sequence motifs belonging to individual subclasses and a close association of kinetoplastid sirtuins to Gram-negative bacterial sirtuins.

Histone demethylases

Amine oxidase (LSD1) and Jumonji-type demethylases are the two known families of histone demethylases. Demethylation of histones by LSD1 happens in an FAD-dependent manner while Jumonji-type demethylases (Tsukada *et al.*, 2006; Yamane *et al.*, 2006) utilize Fe(II) and alpha-ketoglutarate cofactors along with molecular oxygen to demethylate lysines (Yamane *et al.*, 2006) and arginines (Chang *et al.*, 2007) on the histone tail (Elkins *et al.*, 2003; Webby *et al.*, 2009).

Our analysis shows that, interestingly, amine oxidase LSD1 histone demethylases are absent in kinetoplastids. The organism does contain, however, a small zinc finger motif containing protein, zf-LSD1 (LinJ.14.0640), which lacks the amine oxidase and the FAD domains. This protein in *A. thaliana* (NP_849549.1) has been shown to monitor superoxide-dependent signal and negatively regulate plant cell death pathway (Dietrich *et al.*, 1997), and thus, probably is not a histone demethylase.

We also identified four Jumonji-type demethylases in *L. infantum* (Table 1.6). All the four demethylases contain the catalytic JmjC domain. LinJ.30.1250 (a *Leishmania*-specific Jmjd6 enzyme) and LinJ.27.1240 contain an F-box domain tethered N-terminus to the JmjC domain. Sequence-based phylogeny of the Jumonji family proteins from different domains of life with kinetoplastid enzymes shows a clear grouping of the different subfamilies (Fig. 1.7). Thus, *L. infantum* encodes an NO66 (LinJ.31.0250) enzyme which demethylates K4Me3 and K4Me1 on histone H3, a JmjD2 (LinJ.27.1030) enzyme demethylating K9Me3 and K36Me3 on histone H3 and two arginine demethylases (LinJ.30.1250 and LinJ.27.1240) that demethylate R2Me2 and R3Me2 on histone H3.

Epigenetic regulators as druggable targets

Epigenetic signalling key protein families that mediate through the acetylation and methylation of histones include histone deacetylases, protein methyltransferases, lysine demethylases, bromodomain-containing proteins and proteins that bind to methylated histones (Arrowsmith *et al.*, 2012). These protein families are emerging as druggable classes of enzymes. Histone-modifying enzymes (HME) are central actors in the regulation of the epigenetic modification of chromatin and aberrant epigenetic states often associated with cancer. This has led to an interest in HME as targets for therapy. There has been a considerable effort to develop HDAC inhibitors (HDACi), a significant number are in clinical trials, and two of these, suberoylanilide hydroxamic acid (SAHA) and the depsipeptide romidepsin, have been approved by the FDA for use in adults with cutaneous T-cell lymphoma (Johnstone *et al.*, 2002). Most inhibitors that are being developed as anti-cancer agents target class I, II and IV HDACs (Duvic *et al.*, 2007). Broadly, HDACi induces cell death in cancer cells via apoptosis, but they can also act on the cell cycle, on tumour angiogenesis or via the regulation of host cell responses. HDACi have also stimulated interest as anti-parasitic drugs and have been tested against *P. falciparum* (Drummond *et al.*, 2005; Andrews *et al.*, 2009), *Toxoplasma gondii* (Andrews *et al.*, 2008) and the major kinetoplastid parasites (Vanagas *et al.*, 2012; Andrews *et al.*, 2012). In the case of *P. falciparum*, it has been possible to develop HDACi that are significantly more toxic to the parasite than towards human cells. As for targeting parasite epigenetic gene regulation through histone posttranslational modifications, the few studies present in literature have focused exclusively on modulating histone acetylation via the histone acetyltransferase (HAT) inhibitors curcumin (Cui L *et al.*, 2008) or anacardic acid (Prusty *et al.*, 2008), or the histone deacetylase (HDAC) inhibitors nicotinamide (Darkin-Rattray *et al.*, 1996), or apicidin (Andrews *et al.*, 2008), or derivatives of hydroxamic acid (Dow *et al.*, 2008; Agbor-Enoh *et al.*, 2009).

Epigenetic factors such as histone methylation control the developmental progression of malaria parasites during the complex life cycle in the human host. *Plasmodium falciparum* histone lysine methyltransferases are reported to be a potential target class for the development of novel antimalarials (Malmquist *et al.*, 2012).

A number of companies have HDAC inhibitors in various stages of development. Several HDAC inhibitors are in clinical trials for oncology indications. Some of these HDAC inhibitors are orally active which makes therapeutic administration easier. However, there are currently no

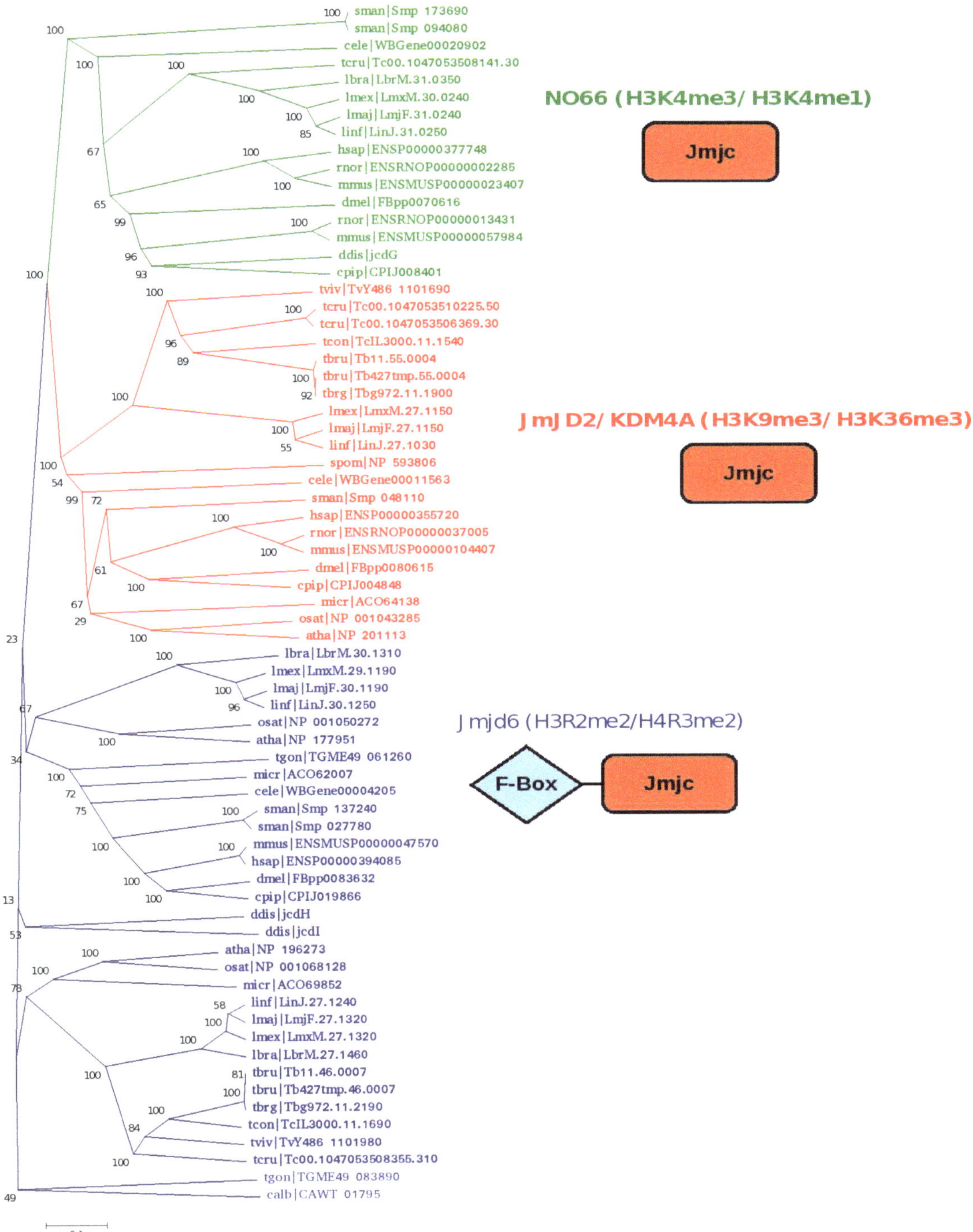

Figure 1.7 Sequence-based phylogeny of the Jumonji family from different domains of life with kinetoplastid Jumonji enzymes. The domain organization, demethylating lysine residues and the methylation states of the different subfamilies are shown.

HDAC inhibitors approved for the treatment of neglected diseases. We believe that it may be possible to exploit differences between human and parasitic HDACs to develop selective inhibitors for neglected disease indications. HDAC homologues are present in human-infecting parasites causing leishmaniasis, human African trypanosomiasis (HAT), schistosomiasis and malaria and are being studied as potential drug targets. We, therefore, consider that the HDACs are promising targets for the development of new drugs.

Conclusion

In this study, the first genome-wide comprehensive survey of the epigenetic regulator proteins in *Leishmania infantum* has been performed. A total of 238 epigenetic regulator proteins, including 18 kinetoplastid-specific methyltransferases were identified in our *in silico* analysis. Three DOT1 methyltransferases were identified as against an earlier report suggesting the presence of only two DOT1 homologues in *T. brucei*.

There were one Chromo domain- and one Tudor domain-containing proteins in *L. infantum*. While the chromodomain is a stand-alone single domain, the Tudor domain-containing protein has staphylococcal-type nuclease (SNc) domains tethered N-terminus and C-terminus to the Tudor domains. There were four PHD type zinc finger proteins, two Zf-CW type zinc fingers, and three PWWP type zinc finger proteins in *L. infantum*. One of the PWWP zinc finger proteins has a UPF0066 domain at the C-terminus. This UPF0066 domain, although recognized as a domain of unknown function, shows a weak similarity to tsaA-like domain suggesting a role in methyltransferase activity. There were five bromodomain-containing proteins. LinJ.14.0360 has a Zf-CW type zinc finger domain which recognizes methylated lysine (MeK) while the bromodomain recognizes acetyl-lysine (KAc). Thus, LinJ.14.0360 could be considered as a bifunctional reader protein. Only the WD40 repeats and Ankyrin repeats have transmembrane helices. 18 ATP-dependent chromatin remodellers belonging to SNF2 family were identified in *L. infantum*. There were two kinetoplastid-specific SNF2 enzymes and three *Leishmania*-specific SNF2 enzymes in *L. infantum*. One of the *Leishmania*-specific SNF2 copy (LinJ.23.1100) is a pseudogene. The study also reveals an abundance of chromatin remodelling proteins that participate in DNA repair in *Leishmania*.

Finally, unlike humans, *L. infantum* contains only the Jumonji-type demethylase and LSD1 type demethylase is completely absent. We have also identified two class I HDACS, two class II HDACs and three class III HDACs (Sirtuins). Further studies will help in understanding the epigenetic regulatory mechanism in *Leishmania* and enable the development of drugs against some of the novel epigenetic regulators.

Materials and methods

Databases and web servers used

Leishmania infantum (version 3.1) protein sequences from the Tritrypdb database (Aslett *et al.*, 2010) (www.tritrypdb.org) are used here. Hidden Markov model (HMM) profiles for all the histone-modifying enzymes were downloaded from the Pfam (version 27.0) database (Finn *et al.*, 2014) from the Sanger Institute (http://xfam.pfam.org). The human homologues of the histone-modifying enzymes were extracted from the HIstome database (the human histone-modifying enzymes infobase) (www.actrec.gov.in/histome/) (Khare *et al.*, 2012). A stand-alone version of the BLAST (Basic Local Alignment Search Tool), WolfPsort (subcellular localization prediction tool) (Horton *et al.*, 2007), ClustalW (Thompson *et al.*, 1994) and MEGA version 5 (Tamura *et al.*, 2011) were used for doing the basic searches, sequence analysis and molecular phylogenetic analysis. Conserved Domains Database (CDD) (Marchler-Bauer *et al.*, 2015) at NCBI (www.ncbi.nlm.nih.gov/Structure/cdd/wrpsb.cgi) was used for verification of protein domain architecture. Orthologues for sequence analysis and phylogeny are extracted from OrthoMCL-DB (Chen *et al.*, 2006) (www.orthomcl.org/orthomcl/).

Search procedure

The hidden Markov model (HMM) profiles from PFAM of the various histone-modifying enzymes were searched using a stand-alone HMM search tool against the *L. infantum* sequence database. The human homologues of the histone-modifying enzymes extracted from the HIstome database were also searched using the stand-alone BLAST search tool against the *L. infantum* sequence database to find the nearest *L. infantum* homologue to the human protein. Subcellular localization prediction of the *L. infantum* protein was made using a stand-alone version of WolfPsort. The domain architecture of the *L. infantum* proteins was then validated using the Conserved Domains Database server at NCBI.

Multiple sequence alignment and phylogeny

Sequence analysis and phylogeny was performed combining a set of orthologues from OrthoMCL-DB webserver. Multiple sequence alignment of these sequences is generated using a stand-alone version of ClustalW using default parameters. These MSAs were used as seed sequences for phylogenetic tree generation using Jones-Taylor-Thornton (JTT) model (Jones *et al.*, 1992). MEGA v5 was used for both analysis and visualization of the phylogenetic trees.

Acknowledgements

Rentala Madhubala is a JC Bose National Fellow and UGC-BSR Faculty Fellow. VSG is supported by UGC-DS Kothari Postdoctoral fellowship. NM is funded by Council for Scientific and Industrial Research (CSIR), India. This work was supported by the University for Potential for Excellence grant (UPE-II, Project ID–11) from the University Grants Commission, Government of India to R. Madhubala and Rohini Muthuswami.

Author contributions

VSG performed the analysis. VSG, NM and RM wrote the manuscript. RMB and RM analysed the problem and reviewed the manuscript. All authors read and approved the final manuscript.

Additional information

The authors declare no competing financial interests.

References

Agbor-Enoh, S., Seudieu, C., Davidson, E., Dritschilo, A., and Jung, M. (2009). Novel inhibitor of Plasmodium histone deacetylase that cures P. berghei-infected mice. Antimicrob. Agents Chemother. 53, 1727–1734. https://dx.doi.org/10.1128/AAC.00729-08.

Alsford, S., and Horn, D. (2011). Elongator protein 3b negatively regulates ribosomal DNA transcription in african trypanosomes. Mol. Cell Biol. *31*, 1822–1832. https://dx.doi.org/10.1128/MCB.01026-10.

Anderson, L., Pierce, R.J., and Verjovski-Almeida, S. (2012). *Schistosoma mansoni* histones: from transcription to chromatin regulation; an in silico analysis. Mol. Biochem. Parasitol. *183*, 105–114. https://dx.doi.org/10.1016/j.molbiopara.2012.03.001.

Andrews, K.T., Tran, T.N., Lucke, A.J., Kahnberg, P., Le, G.T., Boyle, G.M., Gardiner, D.L., Skinner-Adams, T.S., and Fairlie, D.P. (2008). Potent antimalarial activity of histone deacetylase inhibitor analogues. Antimicrob. Agents Chemother. *52*, 1454–1461. https://dx.doi.org/10.1128/AAC.00757-07.

Andrews, K.T., Tran, T.N., Wheatley, N.C., and Fairlie, D.P. (2009). Targeting histone deacetylase inhibitors for anti-malarial therapy. Curr. Top Med. Chem. *9*, 292–308.

Andrews, K.T., Haque, A., and Jones, M.K. (2012). HDAC inhibitors in parasitic diseases. Immunol. Cell Biol. *90*, 66–77. https://dx.doi.org/10.1038/icb.2011.97.

Arrowsmith, C.H., Bountra, C., Fish, P.V., Lee, K., and Schapira, M. (2012). Epigenetic protein families: a new frontier for drug discovery. Nat. Rev. Drug Discov. *11*, 384–400. https://dx.doi.org/10.1038/nrd3674.

Aslett, M., Aurrecoechea, C., Berriman, M., Brestelli, J., Brunk, B.P., Carrington, M., Depledge, D.P., Fischer, S., Gajria, B., Gao, X., *et al.* (2010). TriTrypDB: a functional genomic resource for the Trypanosomatidae. Nucleic Acids Res. *38*, D457–462. https://dx.doi.org/10.1093/nar/gkp851.

Bannister, A.J., and Kouzarides, T. (2011). Regulation of chromatin by histone modifications. Cell Res. *21*, 381–395. https://dx.doi.org/10.1038/cr.2011.22.

Baumann, K. (2015). Post-translational modifications: crotonylation versus acetylation. Nat. Rev. Mol. Cell Biol. *16*, 265. https://dx.doi.org/10.1038/nrm3992.

Berger, S.L. (2007). The complex language of chromatin regulation during transcription. Nature *447*, 407–412. https://dx.doi.org/10.1038/nature05915.

Chang, B., Chen, Y., Zhao, Y., and Bruick, R.K. (2007). JMJD6 is a histone arginine demethylase. Science *318*, 444–447. https://dx.doi.org/10.1126/science.1145801.

Chen, F., Mackey, A.J., Stoeckert, C.J., Jr., and Roos, D.S. (2006). OrthoMCL-DB: querying a comprehensive multi-species collection of ortholog groups. Nucleic Acids Res. *34*, D363–368. https://dx.doi.org/10.1093/nar/gkj123.

Cho, K.S., Elizondo, L.I., and Boerkoel, C.F. (2004). Advances in chromatin remodeling and human disease. Curr. Opin. Genet. Dev. *14*, 308–315. https://dx.doi.org/10.1016/j.gde.2004.04.015.

Corona, D.F., and Tamkun, J.W. (2004). Multiple roles for ISWI in transcription, chromosome organization and DNA replication. Biochim. Biophys. Acta *1677*, 113–119. https://dx.doi.org/10.1016/j.bbaexp.2003.09.018.

Cui, L., Miao, J., Furuya, T., Fan, Q., Li, X., Rathod, P.K., Su, X.Z., and Cui, L. (2008). Histone acetyltransferase inhibitor anacardic acid causes changes in global gene expression during *in vitro* Plasmodium falciparum development. Eukaryot. Cell *7*, 1200–1210. https://dx.doi.org/10.1128/EC.00063-08.

D'Amours, D., Desnoyers, S., D'Silva, I., and Poirier, G.G. (1999). Poly(ADP-ribosyl)ation reactions in the regulation of nuclear functions. Biochem. J. *342 (Pt 2)*, 249–268.

Darkin-Rattray, S.J., Gurnett, A.M., Myers, R.W., Dulski, P.M., Crumley, T.M., Allocco, J.J., Cannova, C., Meinke, P.T., Colletti, S.L., Bednarek, M.A., *et al.* (1996). Apicidin: a novel antiprotozoal agent that inhibits parasite histone deacetylase. Proc. Natl. Acad. Sci. U. S. A. *93*, 13143–13147.

Dhalluin, C., Carlson, J.E., Zeng, L., He, C., Aggarwal, A.K., and Zhou, M.M. (1999). Structure and ligand of a histone acetyltransferase bromodomain. Nature *399*, 491–496. https://dx.doi.org/10.1038/20974.

Dietrich, R.A., Richberg, M.H., Schmidt, R., Dean, C., and Dangl, J.L. (1997). A novel zinc finger protein is encoded by the Arabidopsis LSD1 gene and functions as a negative regulator of plant cell death. Cell *88*, 685–694.

Dow, G.S., Chen, Y., Andrews, K.T., Caridha, D., Gerena, L., Gettayacamin, M., Johnson, J., Li, Q., Melendez, V., Obaldia, N., 3rd, *et al.* (2008). Antimalarial activity of phenylthiazolyl-bearing hydroxamate-based histone deacetylase inhibitors. Antimicrob. Agents Chemother. *52*, 3467–3477. https://dx.doi.org/10.1128/AAC.00439-08.

Drummond, D.C., Noble, C.O., Kirpotin, D.B., Guo, Z., Scott, G.K., and Benz, C.C. (2005). Clinical development of histone deacetylase inhibitors as anticancer agents. Annu. Rev. Pharmacol. Toxicol. *45*, 495–528. https://dx.doi.org/10.1146/annurev.pharmtox.45.120403.095825.

Duvic, M., Talpur, R., Ni, X., Zhang, C., Hazarika, P., Kelly, C., Chiao, J.H., Reilly, J.F., Ricker, J.L., Richon, V.M., *et al.* (2007). Phase 2 trial of oral vorinostat (suberoylanilide hydroxamic acid, SAHA) for refractory cutaneous T-cell lymphoma (CTCL). Blood *109*, 31–39. https://dx.doi.org/10.1182/blood-2006-06025999.

Elkins, J.M., Hewitson, K.S., McNeill, L.A., Seibel, J.F., Schlemminger, I., Pugh, C.W., Ratcliffe, P.J., and Schofield, C.J. (2003). Structure of factor-inhibiting hypoxia-inducible factor (HIF) reveals mechanism of oxidative modification of HIF-1 alpha. J. Biol. Chem. *278*, 1802–1806. https://dx.doi.org/10.1074/jbc.C200644200.

Finn, R.D., Bateman, A., Clements, J., Coggill, P., Eberhardt, R.Y., Eddy, S.R., Heger, A., Hetherington, K., Holm, L., Mistry, J., *et al.* (2014). Pfam: the protein families database. Nucleic Acids Res. *42*, D222–230. https://dx.doi.org/10.1093/nar/gkt1223.

Flaus, A., Martin, D.M., Barton, G.J., and Owen-Hughes, T. (2006). Identification of multiple distinct Snf2 subfamilies with conserved structural motifs. Nucleic Acids Res. *34*, 2887–2905. https://dx.doi.org/10.1093/nar/gkl295.

Fong, H.K., Hurley, J.B., Hopkins, R.S., Miake-Lye, R., Johnson, M.S., Doolittle, R.F., and Simon, M.I. (1986). Repetitive segmental structure of the transducin beta subunit: homology with the CDC4 gene and identification of related mRNAs. Proc. Natl. Acad. Sci. U.S.A. *83*, 2162–2166.

Forouhar, F., Kuzin, A., Seetharaman, J., Lee, I., Zhou, W., Abashidze, M., Chen, Y., Yong, W., Janjua, H., Fang, Y., *et al.* (2007). Functional insights from structural genomics. J. Struct. Funct. Genomics *8*, 37–44. https://dx.doi.org/10.1007/s10969-007-9018-3.

Gassen, A., Brechtefeld, D., Schandry, N., Arteaga-Salas, J.M., Israel, L., Imhof, A., and Janzen, C.J. (2012). DOT1A-dependent H3K76 methylation is required for replication regulation in *Trypanosoma brucei*. Nucleic Acids Res. *40*, 10302–10311. https://dx.doi.org/10.1093/nar/gks801.

Haokip, D.T., Goel, I., Arya, V., Sharma, T., Kumari, R., Priya, R., Singh, M., and Muthuswami, R. (2016). Transcriptional regulation of ATP-dependent chromatin Remodeling factors: Smarcal1 and Brg1 mutually Co-regulate each other. Sci. Rep. *6*, 20532. https://dx.doi.org/10.1038/srep20532.

Hargreaves, D.C., and Crabtree, G.R. (2011). ATP-dependent chromatin remodeling: genetics, genomics and mechanisms. Cell Res. *21*, 396–420. https://dx.doi.org/10.1038/cr.2011.32.

He, F., Umehara, T., Saito, K., Harada, T., Watanabe, S., Yabuki, T., Kigawa, T., Takahashi, M., Kuwasako, K., Tsuda, K., *et al.* (2010). Structural insight into the zinc finger CW domain as a histone modification reader. Structure *18*, 1127–1139. https://dx.doi.org/10.1016/j.str.2010.06.012.

Horton, P., Park, K.J., Obayashi, T., Fujita, N., Harada, H., Adams-Collier, C.J., and Nakai, K. (2007). WoLF PSORT: protein localization predictor. Nucleic Acids Res. *35*, W585–587. https://dx.doi.org/10.1093/nar/gkm259.

Hossain, M.J., Korde, R., Singh, S., Mohmmed, A., Dasaradhi, P.V., Chauhan, V.S., and Malhotra, P. (2008). Tudor domain proteins in protozoan parasites and characterization of Plasmodium falciparum tudor staphylococcal nuclease. Int. J. Parasitol. *38*, 513–526. https://dx.doi.org/10.1016/j.ijpara.2007.09.018.

Ignarski, M., Singh, A., Swart, E.C., Arambasic, M., Sandoval, P.Y., and Nowacki, M. (2014). *Paramecium tetraurelia* chromatin assembly factor-1-like protein PtCAF-1 is involved in RNA-mediated control of DNA elimination. Nucleic Acids Res. *42*, 11952–11964. https://dx.doi.org/10.1093/nar/gku874.

Janzen, C.J., Hake, S.B., Lowell, J.E., and Cross, G.A. (2006). Selective di- or trimethylation of histone H3 lysine 76 by two DOT1 homologs is important for cell cycle regulation in *Trypanosoma brucei*. Mol. Cell *23*, 497–507. https://dx.doi.org/10.1016/j.molcel.2006.06.027.

Johnstone, R.W. (2002). Histone-deacetylase inhibitors: novel drugs for the treatment of cancer. Nat. Rev. Drug Discov. *1*, 287–299. https://dx.doi.org/10.1038/nrd772.

Jones, D.O., Cowell, I.G., and Singh, P.B. (2000). Mammalian chromodomain proteins: their role in genome organisation and expression. Bioessays *22*, 124–137. https://dx.doi.org/10.1002/(S.I.C.I)1521-1878(200002)22:2<124::AID-BIES4>3.0.CO;2-E.

Jones, D.T., Taylor, W.R., and Thornton, J.M. (1992). The rapid generation of mutation data matrices from protein sequences. Comput. Appl. Biosci. *8*, 275–282.

Kernstock, S., Davydova, E., Jakobsson, M., Moen, A., Pettersen, S., Maelandsmo, G.M., Egge-Jacobsen, W., and Falnes, P.O. (2012). Lysine methylation of VCP by a member of a novel human protein methyltransferase family. Nat. Commun. *3*, 1038. https://dx.doi.org/10.1038/ncomms2041.

Khare, S.P., Habib, F., Sharma, R., Gadewal, N., Gupta, S., and Galande, S. (2012). HIstome--a relational knowledgebase of human histone proteins and histone modifying enzymes. Nucleic Acids Res. *40*, D337–342. https://dx.doi.org/10.1093/nar/gkr1125.

Kobza, K., Camporeale, G., Rueckert, B., Kueh, A., Griffin, J.B., Sarath, G., and Zempleni, J. (2005). K4, K9 and K18 in human histone H3 are targets for biotinylation by biotinidase. FEBS J. *272*, 4249–4259. https://dx.doi.org/10.1111/j.1742-4658.2005.04839.x.

Kobza, K., Sarath, G., and Zempleni, J. (2008). Prokaryotic BirA ligase biotinylates K4, K9, K18 and K23 in histone H3. BMB Rep. *41*, 310–315.

Koonin, E.V., Zhou, S., and Lucchesi, J.C. (1995). The chromo superfamily: new members, duplication of the chromo domain and possible role in delivering transcription regulators to chromatin. Nucleic Acids Res. *23*, 4229–4233.

Lu, R., and Wang, G.G. (2013). Tudor: a versatile family of histone methylation 'readers'. Trends Biochem. Sci. *38*, 546–555. https://dx.doi.org/10.1016/j.tibs.2013.08.002.

Malmquist, N.A., Moss, T.A., Mecheri, S., Scherf, A., and Fuchter, M.J. (2012). Small-molecule histone methyltransferase inhibitors display rapid antimalarial activity against all blood stage forms in *Plasmodium falciparum*. Proc. Natl. Acad. Sci. U.S.A. *109*, 16708–16713. https://dx.doi.org/10.1073/pnas.1205414109.

Marchler-Bauer, A., Derbyshire, M.K., Gonzales, N.R., Lu, S., Chitsaz, F., Geer, L.Y., Geer, R.C., He, J., Gwadz, M., Hurwitz, D.I., *et al.* (2015). CDD: NCBI's conserved domain database. Nucleic Acids Res. *43*, D222–226. https://dx.doi.org/10.1093/nar/gku1221.

Martin, C., and Zhang, Y. (2005). The diverse functions of histone lysine methylation. Nat. Rev. Mol. Cell Biol. *6*, 838–849. https://dx.doi.org/10.1038/nrm1761.

Maurer-Stroh, S., Dickens, N.J., Hughes-Davies, L., Kouzarides, T., Eisenhaber, F., and Ponting, C.P. (2003). The Tudor domain 'Royal Family': Tudor, plant Agenet, Chromo, PWWP and MBT domains. Trends Biochem. Sci. *28*, 69–74. https://dx.doi.org/10.1016/S0968-0004(03)00004-5.

Musselman, C.A., and Kutateladze, T.G. (2009). PHD fingers: epigenetic effectors and potential drug targets. Mol. Interv. *9*, 314–323. https://dx.doi.org/10.1124/mi.9.6.7.

Perry, J., and Zhao, Y. (2003). The CW domain, a structural module shared amongst vertebrates, vertebrate-infecting parasites and higher plants. Trends Biochem. Sci. *28*, 576–580. https://dx.doi.org/10.1016/j.tibs.2003.09.007.

Petrucco, S., and Percudani, R. (2008). Structural recognition of DNA by poly(ADP-ribose)polymerase-like zinc finger families. FEBS J. *275*, 883–893. https://dx.doi.org/10.1111/j.1742-4658.2008.06259.x.

Pierce, R.J., Dubois-Abdesselem, F., Lancelot, J., Andrade, L., and Oliveira, G. (2012). Targeting schistosome histone modifying enzymes for drug development. Curr. Pharm. Des. *18*, 3567–3578.

Portela, A., and Esteller, M. (2010). Epigenetic modifications and human disease. Nat. Biotechnol. *28*, 1057–1068. https://dx.doi.org/10.1038/nbt.1685.

Prusty, D., Mehra, P., Srivastava, S., Shivange, A.V., Gupta, A., Roy, N., and Dhar, S.K. (2008). Nicotinamide inhibits *Plasmodium falciparum* Sir2 activity *in vitro* and parasite growth. FEMS Microbiol. Lett. *282*, 266–272. https://dx.doi.org/10.1111/j.1574-6968.2008.01135.x.

Qin, S., and Min, J. (2014). Structure and function of the nucleosome-binding PWWP domain. Trends Biochem. Sci. *39*, 536–547. https://dx.doi.org/10.1016/j.tibs.2014.09.001.

Saunders, L.R., and Verdin, E. (2007). Sirtuins: critical regulators at the crossroads between cancer and aging. Oncogene *26*, 5489–5504. https://dx.doi.org/10.1038/sj.onc.1210616.

Sharma, T., Bansal, R., Haokip, D.T., Goel, I., and Muthuswami, R. (2015). SMARCAL1 negatively regulates C-Myc transcription by altering the conformation of the promoter region. Sci. Rep. *5*, 17910. https://dx.doi.org/10.1038/srep17910.

Suganuma, T., and Workman, J.L. (2011). Signals and combinatorial functions of histone modifications. Annu. Rev. Biochem. *80*, 473–499. https://dx.doi.org/10.1146/annurev-biochem-061809-175347.

Sunil, S., Hossain, M.J., Ramasamy, G., and Malhotra, P. (2008). Transient silencing of *Plasmodium falciparum* Tudor staphylococcal nuclease suggests an essential role for the protein. Biochem. Biophys. Res. Commun. *372*, 373–378. https://dx.doi.org/10.1016/j.bbrc.2008.05.033.

Tamura, K., Peterson, D., Peterson, N., Stecher, G., Nei, M., and Kumar, S. (2011). MEGA5: molecular evolutionary genetics analysis using maximum likelihood, evolutionary distance, and maximum parsimony methods. Mol. Biol. Evol. *28*, 2731–2739. https://dx.doi.org/10.1093/molbev/msr121.

Thompson, J.D., Higgins, D.G., and Gibson, T.J. (1994). CLUSTAL W: improving the sensitivity of progressive multiple sequence alignment through sequence weighting, position-specific gap penalties and weight matrix choice. Nucleic Acids Res. *22*, 4673–4680.

Tsukada, Y., Fang, J., Erdjument-Bromage, H., Warren, M.E., Borchers, C.H., Tempst, P., and Zhang, Y. (2006). Histone demethylation by a family of JmjC domain-containing proteins. Nature *439*, 811–816. https://dx.doi.org/10.1038/nature04433.

Vainio, S., Genest, P.A., ter Riet, B., van Luenen, H., and Borst, P. (2009). Evidence that J-binding protein 2 is a thymidine hydroxylase catalyzing the first step in the biosynthesis of DNA base J. Mol. Biochem. Parasitol. *164*, 157–161. https://dx.doi.org/10.1016/j.molbiopara.2008.12.001.

Van, Hove, J.L., Josefsberg, S., Freehauf, C., Thomas, J.A., Thuy le, P., Barshop, B.A., Woontner, M., Mock, D.M., Chiang, P.W., Spector, E., *et al.* (2008). Management of a patient with holocarboxylase synthetase deficiency. Mol. Genet. Metab. *95*, 201–205. https://dx.doi.org/10.1016/j.ymgme.2008.09.006.

Vanagas, L., Jeffers, V., Bogado, S.S., Dalmasso, M.C., Sullivan, W.J., Jr., and Angel, S.O. (2012). Toxoplasma histone acetylation remodelers as novel drug targets.

Expert Rev. Anti. Infect. Ther. *10*, 1189–1201. https://dx.doi.org/10.1586/eri.12.100.

Verdin, E., Dequiedt, F., and Kasler, H.G. (2003). Class II histone deacetylases: versatile regulators. Trends Genet. *19*, 286–293. https://dx.doi.org/10.1016/S0168-9525(03)00073-8.

Vergnes, B., Sereno, D., Madjidian-Sereno, N., Lemesre, J.L., and Ouaissi, A. (2002). Cytoplasmic SIR2 homologue overexpression promotes survival of *Leishmania* parasites by preventing programmed cell death. Gene *296*, 139–150.

Webby, C.J., Wolf, A., Gromak, N., Dreger, M., Kramer, H., Kessler, B., Nielsen, M.L., Schmitz, C., Butler, D.S., Yates, J.R., 3rd, *et al.* (2009). Jmjd6 catalyses lysyl-hydroxylation of U2AF65, a protein associated with RNA splicing. Science *325*, 90–93. https://dx.doi.org/10.1126/science.1175865.

Yamane, K., Toumazou, C., Tsukada, Y., Erdjument-Bromage, H., Tempst, P., Wong, J., and Zhang, Y. (2006). JHDM2A, a JmjC-containing H3K9 demethylase, facilitates transcription activation by androgen receptor. Cell *125*, 483–495. https://dx.doi.org/10.1016/j.cell.2006.03.027.

Yun, M., Wu, J., Workman, J.L., and Li, B. (2011). Readers of histone modifications. Cell Res. *21*, 564–578. https://dx.doi.org/10.1038/cr.2011.42.

Yusufzai, T., and Kadonaga, J.T. (2008). HARP is an ATP-driven annealing helicase. Science *322*, 748–750. https://dx.doi.org/10.1126/science.1161233.

Yusufzai, T., and Kadonaga, J.T. (2010). Annealing helicase 2 (AH2), a DNA-rewinding motor with an HNH motif. Proc. Natl. Acad. Sci. U S A *107*, 20970–20973. https://dx.doi.org/10.1073/pnas.1011196107.

Role of Hypoxia Inducible Factor-1 in *Leishmania*–Macrophage Interaction: A New Therapeutic Paradigm

2

Amit K. Singh, Vishnu Vivek G., Shalini Saini, Sandhya Sandhya and Chinmay K. Mukhopadhyay*

Special Centre for Molecular Medicine, Jawaharlal Nehru University, New Delhi, India.

*Correspondence: ckm2300@mail.jnu.ac.in

https://doi.org/10.21775/9781910190715.02

Abstract

Intracellular parasites use host components for their survival and growth after invasion within host, while hosts spread out their innate immune defences to deny any advantage to the invading parasite. Research in the last decade has provided evidence of oxygen-sensing mammalian transcription factor hypoxia-inducible factor-1 (HIF-1) as a master controller of innate immune response of phagocytes against various intracellular and extracellular pathogens. In response to most of these pathogens host phagocytes increase transcription of HIF-1α, the regulatory component of HIF-1, to express various effector molecules against invaders. The involvement of NFκB in regulating HIF-1α transcription further strengthened the paradigm. However, more recent evidence has revealed that protozoan parasite *Leishmania donovani* (LD), the causative agent of fatal visceral leishmaniasis, in contrast, can promote and exploit HIF-1 activation for its survival advantage within host macrophages. HIF-1 is a heterodimer of regulatory subunit HIF-1α and constitutive HIF-1β. In conditions of oxygen deficiency or cellular iron depletion, expression of HIF-1α is regulated by a post-translational protein stability mechanism mediated by a family of prolyl hydroxylases (PHDs), while during phagocytic invasion by pathogens HIF-1α is regulated mainly by a transcriptional mechanism. Interestingly, LD activates HIF-1 by both transcriptional and post-translational protein stability mechanisms. This chapter will summarize the detailed mechanism of HIF-1 activation and its potential role in survival advantage of intracellular LD within host macrophages. Owing to its pivotal role in angiogenesis and cancer, HIF-1 is a crucial drug target and matter of intense research. The recent finding of role of host HIF-1 in survival and growth of LD thus presents an opportunity to repurpose the HIF-1 inhibitors as potential drugs against visceral leishmaniasis.

Introduction

Hypoxia-inducible factor-1 (HIF-1) is an oxygen-sensing transcription factor that, under conditions of depleted oxygen, modulates the genes responsible for energy homeostasis, growth and proliferation, iron homeostasis and erythropoiesis for survival of the cell (Semenza, 2001). It is normally expressed in wounds, tumours and atherosclerotic lesions, of which hypoxia is a common feature. HIF-1 comprises an oxygen regulatory subunit HIF-1α and a constitutively expressed subunit HIF-1β. It is now well established that HIF-1α can also be regulated under normoxic conditions by several factors such as infectious agents, iron deprivation, insulin and other hormones (Mukhopadhyay *et al.*, 2010). Recent studies have further revealed the crucial role of HIF-1 in regulating the host innate immune response towards pathogens.

Leishmaniasis represents a group of parasitic diseases caused by intracellular protozoan parasites of the genus *Leishmania*. It is considered as one of the most neglected tropical diseases, with 2 million new cases occurring annually, 1.5 million of which are affected by cutaneous leishmaniasis and around 0.5 million by visceral leishmaniasis. Visceral leishmaniasis (VL), also known as kala-azar, is caused by *Leishmania donovani* (LD). It is the most severe form of leishmaniasis and reported to claim more than 50,000 lives annually. VL is a systemic infection mainly affecting the reticuloendothelial system such as the liver, spleen and lymph nodes. *Leishmania* has a digenetic life cycle. The flagellar promastigote form resides in the gut of the female phlebotomine sandfly and is transmitted to the mammalian host during the sandflies blood meal. Inside the mammalian host it infects the macrophages and resides within the parasitophorous vacuoles (PV), where it is transformed to the non-motile amastigote form. Intracellular *Leishmania* successfully evades the host immune response as the amastigote form is adapted to survive and multiply in the acidic and harsh environment of the PV.

Recent reports established the crucial role of HIF-1 in determining bactericidal capacity of the phagocytes (Zarember and Malech, 2005). It has been revealed that the regulatory subunit HIF-1α is induced by bacterial infection even under normoxia and produces key immune effector molecules, including granule proteases, antimicrobial peptides, nitric oxide and TNF-α for maintaining immunity of host (Peyssonnaux *et al.*, 2005). Although the role of HIF-1 in bacterial infections is well appreciated but its role during infection by protozoan parasites remains far less unexplored. An increased HIF-1α in cutaneous lesions induced by *Leishmania amazonensis* infection was reported first (Arrais-Silva *et al.*, 2005). A further report suggested that HIF-1 activation in mononuclear phagocytes during infection by *Leishmania amazonensis* might be independent of hypoxia (Degrossoli *et al.*, 2007). Recently, it has been revealed that HIF-1 is activated in macrophages by *Leishmania donovani* infection for the survival benefit of the parasite (Singh *et al.*, 2012). This chapter will elaborate the understanding the molecular basis of HIF-1 activation by various stimuli and infections including LD.

Structure of HIF-1

HIF-1 is heterodimer of oxygen regulated HIF-1α subunit and the constitutively expressed HIF-1β subunit. These subunits belong to the basic helix–loop–helix-Per-ARNT-Sim (bHLH–PAS) protein family (Wang *et al.*, 1995). HIF-1β is also known as aryl hydrocarbon nuclear translocator (ARNT). It was identified as the binding partner of the aryl hydrocarbon receptor (Wood *et al.*, 1996). It is expressed constitutively independent of oxygen availability. Expression of HIF-1α is strongly regulated by availability of oxygen. It is rapidly degraded during normoxia because of its very short half life ($t_{1/2} \approx 5$ minutes) (Salceda and Caro, 1997). The capacity of oxygen sensing is dependent on HIF-1α due to its structural attributes. HIF-1α contains ODDD (oxygen-dependent degradation domain) through which it is regulated by oxygen (Masson *et al.*, 2001). It also consists of two transactivation domains, one at the N-terminal (NAD) and the other at the C-terminal (CAD). It is through these two regions HIF-1 interacts with the hypoxia response elements present in targeted genes to cause transcriptional activation (Jiang *et al.*, 1997).

Regulation of HIF-1α

HIF-1α, the oxygen regulated subunit of HIF-1, demonstrates an interesting mechanism of regulation by oxygen. During normoxia the ODDD of HIF-1α gets hydroxylated at two proline residues at positions 402 and 564 by a family of 2-oxoglutarate-dependent dioxygenases namely prolyl hydroxylases (PHDs) (Fandrey *et al.*, 2006). In general, hydroxylation of proline residues is carried out by PHDs in presence of oxygen as well as cofactors Fe^{2+} and ascorbate (Kivirikko *et al.*, 1989). Along with hydroxylation of proline residues at the ODDD of HIF-1α, acetylation of Lysine at position 532 by an acetyl transferase called as arrest-defective-1 (ARD-1) has been observed (Jeong *et al.*, 2002). These post-translational modifications give access to pVHL (Van Hippel Lindau tumour suppressor protein) to bind to HIF-1α. This complex then binds to protein elongin C, elongin B, cullin-2 and Rbx 1 to form E3 ligase complex leading to polyubiquitination of HIF-1α followed by its degradation by the proteasome pathway (Semenza, 2007a).

Under hypoxic conditions, oxygen becomes the rate-limiting factor minimizing the hydroxylation of proline residues on HIF-1α by prolyl hydroxylases, thereby preventing the interaction of pVHL with HIF-1α. As a result, proteolytic degradation of HIF-1α is prevented, subsequently allowing the accumulation of HIF-1α protein during hypoxia. The activity of prolyl hydroxylase also depends on the availability of iron and the reducing agent ascorbic acid. Any situation that affects the availability of iron or ascorbic acid may result in HIF-1α stability (Ke *et al.*, 2006; Pan *et al.*, 2007).

Regulation of HIF-1α regulation during normoxia

HIF-1α was initially attributed to hypoxic conditions, but further studies provided evidence of increased expression of HIF-1α under normoxia by various stimuli such as growth factors and hormones, cytokines and transition metals as well as infectious agents (Keitzmann and Gorlach, 2005; Maxwell and Salnikow, 2004; Werth *et al.*, 2010; Mukhopadhyay *et al.*, 2010).

Factors and mechanism of HIF-1 activation during normoxia

Cellular exposure of insulin was the first normoxic condition reported to activate HIF-1 other than hypoxia mimetics such as cobalt, nickel and iron chelators in hepatic and skeletal muscle cells (Zelzer *et al.*, 1998). Insulin-induced activation of HIF-1 was found to depend on generation of reactive oxygen species (ROS) (Biswas *et al.*, 2007). Later it was reported that ROS-dependent activation of transcription factor Sp1 played a critical role in HIF1α transcription (Biswas *et al.*, 2013).

Hormones and growth factors such as angiotensin II, thrombin and platelet-derived growth factors have been shown to activate HIF-1 in vascular smooth muscle cells (Richard *et al.*, 2000). Recently, it was shown that angiotensin II-mediated H_2O_2 generation depletes intracellular antioxidant ascorbic acid, an essential co-factor of prolyl hydroxylase, resulting in HIF-1α stability (Page *et al.*, 2008). Cytokines such as TNF-α as well as IL-1β can also regulate HIF-1 under normoxic conditions dependent on generation of ROS (Haddad and Land, 2001; Haddad, 2002).

Transition metals such as cobalt, nickel and copper activate HIF-1 in several cell types by depleting cellular ascorbic acid, and thus impairing the prolyl hydroxylase activity (Salnikow *et al.*, 2004). Iron depletion caused by treatment with iron chelators such as desferrioxamine (DFO) also leads to HIF-1 activation by affecting prolyl hydroxylase activity (Hunt *et al.*, 1979; Wang *et al.*, 1993).

Bacterial lipopolysaccharide (LPS) strongly induces HIF-1 in normoxia as shown in differentiated THP-1 cells by increasing HIF-1α protein expression through a translation-dependent pathway (Blouin *et al.*, 2004; Frede *et al.*, 2006). LPS induced ROS generation was detected as critical event for HIF-1 activation (Nishi *et al.*, 2008). This pathway is independent of NFκB mediation but requires the expression of TLR4 in THP-1 cells (Nishi *et al.*, 2008).

Regulation of HIF-1 activity

The mechanism that regulates the transcriptional activity of HIF-1 acts through the modulation of the transactivation domains N-TAD and C-TAD of HIF-1α. To be functionally active these transactivation domains recruit transcriptional co-activators such as CBP/P300, SRC-1 and TIF2 (Arany *et al.*, 1996; Kallio *et al.*, 1998; Carrero *et al.*, 2000). Under normoxic conditions the asparagine residue at position 803 in the C-TAD is hydroxylated by the asparaginyl hydroxylase factor inhibiting HIF-1 (FIH-1) and blocks the interaction of HIF-1α with CBP/P300 (Cockman *et al.*, 2009, Lando *et al.*, 2002b), thereby preventing the transcriptional activation of HIF-1 during normal oxygen tension. FIH-1 is another 2-oxoglutarate-dependent dioxygenase that requires Fe^{2+} and ascorbate as co-factors just like the prolyl hydroxylases (Lando *et al.*, 2002a). During hypoxia or iron deprivation FIH-1 activity is blocked to inhibit the hydroxylation of the asparagine residue at the C-TAD, thus allowing HIF-1α to interact with CBP/P300. HIF-1α then translocates to the nucleus and forms heterodimeric complex with HIF-1β forming active HIF-1 that binds to the HRE of its target genes and induce their transcription (Lando *et al.*, 2002a).

Role of HIF-1 in physiology

To ensure to adapt to conditions during hypoxia and even in specific normoxic conditions, activated HIF-1 binds to the hypoxia response element (HRE; 5′-A/GCGTG-3′) located in the enhancer or promoter regions of more than 1000 target genes (Charpentier *et al.*, 2016). HIF-1 activates genes with varying functions primarily to maintain oxygen and energy homeostasis to sustain important life processes. It is involved in several critical physiological functions but not limited to as described below.

Erythropoiesis

Erythropoietin is a hormone that maintains red blood cell mass by promoting the survival, proliferation and differentiation of erythrocytic progenitors (Jelkmann, 2010; Stockmann and Fandrey, 2006). An increased number of red blood cells ensures increased supply of oxygen during hypoxia (Semenza *et al.*, 1991b). During systemic hypoxia the kidney acts as a highly sensitive oxygen sensor and plays a central role in mediating the hypoxic induction of red blood cell production by erythropoiesis. In response to hypoxia, HIF-1 binds to HRE present in the enhancer region located at the 3′ end of the erythropoietin gene to maintain erythropoiesis (Semenza *et al.*, 1991a).

Iron homeostasis

At the molecular level the roles of iron and oxygen are so much intertwined that deficiencies of either of these have a very similar effect. Lack of both iron and oxygens affect the energy generation process by the electron transport chain (ETC). In an attempt to recover both oxygen and energy homeostasis during hypoxia, HIF-1 activates the expression of major iron homeostasis genes including transferrin, transferrin receptor, ceruloplasmin and haem oxygenase-1 to increase the iron uptake and to promote iron transport required for erythropoiesis to enhance oxygen uptake and its delivery to hypoxic cells (Li and Ginzburg, 2010; Chepelev and Willmore, 2011).

Angiogenesis

In multicellular organisms, supply of oxygen to all metabolizing cells is ensured by a specialized system of blood vessels, as simple diffusion of oxygen would be inadequate to maintain oxygen supply to all cells in solid tissues (Pugh and Ratcliffe, 2003). Angiogenesis is the formation of new blood vessels from pre-existing ones; it is a complex and highly regulated process involving various gene products responsible for proliferation, migration and remodelling of endothelial cells in response to growth factors and cytokines (Bussolino *et al.*, 1997).

As hypoxia acts as the overall inducer of this process (Pugh and Ratcliffe, 2003), HIF-1 activated during hypoxia plays a major role in the transcription of genes involved in angiogenesis, the most important of which is vascular endothelial growth factor (VEGF). VEGF is strongly expressed in areas of tissue injury and in deeper regions of cancerous tissues where oxygen tension is low (Kuo *et al.*, 1999).

HIF-1 also regulates several other genes involved in formation and maintenance of blood vessels, like inducible nitric oxide synthase (iNOS/NOS2) that governs vascular tone (Han *et al.*, 2007).

Glucose metabolism

Under low oxygen tension, cells shift their energy-generating metabolic pathway from the oxygen-dependent tricarboxylic acid cycle (TCA) to the oxygen-independent glycolysis process (Papandreoul *et al.*, 2006). As energy generation via the TCA cycle and oxidative phosphorylation is more efficient than glycolysis alone, cells exposed to hypoxia increase their glucose uptake by inducing glucose transporters like GLUT1 and GLUT3 mediated by HIF-1 to sustain their energy requirement through glycolysis (Ren *et al.*, 2008). Increased glucose uptake is accompanied by HIF-1-mediated increased expression of all the 12 enzymes involved in glycolysis (Semenza *et al.*, 1994). Each step in glycolysis is regulated by substrate and product availability, so increasing the expression of glycolytic enzymes in the cell favours energy generation via glycolysis (Denko, 2008). In hypoxic tissues, HIF-1 activates genes to suppress mitochondrial activity so that scarce oxygen can be utilized for non-energy-producing cellular activities (Semenza, 2009). HIF-1 also blocks mitochondrial biogenesis by directly regulating the expression of MXI1 that displaces MAX and binds to MYC preventing the MYC/MAX complex from activating the transcription of genes involved in mitochondrial biogenesis (Dang *et al.*, 2009).

Role of HIF-1 in infection and innate immunity

Infection, injury, autoimmune diseases or any kind of insult to tissues may lead to inflammation in the affected region. Inflamed regions have some characteristic features such as hypoxia or anoxia (no oxygen), hypoglycaemia (low blood glucose), acidosis (high H^+ concentration) and abundant free oxygen radicals (Saadi *et al.*, 2002). Stressed tissues are infiltrated by the neutrophils and macrophages that act as sentinels and pose as the first line of defence by engulfing the invading micro-organism and removing cellular debris of the damaged tissues (Bosco *et al.*, 2008). HIF-1 plays critical role in resolving such stress situations as it sustains the survival of macrophages and neutrophils by activating genes responsible for oxygen and energy homeostasis (Burke *et al.*, 2003). This enables the phagocytic cells to survive and perform their functions effectively in an anaerobic microenvironment where pathogens multiply during infection (Roiniotis *et al.*, 2009).

HIF-1 plays an important role in regulating innate immunity in response to bacterial infection (Zarember and Malech, 2005). Bacterial infections are known to activate HIF-1 by oxygen-dependent and -independent pathways (Werth *et al.*, 2010). It was found that HIF-1α knockout macrophages showed reduced killing of intracellular bacteria compared to the wild-type macrophages (Peyssonnaux *et al.*, 2008). HIF-1 increasesthe expression of granule proteases, namely neutrophil elastase and cathepsin G, which are important components of innate immune mechanism of the myeloid cells. These proteases either have a direct antimicrobial effect or they act on precursors of inactive antimicrobial peptides (Cole *et al.*, 2001). HIF-1 also increases the expression of cathelicidin-related antimicrobial peptide (CRAMP) (Peyssonnaux *et al.*, 2005). It also up-regulates the expression of inducible nitric oxide synthase (iNOS), which produces nitric oxide (NO), a strong antimicrobial agent that acts against various bacterial species. The increased generation of NO also plays the role of a signalling molecule to enhance the expression of the cytokine TNF-α that subsequently activates the pro-inflammatory response against bacterial infection (Peyssonnaux *et al.*, 2005). In fact, HIF-1 is present in an amplification loop, in which it enhances the production of NO and TNF-α. They in turn increase the stability of HIF-1α in the infection foci (Zhou *et al.*, 2003). It is also reported that activation of HIF-1 is mediated by bacterial lipopolysaccharide LPS by directly up-regulating HIF-1α transcript (Frede *et al.*, 2006). The mechanism shows that LPS-induced activation of NFκB regulates HIF-1α promoter activity, resulting in increased transcript synthesis suggesting a relationship between inflammation, infection and HIF-1 activation (Frede *et al.*, 2006; van Uden *et al.*, 2008).

Crosstalk between NFκB and HIF-1α to mediate innate immune response

Nuclear factor κB (NFκB) is a transcription factor that activates a wide variety of genes involved in critical roles in innate and adaptive immune responses (Liang *et al.*, 2004). Recent studies revealed the complex relationship between NFκB and HIF-1α. Both these transcription factors initiate defensive response during stress situation. Basal activity of NFκB is required to produce sufficient amount of HIF-1α mRNA suggesting an evolutionary conservation of innate immune pathway (Rius *et al.*, 2008). It is also involved in inducing expression of HIF-1α mRNA during bacterial infection or injury. This ensures activation of HIF-1 modulated genes responsible for mounting the innate immune response or the pro-inflammatory response (Rius *et al.*, 2008).

Various pro-inflammatory cytokines (TNF-α, IL-1β) or infectious agents stimulate NFκB activation by different receptors and adaptor proteins but the focal point of NFκB regulation revolves around the phosphorylation and degradation of its inhibitor IκB. NFκB is normally bound to its inhibitor IκB (inhibitor of NFκB) in unstimulated cells. IκB is regulated by IκB kinase (IKK) complex. Upon stimulation IKK phosphorylates IκB and marks it for ubiquitination and subsequent degradation. This sets the NFκB complex free for translocation to the nucleus for subsequent activation of target genes (Scheidereit, 2006; Sun *et al.*, 2008). Phosphorylation of IκB is signalled by various stimulants via reactive oxygen species (ROS). ROS generated during pro-inflammatory responses is an essential partner in NFκB activation (Kaul *et al.*, 1996; Gloire *et al.*, 2006). ROS is also involved in an HIF-1α-stabilizing mechanism (Mukhopadhyay *et al.*, 2010).

There is also evidence that NFκB activation precedes HIF-1α accumulation during hypoxia. NFκB binds to the specific binding site on HIF-1α promoter and activates its transcription during hypoxia so that optimal accumulation of HIF-1α can occur by stabilization for an enhanced innate immune response (Taylor, 2008; van Uden *et al.*, 2008).

HIF-1 and leishmaniasis

Although the role of HIF-1 in bacterial infections is well established, its role in infection by protozoan parasites remains far less explored. An increased HIF-1α expression was first reported in cutaneous lesions induced by *Leishmania amazonensis* infection (Arrais-Silva *et al.*, 2005). A follow-up report suggested that HIF-1α activation in mononuclear phagocytes during infection by *Leishmania amazonensis* could be independent of hypoxia (Degrossoli *et al.*, 2007). The implication of HIF-1 activation in host macrophages by *Leishmania* infection and the molecular mechanism of HIF-1 activation remained mostly unexplored until it was reported that *Leishmania donovani* could utilize host HIF-1 activation for its survival advantage within host macrophages (Singh *et al.*, 2012).

HIF-1 activation is an integral part of the innate immune system as it mediates the pro-inflammatory response by inducing the expression of iNOS, cytokines and several antimicrobial peptides (Nizet and Johnson, 2009); thus, it is intriguing that LD overcomes challenges from host and exploits HIF-1 activation for its own benefit.

Leishmania donovani activates HIF-1 by dual mechanism

HIF-1 activation in general has been reported to occur by two well-defined and distinct mechanisms in response to various stimuli. It occurs either by increased transcription of HIF-1α gene or by enhanced post-translational stability of HIF-1α protein (Rius *et al.*, 2008). *Yersinia enterocolitica* activates HIF-1 by inducing post-translational stabilization of HIF-1α by sequestration iron through siderophores (Hartmann *et al.*, 2008). LD infection is the first report that an infection of pathogen leading to both HIF-1α transcription and post-translation protein stability for activation of HIF-1 (Singh *et al.*, 2012). LD infection simultaneously results into increased expression of HIF-1α transcript both in *in vitro* and *in vivo* infections and HIF-1α protein stability by depletion of intracellular iron (Singh *et al.*, 2012).

Though the precise molecular mechanism of LD-mediated HIF-1α transcript synthesis is not known yet; however, a recent study reported involvement of IRF-5 during *in vivo* LD infection (Hammami *et al.*, 2015). It was further revealed that LD infection in macrophages increased the expression of HIF-1α protein due its enhanced protein stability (Singh *et al.*, 2012). HIF-1α has a very short half-life of around 5 minutes under normoxia (Salceda and Caro, 1997). Newly synthesized HIF-1α is immediately degraded following hydroxylation at specific proline residues 402 and 564 by a family of dioxygenases (Masson *et al.*, 2001). There are three isoforms of this dioxygenase, termed prolyl hydroxylase (PHD) 1, 2 and 3, all having the potential to hydroxylate HIF-1α (Huang *et al.*, 2002). PHD2 is the most active and main rate limiting enzyme controlling the expression of HIF-1α (Berra *et al.*, 2003). An increased mRNA and protein of PHD2 was reported during hypoxia (Epstein *et al.*, 2001), providing a possible mechanism of self-regulation by HIF-1α. PHD2 is mainly localized in the cytoplasm, while PHD1 is localized in the nucleus and PHD3 is found in both cytoplasm and nucleus (Metzen *et al.*, 2003). The stability of HIF-1α is tightly regulated by the PHDs, whereas PHD activity itself is dependent on oxygen level (Kaelin *et al.*, 2008). PHD activity is also dependent on the availability of its co-factors in the cell such as iron and ascorbate (Pan *et al.*, 2007). Infections caused by *Y. enterocolitica* have been reported to stabilize HIF-1α in Peyer's patches mainly by sequestering iron through its siderophores thereby suppressing PHD activity (Hartmann *et al.*, 2008). In a recent report it was also shown that infections caused by *T. gondii* stabilize HIF-1α by suppressing the expression of PHD2 enzyme (Wiley *et al.*, 2010). LD is also observed to suppress the cellular PHD activity to induce HIF-1α stability (Singh *et al.*, 2012). LD-mediated HIF-1α induction is distinctly different from the bacterial or LPS mediated HIF-1 activation. Reports suggest that LPS treatment increases the translation of the elevated HIF-1α mRNA levels via PI3K pathway, a pathway distinct from hypoxia mediated stability of HIF-1α (Blouin *et al.*, 2004). Contrary to this, another report suggests that LPS increases the stability of

HIF-1α by suppressing the mRNA levels of prolyl hydroxylases 2 and 3 (Peyssonaux *et al.*, 2007). In case of infection by LD there was no change in the protein expression of PHD2 (Singh *et al.*, 2012). The same study also found, using a hypoxyprobe marker, that LD infection did not induce hypoxia in the infected macrophages. This suggests that LD infection induces HIF-1α stability by a mechanism not involving hypoxia but still suppression of PHD activity was detected. LD has a unique ability to deplete the labile iron pool (LIP) of the host during its intracellular stay within macrophage (Das *et al.*, 2009), which may explain the molecular mechanism of LD-induced decrease in PHD activity as PHD2 expression remained unaffected in infected macrophage (Singh *et al.*, 2012). Increased HIF-1α mRNA and protein stability due to affected PHD activity strongly suggested utilization of two distinct mechanisms of HIF-1 activation in host macrophages during LD infection.

HIF-1 activation plays an important role in eliminating bacterial infections. It has been reported that macrophages derived from HIF-1α null mice failed to contain bacterial infection (Peyssonnaux *et al.*, 2005). In contrast, intracellular LD failed to survive and multiply within HIF-1α silenced macrophages (Singh *et al.*, 2012). This finding indicates that the ability of LD to survive within macrophages is dependent on its ability to activate HIF-1 to exploit the HIF-1 targeted gene products. However, the precise mechanism by which LD exploits HIF-1 activation for its survival benefit within macrophages is still unknown. It is important to mention here that HIF-1α silencing did not have any influence on the capacity of phagocytosis of the host macrophages as there was no difference in the number of LD that entered in HIF-1α silenced macrophages or control confirming that LD failed to multiply in HIF-1α silenced macrophage. In contrary, when macrophages contained stably expressed HIF-1α mutant (HIF-1α P/A in which Pro402 and Pro564 were mutated to Ala) intracellular LD was grown faster (Singh *et al.*, 2012) further confirming the survival benefit of intracellular LD on activation of host HIF-1. A comparative mechanism of HIF-1 activation during hypoxia, bacterial infection and LD infection in macrophages has been provided in Fig. 2.1.

HIF-1 as possible drug target in visceral form of leishmaniasis

It has now been revealed that HIF-1 activation by LD infection resulted into survival benefits of the intracellular parasite although the precise role of the activated HIF-1 for LD survival still remained elusive. However, this information provides us a novel opportunity to combat intracellular LD in mammalian host by targeting HIF-1. Visceral leishmaniasis (VL) or kala-azar caused by *Leishmania donovani* is a severe form of leishmaniasis and is lethal if left untreated. Ninety per cent of the total worldwide visceral leishmaniasis cases occur in Bangladesh, India, Nepal, Sudan, Ethiopia and Brazil. Recently, it is reported to cause more than 50,000 deaths annually (Chappuis *et al.*, 2007). It

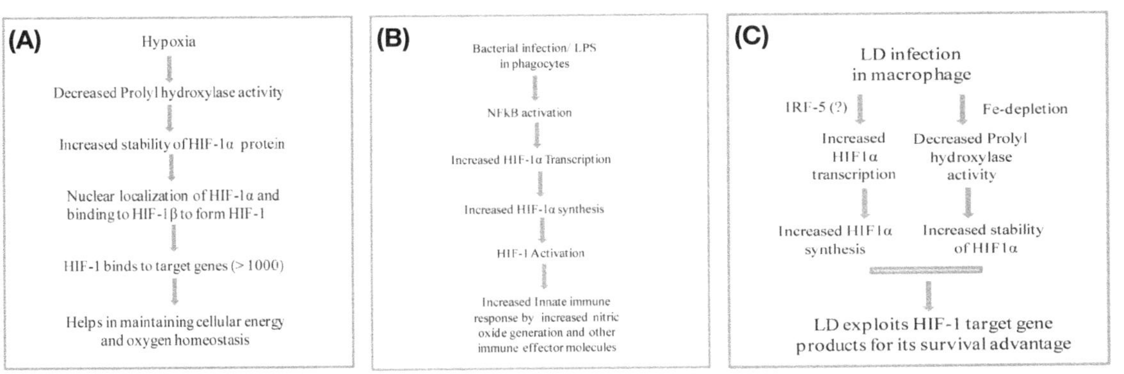

Figure 2.1 A schematic diagram to show mechanisms of HIF-1 activation during (A) hypoxia, (B) bacterial infection and (C) LD infection.

is mostly prevalent in remote and rural areas generally inhabited by poor communities where most of the cases go undetected as reported in studies performed in several clinics in Southern Sudan (Alvar *et al.*, 2006). This disease strikes with multiple impact on the affected people as the increasing cost of diagnosis and treatment affects their day to day living (Boelaert *et al.*, 2010; Thomton *et al.*, 2010). In regions endemic for HIV infection, VL occurs as an opportunistic infection, increasing the mortality risk more than 3-fold (ter Horst *et al.*, 2008). Most alarmingly, the parasite has been reported to develop resistance towards the antimonial drugs that were used once upon a time as first line of treatment (Frezard *et al.*, 2009). Amphotericin B is another effective drug against VL but like antimonials it also causes several life-threatening side-effects (Moore *et al.*, 2010). Miltefosine was initially developed as a cancer treatment but has been proven to be an effective anti-leishmanial drug; however, recent studies have shown that the parasite is developing resistance to this drug (Pérez-Victoria *et al.*, 2003), indicating the need for new drug regimen. Like miltefosine, other anti-cancer compounds were also considered as anti-leishmanials (Fuertes *et al.*, 2008). In cancerous cells HIF-1 activation is a prominent phenomenon leading to activation of genes that play a major role in cancer progression (Kimbro and Simons, 2006, Martin *et al.*, 2011). In recent years many studies and trials are being undertaken to derive HIF-1 inhibitors as drugs against cancer (Onnis *et al.*, 2009; Semenza, 2007b). Since, LD uses HIF-1 for its survival benefit within host macrophages; it will be interesting to study the efficacy of the HIF-1 inhibitors as repurposing drug to treat VL.

Acknowledgements

This work has been supported by Department of Biotechnology, India to CKM. CKM also acknowledges financial support received from a program project on Molecular Parasitology sponsored by Department of Biotechnology, India; ICMR-CAR project to SCMM, JNU and DST-PURSE program to JNU.

References

Alvar, J., Yactayo, S., and Bern, C. (2006). Leishmaniasis and poverty. Trends Parasitol. 22, 552–557. https://doi.org/10.1016/j.pt.2006.09.004.

Arany, Z., Huang, L.E., Eckner, R., Bhattacharya, S., Jiang, C., Goldberg, M.A., Bunn. H.F., and Livingston, D.M. (1996). An essential role for p300/CBP in the cellular response to hypoxia. Proc. Natl. Acad. Sci. U.S.A. *93*, 12969–12973.

Arrais-Silva, W.W., Paffaro, V.A. Jr., Yamada, A.T., and Giorgio, S. (2005). Expression of hypoxia inducible factor-1alpha in the cutaneous lesions of BALB/c mice infected with *Leishmania amazonensis*. Exp. Mol. Pathol. *78*, 49–54. https://doi.org/10.1016/j.yexmp.2004.09.002.

Berra, E., Benizri, E., Ginouvès, A., Volmat, V., Roux, D., and Pouysségur, J. (2003). HIF prolyl-hydroxylase 2 is the key oxygen sensor setting low steady-state levels of HIF-1 alpha in normoxia. EMBO J. *22*, 4082-4090. https://doi.org/10.1093/emboj/cdg392.

Biswas, S., Gupta, M.K., Chattopadhyay, D., and Mukhopadhyay, C.K. (2007). Insulin-induced activation of hypoxia-inducible factor-1 requires generation of reactive oxygen species by NADPH oxidase. Am J Physiol Heart Circ. Physiol. 292, H758-H766. https://doi.org/10.1152/ajpheart.00718.2006.

Biswas, S., Mukherjee, R., Tapryal, N., Singh, A., and Mukhopadhyay, C.K. (2013). Insulin regulates hypoxia-inducible factor-1α transcription by reactive oxygen species sensitive activation of Sp1 in 3T3-L1 preadipocyte. PLOS ONE. 8(4):e62128. https://doi.org/10.1371/journal.pone.0062128.

Blouin, C.C., Pagé, E.L., Soucy, G.M., and Richard, D.E. (2004). Hypoxic gene activation by lipopolysaccharide in macrophages: implication of hypoxia-inducible factor 1alpha. Blood. *103*, 1124–1130. https://doi.org/10.1182/blood-2003-07-2427.

Boelaert, M., Meheus, F., Robays, J., and Lutumba, P. (2010). Socio-economic aspects of neglected diseases: sleeping sickness and visceral leishmaniasis. Ann Trop Med Parasitol. *104*, 535–542. https://doi.org/10.1179/136485910X12786389891641.

Bosco, M.C., Puppo, M., Blengio, F., Fraone, T., Cappello, P., Giovarelli, M., and Varesio, L. (2008). Monocytes and dendritic cells in a hypoxic environment: Spotlights on chemotaxis and migration. Immunobiology *213*, 733–749. https://doi.org/10.1016/j.imbio.2008.07.031.

Burke, B., Giannoudis, A., Corke, K.P., Gill, D., Wells, M., Ziegler-Heitbrock, L., and Lewis, C.E. (2003). Hypoxia-induced gene expression in human macrophages: implications for ischemic tissues and hypoxia-regulated gene therapy. Am J. Pathol. *163*, 1233–1243. https://doi.org/10.1016/S0002-9440(10)63483-9.

Bussolino, F., Mantovani, A., and Persico, G. (1997). Molecular mechanisms of blood vessel formation. Trends Biochem. Sci. *22*, 251–256.

Carrero, P., Okamoto, K., Coumailleau, P., O'Brien, S., Tanaka, H., and Poellinger, L. (2000). Redox-regulated recruitment of the transcriptional co activators CREB – binding protein and SRC-1 to hypoxia-inducible factor 1alpha. Mol Cell Biol. *20*, 402–415.

Charpentier, T., Hammami, A., and Stäger, S. (2016). Hypoxia inducible factor 1α: A critical factor for the immune response to pathogens and *Leishmania*. Cell Immunol. *309*, 42–49. https://doi.org/10.1016/j.cellimm.2016.06.002.

Chappuis, F., Sundar, S., Hailu, A., Ghalib, H., Rijal, S., Peeling, R.W., Alvar, J., and Boelaert, M. (2007). Visceral leishmaniasis: what are the needs for diagnosis, treatment and control? Nat. Rev. Microbiol. *5*, 873–882. https://doi.org/10.1038/nrmicro1748.

Chepelev, N.L., and Willmore, W.G. (2011). Regulation of iron pathways in response to hypoxia. Free Radic. Biol. Med. *50*, 645–666. https://doi.org/10.1016/j.freeradbiomed.2010.12.023.

Cockman, M.E., Webb, J.D., and Ratcliff, P.J. (2009). FIH-dependent asparaginyl hydroxylation of Ankyrin repeat domain-containing proteins. Ann. N. Y. Acad. Sci. *1177*, 9–18. https://doi.org/10.1111/j.1749-6632.2009.05042.x.

Cole, A.M., Shi, J., Ceccarelli, A., Kim, Y.H., Park, A., and Ganz, T. (2001). Inhibition of neutrophil elastase prevents cathelicidin activation and impairs clearance of bacteria from wounds. Blood *97*, 297–304.

Dang, C.V., Le, A., and Gao, P. (2009). MYC-induced cancer cell energy metabolism and therapeutic opportunities. Clin. Cancer Res. *15*, 6479–6483. https://doi.org/10.1158/1078-0432.CCR-09-0889.

Das, N.K., Biswas, S., Solanki, S., and Mukhopadhyay, C.K. (2009). *Leishmania donovani* depletes labile iron pool to exploit iron uptake capacity of macrophage for its intracellular growth. Cell. Microbiol. *11*, 83–94. https://doi.org/10.1111/j.1462-5822.2008.01241.x.

Degrossoli, A., Bosetto, M.C., Lima, C.B., and Giorgio, S. (2007). Expression of hypoxia-inducible factor 1alpha in mononuclear phagocytes infected with *Leishmania amazonensis*. Immunol. Lett. *114*, 119–125. https://doi.org/10.1016/j.imlet.2007.09.009.

Denko, N.C. (2008). Hypoxia, HIF-1 and glucose metabolism in the solid tumour. Nat. Rev. Cancer. *8*, 705–713. https://doi.org/10.1038/nrc2468.

Epstein, A.C., Gleadle. J.M., McNeill, L.A., Hewitson, K.S., O'Rourke, J., Mole, D.R., Mukherji, M., Metzen, E., Wilson, M.I., Dhanda, A., Tian, Y.M., Masson, N., Hamilton, D.L., Jaakkola, P., Barstead, R., Hodgkin, J., Maxwell, P.H., Pugh, C.W., Schofield, C.J., and Ratcliffe, P.J. (2001). *C. elegans* EGL-9 and mammalian homologs define a family of dioxygenases that regulate HIF by prolyl hydroxylation. Cell. *107*, 43–54.

Fandrey, J., Gorr, T.A., and Grassmann, M. (2006). Regulating cellular oxygen sensing by hydroxylation. Cardiovascular Res.*71*, 642–651. https://doi.org/10.1016/j.cardiores.2006.05.005.

Frede, S., Stockmann, C., Freitag, P., and Fandrey, J. (2006). Bacterial lipopolysaccharide induces HIF-1 activation in human monocytes via p44/42 MAPK and NF-kappaB. Biochem. J. *396*, 512–527. https://doi.org/10.1042/BJ20051839.

Frézard, F., Demicheli, C., and Ribeiro, R.R. (2009). Pentavalent antimonials: new perspectives for old drugs. Molecules *14*, 2317–2336. https://doi.org/10.3390/molecules14072317.

Fuertes, M.A., Nguewa, P.A., Castilla, J., Alonso, C., and Pérez, J.M. (2008). Anticancer compounds as leishmanicidal drugs: challenges in chemotherapy and future perspectives. Curr. Med. Chem. *15*, 433–439.

Gloire, G., Legrand-Poels, S., and Piette, J. (2006). NF-κappaB activation by reactive oxygen species: fifteen years later. Biochem Pharmacol. 72, 1493–1505. https://doi.org/10.1016/j.bcp.2006.04.011.

Haddad, J.J. (2002). Recombinant human interleukin (IL)-1 beta mediated regulation of hypoxia-inducible factor-1 alpha (HIF-1alpha) stabilization, nuclear translocation and activation requires an antioxidant/reactive oxygen species (ROS)-sensitive mechanism. European Cytokine Netw. *13*, 250–260.

Haddad, J.J., and Land, S.C. (2001). A non-hypoxic, ROS-sensitive pathway mediates TNF-alpha-dependent regulation of HIF-1 alpha. FEBS Lett. *505*, 269–274.

Hammami, A., Charpentier, T., Smans, M., and Stäger, S. (2015). IRF-5-mediated inflammation limits CD8⁺ T Cell expansion by inducing HIF-1α and impairing dendritic cell functions during *Leishmania* infection. PLOS Pathog. *11*(6), e1004938. https://doi.org/10.1371/journal.ppat.1004938.

Han, Y.L., He, D.L., Luo, Y., Cheng, H.P., and Zhu, G.F. (2007). Over-expression of hypoxia-inducible factor-1 alpha increases angiogenesis of LNCaP cells. Zhonghua Nan Ke Xue. *13*, 988–991.

Hartmann, H., Eltzschig, H.K., Wurz, H., Hantke, K., Rakin, A., Yazdi, A.S., Matteoli, G., Bohn, E., Autenrieth, I.B., Karhausen, J., Neumann, D., Colgan, S.P., and Kempf, V.A. (2008). Hypoxia-independent activation of HIF-1 by enterobacteriaceae and their siderophores. Gastroenterology *134*, 756–767. https://doi.org/10.1053/j.gastro.2007.12.008.

Hunt, J., Richards, R.J., Harwood, R., and Jacobs, A. (1979). The effect of desferrioxamine on fibroblasts and collagen formation in cell cultures. Br. J. Haematol. *41*, 69–76.

Jelkmann, W. (2011). Regulation of erythropoietin production. J. Physiol. *589*, 1251–1258. https://doi.org/10.1113/jphysiol.2010.195057.

Jeong, J.W., Bae, M.K., Ahn, M.Y., Kim, S.H., Sohn, T.K., Bae, M.H., Yoo, M.A., Song, E.J., Lee, K.J., and Kim, K.W. (2002). Regulation and destabilization of HIF-1alpha by ARD1-mediated acetylation. Cell. 27,111, 709–720.

Jiang, B.H., Zheng, J.Z., Leung, S.W., Roe, R., and Semenza, G.L. (1997). Transactivation and inhibitory domains of hypoxia-inducible factor 1alpha. Modulation of transcriptional activity by oxygen tension. J. Biol. Chem. *272*, 19253–19260.

Kaelin, W.G., Jr., and Ratcliffe, P.J. (2008). Oxygen sensing by metazoans: the central role of the HIF hydroxylase pathway. Mol. Cell. *30*, 393–402. https://doi.org/10.1016/j.molcel.2008.04.009.

Kallio, P.J., Okamoto, K., O'Brien, S., Carrero, P., Makino, Y., Tanaka, H., and Poellinger, L. (1998). Signal transduction in hypoxic cells: inducible nuclear translocation and recruitment of the CBP/p300 coactivator by the hypoxia-inducible factor-1alpha. EMBO J. *17*, 6573–6586. https://doi.org/10.1093/emboj/17.22.6573.

Kaul, N., and Forman, H.J. (1996). Activation of NF-κappa B by the respiratory burst of macrophages. Free Radic. Biol. Med. *21*, 401–405.

Ke, Q., and Costa, M. (2006). Hypoxia-inducible factor-1 (HIF-1). Mol Pharmocol. *70*, 1469–1480. https://doi.org/10.1124/mol.106.027029.

Kietzmann, T., and Gorlach, A. (2005). Reactive oxygen species in the control of hypoxia-inducible factor-mediated gene expression. Semin. Cell Dev. Biol. *16*, 474–486. https://doi.org/10.1016/j.semcdb.2005.03.010.

Kimbro, K.S., and Simons, J.W. (2006). Hypoxia-inducible factor-1 in human breast and prostate cancer. Endocr. Relat. Cancer *13*, 739–749. https://doi.org/10.1677/erc.1.00728.

Kivirikko, K.I., Myllylä, R., and Pihlaianiemi, T. (1989). Protein hydroxylation: prolyl 4-hydroxylase, an enzyme with four co-substrates and a multifunctional subunit. FASEB J. *3*, 1609–1617.

Kuo, N.T., Benhayon, D., Przybylski, R.D., Martin, R.J., and LaManna, J.C. (1999). Prolonged hypoxia increases vascular endothelial growth factor mRNA and protein in adult mouse brain. J. Appl. Physiol. *86*, 260–264.

Lando, D., Peet, D.J., Gorman, J.J., Whelan, D.A., Whitelaw, M.L., and Bruick, R.K. (2002a). FIH-1 is an asparaginyl hydroxylase enzyme that regulates the transcriptional activity of hypoxia-inducible factor. Genes Dev. *16*, 1466–1471. https://doi.org/10.1101/gad.991402.

Lando, D., Peet, D.J., Whelan, D.A., Gormann, J.J., and Whitelaw, M.L. (2002b). Asparagine hydroxylation of the HIF transactivation domain a hypoxic switch. Science. *295*, 858–861. https://doi.org/10.1126/science.1068592.

Li, H., and Ginzburg, Y.Z. (2010). Crosstalk between iron metabolism and erythropoiesis. Adv. Hematol. *2010*, 605435–605447. https://doi.org/10.1155/2010/605435.

Liang, Y., Zhou, Y., and Shen, P. (2004). NF-kappaB and its regulation of the immune system. Cell. Mol. Immunol. *1*, 343–350.

Martin, S.K., Diamond, P., Gronthos, S., Peet, D.J., and Zannettino, A.C. (2011). The emerging role of hypoxia, HIF-1 and HIF-2 in multiple myeloma. Leukemia *25*, 1533–1542. https://doi.org/10.1038/leu.2011.122.

Masson, N., William, C., Maxwell, P.H., Pugh, C.W., and Ratcliffe, P.J. (2001). Independent function of two destruction domains in hypoxia-inducible factor–alpha chains activated by prolyl hydroxylation. EMBO J. *20*, 5197–5206. https://doi.org/10.1093/emboj/20.18.5197.

Maxwell, P., and Salnikow, K. (2004). HIF-1: an oxygen and metal responsive transcription factor. Cancer Biol. Ther. *3*, 29–35.

Metzen, E., Berchner-Pfannschmidt, U., Stengel, P., Marxsen, J.H., Stolze, I., Klinger, M., Huang, W.Q., Wotzlaw, C., Hellwig-Burgel, T., Jelkmann, W., Acker, H., and Fandrey, J. (2003). Intracellular localization of human HIF-1 alpha hydroxylases: implications for oxygen sensing. J. Cell Sci. *116*, 1319–1326.

Moore, E.M., and Lockwood, D.N. (2010). Treatment of visceral leishmaniasis. J. Glob. Infect. Dis. 2, 151–158. https://doi.org/10.4103/0974-777X.62883.

Mukhopadhyay, C.K., Biswas, S., and Mukherjee, R. (2010). Reactive oxygen species in the regulation of Hypoxia-Inducible Factor-1 in normoxia. In Methods in Redox signalling, Das, D.K., ed. (New York: Marry Ann Libert), pp. 118–123.

Nishi, K., Oda, T., Takabuchi, S., Oda, S., Fukuda, K., Adachi, T., Semenza, G.L., Shingu, K., and Hirota, K. (2008). LPS induces hypoxia-inducible factor 1 activation in macrophage-differentiated cells in a reactive oxygen species-dependent manner. Antioxid Redox Signal. *10*, 983–995. https://doi.org/10.1089/ars.2007.1825.

Nizet, V., and Johnson, R.S. (2009). Interdependence of hypoxic and innate immune responses. Nat. Rev. Immunol. *9*, 609–617. https://doi.org/10.1038/nri2607.

Onnis, B., Rapisarda, A., and Melillo, G. (2009). Development of HIF-1 inhibitors for cancer therapy. J. Cell. Mol. Med. *13*, 2780–2786. https://doi.org/10.1111/j.1582-4934.2009.00876.x.

Page, E.L., Chan, D.A., Giaccia, A.J., Levine, M., and Richard, D.E. (2008). Hypoxia–inducible factor-1alpha stabilization in non-hypoxic conditions: role of oxidation and intracellular ascorbate depletion. Mol. Biol. Cell *19*, 86–94. https://doi.org/10.1091/mbc.E07-06-0612.

Pan, Y., Mansfield, K.D., Bertozzi, C.C., Rudenko, V., Chan, D.A., Giaccia, A.J., and Simon, M.C. (2007). Multiple factors affecting cellular redox status and energy metabolism modulate hypoxia-inducible factor prolyl hydroxylase activity *in vivo* and *in vitro*. Mol. Cell Biol. *27*, 912–925. https://doi.org/10.1128/MCB.01223-06.

Papandreoul, I., Cairns, R.A., Fontana, L., Lim, A.L., and Denko, N.C. (2006). HIF-1 mediates adaptation to hypoxia by actively downregulating mitochondrial oxygen consumption. Cell Metab. *3*, 187–197. https://doi.org/10.1016/j.cmet.2006.01.012.

Pérez-Victoria, F.J., Castanys, S., and Gamarro, F. (2003). *Leishmania donovani* resistance to miltefosine involves a defective inward translocation of the drug. Antimicrob. Agents Chemother. *47*, 2397–2403.

Peyssonnaux, C., Datta, V., Cramer, T., Doedens, A., Theodorakis, E.A., Gallo, R.L., Hurtado-Ziola, N., Nizet, V., and Johnson, R.S. (2005). HIF-1alpha expression regulates the bactericidal capacity of phagocytes. J. Clin Invest. *115*, 1806–1815. https://doi.org/10.1172/JCI23865.

Peyssonnaux, C., Ceiudo-Martin, P., Doedens, A., Zinkernagel, A.S., Johnson, R.S., and Nizet, V. (2007). Cutting edge: Essential role of hypoxia inducible factor-1alpha in development of lipopolysaccharide-induced sepsis. J. Immunol. *178*, 7516–7519.

Peyssonnaux, C., Boutin, A.T., Zinkernagel, A.S., Datta, V., Nizet, V., and Johnson, R.S. (2008). Critical role of HIF-1 alpha in keratinocyte defence against bacterial infection. J. Invest. Dermatol. *128*, 1964–1968. https://doi.org/10.1038/jid.2008.27.

Pugh, C.W., and Ratcliffe, P.J. (2003). Regulation of angiogenesis by hypoxia: role of the HIF system. Nat Med. *9*, 677–684. https://doi.org/10.1038/nm0603-677.

Ren, B.F., Deng, L.F., Wang, J., Zhu, Y.P., Wei, L., and Zhou, Q. (2008). Hypoxia regulation of facilitated glucose transporter-1 and glucose transporter-3 in mouse chondrocytes mediated by HIF-1alpha. Joint Bone Spine *75*, 176–181. https://doi.org/10.1016/j.jbspin.2007.05.012.

Richard, D.E., Berra, E., and Pouyssegur, J. (2000). Nonhypoxic pathway mediates the induction of hypoxia-inducible factor 1alpha in vascular smooth muscle cells. J. Biol. Chem. *275*, 26765–26771. https://doi.org/10.1074/jbc.M003325200.

Rius, J., Guma, M., Schachtrup, C., Akassoglou, K., Zinkernagel, A.S., Nizet, V., Johnson, R.S., Haddad, G.G., and Karin, M. (2008). NF-κappaB links innate immunity to the hypoxic response through transcriptional regulation of HIF-1 alpha. Nature. *453*, 807–811. https://doi.org/10.1038/nature06905.

Roiniotis, J., Dinh, H., Masendycz, P., Turner, A., Elsegood, C.L., Scholz, G.M., and Hamilton, J.A. (2009). Hypoxia prolongs monocyte/macrophage survival and enhanced glycolysis is associated with their maturation under aerobic conditions. J. Immunol. *182*, 7974–7981. https://doi.org/10.4049/jimmunol.0804216.

Saadi, S., Wrenshall, L.E., and Platt, J.L. (2002). Regional manifestations and control of the immune system. FASEB J. *16*, 849–856. https://doi.org/10.1096/fj.01-0690hyp.

Salceda, S., and Caro, J. (1997). Hypoxia-inducible Factor 1alpha (HIF-1alpha) protein is rapidly degraded by the ubiquitin – proteasome system under normoxic conditions. J. Biol. Chem. *272*, 22642–22647.

Salnikow, K., Donald, S.P., Bruick, R.K., Zhitkovich, A., Phang, J.M., and Kasprzak, K.S. (2004). Depletion of intracellular ascorbate by the carcinogenic metals nickel and cobalt results in the induction of hypoxia stress. J. Biol. Chem. *279*, 40337-40344. https://doi.org/10.1074/jbc.M403057200.

Scheidereit, C. (2006). IkappaB kinase complexes: gateways to NF-kappaB activation and transcription. Oncogene. *25*, 6685–6705. https://doi.org/10.1038/sj.onc.1209934.

Semenza, G.L., Neifelt, M.K., Chi, S.M., and Antonarakis, S.E. (1991a). Hypoxia-inducible nuclear factor bind to an enhancer element located 3' to the human erythropoietin gene. Proc. Natl. Acad. Sci. U.S.A. *88*, 5680–5684.

Semenza, G.L., Koury, S.T., Nejfelt, M.K., Gearhart, J.D., and Antonarakis, S.E. (1991b). Cell-type-specific and hypoxia-inducible expression of the human erythropoietin gene in transgenic mice. Proc. Natl. Acad. Sci. U.S.A. *88*, 8725–8729.

Semenza, G.L., Roth, P.H., Fang, H.M., and Wang, G.L. (1994). Transcriptional regulation of genes encoding glycolytic enzymes by hypoxia-inducible factor 1. J. Biol. Chem. *269*, 23757–23763.

Semenza, G.L. (2001). HIF-1 and mechanisms of hypoxia sensing. Curr. Opin. Cell Biol. *13*, 167–171.

Semenza, G.L. (2007a). Hypoxia-Inducible factor (HIF-1) pathway. Sci STKE. *2007*, cm8. https://doi.org/10.1126/stke.4072007cm8.

Semenza, G.L. (2007b). Evaluation of HIF-1 inhibitors as anticancer agents. Drug Discov. Today *12*, 853–859. https://doi.org/10.1016/j.drudis.2007.08.006.

Semenza, G.L. (2009). Regulation of cancer cell metabolism by hypoxia-inducible factor 1. Semin Cancer Biol. *19*, 12–16. https://doi.org/10.1016/j.semcancer.2008.11.009.

Singh, A.K., Mukhopadhyay, C., Biswas, S., Singh, V.K., and Mukhopadhyay, C.K. (2012). Intracellular pathogen *Leishmania donovani* activates hypoxia inducible factor-1 by dual mechanism for survival advantage within macrophage. PLOS One. 7(6), e38489. https://doi.org/10.1371/journal.pone.0038489.

Stockmann, C., and Fandrey, J. (2006). Hypoxia-induced erythropoietin production: a paradigm for oxygen-regulated gene expression. Clin. Exp. Pharmacol. Physiol. *33*, 968–979. https://doi.org/10.1111/j.1440-1681.2006.04474.x.

Sum, S.C., and Ley, S.C. (2008). New insights into NF-kaapaB regulation and function. Trends Immunol. *29*, 469–478. https://doi.org/10.1016/j.it.2008.07.003.

Taylor, C.T. (2008). Interdependent roles for hypoxia inducible factor and nuclear factor-kappaB in hypoxic inflammation. J. Physiol. *586*, 4055–4059. https://doi.org/10.1113/jphysiol.2008.157669.

ter Horst, R., Collin, S.M., Ritmeijer, K., Bogale, A., and Davidson, R.N. (2008). Concordant HIV infection and visceral leishmaniasis in Ethiopia: the influence of antiretroviral treatment and other factors on outcome. Clin Infect Dis. *46*, 1702–1709. https://doi.org/10.1086/587899.

Thornton, S.J., Wasan, K.M., Piecuch, A., Lynd, L.L., and Wasan, E.K. (2010). Barriers to treatment for visceral leishmaniasis in hyperendemic areas: India, Bangladesh, Nepal, Brazil and Sudan. Drug Dev. Ind. Pharm. *36*, 1312–1319. https://doi.org/10.3109/03639041003796648.

van Uden, P., Kenneth, N.S., and Rocha, S. (2008). Regulation of hypoxia-inducible factor-1alpha by NF-kappaB. Biochem J. *412*, 477–484. https://doi.org/10.1042/BJ20080476.

Wang, G.L., and Semenza, G.L. (1993). Desferrioxamine induces erythropoietin gene expression and hypoxia-inducible factor 1 DNA-binding activity: implications for models of hypoxia signal transduction. Blood. *82*, 3610–3615.

Wang, G.L., Jiang, B.H., Rue, E.A., and Semenza, G.L. (1995). Hypoxia-inducible factor 1 is a basic helix-loop –helix–PAS heterodimer regulated by cellular O_2 tension. Proc. Natl. Acad. Sci. U.S.A. *92*, 5510–5514.

Werth, N., Beerlage, C., Rosenberger, C., Yazdi, A.S., Edelmann, M., Amr, A., Bernhardt, W., von Eiff, C., Becker, K., Schäfer, A., Peschel, A., and Kempf, V.A. (2010). Activation of hypoxia inducible factor 1 is a general phenomenon in infections with human pathogens. PLOS ONE. 5(7), e11576. https://doi.org/10.1371/journal.pone.0011576.

Wiley, M., Sweeney, K.R., Chan, D.A., Brown, K.M., McMurtrey, C., Howard, E.W., Giaccia, A.J., and Blader, I.J. (2010). *Toxoplasma gondii* activates hypoxia-inducible factor (HIF) by stabilizing the HIF-1alpha subunit via type 1 activin-like receptor kinase receptor signaling. J. Biol. Chem. *285*, 26852–26860. https://doi.org/10.1074/jbc.M110.147041.

Wood, S.M., Gleadle, J.M., Pugh, C.W., Hankinson, O., and Ratcliffe, P.J. (1996). The role of the aryl hydrocarbon receptor nuclear translocator (ARNT) in hypoxic induction of gene expression. Studies in ARNT deficient cells. J. Biol. Chem. *271*, 15117–15123.

Zarember, K.A., and Malech, H.l. (2005). HIF-1alpha: a master regulator of innate host defenses? J. Clin. Invest. *115*, 1702–1704. https://doi.org/10.1172/JCI25740.

Zelzer, E., Levy, Y., Kahana, C., Shilo, B.Z., Rubinstein, M., and Cohen, B. (1998). Insulin induces transcription of target genes through the hypoxia-inducible factor HIF-1alpha/ARNT. EMBO J. *17*, 5085–5094. https://doi.org/10.1093/emboj/17.17.5085.

Zhou, J., Fandrey, J., Schümann, J., Tiegs, G., and Brune, B. (2003). NO and TNF-alpha released from activated macrophages stabilize HIF-1alpha in resting tubular LLC-PK1 cells. Am. J. Physiol. Cell Physiol. *284*, 439–446. https://doi.org/10.1152/ajpcell.00294.2002.

Response of B Lymphocytes During *Leishmania* Infection

3

Koushik Mondal[1]* and Syamal Roy[2]

[1]Division of Infectious Diseases and Immunology, CSIR-Indian Institute of Chemical Biology, Kolkata, India.
[2]National Institute of Pharmaceutical Education and Research, Kolkata, India.

*Correspondence: mkoushik9@yahoo.co.in

https://doi.org/10.21775/9781910190715.03

Abstract

Leishmania protozoa affects peripheral B-cell subset and expansion of multiple B-cell clones leading to polyclonal B-cell activation. Both cutaneous and visceral leishmaniasis stimulates polyclonal B-cell activation leading to generation of self or autoantibody and secretion of IgG. The polyclonal B-cell activation also stimulates and activates Transitional B-cell and Marginal Zone B-cell subsets to function as an IL-10 secreting regulatory B-cell during infection. Protozoal parasite mediated generation of hypergammaglobulinaemia and secretion of IL-10 are the major B-cell immunopathological changes responsible for disease progression during leishmaniasis.

Leishmaniasis

Leishmaniasis is a parasitic disease caused by an obligate intracellular protozoa of the genus *Leishmania* and family Trypanosomatidae. The parasite follows a digenetic life cycle: an extracellular flagellated promastigote stage in the gut of female phlebotomine sandfly and intracellular non-flagellated amastigote stage within mammalian cell (Kaye and Scott, 2011). Among the wide spectrum of diseases caused by protozoa, the genus *Leishmania* is responsible for four clinically different types of disease manifestations in humans. These range from symptomatic skin lesions in cutaneous leishmaniasis (CL), caused by *Leishmania major*, *L mexicana* and *L. amazonensis*, to severe tissue destruction and the horrible disfiguring features of mucocutaneous leishmaniasis (MCL), caused by *Leishmania braziliensis* and a fatal visceral form, which needs proper clinical diagnosis; in India this visceral form of leishmaniasis (VL) is popularly known as kala-azar, caused by *Leishmania donovani* and *L. chagasi* (Walker *et al.*, 2014). There is another form of VL, which arise following drug treatment and is known as post-kala-azar dermal leishmaniasis (PKDL). A clinical manifestation of this disease bears a high level of interleukin 10 (IL-10) in skin and blood with pathologically accumulation of highly infected macrophages in the skin (Gasim *et al.*, 1998; Zijlstra *et al.*, 2003).

Immune system elements

Host immune mechanism follows two distinct procedural steps. The first is identification of the pathogen or foreign particle and next step is its elimination. In its defensive strategy host cell use white blood cells and molecules such as antibodies and complement proteins for its protection. The immune response controlled by white blood cell has been categorized as innate immunity and adaptive immunity. During innate or natural immunity, it does not need to recognize the intruder, it responds immediately. There are different types of white blood cells or leucocytes; these are involved in innate and adaptive immune response. The common progenitor called haematopoietic stem cell is the source of leucocyte generation; the same progenitor also gives rise to red blood cells and megakaryocytes. This pluripotent stem cell divides and differentiated into myeloid, lymphoid and erythroid lineage progenitor cells. Granulocytes, monocytes, dendritic cells and mast cells are differentiated from myeloid progenitor cells. Granulocytes contain cytoplasmic granules with reactive substances that kill micro-organisms and take part in inflammation. Having irregularly shaped nuclei with two to five different lobes, granulocytes are also coined as polymorphonuclear leucocytes. Neutrophils are the most abundant granulocytes among white blood cells. During innate immunity neutrophils rapidly mobilize at the site of infection and capture, engulf and kill micro-organisms by the process of phagocytosis. Eosinophils are another type of granulocyte cell that defend the body against helminthic worms and other intestinal parasites. The least abundant granulocyte cell is the basophil, which releases histamines during allergic reactions and also attracts neutrophils and eosinophils at the infected site. Macrophages and monocytes are other types of leucocyte which have phagocytic characteristics. Monocytes are distinguished from granulocytes by having a distinctive indented nucleus and are bigger. Macrophages are stationed at the tissue, whereas their mobile progenitors monocytes, circulate in the blood. The expression of Complement receptor 3 (CR3), a β2 integrin, is important for phagocytic behaviour of macrophages. It is also expressed in polymorphonuclear cells and natural killer cells. Another tissue phagocytic cell is dendritic cell having distinctive star shaped morphology, which also acts as antigen presenting cell (APC) during adaptive immunity. Like basophils, mast cells also release histamine during allergic condition which resides in all connective tissues, but their development is different from basophils.

The lymphoid progenitor gives rise to lymphoid lineage white blood cells which are distinguished into large granular lymphocytes with a granular cytoplasm and with almost no cytoplasm of small lymphocytes. Natural killer cells (NK cells) belong to the large granular lymphocytes and takes part in innate immunity. The important difference between innate and adaptive is that an adaptive immune response is highly specific against the pathogen and 'remembers' the infectious agent, which can be prevented from causing disease later. During adaptive immune response small lymphocytes are involved which are classified as B lymphocytes and T-lymphocytes. Cell surface marker for B lymphocyte is immunoglobulin, whereas T-cell receptor is the cell surface marker for T-lymphocytes. T-cells are further subdivided depending upon their function in immune response into cytotoxic T-cell and helper T-cell. Infected host cells are killed by cytotoxic T-cells which have similar effector functions with NK cells; the difference is former takes part in adaptive immunity and the later in innate immunity. Unlike cytotoxic T-cells, helper T-cells activate B-cells to become plasma cells and also secrete cytokines. The activated B-cells or plasma cells secrete a soluble form of immunoglobulins, called an antibody, which binds to the pathogen and pathogen secreted antigen.

Host immune response

The immune cells which first encounter with the promastigotes are dermal macrophages, and Langerhans cells. They rapidly phagocytose the parasites with the help of CR3-dependent mechanism when they confronted with the parasites (Locksley *et al.*, 1988; Mosser and Edelson, 1985). Langerhans cells are a subset of dendritic cells (DCs) which track and engulf parasites with its pseudopods. These engulfed promastigotes are then metamorphosed to amastigotes. At the site of sandfly bite, there is inflammation and eventually neutrophils and monocytes are recruited there. Parasite engulfed macrophages also secretes MCP1 and CXCL1 which act as chemoattractant for neutrophils and monocytes (Racoosin and Beverley, 1997). These immune cells then try to eliminate the parasites (Fig. 3.1). Neutrophils along with macrophages generates tumour necrosis factor-α (TNF-α) and superoxide anion for parasite elimination (Mougneau *et al.*, 2011). There is also another mechanism adapted by neutrophil to eliminate the parasite. Making of Neutrophil extracellular traps (NET); a fibrous trap made up of DNA, histones and granule proteins is another way out to extinguish *Leishmania* (Guimaraes-Costa *et al.*, 2009). Surprisingly, neutrophil could behave as both protection and dissemination of leishmaniasis. While uptake of infected apoptotic neutrophils by macrophages, parasites can also thrive within the host (Nylén and Gautam, 2010). Parasites induce

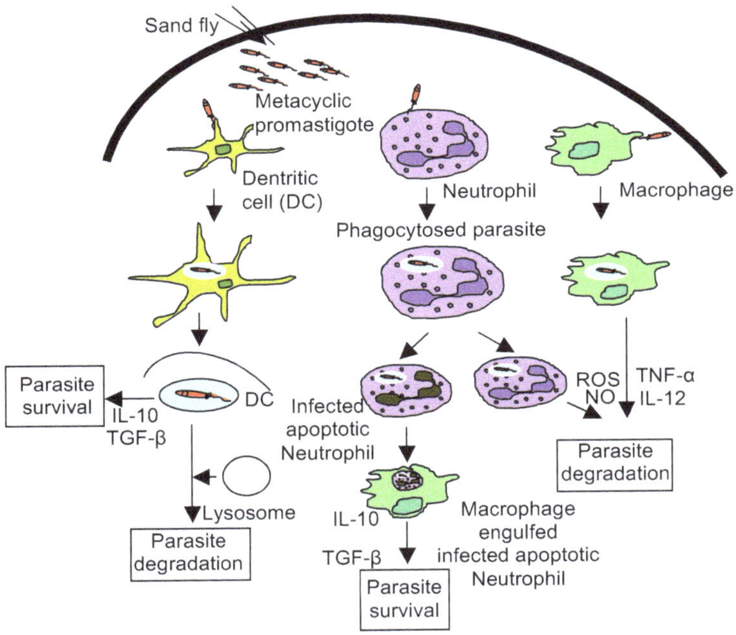

Figure 3.1 Schematic presentation of host immune response by different types of immune cell during *Leishmania* infection. ROS, Reactive oxygen species; NO, nitric oxide; IL-10, interleukin 10; IL-12, interleukin 12; TNF-α, tumour necrosis factor-α; TGF-β, transforming growth factor-β.

several mechanisms within the macrophage for its own survival and establish successful infection. Secretion of TGF-β and synthesis of IL-10 from macrophages and also inhibition of production of NO makes their survival and infection successful within the host (Kima, 2007).

Paradigm of Th1/Th2 response in cutaneous leishmaniasis in mice model

Host also use adaptive immune response against *Leishmania* where Langerhans cells, dermal dendritic cells and lymph node dendritic cells act as antigen-presenting cells (APC)s (Iezzi *et al.*, 2006; Moll *et al.*, 1993; Ritter *et al.*, 2004). This adaptive immune system during *L. major* mediated cutaneous leishmaniasis is different from visceral leishmaniasis. After activation through APC, $CD4^+$ T-cells develop either a Th1 (T-Helper 1) or Th2 response, which expand and differentiate to cytokine-secreting cells. Differentiated Th1 cells secrete IFN-γ, TNF-α and IL-12, which helps parasite elimination, whereas Th2 cells secrete IL-4, IL-10 and IL-13. The cytokines secreted during a Th2 response are responsible for disease progression (Heinzel *et al.*, 1989; Malherbe *et al.*, 2000; Stetson *et al.*, 2002). Apart from helper T-cells, regulatory T-cells ($CD4^+CD25^+$) contribute significantly during *Leishmania* pathogenesis by secreting IL-10 (Bhattacharya *et al.*, 2016). Therefore, disease protection and progression is dependent upon Th1/Th2 response during *Leishmania* infection, in which $CD4^+$ T-helper cells plays significant role.

Visceral leishmaniasis: tissue specific immune response and lack of Th1/Th2 dichotomy

Unlike in CL, where a distinctive Th1/Th2 response leads to either disease protection or progression, the immune response is different in case of VL. Patients infected with VL have been reported to have elevated levels of IL-4 and IL-13; these are te dominant Th2 type of immune response (Sundar *et al.*, 1997). Patients with VL have also been reported to show an increase in the plasma concentrations of pro-inflammatory cytokines such as IL-1, IL-6, IL-8, IL-12, IL-15 ad IFN-γ (Ansari *et al.*, 2006). During the acute phase of infection by VL there is increase in transcript level of IL-10 from spleen and bone marrow (Nylén *et al.*, 2007). Several chronic infectious diseases in humans, such as HIV, tuberculosis, malaria, are correlated with elevated levels of IL-10, which has immunosuppressive effect on antigen presentation pathways in macrophages and DC cells and disrupts T-cell activation, eventually leading to persistence of microbes. The same immunopathological observations have also been reported in chronic VL patients, in whom serum IL-10 is elevated level and accumulation of mRNA in spleen and bone marrow is higher than in controls (Singh *et al.*, 2012). During active VL, there is also increase in plasma TGF-β, at the level of both transcription and translation (Caldas *et al.*, 2005). An increase in TGF-β leads to loss of nitric oxide production, resulting in parasite survival. Progressive parasitic infection has been observed in human patients with varying degrees of parasitic load in the viscera (Faleiro *et al.*, 2014). Interestingly, infection with *Leishmania donovani* in hamster represents a good experimental model, because this animal develops similar immunopathological alterations which follow the characteristics of human disease. The parasite replicates in spleen, liver and bone marrow which eventually leads to death of the host. Compartmentalized immune responses are observed in murine VL, which is absent in case of human and hamster. In liver, amastigotes multiply rapidly and form acute infection which is resolvable. In contrast, chronic infection is established in spleen and bone marrow where they grow slowly and persist for longer time (Ahmed *et al.*, 2003; Squires *et al.*, 1990; Wilson *et al.*, 1996). Apparently, the liver serves as the site of initial parasitic expansion and the spleen serves as a safe harbour for chronic infection.

Immune response in the liver

Within the liver, Kupffer cells (KCs) are the major tissue macrophages which are infected by the parasites (Crocker *et al.*, 1984). Although the infection is rapid here, the liver recovers well from infection, with generation of hepatic granuloma playing a significant role. Following infection, KCs secrete different type of chemokines, including CCL3, CCL2 and CXCL10 (Cotterell *et al.*, 1999). For effective secretion of these chemokines KCs need natural killer T-cells (NKTs), the innate immune cells that secrete IFN-γ in the liver (Svensson *et al.*, 2005). These chemoattractants recruit monocytes and neutrophils in the developing granuloma within the first few days of infection (Cervia *et al.*, 1993; Smelt *et al.*, 2000). After a week of infection both $CD4^+$ and $CD8^+$ T-cells are transported to the granuloma (Kaye *et al.*, 2004). There is also upregulation of MHCII expression during infection which facilitates interaction between $CD4^+$ T-cell and Kupffer cell. The secretion of TNF-α and IFN-γ provides an inflammatory response which generates reactive oxygen intermediates (ROI) and RNI (reactive nitrogen intermediates) by granuloma (Murray and Nathan, 1999).

Immune response in the spleen

The infection in spleen takes longer time to establish with lesser number of parasites which persist in the animal throughout life. The spleen is the largest secondary lymphoid organ where immune response is initiated against blood-borne pathogens and filters foreign materials and damaged red blood cell (Kuper *et al.*, 2002). These functions are carried out by two functionally and morphologically distinct compartments, the red pulp and the white pulp. White pulp is further subcompartmentalized into periarteriolar lymphoid sheath (PALS), the follicles and the marginal zone (MZ). During the initial first month of acute type of infection splenic structure is not altered and there is generation of immune response against the parasite. While travelling through the marginal sinus into the MZ, *Leishmania* parasites in the blood are captured by two specialized type of macrophages in the MZ, the marginal metallophilic macrophages (MMM) and marginal zone macrophages (MZM) and by red pulp macrophages (Gorak *et al.*, 1998; Kraal, 1992). In the marginal zone, DC acquire *Leishmania* antigen and eventually migrate towards T-cell areas of the PALS. The chemokine receptor CCR7 is expressed by mature DC cells and naive T-cells and migrates into the PALS in response to the chemokines CCL19 and CCL21 which are expressed by $gp38^+$ stromal cells and central arterioles. Then DC in the PALS, secretes IL-12 which help to present *Leishmania* antigen to T-cells, resulting in the generation of an antigen-specific T-cell response (Ato *et al.*, 2006). The chronic infection starts during the later stage of infection and parasite persistence in the spleen is associated with severe splenomegaly, resulting in host immunocompromise as a result of the breakdown of splenic architecture. This breakdown in tissue microarchitecture is associated with the secretion of TNF from heavily parasitized macrophages those have migrated to the PALS (Körner *et al.*, 1997). The affect of TNF secretion leads to loss of gp38$^+$ stromal cells and their reticular matrix and subsequently leading to loss of CCL19 and CCL21 on both the stroma and central artioles, leading to reduced cellular recruitment to the PALS (Ato *et al.*, 2002; Nolte *et al.*, 2003; Zlotnik and Yoshie, 2000). There is selective loss of MZM but not MMM by TNF, which positively affects secretion of IL-10 (Engwerda *et al.*, 2002). This induction of IL-10 directly affects down-regulation of CCR7 on the surface of DC cells, which prevents their migration into the PALS (Murphy *et al.*, 2001). This altered function of stromal cell network supports development of regulatory DC from haematopoietic progenitor cells (Svensson *et al.*, 2004). These subsets of regulatory DC cells secrete IL-10, which promotes expansion of IL-10-secreting regulatory T-cells (Tr1) (Wakkach *et al.*, 2003). There is also secretion of IL-27 from regulatory DCs and macrophages. This IL-27, along with T-cell derived IL-21, drives differentiation of Th1 cells into Tr1 cells and also inhibits development of Th17 cells (Spolski *et al.*, 2009). This secretion of IL-10 from Tr1 suppresses antigen presentation, leading to dysfunction of T-cell and secretion of IFN-γ from CD4$^+$ T-cell (Kaye *et al.*, 2004). Thus white pulp is reduced in size and disorganized and there is hypertrophy of the red pulp. Infected macrophages are observed in multiple and unusual locations, such as in the PALS. A pathological change similar to that which occurs in human VL has also been observed in murine spleen, including GC involution and loss of follicular dendritic cell (FDC) networks (Smelt *et al.*, 1997; Zijlstra and el-Hassan, 2001).

Immune response in the bone marrow

Having similarity with the spleen, bone marrow (BM) is also the site of chronic infection during VL, where it infects the stromal macrophages which affects haematopoiesis. There is an increased level of myelopoiesis in infected bone marrow stromal cell due to increase in granulocyte macrophage colony-stimulating factor (GM-CSF) and TNF-α, similar observation of myelopoiesis has also been reported from stromal cell while induction by GM-CSF and TNF-α during *in vitro* condition (Cotterell *et al.*, 2000). In VL patients related changes have been reported in BM, with an increase in lymphocytes and plasma cells, erythroid hyperplasia, marrow granulomas, hypercellular marrows, benign lymphoid nodules and moderate to severe megaloblastosis (Verma and Naseem, 2010; Kumar *et al.*, 2007).

B-cell development and functions of B-cells

B-cell subsets in primary lymphoid organ

B-cells first arise from haematopoietic stem cells in the bone marrow (BM) after birth and within the fetal liver before birth (Hardy and Hayakawa, 2001). The first B-lineage-specific cell is the pro-B-cell, which originates from a multipotent progenitor cell (MPP) through a common lymphoid progenitor cell (CLP). The B-cell then goes through a series of developmental transitions to become a peripheral functional B-cell (Kondo *et al.*, 1997). Stromal cells within the bone marrow play a significant role in these developmental processes, which provide a suitable microenvironment to regulate haematopoiesis. Using multicolour flow cytometry, it is easier to understand functionally heterogeneous different developmental stages of B-cell which have specific phenotypic characteristics (Jackson *et al.*, 2008). This phenotypic characterization of B-cell developmental stages could be identified through differential expression of various cell surface receptors, e.g. B220, CD43, c-kit, heat-stable antigen (HSA) and CD19, which are expressed accordingly towards maturity in bone marrow. Specific surface receptor properties against particular antigens are genetically programmed within the heavy and light chain genes in immunoglobulin. Having this property, early B-cell development is characterized by sequential rearrangement of immunoglobulin (Ig) heavy (H) and light (L) chain gene loci; where combinatorial rearrangement occurs in variable (V), diversity (D) and joining (J) gene segments in the heavy chain locus and the V and J gene segments in L chain locus (Brack *et al.*, 1978). Immunoglobulin heavy chain gene is first rearranged within the bone marrow of mice pro-B-cell (B220$^+$CD43$^+$). In the locus of heavy chain, this rearrangement takes place in two consecutive steps; in the first step, J_H gene is rearranged with the D gene and, in second step, V_H is rearranged with the DJ_H rearrangement. This VDJ recombination is regulated by a combination of complex enzymatic machinery; where recombination activating gene 1 and 2 (RAG1 and RAG2) proteins play a pivotal role which perform cutting and joining of DNA strand near the site of V, D and J segments (Fig. 3.2) (Schatz *et al.*, 1989, Oettinger *et al.*, 1990). This RAG complex initiates DNA double-strand breaks, specifically between a variable region gene-coding segment and an associated recombination signal sequence (RSS). Each RSS is made up of conserved heptamer (5′-CACAGTG-3′) and nonamer (5′-ACAAAAACC-3′) sequences and an intervening spacer sequence of either 12 or 23 base pairs (bp). The 23-spacers RSS flanks V_H and J_H genes, whereas D genes are flanked by 12-spacers RSS. In this process of RAG mediated recombination a gene flanked by 12-RSS generally paired with the gene flanked by 23-RSS, referred as 12/23 rule. This 12/23 rule ensures that the IgH gene is assembled in a correct order. Then this DNA break is repaired by ubiquitously expressed Non-homologous End Joining (NHEJ) proteins, giving rise to precise signal end joint (SJ) and imprecise coding end joint (CJ) (Jung and Alt, 2004). Productive rearrangement of heavy chain leads to pre-B-cell ($B220^+CD43^-$) and formation of pre-B-cell receptor (pre-BCR) at the surface by pairing of μ H chain and surrogate light chain (SLC), where SLC is a heterodimer of two distinct proteins named λ5 and VpreB (Karasuyama *et al.*, 1993). Cytokines secreted from stromal microenvironment plays significant role in this hematopoiesis. One of the major cytokines secreted from stromal cells, interleukin 7 (IL-7), affects both B- and T-cell development. Although, pro-B-cell is not affected by IL-7; but it is an indispensable requirement for pre-B-cell development. Interleukin-7 and transcription factor, pax5 facilitate the commitment for haematopoietic cells to B lineage cell (Corcoran *et al.*, 1998). After assembling of functional pre-BCR, there is degradation of RAG2 and RAG1 expression is suppressed; not having further rearrangement of heavy chain locus. Then pre-B I cells are developed into large pre-B II cells, followed by a burst of proliferation. During this stage they also loss expression of c-kit and CD43. After proliferation, large pre-B II cells are transited to small resting pre-B II cells; and again RAG1:RAG2 are expressed leading to rearrangement at the light chain locus. Both in human and mice κ light chain locus rearranges before the rearrangement of λ light chain. Interestingly, in light chain locus rearrangement takes place only one allele at a time, whereas, in the case of heavy chain genes there is scope of repeated rearrangements of unused V and J gene segments at each allele. There is equal probability of choosing κ gene from either

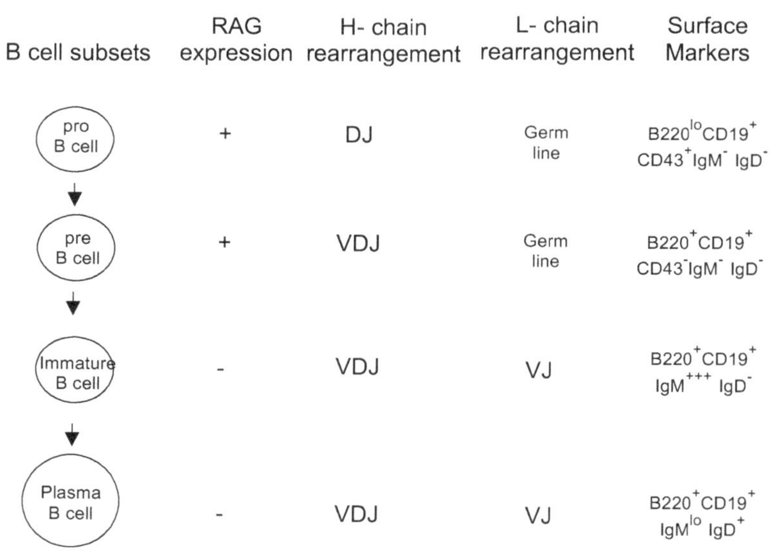

Figure 3.2 Formation of functional BCR by VDJ recombination during B-cell development.

allele, which is a random or independent event; if one is exhausted the cell choose other allele. And if both the alleles are out of κ rearrangement, the cell automatically switches to λ rearrangement. Cells will die if there is no light chain gene rearrangement. Following productive assembling of the light chain gene, the next-level B-cell developmental subset is immature B-cell ($IgM^{+}IgD^{-}$), which expresses IgM on its cell surface. Together with Igα and Igβ, IgM at the cell surface forms a functional B-cell receptor (BCR) complex. These newly expressed receptors, when confronted with a strongly cross-linking antigen, i.e. a self-reactive antigen, halt further development of the B-cell; thus, the B-cell undergoes negative selection. B-cells which do not have self-reactive IgM proceeds towards further development. The next level of development and differentiation then starts at the secondary lymphoid organ, preferentially in the spleen.

Development of transitional B-cells

Immature B-cells enter the spleen through blood from BM as transitional B-cells ($B220^{+}CD93^{+}$) (Allman *et al.*, 2001). Transitional B-cells are further characterized according to the expression level of cell surface markers IgM, IgD, CD21 and CD23 and are divided into three different subsets (Merrell *et al.*, 2006). First immigrated population reside in periarterial lymphatic sheath (PALS) region of spleen as transitional B-cells of type 1 (T1) ($IgM^{high}IgD^{-}CD21^{-}CD23^{-}$) (Liu, 1997), which further develop into transitional B-cells of type 2 (T2). T2 B-cells are observed exclusively in the primary follicles in the spleen (Chung *et al.*, 2003); which retain higher expression level of IgM and are also IgD^{+}, $CD21^{+}$, $CD23^{+}$. The third non proliferating B-cell is characterized as transitional B-cells of type 3 (T3), which resembles T2, with lower expression level of IgM (Su and Rawlings, 2002). Although T2 and T3 populations could be detected in lymph node, T1 is absent there. Due to the absence of CD62L/MEL-14 expression at the surface of T1, this population can not migrate into lymph node (Loder *et al.*, 1999). Having distinct and specific phenotypic characteristics, different Transitional B-cell subsets are engaged in different functional role. T2 B-cell leads the maturation process and give rise to mature naiveB-cells ($IgM^{+}IgD^{+}$), whereas transitional T3 subsets are believed to consist of potentially autoimmune cell and remained as anergic, having competency to secrete autoantibody during the loss of tolerance (Teague *et al.*, 2007).

T-cell-dependent B-cell development

Furthermore, this naive mature B-cell originated from T2 B-cell can proliferate and differentiate into antibody-producing plasma cells and memory B-cells after encountering antigen (Ag) (Sanz *et al.*, 2008). There are three major subsets of mature B-cell: follicular (FO) B-cells, marginal zone (MZ) B-cells and B1 B-cells (Carsetti *et al.*, 2004). FO B-cells are dominant subset and they reside in the lymphoid follicle of spleen and lymph node. Both B1 and MZ B-cells respond to T-cell-independent antigens (TI), whereas FO B-cells respond to protein antigen where $CD4^{+}$ T-cell activation plays significant role for generation of antibody secreting cell which is T-cell-dependent process (TD) (Fairfax *et al.*, 2008). Naive B-cell having BCR eventually migrates towards T zone–B zone border; after having antigen-dependent signalling some cells move towards the T zone–red pulp border, some towards the marginal sinus bridging channels and some migrate to red pulp (Crotty, 2012). Within the extrafollicular region they continue to proliferate and generation of B lymphoblasts and differentiate into short lived plasmablasts that secretes antibody (Nutt *et al.*, 2015). This is known as 'extrafollicular response', which in lymph node occurs in medullary cords and within the bridging channel in spleen (MacLennan *et al.*, 2003). These responses are associated with class switch recombination (CSR) with very limited somatic hypermutation (SHM); as a result antibodies generated from plasmablast are moderate having low affinity against pathogens (Jacob and Kelsoe, 1992).

For long-lived antibody production, activated B-cell needs to migrate to the follicle, where antigen is associated on the surface of follicular dendritic cell (FDC). These migrated B-cells within the follicle proliferates rapidly under the influence of FDC and generates germinal centre (GC) (Nutt and Tarlinton, 2011). Within GC there are dark and light zones, containing centroblasts and non-dividing centrocytes, respectively (Allen *et al.*, 2007). Immunoglobulin variable region gene undergoes SHM in the dark zone centroblasts and subsequently migrates towards light zone where they confront with the antigen (Berek *et al.*, 1991). Ultimately, GC give rise to long-lived high-affinity plasma cells that produce antibody and also memory B-cells (Fig. 3.3). The formation of antibody secreting cell (ASC) is the outcome of cooperation between B-cell and T helper cell in response of protein antigen. In this B-cell activation, BCR performs two distinct functions. At first, it sends signal to the cell interior while binding with the antigen and secondly, it delivers the antigen within the cell where it is degraded and returned to the B-cell surface as peptide attached with major histocompatibility complex II (MHC II) molecules. The helper T-cell then recognizes the peptide: MHC complex which delivers activating signals to B-cell. This T-B immunological synapse is crucial for B-cell biology where Follicular T helper cell (T_{FH}) provides signal that promote hypermutation and class-switch recombination (Crotty, 2015; Mitchison, 2004). Also, T_{FH} contributes pro-mitotic signalling to B-cells through CD40L during induction of proliferation to B-cell. As germinal centre B-cells are sensitively pro-apoptotic, IL-4 secreted from T_{FH} provides pro-survival signals to GC B-cells. These supports from T-cell helps activated germinal centre to secrete Activation Induced Cytidine Deaminase (AID) enzyme for SHM. This DNA damage causing enzyme helps immunoglobulin gene mutation and DNA repair enzymes in the process of affinity maturation. To protect itself against AID mediated damage, GC also secretes Bcl6. The combination of CD40L, IL-4 and IL-21 from T_{FH} cells are the primary contributors during B-cell proliferation, somatic hypermutation and differentiation. Like SHM, CSR is also being regulated by the immune synapse complex of T-B-cell, where AID plays significant role in class switching of constant region in immunoglobulin. In contrast to the RAG mediated recombination signal sequence-specific VDJ recombination, CSR targets large region of repetitive DNA sequences (1–12 kb), known as switch (S) regions which is located upstream of all constant heavy chain (C_H) genes, except C_δ (Dudley *et al.*, 2005). Secretion of various cytokines from $CD4^+$ T-cell regulates this process; class-switching from IgM to IgG is regulated by IL-21 and high ratio of IL-4 to IL-21 is responsible for IgE recombination (Crotty, 2015). Thus, T helper cells provide cytokine and other co-stimulatory signals to B-cells for synthesis of Ig. Interestingly, regulatory T-cell subset behaves differently with B-cell in this mechanism. It has been reported that $Foxp3^+$ regulatory T-cells are present at the site of T-B area border and within GC of human lymphoid tissue and suppress B-cell Ig production and class switch recombination (Lim *et al.*, 2005). As mature plasma cell develops, expression level of early

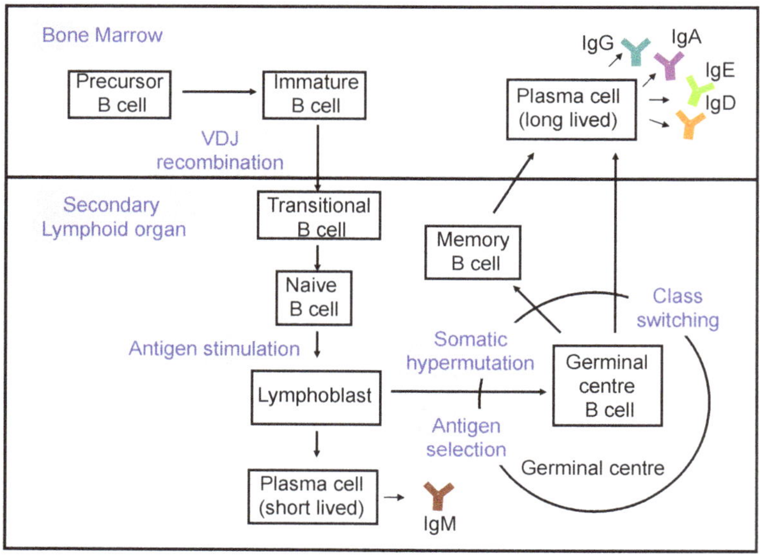

Figure 3.3 Schematic diagram of Plasma B-cell development.

B-cell development cell surface markers (CD19, CD20, CD45 and MHC class II) fall in comparison with naive B-cell and plasmablast (Crotty, 2012; Nutt *et al.*, 2015). Although the expression level of CD138 and CXCR4 gradually rises from plasmablast to mature plasma B-cell, they are absent from naive B-cells (Nutt *et al.*, 2015). There is heterogeneous expression of Blimp1 protein during terminal B-cell differentiation; in spleen, the ratio of $Blimp1^{int}$ to $Blimp1^{high}$ plasma B-cells is 1 : 1. Interestingly, Blimp1 level is low in the short-lived plasmablast, which suggests that Blimp1 plays a significant role in progressive maturity towards the antibody-secreting cell (ASC). In contrast, bone marrow plasma B-cells are all $Blimp1^{high}$; however, Blimp1expression is absent in memory B-cell (Kallies *et al.*, 2004; Nutt *et al.*, 2015).

Memory B-cells (MBC) are classically defined as progenies of GC B-cells which are isotype-switched and mutated BCR having enhanced immune response while reinfection with the pathogens. Memory B-cell has been originally defined as $CD27^+$ B-cell (Agematsu *et al.*, 1997), this B-cell population is further characterized into additional subsets by co-staining with IgD into non-switched MBCs ($CD27^+$ IgD^+), switched MBCs ($CD27^+$ IgD^-) and double negative MBCs ($CD27^-$ IgD^-) (Klein *et al.*, 1998). Differently, co-staining with CD21, MBC is characterized as classical MBCs ($CD27^+CD21^+$), activated MBCs ($CD27^+CD21^-$) and atypical MBCs ($CD27^-CD21^-$) (Moir *et al.*, 2008). Based on these markers, naive B-cells are characterized as $CD27^-$ IgD^+ or $CD27^-CD21^+$. According to the expression of three different cell surface expression markers (CD80, PD-L2 and CD73) five different types of memory B-cells have been characterized in murine model (Bergmann *et al.*, 2013). The lower the expression of these markers, the less likely B-cells are to be memory in nature, whereas an increase in marker expression level causes them to behave more like the memory type of B-cell.

T-cell-independent B-cell development

Marginal zone B-cell and B1 B-cell constitute the first line of defence owing to their anatomical location, and both play an important role in the TI immune response (Martin *et al.*, 2001). MZ B-cells reside outside the marginal sinus in the secondary lymphoid organs and along with TI-dependent immune response, it also take part in T-cell-dependent process. TI immune response is divided into two categories, TI-1 and TI-2. TI-1 antigens are lipopolysachharide (LPS), CpG, poly-IC and induce polyclonal B-cell activation using Toll-like receptors, whereas polysaccharides with the help of B-cell receptor activate the TI-2 response (Engel *et al.*, 2011). Apart from their variable developmental ontogeny, FO B-cells, MZ B-cells and B1 B-cells are further characterized according to the expression of different surface markers of these cells; FO B-cells are $B220^{high}IgM^{low}IgD^{high}CD21^{inter}CD23^{high}$, whereas MZ B-cells are $B220^{high}IgM^{high}IgD^{low}$ $CD21^{high}CD23^{low}$. However, lower expression of B220 and IgD are observed in B1 B-cells, which secretes higher level of IgM. CD43 is absent in both FO and MZ B-cells, but present in B1 B-cells. B1 B-cells are $CD19^{high}$, whereas FO and MZ B-cells show $CD19^{mid}$ (Baumgarth, 2011; Fairfax *et al.*, 2008). The developmental ontogeny of peritoneal and splenic B1 cells are controversial as well as phenotypic characteristics of peritoneal and splenic B1 cells also differ according to the cell surface marker expression. Expression of surface IgM, CD11b and CD80 are much higher in peritoneal B1 cell in comparison with splenic B1 cell (Tumang *et al.*, 2004). B1 cells are further subdivided into B1a and B1b, respectively according to CD5 positive and negative expression (Baumgarth, 2011; Hardy, 2006). During B1a cell development CD19 plays significant role (Haas *et al.*, 2005), whereas IL-9 critically affect B1b cell development (Knoops *et al.*, 2004).

B-cell pathogenicity during *Leishmania* infection

Polyclonal B-cell activation during leishmaniasis

During infections by bacteria, viruses and parasites different clones of B-cell are proliferated and differentiated to antibody-secreting cells which are not specific for their antigen. This mechanism is known as polyclonal B-cell activation (PBA). Micro-organisms and their secretory molecules have the ability to induce proliferation of multiple B-cell clones and up regulate B-cell surface markers such as MHC class II, CD69, CD25, CD80 and CD86 (Nashar *et al.*, 1997). Due to polyclonal activation, B-cell secretes antibodies that are non-specific and recognize both homologous and heterologous antigens such as myosin, actin, myoglobin, thyroglobulin, DNA (Hunziker *et al.*, 2003; Montes *et al.*, 1999). This polyclonal B-cell activation is a common pathological phenomenon during infection, and leads to hypergammaglobulinaemia and secretion of autoantibodies (Montes *et al.*, 1996; Silva *et al.*, 2003). B-cell dysfunction and generation of hypergammaglobulinaemia with natural antibodies of nonspecific self antigens (tubulin, myosin, myoglobin, actin, vimentin) have been observed during human VL (Bohme *et al.*, 1986; Konstantinos *et al.*, 2009) (Fig. 3.4). In hamsters both PBA and antigen-specific impairment of T-cells have been observed during *Leishmania donovani* infection (Bunn-Moreno *et al.*, 1985). Activation of B-cells during *Leishmania major*-infected cutaneous leishmaniasis (CL) has been observed with increase in CD69 surface marker expression in murine splenic B-cell and also increase in IgM secreting B-cells (Cordeiro-da-Silva *et al.*, 2001). This information has been supported while using anti-IgM to treat CL (Sacks *et al.*, 1984). Disease condition is exacerbated due to secretion of antibodies against heterologous antigens such as myosin, thyroglobulin, DNA and keyhole limpet hemocyanin instead of *Leishmania*-specific antigen because of PBA. As in CL, B-cell activation has also been noticed during murine visceral leishmaniasis with increase in B-cell surface marker CD69 (Deak *et al.*, 2010). In murine VL there is a marked B-cell expansion in the draining lymph nodes and increase in serum IgM of non-leishmanial antigens (TNP, OVA, chromatin) against soluble *Leishmania* antigen (SLA). Exposure with *Leishmania infantum* amastigotes and purified human tonsillar B-cells leads to B-cell activation. Significant increase in B-cell surface markers of CD25, CD69, CD80, CD83 and CD86 have been observed during this interaction between amastigotes and purified B-cell (Andreani *et al.*, 2015). However, sera from Indian kala-azar (KA) and post-KA dermal leishmaniasis (PKDL) patients show highest level of IgG and its subclasses. Using enzyme-linked immunosorbent assay (ELISA) and immunoblot experiment it has been observed that IgG subclasses antibodies are prevalent in patient sera and follows the order as IgG1>IgG2>IgG3>IgG4 (Ghosh *et al.*, 1995). After successful treatment of VL patients IgG level declines and there is development of delayed type hypersensitivity (DTH) responses to leishmanial antigen. This increase in IgG1 induces secretion of

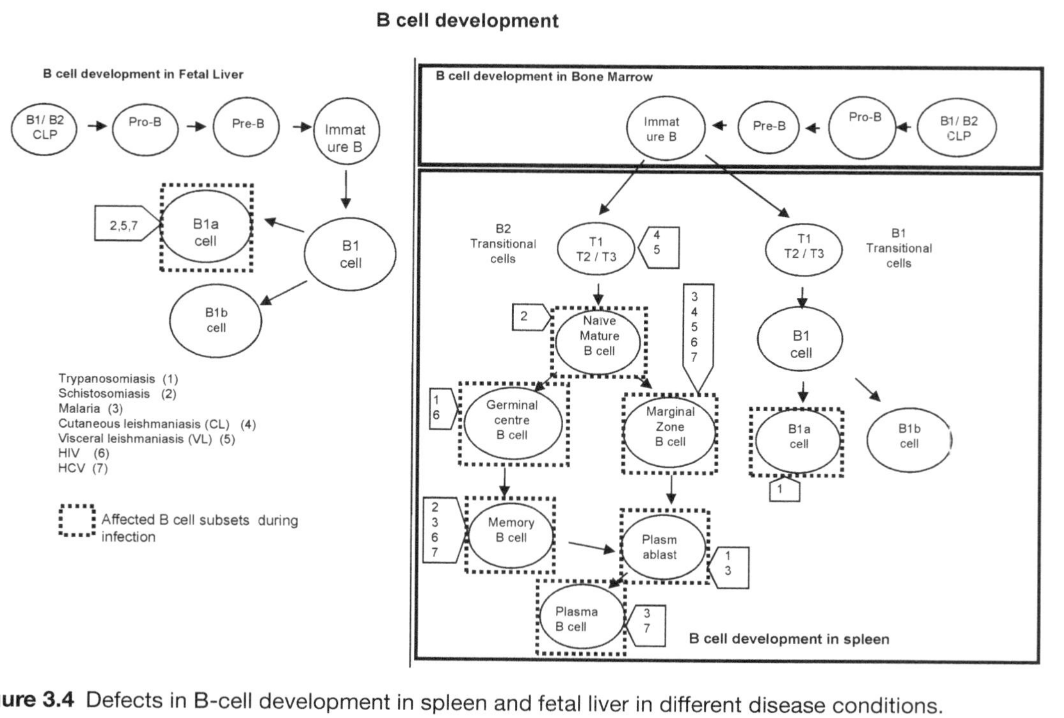

Figure 3.4 Defects in B-cell development in spleen and fetal liver in different disease conditions.

IL-10 from murine macrophage during *Leishmania major* infection leading towards pathological condition (Miles *et al.*, 2005). In murine model of cutaneous leishmaniasis it has been reported that increase in IgG contributes to disease progression; whereas increase in IgM is associated with disease exacerbation in VL.

Occurrence of polyclonal B-cell activation is not specific for *Leishmania* infection only; other protozoan parasites, bacteria and virus also manifest same pathological features in their respective host. Infection with *Trypanosoma cruzi* leads to PBA and secretion of both IgM and IgG, with increase in extrafollicular plasmablasts and germinal centre B-cell (Bermejo *et al.*, 2011). Same pathological condition in splenic B-cells also been reported while *Plamodium chabaudi chabaudi* infection where vigorous growth of plasmablasts and unconventional distribution of plasma cell in the periarteriolar sheath of white pulp leads to delayed parasite-specific antibody response (Achtman *et al.*, 2003). Dysfunction of B-cell induces secretion of IgG2a and IgG2b isotypes during infection with *T. congolens* (Morrison *et al.*, 1978); similarly with *Leishmania*-infected patients serum contains hypergammaglobulinaemia. Another splenic B-cell subset CD5$^+$ B-cells, are also involved in *Trypanosoma* infection as Xid mice lacking CD5$^+$ B-cells are resistant to *Trypanosoma* infection where serum contains less amount of parasite-specific and non-specific antibodies (Minoprio *et al.*, 1991). There is marked expansion and activation of B1 cell in human VL patients-infected with *Leishmania infantum* with increased level of natural autoantibodies in sera which falls after treatment (Louzir *et al.*, 1994). Contrary with this observation, CD5$^+$ depleted peritoneal B-cell has minimal affect of *L. major* infection in both C57Bl/6 and BALB/c mice (Babai *et al.*, 1999). In support of involvement of CD5$^+$ peripheral B-cell in autoimmunity Zuckerman *et al.* noticed expansion of this specific B-cell subset in HCV patients (Zuckerman *et al.*, 2002). Autoimmunity has also been reported by other B-cell in HCV patients. Interestingly, during viral infection, functionally anergic marginal zone-like B-cells (CD21$^{-/low}$IgM$^+$CD27$^+$) are expanded leading to chronic antigenic stimulation that affect on loss of tolerance (Terrier *et al.*, 2011). However, Schistosoma-infected patients show higher level of IgG4 in their serum and there is expansion of atypical Memory B-cell (CD27$^-$CD21$^-$) and loss of naive B-cell (CD27$^-$CD21$^+$) population (Labuda *et al.*, 2013). Whereas in experimental murine schistosomiasis there is increase in B-cell number and polyclonal B-cell activation induce secretion of IgE (Oliveira *et al.*, 2005). There is induction of Th2 dominated immune response by the parasite which secretes both IL-4 and IL-5. This secretion of IL-5 induce class switching and converts B1 (CD23$^-$)-IgE and B2 (CD23$^+$)-IgE B-cells to secrete IgE.

Microbial cell membrane component, cytosol and secretory proteins act as polyclonal activators. Cell surface proteins proline racemase and trans-sialidase (Gao *et al.*, 2002); soluble proteins malate dehydrogenase (MDH) and glutamate dehydrogenase (GDH) act as polyclonal activators during *T. cruzi* infection (Montes *et al.*, 2002). Activation and differentiation of multiple B-cell clones to secrete antibodies are also induced by ribosomal protein S3a homologue of *L. major* (Cordeiro-da-Silva *et al.*, 2001), erythrocyte membrane protein of *Plasmodium falciparum* (Donati *et al.*, 2006), bacterial protein staphylococcal protein A (SpA) (Anderson *et al.*, 2006), viral protein HIV-1 glycoprotein gp120 (He *et al.*, 2006) and influenza haemagglutinin (Rott *et al.*, 1996). Not only the microbial protein has the ability of polyclonal B-cell activation, other microbial product has the ability to trigger PBA. Among these, N-glycans from lactate dehydrogenase elevating virus (LDV) (Plagemann *et al.*, 2000), LPS from *Escherichia coli* and *Brucella abortus*, oligodeoxynucleotide containing CpG motifs (ODN) from *E. coli* and *T. cruzi* have the ability to differentiate multiple B-cell clone and in turn secretes IgG and different types of cytokines (He *et al.*, 2004; Koga *et al.*, 1985; Moreno *et al.*, 1981; Neujhar *et al.*, 1999).

Developmental deregulation of B-cells during pathological condition

Microbial pathogens not only interact directly with innate B-cells, but also modify their microenvironment and change the humoral response (Fig. 3.5). Infection with malaria results in substantial changes in blood B-cells. A reduction in the number of circulating MZ B-cells in *Plasmodium*-infected patients correlates with a lower level of parasite-specific plasma IgG (Asito *et al.*, 2011; Requena *et al.*,

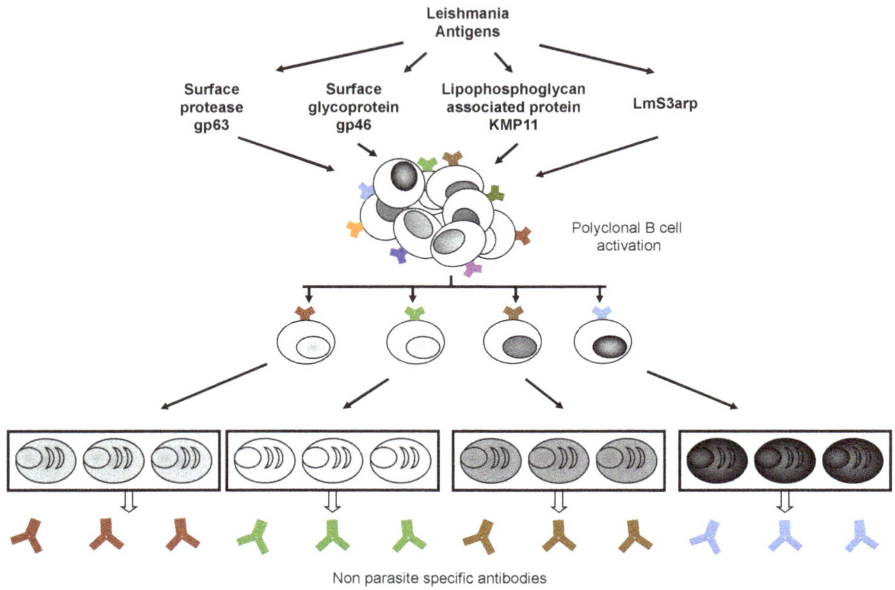

Figure 3.5 *Leishmania* antigen-induced polyclonal B-cell activation.

2014) during chronic infection whereas an increase in atypical memory B-cells during malaria infection (SIgG/A^{+}CD27^{-}CD21^{-}) enriches poly-reactive B-cells, which recognize different *Plasmodium*-associated Ags (Muellenbeck *et al.*, 2013). In case of schistosomiasis both atypical and activated memory B-cells (IgD^{-}CD27^{+}CD21^{-}) are increased but decrease level of naive B-cell (IgM^{+}IgD^{hi}CD27^{-}CD21^{int}) has been reported (Nduati *et al.*, 2011). Resting memory B-cell (IgD^{-}CD27^{+}CD21^{+}) and MZ B-cell remain unchanged in *Schistosoma* infection but in case of malaria resting memory B-cell decreases. The puzzling topic in B-cell biology while HIV infection is the inefficiency of HIV-induced Ab response where both innate and virus-specific antibody response are impaired. During chronic HIV infection, it has been reported that there is loss of circulating MZ like B-cell associated with pneumococcal Ags (Hart *et al.*, 2007). Similarly, HIV-infected macaques show reduce proportions of MZ B-cell not only in blood but also in spleen and lymph node (Peruchon *et al.*, 2009). This loss of MZ B-cell substantially increases plasma cell B-cell and increases circulating IgM and IgG levels. Interestingly striking affect has been observed in memory B-cell while HIV infection. Increase in both activated and atypical memory B-cell and decrease in resting memory B-cell in the blood has been observed in chronically HIV-infected patients (Moir and Fauci, 2009), which in sharp contrast with normal individual where predominant fraction is resting memory B-cell with low percentage of activated and atypical memory B-cell (Good *et al.*, 2009). During infection, FcRL4 and TGF-β1 are expressed in peripheral B-cell and low surface expression of CD80. All of these factors inhibit activation and proliferation of B-cell by dysfunction of BCR responses and co-stimulation abilities of B-cells, which in turn affects immunoglobulin production, isotype switching, affinity maturation and generation of plasma cells (Jelicic *et al.*, 2013). Further decrease in IL-21 secretion together with altered expression of CCL19 and CXCL13 during infection of T-follicular helper (Tfh) cell with HIV-1 disrupts germinal centre formation (Ruffin *et al.*, 2012). Also there is reduction of follicular dendritic cell (FDC) network. So disruption of GC affects on plasma B-cell and memory B-cell during HIV infection. During chronic infection with HCV, dysfunction of B-cell is associated with IgG production with delayed onset Ab response and loss of resting memory B-cell with increase in atypical memory B-cell (Chen *et al.*, 1999).

The diseases of humoral immunity occur when one or more of its components are defective. Immunoglobulin is a primary product of B-cell and its production proceeds through the complex mechanism of VDJ gene rearrangement which is regulated by RAG1/RAG2 (recombination activating gene 1/2). RAG also controls the rearrangement of T-cell receptor; dysfunction of RAG prevents production of both B-cells and T-cells, which leads to severe combined immunodeficiency (SCID). Omen syndrome arises from incomplete block in VDJ rearrangement (Notarangelo *et al.*, 2009). Failure to create immunoglobulin is also associated with increased radiosensitivity; this class of SCID is called radiation-sensitive SCID (RS-SCID). Defects in the gene of Artemis, DNA protein kinase catalytic subunit (DNA-PKs) and DNA ligase IV (Genery *et al.*, 2000; Moshous *et al.*, 2001) causes RS-SCID. This type of defect is associated with DNA repair mechanism while non-homologous end joining (NHEJ) recombination during VDJ rearrangement, also lead to malignant transformation in lymphocyte. The failure to transduce signals through the pre-BCR inhibits B-cell development, X-linked agammaglobulinaemia (XLA). Mutations in Bruton's tyrosine kinase (BTK) and B-cell linker protein (BLNK) result in the arrest of B-cell development at the pre-B-cell stage, yielding an agammaglobulinemic state (Conley *et al.*, 1998; Minegishi *et al.*, 1999). Patients with loss of function mutations in CD40L affect B-cell class switching, have high serum levels of IgM and suffer from recurrent infections (Hyper-IgM syndrome) (Facchetti *et al.*, 1995). In this pathological condition, there is a severe reduction in circulating levels of all antibody isotypes except IgM and patients are highly susceptible to pyogenic extracellular bacteria. A very similar syndrome has also been observed due to the mutation in NEMO protein which regulates signalling cascade in the downstream of CD40: CD40L interaction (Mancini *et al.*, 2008). NEMO deficiency permits the production of pathogenic IgM autoantibodies that are frequently directed against haematopoietic cells including red blood cells, white blood cells, and platelets. The patient having IgA deficiency is one type of common variable immunodeficiencies (CVID), which is caused by mutation in BAFF (B-cell activating factor of the tumour necrosis family) and TACI (transmembrane activator and calcium-modulator and cyclophilin ligand interactor) which provides survival signal for B-cell activation and class switching (Castigli and Geha, 2006).

B-cell developmental deregulation also associated with B-cell malignancies which occur in different stages of development. These B-cell neoplasms occur due to chromosomal translocations and gene mutations. B-cell acute lymphocytic leukaemia (B-ALLs) which is mainly paediatric leukaemia, arise during pro-B and pre-B-cell stage and usually involve a breakpoint cluster region (*BCR*)–*ABL1* translocation or mutations affecting one or more of the runt-related transcription factor 1 (*RUNX1*), pre-B-cell leukaemia homeobox 1 (*PBX1*), mixed-lineage leukaemia (*MLL*), protein tyrosine phosphatase non-receptor type 11 (*PTPN11*) and *RAS* genes (Zhang *et al.*, 2011). The high risk of B-ALL is associated with the $CD34^{+}CD19^{-}$ pro-B-cell subset, whereas $CD34^{+}CD19^{+}$ pre-B-cell has standard risk of ALL (le Viseur *et al.*, 2008). B-cell chronic lymphocytic leukaemia (B-CLL) occurs in the follicle of spleen or lymph node (Kuppers, 2005) and an unmutated type of B-CLL derives from mature CD5-positive B-cells (Seifert *et al.*, 2012). However, a mutated type of B-CLL, where there is hypermutation of the immunoglobulin heavy-chain variable gene segment, GC-derived memory B-cells are affected (Oppezzo *et al.*, 2002). Mantle cell lymphoma (MCL) also arises from circulating mature B-cells which do not enter the follicular germinal centre (Campo and Rule, 2015). In the secondary lymphoid organ other different types of lymphoma arise from differentiated subsets of mature B-cells. The splenic marginal zone lymphoma (SMZL) and mucosa-associated lymphoid tissue (MALT) lymphoma are indolent malignancies derived from marginal zone B-cells (Bende *et al.*, 2009). The germinal centre B-cell malignancies are follicular lymphoma, diffuse large B-cell lymphoma (DLBCL) and Burkitt's lymphoma. A GC-derived plasma cell malignancy is multiple myeloma which persists in the bone marrow and that is dependent on contact of stromal cell and cytokines such as IL-6 (Rickert, 2013).

Regulatory B-cells

B-cells, in addition to their conventional role of secreting antibody, also play a regulatory role by secreting cytokines to dampen the pathological condition due to autoimmune disorders and inflammation (Yang *et al.*, 2013; Wolf *et al.*, 1996).

Mizoguchi and Bhan first coined the term 'regulatory B-cell', which secretes regulatory cytokines such as IL-10 and TGF-β (Mizoguchi *et al.*, 2000). Although there are not any specific cell surface marker or set of markers for identification of IL-10 secreting regulatory B-cells; these cells are characterized as B10 cells with their ability to produce and secrete IL-10 following appropriate stimulation (Fillatreau *et al.*, 2002). In mouse, B10 cells are mainly localized in spleen, peritoneal cavity, peripheral and mesenteric lymph node and in blood their frequency is very low (Lykken *et al.*, 2015). B10 cells residing within spleen are phenotypically characterized as $CD1d^{high}CD5^{+}CD19^{+}$ subset (Tedder, 2015) and about 70–80% decrease of IL-10 secreting Breg cell has been noticed in CD19-deficient mice (Watanabe *et al.*, 2010). In contrast, over expression of CD19 increase IL-10 production (Yanaba *et al.*, 2009). The loss of CD22 also results in increase in B10 cell numbers, as CD22 antagonistically effect on CD19 and BCR signalling (Poe *et al.*, 2011; Yanaba *et al.*, 2008). As there are not any specific phenotypic marker for IL-10 secretion, functionally other peripheral B-cell subsets such as MZ B-cell ($CD21^{high}CD23^{-}IgM^{high}CD1d^{high}$) (Gray *et al.*, 2007), Transitional cell T2-MZ precursor B-cell ($CD21^{high}CD23^{+}IgM^{high}CD1d^{+}IgD^{+}$) (Evans *et al.*, 2007), B1a($CD5^{+}$) (O'Garra *et al.*, 1992), plasmablasts ($CD138^{+}CD44^{high}$) (Matsumoto *et al.*, 2014), plasma B-cells ($CD138^{high}IgM^{+}TACI^{+}CXCR4^{+}CD1d^{high}Tim^{int}$) (Shen *et al.*, 2014) secretes IL-10 during different pathological condition. Most abundant peritoneal cavity B10 cell belongs to $CD5^{+}CD11b^{+}$ B1a subset; also at lower frequencies in $CD5^{-}CD11b^{+}$ B1b and $CD5^{-}CD11b^{-}$ B2 subsets (Maseda *et al.*, 2013). It is difficult to identify IL-10 secreting B-cell (B10) during *in vivo* condition, as their number is very low. B-cells that are functionally programmed to secrete IL-10 are labelled as progenitor B10 cells ($B10_{pro}$) and phenotypically characterized as $CD1d^{-}CD5^{-}$ in adult blood and lymph node, whereas $CD1d^{-}CD5^{+}$ in neonatal spleen and adult peritoneal cavity B-cell subset in mice (Kalampokis *et al.*, 2013). To identify functionally IL-10 competent IL-10 cell, during *ex vivo* condition $B10_{pro}$ cells are stimulated with LPS, phorbol 12-myristate 13-acetate (PMA) and ionomycin and are converted into IL-10 secreting B10 cells (Maseda *et al.*, 2012) (Fig. 3.6).

Possible role of regulatory B-cell during pathological condition

Regulatory B-cell has been discovered during understanding of inflammation and autoimmunity which has critical role during various diseased conditions. During systemic lupus erythematosus (SLE) there is impairment of $CD24^{hi}CD38^{hi}$ regulatory B-cells in peripheral blood. Applying CD40 stimulation in these regulatory B-cells, they suppress IFN-γ and TNF-α production from $CD4^{+}$ T-cells in normal condition; but this inhibition is absent from SLE

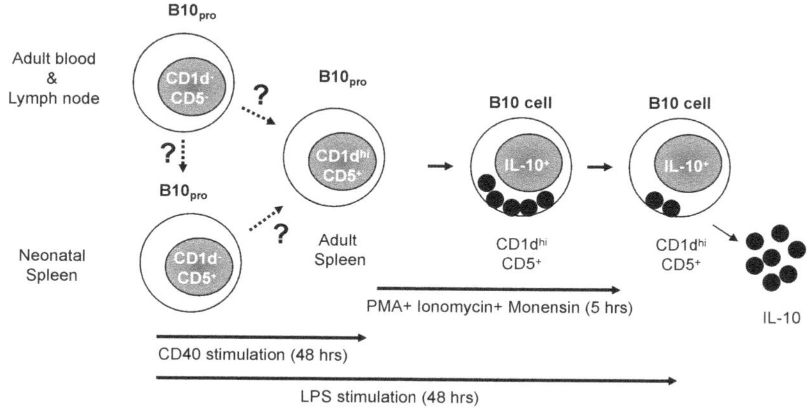

Figure 3.6 Model of potential developmental pathway of IL-10-secreting B10 cell.

patients as there is less production of IL-10 with CD40 stimulation (Blair *et al.*, 2010). The similar observation has been reported with pemphigus patients, where activated B-cells act as positive regulator to secrete autoantibodies (Zhu *et al.*, 2015). In contrast to SLE and pemphigus patients, $CD24^{hi}CD38^{hi}$ regulatory B-cells from peripheral blood of Rheumatoid arthritis (RA) patient fail to suppress T_h17 differentiation. Their frequency also decreases during diseased condition (Flores-Borja *et al.*, 2013). These $CD24^{hi}CD38^{hi}$ B-cell subsets are transitional B-cell which functionally acts as regulatory B-cell during autoimmune disorder. The heterogeneous phenotypic population of regulatory B-cell has been observed in autoimmune diseases. The SLE patients have higher frequencies of IL-10 secretion from peripheral blood $CD5^+$ B-cell than the normal controls in the presence or absence of PMA, ionomycin and LPS (Amel Kashipaz *et al.*, 2003). In the mouse model of inflammatory bowel disease (IBD) $CD1d^{hi}CD5^+$ regulatory B-cells from spleen and peritoneal cavity have suppressive activity in different experimental mouse colitis. This suppressive activity during intestinal inflammation by regulatory B-cell in IBD mouse model is IL-10 dependent (Maseda *et al.*, 2013). In lupus patient, $CD154^+$ B-cells from peripheral blood also secretes IL-10, while stimulating with *Staphyloccocal aureus* Cowan 1 (SAC1) (Diaz-Alderete *et al.*, 2004). IL-10 contributes a complex biology and multiple roles in autoimmune biology. The secretion of IL-10 by hyperactive B-cell regulates pathogenesis in systemic lupus erythematosus, whereas a relative deficiency of B-cell secreted IL-10 is associated with dysfunctional immune regulation in multiple sclerosis (MS). In normal human B-cell subsets IL-10 is produced by naive B-cell, whereas memory B-cell secretes pro-inflammatory cytokines. During MS there is an *in vivo* switch of cytokine program that transit naive B-cell pool towards memory B-cell pool and dysfunction immune response (Duddy *et al.*, 2007).

Regulatory B-cells also plays significant role in cancer. In the mouse 4T1 model of breast cancer, phenotypic subset $CD19^+B220^+CD25^+$ B2 lymphocyte behaves as regulatory B-cell which progress murine breast cancer cells to metastasise into lungs (Olkhanud *et al.*, 2011). Interestingly, CD25 is highly expressed on all activated T-cells, B-cells and in the thymic Tregs and this novel subpopulation of Bregs is termed as tBregs whose population significantly rises in peripheral blood and secondary lymphoid organs. Both in murine model of 4T1 and in the human cancers *in vitro* there is generation of tBregs which secretes TGF-β (Olkhanud *et al.*, 2011). In the absence of Bregs 4T1 tumours can not metastasize into the lungs efficiently, due to inhibition of conversion into Tregs from $CD4^+$ T-cells. It has been reported that anti-CD20 antibody reduction is one of the effective therapy for treating non Hodgkin lymphomas and CLL, but during relapsed condition the poor response of this therapy is due to the presence of Bregs and its secretion of IL-10 (Horikawa *et al.*, 2011). In skin cancer, secretion of IL-10 from $CD19^+CD21^+$ regulatory B-cell has been established through TNF-α knockout model in mice where significant reduction of Breg cell has been observed (Schioppa *et al.*, 2011).

Apart from its significant involvement in autoimmunity and cancer, regulatory B-cell plays its role during different infectious conditions. During acute infection malaria patients show increased number of immature Transitional B-cell population in blood that could behave as IL-10 regulatory B-cell (Nduati *et al.*, 2011). Increase in IL-10 secretion has been reported in Schistosoma infection which suppresses IFN-γ and IL-17 from effector T-cell (Labuda *et al.*, 2013). The major source of this IL-10 secretion is $CD1d^{hi}$ regulatory B-cells (Breg) from peripheral blood mononuclear cell (PBMC), whereas $CD24^{hi}CD27^+$ Breg cell secretes TGF-β (van der Vlugt *et al.*, 2014). However, during *Mycobacterium* infection $CD1d^{hi}CD5^+$ B-cell subset inhibits Th17/22 not by supplying IL-10 or TGF-β but through development of cognate interactions (Zhang *et al.*, 2012). Regulatory B-cell also plays significant role by secreting IL-10 during HIV infection. There is inhibition of $CD8^+$ T-cell proliferation and the HIV-specific cytotoxic response in antiretroviral-treated or untreated HIV-infected patients by the secretion of IL-10. B-cell phenotype $CD19^+CD38^{hi}CD24^{hi}PD\text{-}L1^+$ ($CD27^-$) is the source of this IL-10 during HIV infection (Siewe *et al.*, 2013). Here, interaction between PD-L1/PD1 has critical role in the exhaustion of $CD8^+$-T-cell. Similar immunopathological feature has also been noticed in case of HCV infection (Das *et al.*, 2012).

Secretion of IL-10 from B-lymphocytes during *Leishmania* infection

Interleukin 10 (IL-10) belongs to anti-inflammatory cytokine (Moore *et al.*, 2001), which plays a significant role against inflammation and autoimmune disorders and loss of IL-10 leads to exacerbation of disease in mutant mice (Mauri *et al.*, 2003). Initially this cytokine has been identified as a cytokine synthesis inhibitory factor (CSIF), the product which is secreted from Th2 cells inhibit production of IFN-γ and TNF-α from Th1 cells (Fiorentino *et al.*, 1989). The source of IL-10 secretion is much diversified and it targets multiple immune cells to check their pro-inflammatory response. The activation of conventional dendritic cell is impaired by IL-10, which inhibits the up-regulation of co-stimulatory molecules and reduces production of IL-12. This indirectly suppresses the proliferation and IFN-γ secretion of T-cells and NK cells (Redpath *et al.*, 2014). It has been reported that macrophage destruction is impaired due to IL-10 while infection with intracellular and extracellular parasites. During infection, it reduces production of IL-12 from parasite induced dendritic cells (Groux *et al.*, 1999). IL-10 further reduces TNF-α production and prevents generation of reactive oxygen species and nitric oxide from macrophages and dendritic cell while infection with the parasite. This reduces toxicity at the local site of infection and parasite killing (Bogdan *et al.*, 1991; Gazzinelli *et al.*, 1992; Oswald *et al.*, 1992). Although, suppression of pro-inflammatory cytokines by IL-10 hamper parasite killing, alternatively it restricts tissue damage from those pro-inflammatory cytokines. (Chen *et al.*, 2012). IL-10 secretion is the healing strategy of the host against inflammation and acts as a balancing factor between protection and pathology during microbial infections. Pathogen tactfully manipulates or hijacks the host defence mechanism for their survival and infection. The granulomatous hypersensitivity reactions that set in spleen during kinetoplastid infection can be deleterious for the host. However, the tissue damage is substantially prevented by IL-10, an anti-inflammatory cytokine that is secreted from Th2 cells, Treg cells, M2 macrophages, regulatory macrophages, immature dendritic cells and regulatory B-cells (Hunter *et al.*, 1997; Murphy *et al.*, 2001). Whereas, IL-10 mediated protection against immunopathology does not compromise protective immunity during Plasmodium and Toxoplasma infection (Jankovic *et al.*, 2007; Freitas do Rosario *et al.*, 2012). However during helminth infection protective immunity development is disturbed by IL-10 and it does not effect on host immunopathology (Hesse *et al.*, 2004). The source of IL-10 secretion is versatile like macrophages (Kane and Mosser, 2001), dendritic cells (Poncini *et al.*, 2008), B-cells (Ronet *et al.*, 2010) and other subsets of $CD4^+$ and $CD8^+$ T-cells (Hesse *et al.*, 2004; Hoffmann *et al.*, 2010) and also varies accordingly with different microbial infections. During *Leishmania* infection secretion of IL-10 from macrophage (Mukherjee *et al.*, 2014) and T regulatory cell (Guha *et al.*, 2014) are very well known immunopathological feature associated with parasitic infection. Interestingly, antimony drug resistance property of *Leishmania* is associated with higher level of IL-10 secretion from host macrophage which makes IL-10 a significant pathological factor in leishmaniasis (Berman *et al.*, 1985; Mukherjee *et al.*, 2015). Among B-cells, major IL-10 producing cell is regulatory B-cell (Breg) which specifically known as B10 cell and in autoimmunity it is the major source of IL-10 secretion. During *in vitro* condition purified splenic B-cell from BALB/c mice secretes IL-10 during interaction with *Leishmania major* parasite, whereas draining lymph node B-cell secretes IL-10 during *in vivo* condition of *L. major*-infected mice. The phenotypic characterization of these IL-10 producing cells have been identified as $CD1d^+CD5^+$ B-cells both *in vitro* and *in vivo* condition (Table 3.1). This *in vitro* leishmanial parasite interacted IL-10 secreting $CD5^+CD1d^+$ B-cells are $CD21^{low}$ and $CD23^{low}$; whereas CpG stimulated IL-10-secreting B-cells are characterized as $CD5^+CD1d^-$ $CD21^{high}CD23^{high}$ (Ronet *et al.*, 2010). Interestingly, these phenotypic markers overlap some common subsets of B-cell, like $CD5^+$ B1a B-cell, $CD1d^+CD23^+IgM^+IgD^+$ Transitional B-cell T2 and $CD19^+CD1d^+CD23^{+/-}IgM^+IgD^{+/-}$ Marginal Zone B-cell. These different subsets of B-cell could behave as regulatory B-cell during autoimmunity and infectious conditions, accordingly. However IL-10 secreting B-cell during interaction of *Leishmania donovani* amastigote and purified splenic B-cell from C57BL6 mice are phenotypically $CD21^+$ and $CD5^+$. Among these IL-10 secreting B-cell subsets, one population resembles MZ B-cell

Table 3.1 Source of IL-10, TGF-β and IL-6 from different subsets of B-cells in disease conditions

Diseased condition	Host	Source	Experimental condition	Phenotypic characterization	Cytokine production		
					IL-10	IL-6	TGF-β
Cutaneous leishmaniasis	Mouse	Draining lymph node	*In vivo*	$CD1d^{+}CD5^{+}$	+	–	–
Cutaneous leishmaniasis	Mouse	Spleen	*In vitro*	$CD1d^{+}CD5^{+}CD21^{low}CD23^{low}$	+	–	–
				$CD19^{+}CD21^{+}CD1d^{+}CD5^{+}CD23^{hi}$	+	–	–
Visceral leishmaniasis	Mouse	Spleen	*In vitro*	$CD19^{+}CD21^{+}CD1d^{+/-}CD5^{+}CD23^{low}$	+	–	–
Visceral leishmaniasis	Human	Tonsillar tissue	*In vitro*	$CD19^{+}CD24^{+}CD27^{-}$	+	–	–
				$CD19^{+}CD1d^{hi}$	+	–	–
Schistosomiasis	Human	PBMC	*In vivo*	$CD24^{hi}CD27^{-}$	–	–	+
HIV-1/SIV; HBV	Human	PBMC	*Ex vivo*	$CD19^{hi}CD38^{hi}CD24^{hi}CD27^{-}$	+	–	–
Bacterial infection	Mouse	Spleen	*In vitro*	$B220^{+}CD21/35^{hi}CD23^{-}$	–	+	–

($CD19^{+}CD21^{+}CD1d^{+/-}CD5^{+}CD23^{low}$) and another population has similarity with Breg cell ($CD19^{+}CD21^{+}CD1d^{+}CD5^{+}CD23^{high}$). The significance of MZ like B-cell secretes IL-10 has been established during *Leishmania* infection, as depletion of this specific B-cell leads to dramatic fall of IL-10 secretion. The only source of IL-10 secretion from MZ B-cell depleted group is regulatory B-cell. This stimulation of IL-10 secretion is associated with MyD88-dependent pathway (Bankoti *et al.*, 2012); where B-cell activation is triggered by endosomal TLR (Toll-like receptor) while interaction with *Leishmania donovani* amastigotes and B-cells (Silva-Barrrios *et al.*, 2016). It has been observed that expansion of B10 cell during *in vivo* condition is positively regulated due to overexpression of CD19, CD22 deficiency and ectopic CD40L expressions on B-cells (Yanaba *et al.*, 2009). An adaptor molecule MyD88 plays significant role in TLR signalling pathway and MyD88 is a common adapter protein which is shared by all TLRs. The role of MyD88 has been established during cutaneous leishmaniasis in C57BL/6 mouse model. The mutant $MyD88^{-/-}$ mice shows greater number of cutaneous lesions than the wild-type mice, the same pattern has been observed in infected *Leishmania* susceptible strain BALB/c mice. This loss of MyD88 leads to increase in IL-4 and decrease level of IFN-γ and IL-12(p40) cytokines from dendritic cells (Muraille *et al.*, 2003). More specifically, the efficient control against *Leishmania major* infection is regulated by TLR-4-dependent pathway. In this case deletion of TLR-4 leads to higher level secretion of IL-10 from lymph node, but not IL-4 cytokine. A higher level of IL-6 and IL-13 secretion has also been reported from these infected TLR-4 deleted mice compare to the wild-type animals (Kropf *et al.*, 2004). So, several studies have highlighted TLR-MyD88 pathway activates Th1 response against different disease model, including *Leishmania*. In this pathway, transcription factor interferon regulatory factor 5 (IRF-5) is activated by TLR-7 and TLR-9 and plays role in innate antiviral immune response (Schoenemeyer *et al.*, 2005). While observing similar deletion affect on cytokine status in MyD88 and TLR-4 mice while *Leishmania* infection, $Irf5^{-/-}$ mice also shows increase in IL-4 and IL-10 response and concomitant reduction in iNOS expression with smaller liver granulomas during *Leishmania donovani* infection. Also Th1 response is hampered through TLR-7 mediated Irf-5 activation during chronic infection in spleen (Paun *et al.*, 2011). During visceral leishmaniasis activation of NK cell and release of IFN-γ requires TLR-9, myeloid dendritic cell and IL-12 (Schleicher *et al.*, 2007). For activation of NK cell IL-12 and IFN-α/β are necessary, while in this case IFN-α/β receptor (IFNAR) plays minor role in NK cell activation. *Leishmania donovani* amastigote also activates B-cells through endosomal TLR activation where TLR3, TLR7 and TLR9 are involved (Silva-Barrrios *et al.*, 2016; Honda *et al.*, 2016). Unlike NK cell activation, B-cell activation by TLR is controlled through IFNAR. Disease exacerbation during VL is regulated through TLR mediated B-cell activation which secretes IL-10 and IFN-I and generation of hypergammaglobulinaemia. In human B-cell, TLR induced IL-10 production requires activation of STAT3 and ERK (Liu *et al.*, 2014). Furthermore, when stimulated with IFN-α, a member of the type I interferon family, the IL-10 production is regulated through TLR7/8, but not through TLR9. The exact role or mechanism of IL-10-mediated disease progression is not quite clear. It has been observed that blockage of IL-10 receptor leads to CD4 and CD8 expansion and interaction between *Leishmania* parasite and MZB-cell is responsible for suppression of effector T-cell. Interestingly, genetically modified IgM transmembrane domain (μMT) mutant mice display higher cytotoxic potentiality by CD8 T-cells and increase in parasite-specific IFN-γ^+-CD4 T-cells while infection with *L. donovani* in compare to parasite-infected wild-type mice. Formation of hepatic granuloma and presence of increased number of neutrophils in the liver has been observed in μMT mutant mice during parasite infection (Smelt *et al.*, 2000). The Th1/Th2 immune responses in μMT mutant mice do not clearly establish its resistance to *Leishmania* infection. In contrast, Darrah *et al.* (2010) have reported that an increase in IL-10 disturbs Th1/Th2 balance by decreasing IL-12 secretion from dendritic cells and macrophages during *Leishmania* infection. Similarly, μMT mice show misbalance of Th1/Th2 immune response during infection with *Plasmodium chabaudi chabaudi* (Langhorne *et al.*, 1998). In case of Chlamydia infection, this mutant mice fails to produce delayed type hypersensitivity response (Yang and Brunham, 1998). Suppression of NK1.1$^+$ cells through MyD88–IL-10 axis by MZ B-cell has also been reported during *Leishmania* infection. This observation is well studied in *Salmonella* infection; where splenic CD19$^+$CD138$^+$ B-cell secretes IL-10 during bacterial infection through MyD88-dependent pathway (Neves *et al.*, 2010). Furthermore, Fillatreau group has also noticed increase infiltration of neutrophil and natural killer cell in Salmonella-infected B-cell-specific IL-10$^{-/-}$ mice liver. Similar observation in case of B-cell-specific MyD88 mutant mice confirms B-cell-MyD88-IL-10 mediated immunopathological deregulation leading to inhibition of effective host defence against Salmonella infection. Thus, IL-10 secretion could affect host immunopathology in larger context which is complex to understand still now. Not only in mice but also from human tonsillar tissue purified B-cell secretes IL-10 during their interaction with *Leishmania infantum* amastigote. Parasite interaction triggers regulatory function of B-cell and induce CD19$^+$CD24$^+$CD27$^-$ B-cell to secrete IL-10 (Andreani *et al.*, 2015). Similar to murine B-cell activation mediated suppression of CD4$^+$ T-cell, activation of human tonsillar B-cell leads to suppression of CD4$^+$ T-cell activation, proliferation and function during *Leishmania* and parasite interaction. Thus generation of hypergammaglobulinaemia and disbalance of various cytokines, specifically IL-10 could be the reason behind *Leishmania* pathogenesis.

Acknowledgements

We acknowledge research grant supported by CSIR under CSIR-Pool Scientist scheme (to K.M) and J.C. Bose Fellowship (to S.R) (SB/S2/JCB-65/2014).

References

Achtman, A.H., Khan, M., MacLennan, I.C., and Langhorne, J. (2003). *Plasmodium chabaudi chabaudi* infection in mice induces strong B-cell responses and striking but temporary changes in splenic cell distribution. J. Immunol. *171*, 317–324.

Agematsu, K., Nagumo, H., Yang, F.C., Nakazawa, T., Fukushima, K., Ito, S., Sugita, K., Mori, T., Kobata, T., Morimoto, C., *et al.* (1997). B-cell subpopulations separated by CD27 and crucial collaboration of CD27+ B-cells and helper T-cells in immunoglobulin production. Eur. J. Immunol. *27*, 2073–2079. https://doi.org/10.1002/eji.1830270835.

Ahmed, S., Colmenares, M., Soong, L., Goldsmith-Pestana, K., Mustnermann, L., Molina, R., and McMahon-Pratt, D. (2003). Intradermal infection model for pathogenesis and vaccine studies of murine visceral leishmaniasis. Infect Immun.71, 401–410. https://doi.org/10.1038/nri3803 10.1128/IAI.71.1.401-410.2003.

Allen, C.D., Okada, T., Tang, H.L., and Cyster, J.G. (2007). Imaging of germinal center selection events during affinity maturation. Science *315*, 528–531.

Allman, D., Lindsley, R.C., DeMuth, W., Rudd, K., Shinton, S.A., and Hardy, R.R. (2001). Resolution of three non proliferative immature splenic B-cell subsets reveals multiple selection points during peripheral B-cell maturation. J Immunol. *167*, 6834–6840. https://doi.org/10.4049/jimmunol.167.12.6834.

Amel Kashipaz, M.R., Huggins, M.L., Lanyon, P., Robins, A., Powell, R.J., and Todd, I. (2003). Assessment of Be1 and Be2 cells in systemic lupus erythematosus indicates elevated interleukin-10 producing CD5$^+$ B-cells. Lupus *12*, 356–363. https://doi.org/10.1191/0961203303lu338oa.

Anderson, A.L., Sporici, R., Lambris, J., Larosa, D., and Levinson, A.I. (2006). Pathogenesis of B-cell superantigen-induced immune complex-mediated inflammation. Infect. Immun. *74,* 1196–1203.

Andreani, G., Ouellet, M., Menasria, R., Gomez, A.M., Barat, C., and Tremblay, M.J. (2015). *Leishmania infantum* amastigotes trigger a subpopulation of human B-cells with an immunoregulatory phenotype. PLOS Negl. Trop. Dis 9(2): e0003543. 1-26. https://doi.org/10.1038/nri3803 10.1371/journal.pntd.0003543.

Ansari, N.A., Saluja, S., and Salotra, P. (2006). Elevated levels of interferon-gamma, interleukin-10, and interleukin-6 during active disease in Indian kala azar. Clin. Immunol. *119,* 339–345.

Asito, A.S., Piriou, E., Jura, W.G., Ouma, C., Odada, P.S., Ogola, S., Fiore, N., and Rochford, R. (2011). Suppression of circulating IgD+CD27+ memory B-cells in infants living in a malaria-endemic region of Kenya. Malar. J. *10,* 362. https://doi.org/10.1186/1475-2875-10-362.

Ato, M., Maroof, A., Zubairi, S., Nakano, H., Kakiuchi, T., and Kaye, P.M. (2006). Loss of dendritic cell migration and impaired resistance to *Leishmania donovani* infection in mice deficient in CCL19 and CCL21. J. Immunol. *176,* 5486–5493.

Ato, M., Stäger, S., Engwerda, C.R., and Kaye, P.M. (2002). Defective CCR7 expression on dendritic cells contributes to the development of visceral leishmaniasis. Nat. Immunol. *3,* 1185–1191. https://doi.org/10.1038/ni861.

Babai, B., Louzir, H., Cazenave, P.A., and Dellagi, K. (1999). Depletion of peritoneal $CD5^+$ B-cells has no effect on the course of *Leishmania major* infection in susceptible and resistant mice. Clin. Exp. Immunol. *117,* 123–129.

Bankoti, R., Gupta, K., Levchenko, A., and Stäger, S. (2012). Marginal zone B-cells regulate antigen-specific T-cell responses during infection. J. Immunol. *188,* 3961–3971. https://doi.org/10.4049/jimmunol.1102880.

Baumgarth, N. (2011). The double life of a B-1 cell: self-reactivity selects for protective effector functions. Nat. Rev. Immunol. *11,* 34–46. https://doi.org/10.1038/nri2901.

Bende, R.J., van Maldegem, F., and van Noesel, C.J. (2009). Chronic inflammatory disease, lymphoid tissue neogenesis and extranodal marginal zone B-cell lymphomas. Haematology *94,* 1109–1123. https://doi.org/10.1038/nri3803 10.3324/haematol.2009.005983.

Berek, C., Berger, A., and Apel, M. (1991). Maturation of the immune response in germinal centers. Cell *67,* 1121–1129.

Bergmann, B., Grimsholm, O., Thorarinsdottir, K., Ren, W., Jirholt, P., Gjertsson, I., and Mårtensson, I.L. (2013). Memory B-cells in mouse models. Scand. J. Immunol. *78,* 149–156. https://doi.org/10.1111/sji.12073.

Berman, J.D., Waddell, D., and Hanson, B.D. (1985). Biochemical mechanisms of the antileishmanial activity of sodium stibogluconate. Antimicrob. Agents Chemother. *27,* 916–920.

Bermejo, D.A., Vesely, M.C.A., Khan, M., Acosta-Rodríguez, E.V., Montes, C.L., Merino, M.C., Toellner, K.M., Mohr, E., Taylor, D., Cunningham, A.F., and Gruppi, A.(2011). *Trypanosoma cruzi* infection induces a massive extrafollicular and follicular splenic B-cell response which is a high source of non-parasite-specific antibodies. Immunol.132(1),123–133. https://doi.org/10.1111/j.1365-2567.2010.03347.x.

Bhattacharya, P., Ghosh, S., Ejazi, S.A., Rahaman, M., Pandey, K., Ravi Das, V.N., Das, P., Goswami, R.P., Saha, B., and Ali, N. (2016). Induction of IL-10 and TGFβ from $CD4^+$CD25+FoxP3+ T cells correlates with parasite load in Indian kala-azar patients infected with *Leishmania donovani.* PLOS Negl. Trop. Dis *10,* e0004422. https://doi.org/10.1371/journal.pntd.0004422.

Blair, P.A., Noreña, L.Y., Flores-Borja, F., Rawlings, D.J., Isenberg, D.A., Ehrenstein, M.R., and Mauri, C. (2010). CD19(+) CD24(hi)CD38(hi) B-cells exhibit regulatory capacity in healthy individuals but are functionally impaired in systemic lupus erythematosus patients. Immunity. *32,* 129-40. https://doi.org/10.1038/nri3803 10.1016/j.immuni.2009.11.009.

Bogdan, C., Vodovotz, Y., and Nathan, C. (1991). Macrophage deactivation by interleukin 10. J. Exp. Med. *174,* 1549–1555.

Bohme, M.W.J., Evans, D.A., Miles, M.A., and Holborowi, E.J. (1986). Occurrence of autoantibodies to intermediate filament proteins in human visceral leishmaniasis and their induction by experimental polyclonal B-cell activation. Immunology *59,* 583-588. (PMC 1453314).

Brack, C., Hirama, M., Lenhard-Schuller, R., and Tonegawa, S. (1978). A complete immunoglobulin gene is created by somatic recombination. Cell. 15,1-14. https://dx.doi.org/10.1016/0092-8674(78)90078-8.

Bunn-Moreno, M.M., Madeira, E.D., Miller, K., Menezes, J.A., and Campos-Neto, A. (1985). Hypergammaglobulinaemia in *Leishmania donovani* infected hamsters: possible association with a polyclonal activator of B-cells and with suppression of T-cell function. Clin. Exp. Immunol. *59,* 427–434 (PMC 1577132).

Caldas, A., Favali, C., Aquino, D., Vinhas, V., van Weyenbergh, J., Brodskyn, C., Costa, J., Barral-Netto, M., and Barral, A. (2005). Balance of IL-10 and Interferon-gamma plasma levels in human visceral leishmaniasis: Implications in the pathogenesis. BMC Infect Dis. 5, 113. https://doi.org/10.1186/1471-2334-5-113.

Campo, E., and Rule, S. (2015). Mantle cell lymphoma: evolving management strategies. Blood *125,* 48–55. https://doi.org/10.1182/blood-2014-05-521898.

Carsetti, R., Rosado, M.M., and Wardmann, H. (2004). Peripheral development of B-cells in mouse and man. Immunol. Rev. *197,* 179–191.

Castigli, E., and Geha, R.S. (2006). Molecular basis of common variable immunodeficiency. J. Allergy Clin. Immunol. *117,* 740–746.

Cervia, J.S., Rosen, H., and Murray, H.W. (1993). Effector role of blood monocytes in experimental visceral leishmaniasis. Infect. Immun. *61,* 1330–1333.

Chen, F., Liu, Z., Wu, W., Rozo, C., Bowdridge, S., Millman, A., Van Rooijen, N., Urban, J.F., Wynn, T.A., and Gause, W.C. (2012). An essential role for TH2-type responses in limiting acute tissue damage during experimental helminth infection. Nat. Med. *18,* 260–266. https://doi.org/10.1038/nm.2628.

Chen, M., Sällberg, M., Sönnerborg, A., Weiland, O., Mattsson, L., Jin, L., Birkett, A., Peterson, D., and Milich, D.R. (1999). Limited humoral immunity in hepatitis C virus infection. Gastroenterology *116*, 135–143.

Chung, J.B., Silverman, M., and Monroe, J.G. (2003). Transitional B-cells: step by step towards immune competence. Trends Immunol. 24(6), 342–348. https://dx.doi.org/10.1016/S1471-4906(03)00119-4.

Conley, M.E., Mathias, D., Treadaway, J., Minegishi, Y., and Rohrer, J. (1998). Mutations in btk in patients with presumed X-linked agammaglobulinemia. Am. J. Hum. Genet. *62*, 1034–1043.

Corcoran, A.E., Riddell, A., Krooshoop, D., and Venkitaraman, A.R. (1998). Impaired immunoglobulin gene rearrangement in mice lacking the IL-7 receptor. Nature *391*, 904–907. https://doi.org/10.1038/36122.

Cordeiro-Da-Silva, A., Borges, M.C., Guilvard, E., and Ouaissi, A. (2001). Dual role of the *Leishmania major* ribosomal protein S3a homologue in regulation of T- and B-cell activation. Infect. Immun. *69*, 6588–6596. https://doi.org/10.1128/IAI.69.11.6588-6596.2001.

Cotterell, S., Engwerda, C., and Kaye, P., (2000). *Leishmania donovani* infection of bone marrow stromal macrophages selectively enhances myelopoiesis, by a mechanism involving GM-CSF and TNF-α. Blood. 95, 1642–1651 (PMID 10688819).

Cotterell, S.E., Engwerda, C.R., and Kaye, P.M. (1999). *Leishmania donovani* infection initiates T-cell-independent chemokine responses, which are subsequently amplified in a T-cell-dependent manner. Eur. J. Immunol. *29*, 203–214 (PMID 9933102).

Crocker, P.R., Blackwell, J.M., and Bradley, D.J. (1984). Expression of the natural resistance gene Lsh in resident liver macrophages. Infect. Immun. *43*, 1033–1040.

Crotty, S. (2012). The 1-1-1 fallacy. Immunol. Rev. *247*, 133–142. https://doi.org/10.1111/j.1600-065X.2012.01117.x.

Crotty, S. (2015). A brief history of T-cell help to B-cells. Nat. Rev. Immunol. *15*, 185–189. https://doi.org/10.1038/nri3803.

Darrah, P.A., Hegde, S.T., Patel, D.T., Lindsay, R.W.B., Chen, L., Roederer, M., and Seder, R.A. (2010). IL-10 production differentially influences the magnitude, quality, and protective capacity of Th1 responses depending on the vaccine platform. J Exp Med. 207(7), 1421–1433. https://doi.org/10.1084/jem.20092532.

Das, A., Ellis, G., Pallant, C., Lopes, A.R., Khanna, P., Peppa, D., Chen, A., Blair, P., Dusheiko, G., Gill, U., *et al.* (2012). IL-10-producing regulatory B-cells in the pathogenesis of chronic hepatitis B virus infection. J. Immunol. *189*, 3925–3935. https://doi.org/10.4049/jimmunol.1103139.

Deak, E., Jayakumar, A., Cho, K.W., Goldsmith-Pestana, K., Dondji, B., Lambris, J.D., and McMahon-Pratt, D. (2010). Murine visceral leishmaniasis: IgM and polyclonal B-cell activation lead to disease exacerbation. Eur. J. Immunol. *40*, 1355–1368. https://doi.org/10.1002/eji.200939455.

Díaz-Alderete, A., Crispin, J.C., Vargas-Rojas, M.I., and Alcocer-Varela, J. (2004). IL-10 production in B-cells is confined to CD154+ cells in patients with systemic lupus erythematosus. J. Autoimmun. *23*, 379–383.

Donati, D., Mok, B., Chêne, A., Xu, H., Thangarajh, M., Glas, R., Chen, Q., Wahlgren, M., and Bejarano, M.T. (2006). Increased B-cell survival and preferential activation of the memory compartment by a malaria polyclonal B-cell activator. J. Immunol. *177*, 3035–3044.

Duddy, M., Niino, M., Adatia, F., Hebert, S., Freedman, M., Atkins, H., Kim, H.J., and Bar-or, A.(2007). Distinct effector cytokine profiles of memory and naive human B-cell subsets and implication in multiple sclerosis. J Immunol. *178*, 6092–6099. https://doi.org/10.4049/jimmunol.178.10.6092.

Dudley, D.D., Chaudhuri, J., Bassing, C.H., and Alt, F.W. (2005). Mechanism and control of V(D)J recombination versus class switch recombination: similarities and differences. Adv. Immunol. *86*, 43–112.

Engel, P., Gómez-Puerta, J.A., Ramos-Casals, M., Lozano, F., and Bosch, X. (2011). Therapeutic targeting of B-cells for rheumatic autoimmune diseases. Pharmacol. Rev. *63*, 127–156. https://doi.org/10.1124/pr.109.002006.

Engwerda, C.R., Ato, M., Cotterell, S.E., Mynott, T.L., Tschannerl, A., Gorak-Stolinska, P.M., and Kaye, P.M. (2002). A role for tumor necrosis factor-alpha in remodeling the splenic marginal zone during *Leishmania donovani* infection. Am. J. Pathol. 161, 429–437. (PMC1850733).

Evans, J.G., Chavez-Rueda, K.A., Eddaoudi, A., Meyer-Bahlburg, A., Rawlings, D.J., Ehrenstein, M.R., and Mauri, C. (2007). Novel suppressive function of transitional 2 B-cells in experimental arthritis. J. Immunol. *178*, 7868–7878.

Facchetti, F., Appiani, C., Salvi, L., Levy, J., and Notarangelo, L.D. (1995). Immunohistologic analysis of ineffective CD40-CD40 ligand interaction in lymphoid tissues from patients with X-linked immunodeficiency with hyper-IgM. Abortive germinal center cell reaction and severe depletion of follicular dendritic cells. J. Immunol. 154, 6624–6633 (PMID 7539026).

Fairfax, K.A., Kallies, A., Nutt, S.L., and Tarlinton, D.M. (2008). Plasma cell development: from B-cell subsets to long-term survival niches. Semin. Immunol. *20*, 49–58. https://doi.org/10.1016/j.smim.2007.12.002.

Faleiro, R.J., Kumar, R., Hafner, L.M., and Engwerda, C.R. (2014). Immune regulation during chronic visceral leishmaniasis. PLOS Negl. Trop. Dis *8*, e2914. https://doi.org/10.1371/journal.pntd.0002914.

Fillatreau, S., Sweenie, C.H., McGeachy, M.J., Gray, D., and Anderton, S.M. (2002). B-cells regulate autoimmunity by provision of IL-10. Nat. Immunol. 3, 944–950. https://doi.org/10.1038/ni833.

Fiorentino, D.F., Bond, M.W., and Mosmann, T.R. (1989). Two types of mouse T helper cell. IV. Th2 clones secrete a factor that inhibits cytokine production by Th1 clones. J. Exp. Med. *170*, 2081–2095.

Flores-Borja, F., Bosma, A., Ng, D., Reddy, V., Ehrenstein, M.R., Isenberg, D.A., and Mauri, C. (2013). CD19+CD24hiCD38hi B-cells maintain regulatory T-cells while limiting TH1 and TH17 differentiation. Sci. Transl. Med. 5, 173ra23. https://doi.org/10.1126/scitranslmed.3005407.

Freitas, do Rosario, A.P., Lamb, T., Spence, P., Stephens, R., Lang, A., Roers, A., Muller, W., O'Garra, A., and Langhorne, J. (2012). IL-27 promotes IL-10 production by effector Th1 CD4+ T-cells: a critical mechanism for protection from severe immunopathology during malaria infection. J. Immunol. 188, 1178–1190. https://doi.org/10.4049/jimmunol.1102755.

Gao, W., Wortis, H.H., and Pereira, M.A. (2002). The *Trypanosoma cruzi* trans-sialidase is a T-cell-independent B-cell mitogen and an inducer of non-specific Ig secretion. Int. Immunol. *14*, 299–308.

Gasim, S., el-Hassan, A.M., Khalil, E.A., Ismail, A., Kadaru, A.M., Kharazmi, A., and Theander, T.G. (1998). High levels of plasma IL-10 and expression of IL-10 by keratinocytes during visceral leishmaniasis predict subsequent development of post-kala-azar dermal leishmaniasis. Clin. Exp. Immunol. *111*, 64–69 (PMC1904865).

Gazzinelli, R.T., Oswald, I.P., James, S.L., and Sher, A. (1992). IL-10 inhibits parasite killing and nitrogen oxide production by IFN-gamma-activated macrophages. J. Immunol. *148*, 1792–1796.

Genery, A.R., Cant, A.J., and Jeggo, P.A. (2000). Immunodeficiency associated with DNA repair defects. Clin. Exp. Immunol. 121,1-7. https://doi.org/10.1046/j.1365-2249.2000.01257.x.

Ghosh, A.K., Dasgupta, S., and Ghose, A. (1995). Immunoglobulin G subclass-specific antileishmanial antibody responses in Indian Kala-azar and post-Kala-azar dermal leishmaniasis. Clin. Diagn. Lab. Immunol. *2*, 291–296 (PMC170148)

Good, K.L., Avery, D.T., and Tangye, S.G. (2009). Resting human memory B-cells are intrinsically programmed for enhanced survival and responsiveness to diverse stimuli compared to naive B-cells. J. Immunol. *182*, 890–901. https://doi.org/10.4049/jimmunol.182.2.890.

Gorak, P.M., Engwerda, C.R., and Kaye, P.M. (1998). Dendritic cells, but not macrophages, produce IL-12 immediately following *Leishmania donovani* infection. Eur J Immunol. *28*, 687–695 (PMID 9521079).

Gray, M., Miles, K., Salter, D., Gray, D., and Savill, J. (2007). Apoptotic cells protect mice from autoimmune inflammation by the induction of regulatory B-cells. Proc. Natl. Acad. Sci. U.S.A. *104*, 14080–14085.

Groux, H., Cottrez, F., Rouleau, M., Mauze, S., Antonenko, S., Hurst, S., McNeil, T., Bigler, M., Roncarolo, M.G., and Coffman, R.L. (1999). A transgenic model to analyze the immunoregulatory role of IL-10 secreted by antigen-presenting cells. J. Immunol. *162*, 1723–1729.

Guha, R., Das, S., Ghosh, J., Sundar, S., Dujardin, J.C., and Roy, S. (2014). Antimony resistant *Leishmania donovani* but not sensitive ones drives greater frequency of potent T-regulatory cells upon interaction with human PBMCs: role of IL-10 and TGF-β in early immune response. PLOS Negl. Trop. Dis. 8(7):e2995. https://doi.org/10.1371/journal.pntd.0002995.

Guimaraes-Costa, A.B., Nascimento, M.T., Froment, G.S., Soares, R.P., Morgado, F.N., Conceição-Silva, F., and Saraiva, E.M. (2009). *Leishmania amazonensis* promastigotes induce and are killed by neutrophil extracellular traps. Proc Natl Acad Sci USA. *106*, 6748–6753. https://doi.org/10.1073/pnas.0900226106.

Haas, K.M., Poe, J.C., Steeber, D.A., and Tedder, T.F. (2005). B-1a and B-1b cells exhibit distinct developmental requirements and have unique functional roles in innate and adaptive immunity to *S. pneumoniae*. Immunity 23, 7–18. https://doi.org/10.1016/j.immuni.2005.04.011.

Hardy, R.R., and Hayakawa, K. (2001). B-cell development pathways. Annu. Rev. Immunol. *19*, 595–621.

Hardy, R.R. (2006). B-1 B-cell development. J. Immunol. *177*, 2749–2754.

Hart, M., Steel, A., Clark, S.A., Moyle, G., Nelson, M., Henderson, D.C., Wilson, R., Gotch, F., Gazzard, B., and Kelleher, P. (2007). Loss of discrete memory B-cell subsets is associated with impaired immunization responses in HIV-1 infection and may be a risk factor for invasive pneumococcal disease. J. Immunol. *178*, 8212–8220. https://doi.org/10.4049/jimmunol.178.12.8212.

He, B., Qiao, X., and Cerutti, A. (2004). CpG DNA induces IgG class switch DNA recombination by activating human B-cells through an innate pathway that requires TLR9 and cooperates with IL-10. J. Immunol. *173*, 4479–4491.

He, B., Qiao, X., Klasse, P.J., Chiu, A., Chadburn, A., Knowles, D.M., Moore, J.P., and Cerutti, A. (2006). HIV-1 envelope triggers polyclonal Ig class switch recombination through a CD40-independent mechanism involving BAFF and C-type lectin receptors. J. Immunol. *176*, 3931–3941. https://doi.org/10.4049/jimmunol.176.7.3931.

Heinzel, F.P., Sadick, M.D., Holaday, B.J., Coffman, R.L., and Locksley, R.M. (1989). Reciprocal expression of interferon gamma or interleukin 4 during the resolution or progression of murine leishmaniasis. Evidence for expansion of distinct helper T-cell subsets. J. Exp. Med. *169*, 59–72.

Hesse, M., Piccirillo, C.A., Belkaid, Y., Prufer, J., Mentink-Kane, M., Leusink, M., Cheever, A.W., Shevach, E.M., and Wynn, T.A. (2004). The pathogenesis of schistosomiasis is controlled by cooperating IL-10-producing innate effector and regulatory T-cells. J. Immunol. *172*, 3157–3166.

Hoffmann, K.F., Cheever, A.W., and Wynn, T.A. (2000). IL-10 and the dangers of immune polarization: excessive type 1 and type 2 cytokine responses induce distinct forms of lethal immunopathology in murine schistosomiasis. J. Immunol. *164*, 6406–6416. https://doi.org/10.4049/jimmunol.164.12.6406.

Honda, S., Sato, K., Totsuka, N., Fujiyama, S., Fujimoto, M., Miyake, K., Nakahashi-Oda, C., Tahara-Hanaoka, S., Shibuya, K., and Shibuya, A. (2016). Marginal zone B-cells exacerbate endotoxic shock via interleukin-6 secretion induced by Fcα/μR-coupled TLR4 signalling. Nat. Commun. 7, 11498. https://doi.org/10.1038/ncomms11498.

Horikawa, M., Minard-Colin, V., Matsushita, T., and Tedder, T.F. (2011). Regulatory B-cell production of IL-10 inhibits lymphoma depletion during CD20 immunotherapy in mice. J. Clin. Invest. *121*, 4268–4280. https://doi.org/10.1172/JCI59266.

Hunter, C.A., Ellis-Neyes, L.A., Slifer, T., Kanaly, S., Grünig, G., Fort, M., Rennick, D., and Araujo, F.G. (1997). IL-10 is required to prevent immune hyperactivity during infection with *Trypanosoma cruzi*. J. Immunol. *158*, 3311–3316.

Hunziker, L., Recher, M., Macpherson, A.J., Ciurea, A., Freigang, S., Hengartner, H., and Zinkernagel, R.M.

(2003). Hypergammaglobulinemia and autoantibody induction mechanisms in viral infections. Nat. Immunol. *4*, 343–349. https://doi.org/10.1038/ni911.

Iezzi, G., Frohlich, A., Ernst, B., Ampenberger, F., Saeland, S., Glaichenhaus, N., and Kopf, M. (2006). Lymph node resident rather than skin-derived dendritic cells initiate specific T-cell responses after *Leishmania major* infection. J. Immunol. *177*, 1250–1256. https://doi.org/10.4049/jimmunol.177.2.1250.

Jackson, S.M., Wilson, P.C., James, J.A., and Capra, J.D. (2008). Human B-cell subsets. Adv. Immunol. *98*, 151–224. https://doi.org/10.1016/S0065-2776(08)00405-7.

Jacob, J., and Kelsoe, G. (1992). *In situ* studies of the primary immune response to (4-hydroxy-3-nitrophenyl) acetyl. II. A common clonal origin for periarteriolar lymphoid sheath-associated foci and germinal centers. J. Exp. Med. *176*, 679–687. https://doi.org/10.1084/jem.176.3.679.

Jankovic, D., Kullberg, M.C., Feng, C.G., Goldszmid, R.S., Collazo, C.M., Wilson, M., Wynn, T.A., Kamanaka, M., Flavell, R.A., and Sher, A. (2007). Conventional T-bet(+)Foxp3(-) Th1 cells are the major source of host-protective regulatory IL-10 during intracellular protozoan infection. J. Exp. Med. 204, 273–283. https://doi.org/10.1084/jem.20062175.

Jelicic, K., Cimbro, R., Nawaz, F., Huang, d.a.W., Zheng, X., Yang, J., Lempicki, R.A., Pascuccio, M., Van Ryk, D., Schwing, C., *et al.* (2013). The HIV-1 envelope protein gp120 impairs B-cell proliferation by inducing TGF-β1 production and FcRL4 expression. Nat. Immunol. *14*, 1256–1265. https://doi.org/10.1038/ni.2746.

Jung, D., and Alt, F.W. (2004). Unraveling V(D)J recombination; insights into gene regulation. Cell *116*, 299–311.

Kalampokis, I., Yoshizaki, A., and Tedder, T.F. (2013). IL-10-producing regulatory B-cells (B10 cells) in autoimmune disease. Arthritis Res. Ther. 15 (Suppl. 1), S1. https://doi.org/10.1186/ar3907.

Kallies, A., Hasbold, J., Tarlinton, D.M., Dietrich, W., Corcoran, L.M., Hodgkin, P.D., and Nutt, S.L. (2004). Plasma cell ontogeny defined by quantitative changes in blimp-1 expression. J. Exp. Med. *200*, 967–977.

Kane, M.M., and Mosser, D.M. (2001). The role of IL-10 in promoting disease progression in leishmaniasis. J. Immunol. *166*, 1141–1147.

Karasuyama, H., Rolink, A., and Melchers, F. (1993). A complex of glycoproteins is associated with VpreB/lambda 5 surrogate light chain on the surface of mu heavy chain-negative early precursor B-cell lines. J. Exp. Med. *178*, 469–478.

Kaye, P., and Scott, P. (2011). Leishmaniasis: complexity at the host-pathogen interface. Nat. Rev. Microbiol. *9*, 604–615. https://doi.org/10.1038/nrmicro2608.

Kaye, P.M., Svensson, M., Ato, M., Maroof, A., Polley, R., Stager, S., Zubairi, S., and Engwerda, C.R. (2004). The immunopathology of experimental visceral leishmaniasis. Immunol. Rev. *201*, 239–253. https://doi.org/10.1111/j.0105-2896.2004.00188.x.

Kima, P.E. (2007). The amastigote forms of *Leishmania* are experts at exploiting host cell processes to establish infection and persist. Int. J. Parasitol. 37, 1087–1096.

Klein, U., Rajewsky, K., and Kuppers, R. (1998). Human immunoglobulin (Ig)M+IgD+ peripheral blood B-cells expressing the CD27 cell surface antigen carry somatically mutated variable region genes: CD27 as a general marker for somatically mutated (memory) B-cells. J. Exp. Med. *188*, 1679-1689 (PMC2212515)

Knoops, L., Louahed, J., and Renauld, J.C. (2004). IL-9-induced expansion of B-1b cells restores numbers but not function of B-1 lymphocytes in xid mice. J. Immunol. *172*, 6101–6106.

Körner, H., Cook, M., Riminton, D.S., Lemckert, F.A., Hoek, R.M., Ledermann, B., Köntgen, F., Fazekas de St Groth, B., and Sedgwick, J.D. (1997). Distinct roles for lymphotoxin-alpha and tumor necrosis factor in organogenesis and spatial organization of lymphoid tissue. Eur. J. Immunol. *27*, 2600–2609. https://doi.org/10.1002/eji.1830271020.

Koga, T., Nishihara, T., Fujiwara, T., Nisizawa T., Okahashi, N., Noguchi, T., and Hamada, S. (1985). Biochemical and immunobiological properties of lipopolysaccharide (LPS) from *Bacteroides gingivalis* and comparison with LPS from *Escherichia coli*. Infect. Immun. *47*, 638–647.

Kondo, M., Weissman, I.L., and Akashi K. (1997). Identification of clonogenic common lymphoid progenitors in mouse bone marrow. Cell *91*, 661–672.

Konstantinos, P., Makaritsis, K.P., Gatselis, N.K., Ioannou, M., Petinaki, E., and Dalekos, G.N. (2009). Polyclonal hypergammaglobulinemia and high smooth-muscle autoantibody titers with specificity against filamentous actin: consider visceral leishmaniasis, not just autoimmune hepatitis. Int. J. Infect. Dis. *13*, e157–e160. https://doi.org/10.1016/j.ijid.2008.08.011.

Kraal, G. (1992). Cells in the marginal zone of the spleen. Int. Rev. Cytol. *132*, 31–74.

Kropf, P., Freudenberg, M.A., Modolell, M., Price, H.P., Herath, S., Antoniazi, S., Galanos, C., Smith, D.F., and Mu¨ller, I. (2004). Toll-Like Receptor 4 contributes to efficient control of infection with the protozoan parasite *Leishmania major*. Infect. Immun. 72, 1920–1928 (PMC375159)

Kumar, P.V., Vasei, M., Sadeghipour, A., Sadeghi, E., Soleimanpour, H., Mousavi, A., Tabatabaei, A.H., and Rizvi, M.M. (2007). Visceral leishmaniasis bone marrow biopsy findings. J. Pediatr. Hematol. Oncol. *29*, 77–80. https://doi.org/10.1097/MPH.0b013e31803076a8.

Kuper, D.F., de Heer, E., Van Loveren, H., and Vos, J.G. (2002). Immune system. In Handbook of Toxicologic Pathology (Haschek, W. M., Rousseaux, C. G. and Wallig, M. A., eds.), Vol. 2, pp. 585–646. Academic Press, San Diego.

Kuppers, R. (2005). Mechanisms of B-cell lymphoma pathogenesis. Nat Rev. Cancer 5, 251–262. https://doi.org/10.1038/nrc1589.

Labuda, L.A., Ateba-Ngoa, U., Feugap, E.N., Heeringa, J.J., van der Vlugt, L.E., Pires, R.B., Mewono, L., Kremsner, P.G., van Zelm, M.C., Adegnika, A.A., *et al.* (2013). Alterations in peripheral blood B-cell subsets and dynamics of B-cell responses during human schistosomiasis. PLOS Negl. Trop. Dis 7, e2094. https://doi.org/10.1371/journal.pntd.0002094.

Langhorne, J., Cross, C., Seixas, E., Li, C., and von der Weid, T. (1998). A role for B-cells in the development of T-cell helper function in a malaria infection in mice. Proc. Natl. Acad. Sci. U.S.A. *95*, 1730–1734.

le Viseur, C., Hotfilder, M., Bomken, S., Wilson, K., Röttgers, S., Schrauder, A., Rosemann, A., Irving, J., Stam, R.W., Shultz, L.D., et al. (2008). In childhood acute lymphoblastic leukemia, blasts at different stages of immunophenotypic maturation have stem cell properties. Cancer Cell. *14*, 47–58. https://doi.org/10.1016/j.ccr.2008.05.015.

Lim, H.W., Hillsamer, P., Banham, A.H., and Kim, C.H. (2005). Cutting edge: direct suppression of B-cells by CD4⁺ CD25⁺ regulatory T-cells. J. Immunol. *175*, 4180–4183.

Liu, B.S., Cao, Y., Huizinga, T.W., Hafler, D.A., and Toes, R.E. (2014). TLR-mediated STAT3 and ERK activation controls IL-10 secretion by human B-cells. Eur. J. Immunol. *44*, 2121–2129. https://doi.org/10.1002/eji.201344341.

Liu, Y.J. (1997). Sites of B lymphocyte selection, activation, and tolerance in spleen. J. Exp. Med. *186*, 625–629.

Locksley, R.M., Heinzel, F.P., Fankhauser, J.E., Nelson, C.S., and Sadick, M.D. (1988). Cutaneous host defense in leishmaniasis: interaction of isolated dermal macrophages and epidermal Langerhans cells with the insect-stage promastigote. Infect. Immun. 56:336–342. (PMC259285)

Loder, F., Mutschler, B., Ray, R.J., Paige, C.J., Sideras, P., Torres, R., Lamers, M.C., and Carsetti, R. (1999). B-cell development in the spleen takes place in discrete steps and is determined by the quality of B-cell receptor-derived signals. J. Exp. Med. *190*, 75–89.

Louzir, H., Belal-Kacemi, L., Sassi, A., Laouini, D., Ben Ismail, R., and Dellagi, K. (1994). Natural autoantibodies, IgG antibodies to tetanus toxoid and CD5⁺ B-cells in patients with Mediterranean visceral leishmaniasis. Clin. Exp. Immunol. *95*, 479–484. https://doi.org/10.1111/j.1365-2249.1994.tb07022.x.

Lykken, J.M., Candando, K.M., and Tedder, T.F. (2015). Regulatory B10 cell development and function. Int. Immunol. *27*, 471–477. https://doi.org/10.1093/intimm/dxv046.

MacLennan, I.C.M., Toellner, K.M., Cunningham, A.F., Serre, K., Sze, D.M.Y., Zúñiga, E., Cook, M.C., and Vinuesa, C.G. (2003). Extrafollicular antibody responses. Immunol. Rev. *194*, 8–18. https://doi.org/10.1034/j.1600-065X.2003.00058.x.

Malherbe, L., Filippi, C., Julia, V., Foucras, G., Moro, M., Appel, H., Wucherpfennig, K., Guéry, J.C., and Glaichenhaus, N. (2000). Selective activation and expansion of high-affinity CD4⁺ T-cells in resistant mice upon infection with *Leishmania major*. Immunity *13*, 771–782.

Mancini, A.J., Lawley, L.P., and Uzel, G. (2008). X-linked ectodermal dysplasia with immunodeficiency caused by NEMO mutation: early recognition and diagnosis. Arch. Dermatol. *144*, 342–346. https://doi.org/10.1001/archderm.144.3.342.

Martin, F., Oliver, A.M., and Kearney, J.F. (2001). Marginal zone and B1 B-cells unite in the early response against T-independent blood-borne particulate antigens. Immunity *14*, 617–629. https://dx.doi.org/10.1016/S1074-7613(01)00129-7.

Maseda, D., Candando, K.M., Smith, S.H., Kalampokis, I., Weaver, C.T., Plevy, S.E., Poe, J.C., and Tedder, T.F. (2013). Peritoneal cavity regulatory B-cells (B10 cells) modulate IFN-γ+CD4⁺ T-cell numbers during colitis development in mice. J. Immunol. *191*, 2780–2795. https://doi.org/10.4049/jimmunol.1300649.

Maseda, D., Smith, S.H., DiLillo, D.J., Bryant, J.M., Candando, K.M., Weaver, C.T., and Tedder, T.F. (2012). Regulatory B10 cells differentiate into antibody-secreting cells after transient IL-10 production *in vivo*. J. Immunol. *188*, 1036–1048. https://doi.org/10.4049/jimmunol.1102500.

Matsumoto, M., Baba, A., Yokota, T., Nishikawa, H., Ohkawa, Y., Kayama, H., Kallies, A., Nutt, S.L., Sakaguchi, S., Takeda, K., Kurosaki, T., and Baba, Y.(2014). Interleukin-10-producing plasmablasts exert regulatory function in autoimmune inflammation. Immunity *41*, 104–1051. https://doi.org/10.1016/j.immuni.2014.10.016.

Mauri, C., Gray, D., Mushtaq, N., and Londei, M. (2003). Prevention of arthritis by interleukin 10-producing B-cells. J. Exp. Med. *197*, 489–501.

Merrell, K.T., Benschop, R.J., Gauld, S.B., Aviszus, K., Decote-Ricardo, D., Wysocki, L.J., and Cambier, J.C. (2006). Identification of anergic B-cells within a wild-type repertoire. Immunity *25*, 953–962.

Miles, S.A., Conrad, S.M., Alves, R.G., Jeronimo, S.M., and Mosser, D.M. (2005). A role for IgG immune complexes during infection with the intracellular pathogen *Leishmania*. J. Exp. Med. *201*, 747–754.

Minegishi, Y., Rohrer, J., Coustan-Smith, E., Lederman, H.M., Pappu, R., Campana, D., Chan, A.C., and Conley, M.E. (1999). An essential role for BLNK in human B-cell development. Science *286*, 1954–1957.

Minoprio, P., Coutinho, A., Spinella, S., and Hontebeyrie-Joskowicz, M. (1991). Xid immunodeficiency imparts increased parasite clearance and resistance to pathology in experimental Chagas' disease. Int. Immunol. 3, 427–433. (PMID 1911532).

Mitchison, N.A. (2004). T-cell-B-cell cooperation. Nat. Rev. Immunol. *4*, 308–312. https://doi.org/10.1038/nri1334.

Mizoguchi, E., Mizoguchi, A., Preffer, F.I., and Bhan, A.K. (2000). Regulatory role of mature B-cells in a murine model of inflammatory bowel disease. Int. Immunol. *12*, 597–605.

Moir, S., and Fauci, A.S. (2009). B-cells in HIV infection and disease. Nat. Rev. Immunol. *9*, 235–245. https://doi.org/10.1038/nri2524.

Moir, S., Ho, J., Malaspina, A., Wang, W., DiPoto, A.C., O'Shea, M.A., Roby, G., Kottilil, S., Arthos, J., Proschan, M.A., *et al.* (2008). Evidence for HIV-associated B-cell exhaustion in a dysfunctional memory B-cell compartment in HIV-infected viremic individuals. J. Exp. Med. *205*, 1797–1805. https://doi.org/10.1084/jem.20072683.

Moll, H., Fuchs, H., Blank, C., and Röllinghoff, M. (1993). Langerhans cells transport *Leishmania major* from the infected skin to the draining lymph node for presentation to antigen-specific T-cells. Eur. J. Immunol. *23*, 1595–1601. https://doi.org/10.1002/eji.1830230730.

Montes, C.L., Vottero-Cima, E., and Gruppi, A. (1996). *Trypanosoma cruzi* cytosolic alkaline antigens (FI) induce polyclonal activation in murine normal B-cells. Scand. J. Immunol. 44, 93–100. https://doi.org/10.1046/j.1365-3083.1996.d01-285.x.

Montes, C.L., Zuniga, E., Minoprio, P., Vottero-Cima, E., and Gruppi, A. (1999). A *Trypanosoma cruzi* alkaline antigen induces polyclonal B-cell activation of normal murine spleen cells by T-cell-independent, BCR directed stimulation. Scand. J. Immunol. 50, 159–166. https://doi.org/10.1046/j.1365-3083.1999.00577.x.

Montes, C.L., Zuniga, E.I., Vazquez, J., Arce, C., and Gruppi, A. (2002). *Trypanosoma cruzi* mitochondrial malate dehydrogenase triggers polyclonal B-cell activation. Clin. Exp. Immunol. 127, 27–36. https://doi.org/10.1046/j.1365-2249.2002.01746.x.

Moore, K.W., de Waal Malefyt, R., Coffman, R.L., and O'Garra, A. (2001). Interleukin-10 and the interleukin-10 receptor. Annu. Rev. Immunol. *19*, 683–765.

Moreno, E., Berman, D.T., and Boettcher, L.A. (1981). Biological activities of *Brucella abortus* lipopolysaccharides. Infect. Immun. *31*, 362–370.

Morrison, W.I., Roelants, G.E., Mayor-Withey, K.S., and Murray, M. (1978). Susceptibility of inbred strains of mice to *Trypanosoma congolense*: correlation with changes in spleen lymphocyte populations. Clin. Exp. Immunol. 32(1), 25–40. (PMC1541310).

Moshous, D., Callebaut, I., de Chasseval, R., Corneo, B., Cavazzana-Calvo, M., Le-Diest, F., Tezcan, I., Sanal, O., Bertrand, Y., Philippe, N., Fischer, A., de Villartay, J.P. (2001). Artemis, a novel DNA double strand break repair/ V(D)J recombination protein, is mutated in human severe combined immunodeficiency. Cell. 105, 177-186. https://doi.org/10.1016/S0092-8674(01)00309-9.

Mosser, D.M., and Edelson, P.J. (1985). The mouse macrophage receptor for C3bi (CR3) is a major mechanism in the phagocytosis of *Leishmania* promastigotes. J. Immunol. *135*, 2785–2789.

Mougneau, E., Bihl, F., and Glaichenhaus, N. (2011). Cell biology and immunology of *Leishmania*. Immunol. Rev. *240*, 286–296. https://doi.org/10.1111/j.1600-065X.2010.00983.x.

Muellenbeck, M.F., Ueberheide, B., Amulic, B., Epp, A., Fenyo, D., Busse, C.E., Esen, M., Theisen, M., Mordmüller, B., and Wardemann, H.(2013). Atypical and classical memory B-cells produce *Plasmodium falciparum* neutralizing antibodies. J Exp Med. *210*(2), 389–99. https://doi.org/10.1084/jem.20121970.

Mukherjee, B., Paul, J., Mukherjee, S., Mukhopadhyay, R., Das, S., Naskar, K., Sundar, S., Dujardin, J.C., Saha, B., and Roy, S. (2015). Antimony-resistant *Leishmania donovani* exploits miR-466i to deactivate host MyD88 for regulating IL-10/IL-12 levels during early hours of infection. J. Immunol. *195*, 2731–2742. https://doi.org/10.4049/jimmunol.1402585.

Mukherjee, S., Mukherjee, B., Mukhopadhyay, R., Naskar, K., Sundar, S., Dujardin, J.-C., and Roy, S. (2014). Imipramine exploits histone deacetylase 11 to increase the IL-12/IL-10 ratio in macrophages infected with antimony-resistant *Leishmania donovani* and clears organ parasites in experimental infection. J. Immunol. 193, 4083–4094. https://doi.org/10.4049/jimmunol.1400710.

Muraille, E., De Trez, C., Brait, M., De Baetselier, P., Leo, O., and Carlier, Y. (2003). Genetically resistant mice lacking MyD88-adapter protein display a high susceptibility to *Leishmania major* infection associated with a polarized Th2 response. J. Immunol. 170, 4237–4241. https://doi.org/10.4049/jimmunol.170.8.4237.

Murphy, M.L., Wille, U., Villegas, E.N., Hunter, C.A., and Farrell, J.P. (2001). IL-10 mediates susceptibility to *Leishmania donovani* infection. Eur. J. Immunol. *31*, 2848–2856.

Murray, H.W., and Nathan, C.F. (1999). Macrophage microbicidal mechanisms *in vivo*: reactive nitrogen versus oxygen intermediates in the killing of intracellular visceral *Leishmania donovani*. J. Exp. Med. 189, 741–746. https://doi.org/10.1084/jem.189.4.741.

Nashar, T.O., Hirst, T.R., and Williams, N.A. (1997). Modulation of B-cell activation by the B subunit of *Escherichia coli* enterotoxin: receptor interaction up-regulates MHC class II, B7, CD40, CD25 and ICAM-1. Immunology *91*, 572–578.

Nduati, E., Gwela, A., Karanja, H., Mugyenyi, C., Langhorne, J., Marsh, K., and Urban, B.C. (2011). The plasma concentration of the B-cell activating factor is increased in children with acute malaria. J. Infect. Dis. *204*, 962–970. https://doi.org/10.1093/infdis/jir438.

Neujahr, D.C., Reich, C.F., and Pisetsky, D.S. (1999). Immunostimulatory properties of genomic DNA from different bacterial species. Immunobiology *200*, 106–119.

Neves, P., Lampropoulou, V., Calderon-Gomez, E., Roch, T., Stervbo, U., Shen, P., Kühl, A.A., Loddenkemper, C., Haury, M., Nedospasov, S.A., Kaufmann, S.H.E., Steinhoff, U., Calado, D.P., and Fillatreau, S. (2010). Signaling via the MyD88 adaptor protein in B-cells suppresses protective immunity during *Salmonella typhimurium* infection. Immunity 33, 777–790. https://doi.org/10.1016/j.immuni.2010.10.016.

Nolte, M.A., Belien, J.A., Schadee-Eestermans, I., Jansen, W., Unger, W.W., van Rooijen, N., Kraal, G., and Mebius, R.E.(2003). A conduit system distributes chemokines and small blood-borne molecules through the splenic white pulp. J Exp Med.198, 505–512. https://doi.org/10.1084/jem.20021801.

Notarangelo, L.D., Fischer, A., Geha, R.S., Casanova, J.L., Chapel, H., Conley, M.E., Cunningham-Rundles, C., Etzioni, A., Hammartröm, L., Nonoyama, S., *et al.* (2009). Primary immunodeficiencies: 2009 update. J. Allergy Clin. Immunol. *124*, 1161–1178. https://doi.org/10.1016/j.jaci.2009.10.013.

Nutt, S.L., and Tarlinton, D.M. (2011). Germinal center B and follicular helper T-cells: siblings, cousins or just good friends? Nat. Immunol. *12*, 472–477.

Nutt, S.L., Hodgkin, P.D., Tarlinton, D.M., and Corcoran, L.M. (2015). The generation of antibody-secreting plasma cells. Nat. Rev. Immunol. *15*, 160–171. https://doi.org/10.1038/nri3795.

Nylén, S., and Gautam, S. (2010). Immunological perspectives of leishmaniasis. J. Glob. Infect Dis. *2*, 135–146. https://doi.org/10.4103/0974-777X.62876.

Nylén, S., Maurya, R., Eidsmo, L., Manandhar, K.D., Sundar, S., and Sacks, D. (2007). Splenic accumulation of IL-10 mRNA in T-cells distinct from CD4+CD25+ (Foxp3) regulatory T-cells in human visceral leishmaniasis. J. Exp. Med. *204*, 805–817.

O'Garra, A., Chang, R., Go, N., Hastings, R., Haughton, G., and Howard, M. (1992). Ly-1 B (B-1) cells are the main source of B-cell derived Interleukin-10. Eur. J. Immunol. *22*, 711–717. https://doi.org/10.1002/eji.1830220314.

Oettinger, M.A., Schatz, D.G., Gorka, C., and Baltimore, D. (1990). RAG-1 and RAG-2, adjacent genes that synergistically activate V(D)J recombination. Science *248*, 1517–1523.

Oliveira, F.L., Aguiar, A.M., Borojevic, R., and El-Cheikh, M.C. (2005). IgE expression on the surface of B1 and B2 lymphocytes in experimental murine schistosomiasis. Braz. J. Med. Biol. Res. *38*, 1033–1042.

Olkhanud, P.B., Damdinsuren, B., Bodogai, M., Gress, R.E., Sen, R., Wejksza, K., Malchinkhuu, E., Wersto, R.P., and Biragyn, A.(2011). Tumor-evoked regulatory B-cells promote breast cancer metastasis by converting resting CD4+ T-cells to T-regulatory cells. Cancer Res. 71(10), 3505–3515. https://doi.org/10.1158/0008-5472.CAN-10-4316.

Olkhanud, P.B., Rochman, Y., Bodogai, M., Malchinkhuu, E., Wejksza, K., Xu, M., Gress, R.E., Hesdorffer, C., Leonard, W.J., and Biragyn, A. (2011). Thymic stromal lymphopoietin is a key mediator of breast cancer progression. J. Immunol. *186*, 5656–5662. https://doi.org/10.4049/jimmunol.1100463.

Oppezzo, P., Magnac, C., Bianchi, S., Vuillier, F., Tiscornia, A., Dumas, G., Payelle-Brogard, B., Ajchenbaum-Cymbalista, F., Dighiero, G., and Pritsch, O.(2002). Do CLL B-cells correspond to naive or memory B-lymphocytes? Evidence for an active Ig switch unrelated to phenotype expression and Ig mutational pattern in B-CLL cells. Leukemia 16, 2438–2446. https://doi.org/10.1038/sj.leu.2402731.

Oswald, I.P., Wynn, T.A., Sher, A., and James, S.L.(1992). Interleukin 10 inhibits macrophage microbicidal activity by blocking the endogenous production of tumor necrosis factor alpha required as a costimulatory factor for interferon gamma-induced activation. Proc. Natl. Acad. Sci. U.S.A. 89, 8676–8680.(PMC49983).

Paun, A., Bankoti, R., Joshi, T., Pitha, P.M., and Stäger, S. (2011). Critical role of IRF-5 in the development of T helper 1 responses to *Leishmania donovani* infection. PLOS Pathog. 7, e1001246. https://doi.org/10.1371/journal.ppat.1001246.

Peruchon, S., Chaoul, N., Burelout, C., Delache, B., Brochard, P., Laurent, P., Cognasse, F., Prévot, S., Garraud, O., Le Grand, R., *et al.* (2009). Tissue-specific B-cell dysfunction and generalized memory B-cell loss during acute SIV infection. PLOS ONE *4*, e5966. https://doi.org/10.1371/journal.pone.0005966.

Plagemann, P.G., Jones, Q.A., and Cafruny, W.A. (2000). N-glycans on the short ectodomain of the primary envelope glycoprotein play a major role in the polyclonal activation of B-cells by lactate dehydrogenase-elevating virus. J. Gen. Virol. *81*, 2167–2175. https://doi.org/10.1099/0022-1317-81-9-2167.

Poe, J.C., Smith, S.H., Haas, K.M., Yanaba, K., Tsubata, T., Matsushita, T., and Tedder, T.F. (2011). Amplified B lymphocyte CD40 signaling drives regulatory B10 cell expansion in mice. PLOS ONE 6, e22464. https://doi.org/10.1371/journal.pone.0022464.

Poncini, C.V., Alba Soto, C.D., Batalla, E., Solana, M.E., and González Cappa, S.M. (2008). *Trypanosoma cruzi* induces regulatory dendritic cells *in vitro*. Infect. Immun. *76*, 2633–2641. https://doi.org/10.1128/IAI.01298-07.

Racoosin, E.L., and Beverley, S.M. (1997). *Leishmania major*: promastigotes induce expression of a subset of chemokine genes in murine macrophages. Exp. Parasitol. *85*, 283–295.

Redpath, S.A., Fonseca, N.M., and Perona-Wright, G. (2014). Protection and pathology during parasite infection: IL-10 strikes the balance. Parasite Immunol. *36*, 233–252. https://doi.org/10.1111/pim.12113.

Requena, P., Campo, J.J., Umbers, A.J., Ome, M., Wangnapi, R., Barrios, D., Robinson, L.J., Samol, P., Rosanas-Urgell, A., Ubillos, I., Mayor, A., López, M., de Lazzari, E., Arévalo-Herrera, M., Fernández-Becerra, C., del Portillo, H., Chitnis, C.E., Siba, P.M., Bardají, A., Mueller, I., Rogerson, S., Menéndez, C., and Dobaño, C.(2014). Pregnancy and malaria exposure are associated with changes in the B-cell pool and in plasma eotaxin levels. J Immunol. 193(6), 2971–83. 10.4049/jimmunol.1401037.

Rickert, R.C. (2013). New insights into pre-BCR and BCR signalling with relevance to B-cell malignancies. Nat. Rev. Immunol. *13*, 578–591. https://doi.org/10.1038/nri3487.

Ritter, U., Meissner, A., Scheidig, C., and Korner, H. (2004). CD8 alpha- and Langerin-negative dendritic cells, but not Langerhans cells, act as principal antigen-presenting cells in leishmaniasis. Eur. J. Immunol. 34, 1542–1550.

Ronet, C., Hauyon-La Torre, Y., Revaz-Breton, M., Mastelic, B., Tacchini-Cottier, F., Louis, J., and Launois, P. (2010). Regulatory B-cells shape the development of Th2 immune responses in BALB/c mice infected with *Leishmania major* through IL-10 production. J. Immunol. *184*, 886–894. https://doi.org/10.4049/jimmunol.0901114.

Rott, O., Charreire, J., and Cash, E. (1996). Influenza A virus hemagglutinin is a B-cell-superstimulatory lectin. Med. Microbiol. Immunol. *184*, 185–193.

Ruffin, N., Thang, P.H., Rethi, B., Nilsson, A., and Chiodi, F. (2011). The impact of inflammation and immune activation on B-cell differentiation during HIV-1 infection. Front. Immunol. *2*, 90. https://doi.org/10.3389/fimmu.2011.00090.

Sacks, D.L., Scott, P.A., Asofsky, R., and Sher, F.A. (1984). Cutaneous leishmaniasis in anti-IgM-treated mice: enhanced resistance due to functional depletion of a B-cell-dependent T-cell involved in the suppressor pathway. J. Immunol. *132*, 2072-2077 (PMID 6230400).

Sanz, I., Wei, C., Lee, F.E., and Anolik, J. (2008). Phenotypic and functional heterogeneity of human memory B-cells. Semin. Immunol. *20*, 67–82. https://doi.org/10.1016/j.smim.2007.12.006.

Schatz, D.G., Oettinger, M.A., and Baltimore, D. (1989). The V(D)J recombination activating gene, RAG-1. Cell *59*, 1035–1048.

Schioppa, T., Moore, R., Thompson, R.G., Rosser, E.C., Kulbe, H., Nedospasov, S., Mauri, C., Coussens, L.M., and Balkwill, F.R. (2011). B regulatory cells and the tumor-promoting actions of TNF-α during squamous carcinogenesis. Proc. Natl. Acad. Sci. U.S.A. *108*, 10662–10667. https://doi.org/10.1073/pnas.1100994108.

Schleicher, U., Liese, J., Knippertz, I., Kurzmann, C., Hesse, A., Heit, A., Fischer, J.A., Weiss, S., Kalinke, U., Kunz, S., *et al.* (2007). NK cell activation in visceral leishmaniasis requires TLR9, myeloid DCs, and IL-12, but is independent of plasmacytoid DCs. J. Exp. Med. *204*, 893–906.

Schoenemeyer, A., Barnes, B.J., Mancl, M.E., Latz, E., Goutagny, N., Pitha, P.M., Fitzgerald, K.A., and Golenbock, D.T. (2005). The interferon regulatory factor, IRF5, is a central mediator of toll-like receptor 7 signaling. J. Biol. Chem. *280*, 17005–17012.

Seifert, M., Sellmann, L., Bloehdorn, J., Wein, F., Stilgenbauer, S., Dürig, J., and Küppers, R. (2012). Cellular origin and pathophysiology of chronic lymphocytic leukemia. J. Exp. Med. *209*, 2183–2198. https://doi.org/10.1084/jem.20120833.

Shen, P., Roch, T., Lampropoulou, V., O'Connor, R.A., Stervbo, U., Hilgenberg, E., Ries, S., Dang, V.D., Jaimes, Y., Daridon, C., Li, R., Jouneau, L., Boudinot, P., Wilantri, S., Sakwa, I., Miyazaki, Y., Leech, M.D., McPherson, R.C., Wirtz, S., Neurath, M., Hoehlig, K., Meinl, E., Grützkau, A., Grün, J.R., Horn, K., Kühl, A.A., Dörner, T., Bar-Or, A., Kaufmann, S.H., Anderton, S.M., and Fillatreau, S.(2014). IL-35-producing B-cells are critical regulators of immunity during autoimmune and infectious diseases. Nature 507(7492), 366–370. https://doi.org/10.1038/nature12979.

Siewe, B., Stapleton, J.T., Martinson, J., Keshavarzian, A., Kazmi, N., Demarais, P.M., French, A.L., and Landay, A. (2013). Regulatory B-cell frequency correlates with markers of HIV disease progression and attenuates anti-HIV CD8+ T-cell function *in vitro*. J. Leukoc. Biol. *93*, 811–818. https://doi.org/10.1189/jlb.0912436.

Silva, E.E., Ramos, O.P., Bauab, T.M., Falcao, D.P., and de Medeiros, B.M. (2003). *Yersinia enterocolitica* O:3 isolated from patients with or without reactive arthritis induces polyclonal activation of B-cells and autoantibody production *in vivo*. Autoimmunity 36, 261–268. (PMID 14567555)

Silva-Barrios, S., Smans, M., Duerr, C.U., Qureshi, S.T., Fritz, J.H., Descoteaux, A., and Stager, S. (2016). Innate immune B-cell activation by *Leishmania donovani* exacerbates disease and mediates hypergammaglobulinemia. Cell Reports *15*, 2427–2437. https://doi.org/10.1016/j.celrep.2016.05.028.

Singh, O.P., Gidwani, K., Kumar, R., Nylén, S., Jones, S.L., Boelaert, M., Sacks, D., and Sundar, S. (2012). Reassessment of immune correlates in human visceral leishmaniasis as defined by cytokine release in whole blood. Clin. Vaccine Immunol. *19*, 961–966. https://doi.org/10.1128/CVI.00143-12.

Smelt, S.C., Cotterell, S.E., Engwerda, C.R., and Kaye, P.M. (2000). B-cell-deficient mice are highly resistant to *Leishmania donovani* infection, but develop neutrophil-mediated tissue pathology. J. Immunol. *164*, 3681–3688. https://doi.org/10.4049/jimmunol.164.7.3681.

Smelt, S.C., Engwerda, C.R., McCrossen, M., and Kaye, P.M. (1997). Destruction of follicular dendritic cells during chronic visceral leishmaniasis. J. Immunol. *158*, 3813–3821.

Spolski, R., Kim, H.P., Zhu, W., Levy, D.E., and Leonard, W.J. (2009). IL-21 mediates suppressive effects via its induction of IL-10. J. Immunol. *182*, 2859–2867. https://doi.org/10.4049/jimmunol.0802978.

Squires, K.E., Kirsch, M., Silverstein, S.C., Acosta, A., McElraith, M.J., and Murray, H.W. (1990). Defect in the tissue cellular immune response: experimental visceral leishmaniasis in euthymic C57BL/6 ep/ep mice. Infect. Immun. 58, 3893–3898 (PMC313751).

Stetson, D.B., Mohrs, M., Mallet-Designe, V., Teyton, L., and Locksley, R.M. (2002). Rapid expansion and IL-4 expression by *Leishmania*-specific naive helper T-cells in vivo. Immunity 17, 191–200.

Su, T.T., and Rawlings, D.J. (2002). Transitional B lymphocyte subsets operate as distinct checkpoints in murine splenic B-cell development. J. Immunol. *168*, 2101–2110.

Sundar, S., Reed, S.G., Sharma, S., Mehrotra, A., and Murray, H.W. (1997). Circulating T helper 1 (Th1) cell- and Th2 cell-associated cytokines in Indian patients with visceral leishmaniasis. Am. J. Trop. Med. Hyg. *56*, 522–525.

Svensson, M., Maroof, A., Ato, M., and Kaye, P.M. (2004). Stromal cells direct local differentiation of regulatory dendritic cells. Immunity *21*, 805–816.

Svensson, M., Zubairi, S., Maroof, A., Kazi, F., Taniguchi, M., and Kaye, P.M. (2005). Invariant NKT-cells are essential for the regulation of hepatic CXCL10 gene expression during *Leishmania donovani* infection. Infect. Immun. 73, 7541–7547.

Teague, B.N., Pan, Y., Mudd, P.A., Nakken, B., Zhang, Q., Szodoray, P., Kim-Howard, X., Wilson, P.C., and Farris, A.D. (2007). Cutting edge: Transitional T3 B-cells do not give rise to mature B-cells, have undergone selection, and are reduced in murine lupus. J. Immunol. *178*, 7511–7515.

Tedder, T.F. (2015). B10 cells: a functionally defined regulatory B-cell subset. J. Immunol. *194*, 1395–1401. https://doi.org/10.4049/jimmunol.1401329.

Terrier, B., Joly, F., Vazquez, T., Benech, P., Rosenzwajg, M., Carpentier, W., Garrido, M., Ghillani-Dalbin, P., Klatzmann, D., Cacoub, P., *et al.* (2011). Expansion of functionally anergic CD21-/low marginal zone-like B-cell clones in hepatitis C virus infection-related autoimmunity. J. Immunol. *187*, 6550–6563. https://doi.org/10.4049/jimmunol.1102022.

Tumang, J.R., Hastings, W.D., Bai, C., and Rothstein, T.L. (2004). Peritoneal and splenic B-1 cells are separable by phenotypic, functional, and transcriptomic characteristics. Eur. J. Immunol. *34*, 2158–2167. https://doi.org/10.1002/eji.200424819.

van der Vlugt, L.E., Zinsou, J.F., Ozir-Fazalalikhan, A., Kremsner, P.G., Yazdanbakhsh, M., Adegnika, A.A., and Smits, H.H. (2014). Interleukin 10 (IL-10)-producing CD1dhi regulatory B-cells from *Schistosoma haematobium*-infected individuals induce IL-10-positive T-cells and suppress effector T-cell cytokines. J. Infect. Dis. *210*, 1207–1216. https://doi.org/10.1093/infdis/jiu257.

Varma, N., and Naseem, S. (2010). Hematologic changes in visceral leishmaniasis/kala azar. Indian J. Hematol.

Blood Transfus. *26,* 78–82. https://doi.org/10.1007/s12288-010-0027-1.

Wakkach, A., Fournier, N., Brun, V., Breittmayer, J.P., Cottrez, F., and Groux, H. (2003). Characterization of dendritic cells that induce tolerance and T regulatory 1 cell differentiation *in vivo.* Immunity *18,* 605–617.

Walker, D.M., Oghumu, S., Gupta, G., McGwire, B.S., Drew, M.E., and Satoskar, A.R. (2014). Mechanisms of cellular invasion by intracellular parasites. Cell. Mol. Life Sci. *71,* 1245–1263. https://doi.org/10.1007/s00018-013-1491-1.

Watanabe, R., Ishiura, N., Nakashima, H., Kuwano, Y., Okochi, H., Tamaki, K., Sato, S., Tedder, T.F., and Fujimoto, M. (2010). Regulatory B-cells (B10 cells) have a suppressive role in murine lupus: CD19 and B10 cell deficiency exacerbates systemic autoimmunity. J. Immunol. *184,* 4801–4809. https://doi.org/10.4049/jimmunol.0902385.

Wilson, M.E., Sandor, M., Blum, A.M., Young, B.M., Metwali, A., Elliott, D., Lynch, R.G., and Weinstock, J.V. (1996). Local suppression of IFN-γ in hepatic granulomas correlates with tissue-specific replication of *Leishmania chagasi.* J. Immunol. *156,* 2231–2239 (PMID 8690913).

Wolf, S.D., Dittel, B.N., Hardardottir, F., and Janeway, C.A. (1996). Experimental autoimmune encephalomyelitis induction in genetically B-cell-deficient mice. J. Exp. Med. *184,* 2271–2278.

Yanaba, K., Bouaziz, J.D., Haas, K.M., Poe, J.C., Fujimoto, M., and Tedder, T.F. (2008). A regulatory B-cell subset with a unique CD1dhiCD5$^+$ phenotype controls T-cell-dependent inflammatory responses. Immunity *28,* 639–650. https://doi.org/10.1016/j.immuni.2008.03.017.

Yanaba, K., Bouaziz, J.D., Matsushita, T., Tsubata, T., and Tedder, T.F. (2009). The development and function of regulatory B-cells expressing IL-10 (B10 cells) requires antigen receptor diversity and TLR signals. J. Immunol. *182,* 7459–7472. https://doi.org/10.4049/jimmunol.0900270.

Yang, X., and Brunham, R.C. (1998). Gene knockout B-cell-deficient mice demonstrate that B-cells play an important role in the initiation of T-cell responses to *Chlamydia trachomatis* (mouse pneumonitis) lung infection. J. Immunol. *161,* 1439–1446.

Yang, M., Rui, K., Wang, S., and Lu, L. (2013). Regulatory B-cells in autoimmune diseases. Cell. Mol. Immunol. *10,* 122–132. https://doi.org/10.1038/cmi.2012.60.

Zhang, J., Mullighan, C.G., Harvey, R.C., Wu, G., Chen, X., Edmonson, M., Buetow, K.H., Carroll, W.L., Chen, I.M., Devidas, M., Gerhard. D.S., Loh, M.L., Reaman, G.H., Relling, M.V., Camitta, B.M., Bowman, W.P., Smith, M.A., Willman, C.L., Downing, J.R., and Hunger, S.P. (2011). Key pathways are frequently mutated in high-risk childhood acute lymphoblastic leukemia: a report from the Children's Oncology Group. Blood *118,* 3080–3087. https://doi.org/10.1182/blood-2011-03-341412.

Zhang, M., Zheng, X., Zhang, J., Zhu, Y., Zhu, X., Liu, H., Zeng, M., Graner, M.W., Zhou, B., and Chen, X. (2012). CD19(+)CD1d(+)CD5(+) B-cell frequencies are increased in patients with tuberculosis and suppress Th17 responses. Cell Immunol. *274*(1–2), 89–97. https://doi.org/10.1016/j.cellimm.2012.01.007.

Zhu, H.Q., Xu, R.C., Chen, Y.Y., Yuan, H.J., Cao, H., Zhao, X.Q., Zheng, J., Wang, Y., and Pan, M. (2015). Impaired function of CD19(+) CD24(hi) CD38(hi) regulatory B-cells in patients with pemphigus. Br. J. Dermatol. *172,* 101–110. https://doi.org/10.1111/bjd.13192.

Zijlstra, E.E., and el-Hassan, A.M. (2001). Leishmaniasis in Sudan. Visceral leishmaniasis. Trans. R. Soc. Trop. Med. Hyg. 95 (Suppl. 1), S27–58.

Zijlstra, E.E., Musa, A.M., Khalil, E.A., el-Hassan, I.M., and el-Hassan, A.M. (2003). Post-kala-azar dermal leishmaniasis. Lancet Infect Dis. 3: 87–98. https://doi.org/10.1016/S1473-3099(03)00517-6.

Zlotnik, A., and Yoshie, O. (2000). Chemokines: a new classification system and their role in immunity. Immunity *12,* 121–127.

Zuckerman, E., Slobodin, G., Kessel, A., Sabo, E., Yeshurun, D., Halas, K., and Toubi, E. (2002). Peripheral B-cell CD5 expansion and CD81 overexpression and their association with disease severity and autoimmune markers in chronic hepatitis C virus infection. Clin. Exp. Immunol. *128,* 353–358.

Cellular Defence of the *Leishmania* Parasite

4

Sanchita Das and Chandrima Shaha*

Cell Death and Differentiation Research Laboratory, National Institute of Immunology, New Delhi, India.

*Correspondence: cshaha@nii.ac.in

https://doi.org/10.21775/9781910190715.04

Abstract

One of the major defence approaches of host against a pathogen is to generate ROS (reactive oxygen species) and RNS (reactive nitrogen species) to eliminate infections. How the pathogen overcomes this stress response is the key to successful invasion and sustenance of infection. *Leishmania* spp. is accountable for the disease leishmaniasis, which manifests in a cutaneous or a visceral form and survives within macrophage phagosomes, the very compartments that are supposed to eliminate them. *Leishmania* spp. lack catalases but have evolved several important defensive armories. These include the peroxiredoxins, thiols, ascorbate peroxidase, selenoproteins, kinases and phosphatases. Available data on the variety of defence systems are discussed.

Introduction

*Leishmania*sis is a spectrum of manifestations caused by 17 diverse species of *Leishmania* parasites propagated by different species of region-specific sandfly vectors. Based on clinical symptoms, *Leishmania*ses has been traditionally classified in three major forms (Handman, 2001), the often fatal visceral leishmaniasis (VL), the cutaneous leishmaniasis (CL) and mucocutaneous leishmaniasis (MCL) that are not fatal but causes extensive morbidity in a large number of people in endemic areas (Peters *et al.*, 1983; Prasad, 1999). Epidemiological changes have provoked spread of the disease along with acquired immune deficiency syndrome.

Distribution

In Asia, the geographical distribution of leishmaniasis includes the nations of Bangladesh, India, Maldives, Nepal, and Sri Lanka. Among them India is solely accountable for 75% of the VL cases, primarily in the state of Bihar (Muzaffarpur), as well as in some neighbouring districts in Uttar Pradesh and in West Bengal (Croft *et al.*, 2006; Desjeux, 2004; Thakur, 2003). Since VL is mainly a rural disease predominantly affecting the poor, and poverty is a key determinant of this disease, so it is grouped, among the neglected tropical infections. The neglected tropical diseases (NTDs) are the most common diseases of the world's poorest people living in Africa, Asia, and the Americas. The global load of leishmaniasis remains unaffected for past several years but has caused the loss of 2.4 million disability-adjusted years (Croft *et al.*, 2006; Desjeux, 2004; Thakur, 2003). Treatment of VL has become more challenging due to higher numbers of human immunodeficiency virus (HIV) co-infections and due to the effects of human migration and resettlement.

Life cycle

In India, VL is caused by *L. donovani*. Indian VL is anthroponotic and is transmitted chiefly through the bites of the female sandfly, *P. argentipes*. During its life cycle, the parasite exists as two forms, the motile promastigotes, which are elongated extracellular forms, and ranges in size from 2 to

20 μm. Promastigotes exist primarily in the gut of the sandfly and amastigotes are intracellular, immotile with no flagella ranging in size from 2 to 5 μm. Amastigotes multiply within macrophage phagolysosomes of the vertebrate host (Handman, 1999). Following sandfly bite, some flagellates entering the circulation are destroyed while others enter the cells of the reticuloendothelial system. They change into amastigotes which multiply by binary fission until the host cell is packed with the parasites and eventually ruptures releasing the amastigotes into circulation.

Treatment

Advanced diagnostic approaches, vector and reservoir control and search for new drugs are being addressed for decades to improve the situation. A number of potential vaccine candidates have been identified, like various protein subunits and DNA antigens but no trial in humans has taken place. The absence of a vaccine for leishmaniasis makes chemotherapeutic strategies as the only choice. Treatment of VL currently uses the pentavalent antimonials, amphotericin B and its lipid formulations (AmBisome®, Abelcet, Amphocil). In addition, pentamidine, miltefosine, paromomycin and sitamaquine (Mukhopadhyay *et al.*, 2001; Singh *et al.*, 1995; Sundar *et al.*, 1995, 1997) are also used. In addition to the above, currently, combination therapy is emerging as a new approach.

Challenges in the field of treatment

Despite impressive advances in science, technology and medicines, we have until now not been fully successful in fighting this disease that particularly affects the poor. Although drug management in leishmaniasis has evolved rapidly and with success, but the obstacle continue to limit the impact of these advances in regions of endemicity (Murray, 2001). The major challenges in the field are due to increased drug resistance and immune-variation among patients. The main limitation to success of drugs is the incomplete knowledge on drug resistance in the *Leishmania* parasite. Currently only a few drugs are available such as Sb(V), amphotericin B, miltefosine and pentamidine. Disparity in the ability of drugs to treat leishmaniasis is often due to the differences in drug sensitivity of the *Leishmania* species, the immune status of the patient, or the pharmacokinetic properties of the drug.

Parasite's defence arsenal

The parasite's major arsenal has evolved against the huge oxidative stress triggered during infection. There are several important defensive armories present in the parasite to defend themselves against the stressful environment of the host.

Thiols

Trypanothione [N1,N8-bis(glutathionyl)-spermidine adduct] is the redox mediator in trypanosomatids comprising the genera *Crithidia*, *Leishmania* and *Trypanosoma*. Trypanothione is manufactured *de novo* from glutathione, spermidine and ATP (Henderson *et al.*, 1990; Krauth-Siegel and Comini, 2008). It is in addition renewed by trypanothione reductase from its oxidized cyclic disulfide form produced during its activity in the peroxide reducing enzyme cascade that also contains dithiol proteins such as thioredoxin and tryparedoxin. It is the primary thiol in trypanosomatids simulating glutathione. Even though Kinetoplastida lack glutathione reductase, *Leishmania* still has a good amount of amount of glutathione from which trypanothione can be produced (Krauth-Siegel and Comini, 2008). Increased levels of trypanothione have been linked with metal resistance in different *Leishmania* species (Legare *et al.*, 1997).

Ovothiol A

Ovothiol A, a non-enzymatic scavenger of hydrogen peroxide, is comparatively less efficient than trypanothione (Ariyanayagam and Fairlamb, 2001). In late logarithmic and stationary phases of growth of the parasites, ovothiol A can be higher than trypanothione in quantity. There is a difference in ovothiol pool in several species of *Leishmania* as ovothiol A is present in *L. major* but absent in *L. donovani*. The role of ovothiol is not prominent in the parasite because the parasite expresses a very active system of trypanothione peroxidase (Ariyanayagam and Fairlamb, 2001).

Trypanothione reductase and tryparedoxin

Trypanothione reductase (TR), a flavoprotein oxidoreductase, is the central component to the unique thiol-redox system. It is responsible for maintaining trypanothione in its reduced form and is a potential target for the chemotherapy of trypanosomatid infections in several studies (Dumas *et al.*, 1997; Khan, 2007). Mutants of *L. donovani* and *L. major* possessing only one wild-type TR allele have lower TR activity than wild-type cells carrying two copies of the TR gene. Significantly, these mutants show reduced infectivity with a significantly decreased capacity to survive intracellularly within macrophages (Dumas *et al.*, 1997). The proximal reaction partner for trypanothione is Tryparedoxin and serves as a reductant for different types of thiol peroxidases. They have the CxxC motif. Functionally, TXNs are protein disulphide reductants.

Tryparedoxin peroxidases

The peroxiredoxins are the major defence system of the *Leishmania* parasite. The members of the family, the peroxidases are functional in both the promastigotes and amastigotes. They reduce H_2O_2 and alkyl hydroperoxides to water and alcohol respectively while using the reducing equivalents provided by thiol proteins (Rhee *et al.*, 2001). These enzymes are abundant, present in a variety of phyla including both protozoans and helminths. They reduce hydrogen peroxide, peroxynitrite and a wide range of organic hydroperoxides (Castro *et al.*, 2002) thus reducing cellular stress. They appear to be fairly promiscuous with respect to the hydroperoxide substrate; the specificities for the donor substrate vary considerably between the subfamilies. Because of their ability to eliminate a variety of toxic species generated by the host cell machinery to eliminate pathogens, they have been implicated in the virulence of mycobacteria and trypanosomatids. They have also been designated as potential drug targets as these enzymes are not present in the mammalian host. Peroxiredoxins use redox-active Cysteine to reduce peroxides and exist as the 1-Cys and the 2-Cys peroxiredoxins, based on the number of cysteinyl residues directly involved in catalysis. The TXNPxs of Trypanosomatidae belong to the 2-Cys peroxiredoxin based on the sequence homology to higher eukaryotic peroxiredoxins and presence of two catalytic cysteine residues with associated motifs (Flohé *et al.*, 1999).

Pronounced differences exist between the antioxidant machinery of trypanosomatids and other eukaryotes. Trypanosomatids do not express essential anti-oxidant enzymes such as catalase and selenium-containing glutathione peroxidases (Castro *et al.*, 2002). As much as 70% of their glutathione is converted to trypanothione. The component accepting the reduction equivalents from trypanothione is tryparedoxin (TXN), which is related to the Thioredoxin family. Tryparedoxin itself is a substrate for the tryparedoxin peroxidase (TXNPx) which reduces H_2O_2 and organic hydroperoxides. All trypanosomatid organisms studied so far possess 2-Cys peroxiredoxins and 1-Cys peroxiredoxins are not present in the genomes of *T. brucei*, *T. cruzi*, and *L. major* (Harder *et al.*, 2006).

There are two types of 2-Cys peroxiredoxins, one located in the cytosol (cytosolic tryparedoxin peroxidase) (cTXNPx) and the other in the mitochondria (mitochondrial tryparedoxin peroxidase) (mTXNPx) (Fig. 4.1). The trypanosomatid genomes encode multiple and almost nearly identical copies for cytosolic proteins and on another chromosome a single copy gene for a mitochondrial 2-Cys peroxiredoxin. Homologous proteins of cytosolic as well as mitochondrial peroxidase have been identified in *L. major* (Levick *et al.*, 1998), *T. brucei* (Tetaud *et al.*, 2001), *L. infantum* (Castro *et al.*, 2002) and *L. donovani* (Iyer *et al.*, 2008). The cytosolic enzymes express two classical VCP motifs. The mitochondrial enzyme has an N-terminal mitochondrial pre-sequence and an IPC motif as the second redox centre. This motif is similar to the LPC sequence in yeast TSA I and II and is not a general feature of mitochondrial 2-Cys-peroxiredoxins. A 226 amino acid coding gene code for the mitochondrial peroxidase in *L. major* and *L. infantum*. Studies on developmentally induced changes using microarray analysis (Holzer *et al.*, 2006) as well as proteomic analysis (Bente *et al.*, 2003) on lesion-derived from promastigotes, and axenic amastigotes in various *Leishmania* spp. have shown different profiles of expression for the protein.

Both the cytosolic and the mitochondrial forms of the tryparedoxin peroxidases exist as decamers (Alphey *et al.*, 2000). The three-dimensional structure of the two enzymes are very similar although they share about 52% primary sequence identity. Studies from this laboratory have shown that overexpression of the cytosolic enzyme rescues parasites from oxidative and drug-induced stress. The

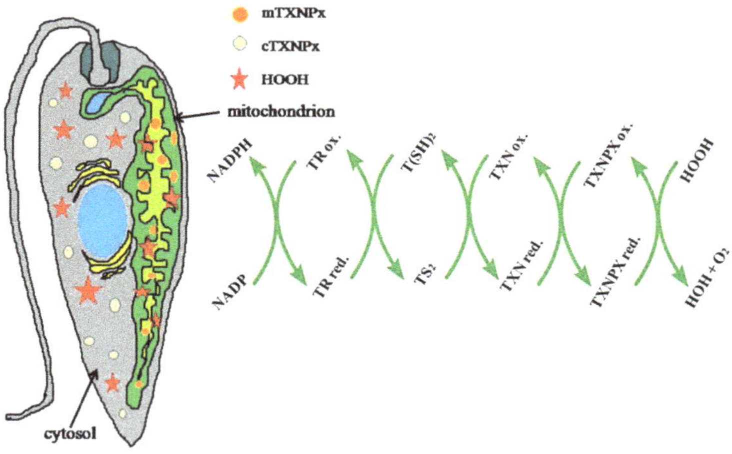

Figure 4.1 Dual TXNPx system in Leishmania. Cytosolic tryparedoxin peroxidase (cTXNPx), and mitochondrial tryparedoxin peroxidase (mTXNPx) for H_2O_2 reduction. Catalytic pathway for detoxification of peroxides by the trypanothione-dependent peroxidoxin system in trypanosomatids is shown where trypanothione is used. trypanothione reductase (TR), reduced trypanothione [$T(SH)_2$], oxidized trypanothione [TS_2], tryparedoxin (TXN), tryparedoxin peroxidase (TXNPx), oxidized (ox.), reduced (red.).

parasites are more sensitive to the combined stress of hydrogen peroxide and nitric oxide that can be overcome through overexpression of the cTXNPx. The elimination of peroxides by the overexpressed enzyme prevents entry of extracellular calcium and release of intracellular calcium induced by the oxidative stress, reducing the forces capable of precipitating cell death consequently increasing cell survival (Iyer *et al.*, 2008). Parasites overexpressing the cTXNPx could infect macrophages *in vitro* in higher numbers as compared to only vector transfected parasites (Iyer *et al.*, 2008), showing that the presence of excess cTXNPx rendered the parasites more capable of combating host defence. Data from this laboratory show that overexpression of the mitochondrial enzyme shows similar effects. The mitochondrial enzyme contains a 30-amino-acid-long mitochondrial targeting sequence (MTS). Our studies have shown that this sequence is essential for transport to the mitochondria and is cleaved upon entry into the organelle (Aich and Shaha, 2013). Interestingly, the MTS contains a calmodulin-binding sequence and *in silico* studies show that calmodulin binds to the MTS. Our investigations using site directed mutagenesis studies clearly demonstrate that substitution of specific residues in place of calmodulin-binding amino acids, impedes the translocation of the protein to the mitochondria. The calmodulin essentially helps HSP70 to bind to the protein for translocation to the mitochondria (Aich and Shaha, 2013). Data from the laboratory show that if translocation of the protein is impeded and bulk of the enzyme resides within the cytosol, the parasites become sensitive to mitochondria generated oxidative stress produced by inhibition of respiratory chain complexes. Interestingly, it has been reported that the mitochondrial enzyme by itself can act as a chaperone and mTXNPx can be a determinant for pathogenicity that is independent of the peroxidative activity of the enzyme (Castro *et al.*, 2011).

Therefore, in a nutshell, these two most essential enzymes of the *Leishmania* defence system, the tryparedoxin peroxidases, cytosolic as well as mitochondrial one are responsible for eliminating peroxides thus protecting the cells from ROS and may also have other functions unrelated to their anti-oxidant activity. Importantly, overexpression of both enzymes rendering the parasites resistant to drug-induced death and increase in the enzyme levels upon relevant oxidative stress in wild-type cells suggests that a close look is necessary at clinical isolates of parasites to determine if they are linked to drug resistance (Wyllie *et al.*, 2010).

Arginases

Internalization of parasites is related to phagocytosis of promastigotes. Further phagocytosis guide the macrophage activation via classical and alternative ways which further outcome in differential L-arginine metabolism through two key enzyme: inducible nitric oxide synthase (iNOS) and arginase (Iniesta *et al.*, 2001; Kropf *et al.*, 2003). Classical activation of macrophages occurs via iNOS which oxidizes L-arginine to nitric oxide (NO), while arginase hydrolyses L-arginine to urea and ornithine through alternative macrophage activation (Gordon, 2003). Ornithine as a result actively participates in synthesis of polyamines, the essential nutrients for growth and proliferation of *Leishmania* parasites (Iniesta *et al.*, 2001, 2002; Kropf *et al.*, 2005). NO has an indispensable role in parasite removal. Since the two pathways contend for arginine, so activation of one pathway slows the other. For example, hydroxyl arginine being a intermediate of classical activation pathway is a powerful arginase inhibitor and treatment of *L. major*-infected mice with its synthetic analogue Nω-hydroxy-L-arginine (NOHA), causes a significant reduction in lesion size and as well as parasite burden (Kropf *et al.*, 2005). The *Leishmania* parasites translate their own arginase which alters infectivity. Also *L. major* null mutants for arginase have reduced capability to infect macrophages both *in vitro* as well as *in vivo* (Muleme *et al.*, 2009). It seems that *Leishmania*-encoded arginase increases progression of the disease by enhancing the host arginase activity.

Kinases and phosphatases

Host macrophages, neutrophils and dendritic cells engulf parasites and start immune responses against them through multiple signalling pathways. The kinases mediating phosphorylation and dephosphorylation processes play an important role in this process. *Leishmania* parasites can modulate macrophage signalling and antimicrobial function. They possess surface protein kinases, which phosphorylate members of complement system thus inactivating cellular cascades and therefore avoid the innate immune responses for a safe environment for parasite proliferation. Forget and coworkers (2001) demonstrated *in vivo* inhibition of host PTP (protein tyrosine phosphatases) controlling disease progression by NO production (Forget *et al.*, 2001). *Leishmania* parasites modulate host's serine/threonine phosphatase PP2 and MAPK phosphatases MKP1 and MKP3 during murine leishmaniasis. Kar and coworkers (2010) demonstrated an enhancement of two PKC isoforms, PKCε and PKCζ, that were implicated in the up-regulation of DSP (dual specific phosphatases) and STP (serine threonine phosphatases) expression along with activity which inhibited macrophage leishmanicidal effects along with higher IL-10 production. Arsenic-based compounds used against leishmaniasis target PTPs. Therefore, kinases and phosphatases are valuable components of *Leishmania* defence against the innate immune system.

Modulation of host cytokines

Leishmania is well known for immunomodulating host immune responses. The T helper cell type 1 (Th1) response is essential to challenge *Leishmania* invasion, while the Th2 response favours development of the disease. Parasites can inhibit the activation of several inflammatory cytokines such as IL-12 (involved in T-cell activation), IFN-γ, IL-1 and TNFα that strengthens parasite survival. *L. donovani* infection inhibits IL-1β secretion, LPG can also repress IL-1β through promoter repression sequence (Hatzigeorgiou *et al.*, 1996; Reiner and Malemud, 1985; Reiner *et al.*, 1990a). *L. donovani* are inferior activator of pro-inflammatory reactions as compared to *L. major* (Matte *et al.*, 2001). Both the species induces heterologous population of host inflammatory cells such as neutrophils and monocytes/macrophages which are effective in controlling infections. IL-12 being a lead player in regulation of cellular immune responses (T-cell activation and IFN-γ secretion), is inhibited by *L. donovani*, *L. major*, and *L. mexicana* amastigotes for securing a safe environment for parasites (Carrera *et al.*, 1996; Weinheber *et al.*, 1998). Cytokine inhibition is further augmented by production of immunosuppressive signalling molecules, such as arachidonic acid metabolites (like PGE2) and Th2 stimulating cytokines such as TGF-β and IL-10 (via interaction with Fcγ receptor) (Matte *et al.*, 2001; Reiner and Malemud, 1984; Reiner and Malemud, 1985b). As a result decreased expression of iNOS and reduced activity of NK cells has been observed. Th2 being a disease progression pathway is also down-regulate the Th1 pathway (microbicidal effects) by suppressing several key players of Th1 pathway such as IL-1, IL-12 and TNFα. Whereas prostaglandin (PGE2) favours parasite survival by inhibiting TNFα, IL-1 and ROS.

Selenoproteins

Selenocysteine (Sec-U) is found in bacteria, archaea and eukarya (Bock *et al.*, 1991; Forchhammer and Böck, 1991). Recently, several selenoprotein families expressing antioxidant properties such as glutathione peroxidase and thioredoxin glutathione reductase (TGR) have been reported to be required for function in flatworms (Maiorino *et al.*, 1996; Williams *et al.*, 1992). Selenoproteins from protozoan parasites are not explored exhaustively. Selenophosphate synthetase has been characterized in *P. falciparum* and other plasmodia. Several genes like in *Leishmania major* (accession no. AAG35734), *Trypanosoma cruzi* (XM_805940) and *Trypanosoma brucei* (XP_823164) were reported. They are a homologue of selenophosphate synthetase, an enzyme that generates selenophosphate, a selenium donor compound used for biosynthesis of Sec (Jayakumar *et al.*, 2004; Lobanov *et al.*, 2006). Functional aspects need to be explored.

Ascorbate peroxidases

Although catalase and selenium-containing glutathione peroxidases are not present in the parasite, ascorbate peroxidase (LmAPX) from *L. major* presents itself to be a potential candidate for scavenging of ROS and has been shown to be central to the redox defence system of *Leishmania* (Dolai et al., 2009). Ascorbate peroxidase is a haem peroxidase identified in the *Leishmania* parasite. This enzyme is localized with to the inner mitochondrial membrane. Overexpression of this enzyme in *L. major* bestows tolerance to oxidative stress-mediated cardiolipin oxidation and thus protects the parasites from protein damage. Parasites with double knockout of this enzyme shows higher intracellular hydrogen peroxide as compared to wild-type parasites. Protection against host cell induced apoptosis is also accorded by ascorbate peroxidase in *Leishmania* (Dolai *et al.*, 2009; Pal *et al.*, 2010).

NADPH oxidase and iNOS expression

Amastigotes resistant to hydrolytic environment succeed well in the phagolysosomal compartment of host macrophages. *Leishmania* promastigotes inhibit phagolysosome biogenesis via its membrane located LPG, which produces periphagosomal accumulation of F-actin with reduced assembly of nicotinamide adenine dinucleotide phosphate (NADPH) oxidase complex and exclusion of vesicular proton ATPase from phagosomes. Amastigote harbouring vacuoles are composed of endoplasmic reticulum such as calnexin and Sec 22b as well as endocytic pathway components such as Rab7, LAMP-1 and LAMP-2 (Antoine *et al.*, 1990, 1998; Vinet *et al.*, 2009). Proteolysis within the phagosome is key to a competent antigen processing and presentation, which is again inhibited by the parasites. The NADPH oxidase complex disrupt antigen presentation (Mantegazza *et al.*, 2008; Rybicka *et al.*, 2012; Savina *et al.*, 2009) by *L. donovani* promastigotes contributing to tricking the immune system.

For NADPH oxidase complex to function requires the cytosolic phosphorylated p47phox and p40/p67 phox heterodimers which associate to form p47/p67/p40phox hetero-trimers before translocation to the membrane, where they interact with membrane-associated flavour cytochrome b558 (DeLeo *et al.*, 1999; El-Benna *et al.*, 2009). To avoid ROS exposure, amastigotes evade the phosphorylation of cytosolic p47 phox, which is necessary for NADPH oxidase activation during phagocytosis (Lodge and Descoteaux, 2006). Thus inhibition of phagolysosome biogenesis which is used for defence by the parasites, is in conflict with the proteolytic activity of the phagosome.

Secretory proteins of *Leishmania*

Secretory proteins in *Leishmania* are supposed to be the key mediators for host–parasite interactions. *Leishmania* secretes plethora of proteins in their environment in order to facilitate the process of infection (Fig. 4.2) (Cuervo *et al.*, 2009; Lambertz *et al.*, 2012). These proteins having serve a dual role in intracellular as well as extracellular events. Among the secretome, a large fraction of proteins belongs to HSP group of proteins such as HSP70, HSP60 etc. which have a known role in protein folding assistance, but their secretory motive in *Leishmania* is not clear (Evdonin *et al.*, 2006; Lancaster and Febbraio, 2005; Requena *et al.*, 2015). In *Toxoplasma gondii* tachyzoites the secretory HSP known to regulate nitric oxide production, immunomodulation and host signalling regulation (Dobbin *et al.*, 2002). Proteins with antioxidative

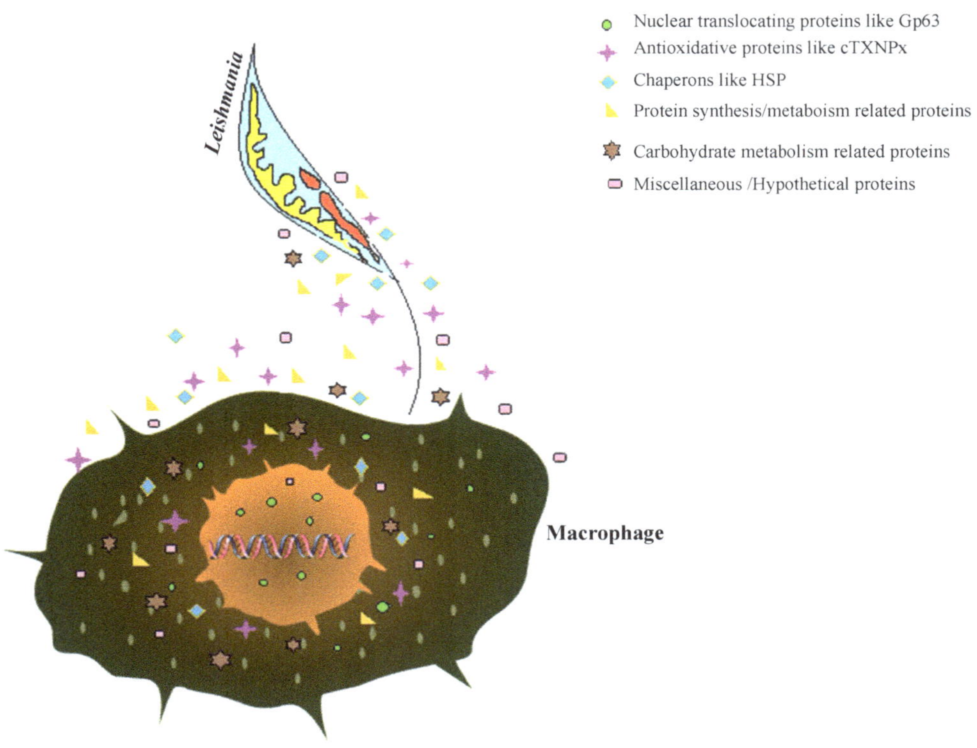

Figure 4.2 Functional categories of secretory proteins of *Leishmania*: *Leishmania* secretes several diverse group of proteins to facilitate the survival of *Leishmania* in macrophages. The secretory fraction of proteins includes chaperons, antioxidative proteins, proteins with nuclear translocation potential, carbohydrate/protein/DNA metabolism proteins and several unidentified proteins. Despite the known functional role of these proteins their secretory mode as well as function are still to be identified.

properties also exist in a good fraction in the secretome of *Leishmania*. Tryparedoxin peroxidase a well characterized protein for their antioxidative properties in the parasites in the cytosol, whereas the motive behind its secretory presence is still to be deciphered. Apart from this, a fraction of proteins involved in carbohydrate, protein and nucleic acid metabolism are also found in the secretome of *Leishmania* (Cuervo *et al.*, 2009).

Host manipulation by parasite

Combating the sandfly's environment

The first host of *Leishmania* is the midgut of sandfly vector. *Leishmania* promastigotes express large amount of glycoconjugates such as GP63, liposhosphoglycan (LPG) on their surface that make them resistant to hydrolytic enzymes of promastigotes (Brittingham *et al.*, 1995; Späth *et al.*, 2003). LPG takes part in attachment to the lectins of sandfly midgut to avoid elimination through excretion. The modification of parasite continues during its differentiation to metacyclic form, the extensive modification involves the modification in expression of LPG. This altered expression of LPG leads to migration of parasites from the midgut of epithelium, at which point they migrate to pharynx for the search of vertebrate host (Olivier *et al.*, 2005).

Within the bloodstream

After transmission to the vertebrate host, the parasite survives by resisting the mammalian complement system. The transformed expression of LPG plays an important role in resisting the cell lysis (Puentes *et al.*, 1988, 1989). The complement system is important and factor C_{3b} is deposited on the cell surface of the parasite but in metacyclic promastigotes, the longer LPG molecules prevents insertion of the C_{5b-9} membrane attack complex into the parasite's membrane. Whereas in *L. donovani* C_{3bi} cannot be deposited on the promastigote surface and thus is not able to play a role in formation of C_5 convertase. The metalloprotease gp63 on the promastigotes helps to prevent complement-mediated lysis and augment promastigote uptake by cleaving C_{3b} to C_{3bi}. Parasites are also known to interact with several serum proteins to activate complement system and facilitate their uptake into host macrophages. Parasite survival is also facilitated by maxadilan (peptide of sandfly saliva) which is a selective agonist of pituitary adenylate-cyclase-activating polypeptide type1 receptor (Andrade *et al.*, 2005, 2007). Maxadilan inhibits LPS-stimulated macrophages from making TNFα and at the same time it reduces the NO production ability. Apart from promastigotes, amastigotes also survive in the bloodstream after being released from macrophages until they find the new macrophage. Proteophosphoglycan (PPG) do the same work for them. PPG activates the complement away from amastigotes to initiate their entry into the new macrophages (Peters *et al.*, 1995, 1997).

Interaction with macrophages

Parasites efficiently escape the humoral immune response of host during their stay in the phagolysosomal compartments of macrophages. Opsonization with C3b and C3bi, which bind to the macrophage receptors CR1 and CR3 respectively, is the main way in which metacyclic promastigotes enter the macrophage (Hawn *et al.*, 2002; Hermoso *et al.*, 1991). Since CR1 and CR3 promote phagocytosis without the oxidative burst, this may be a smart way of survival. Also IL-12 induced cell mediated immunity is inhibited by CR3 receptor so use of CR3 receptor by *Leishmania* is of great advantage. Parasites can enter through Langerhan's cells in the epidermis where the amastigotes transformation can occur (Cáceres-Dittmar *et al.*, 1992; Kautz-Neu *et al.*, 2011). Absence of inducible nitric oxide synthase (iNOS and NOS2) in these cells are advantageous for their survival. Once inside the macrophage, parasites employ a variety of adaptive mechanisms to survive the harsh conditions there. Like *L. donovani* promastigotes can inhibit phagosome–endosome fusion by reducing the fusogenic properties of the membrane, through creating steric repulsion between phagosome and endosome membrane (Desjardins and Descoteaux, 1997). *L. mexicana* amastigotes secrete large amount of PPG in the parasitophorous vacuole possibly to support survival (Ilg *et al.*, 1995). Another way of survival in the host is through inhibiting the hydrolytic enzymes of the host. LPG acts as a degradation barrier because of its highly anionic nature and its unique galactose-β1,4-mannose linkages within the repeating units. The gp63 also exhibits degradation activities of lysosomal enzymes. Binding of calcium to LPG repeating units near the phosphate groups alter the calcium mobilization resulting in defective PKC activation (Culley *et al.*, 1996).

Host signalling pathways modulation

During *Leishmania* infection, weak sensitivity to IFN-γ, LPS and PKC activators has been reported (Carrera *et al.*, 1996; Kima and Soong, 2013). *L. donovani* has been shown to impair tyrosine phosphorylation and activation of JAK1, JAK2 and STAT1 in response to IFN-γ, possibly involving the activation of the cellular protein tyrosine phosphatases SHP-1 (Blanchette *et al.*, 1999; Nandan and Reiner, 1995). As a result dephosphorylation of MAP kinase 1 and 2 occur further activating transcription factor Elk-1 to regulate gene expression. LPG possibly interact with PKC domains containing the diacylglycerol, calcium, and phopholipid-binding sites, or it may interfere the activation of PKC via inhibiting its incorporation into the membrane (Lambertz *et al.*, 2012; McNeely and Turco, 1987). MARCKs (myristolated alanine-rich C kinase substrate) and MRP (MARCKs-related protein) are PKC substrates associated with components of the cellular cytoskeleton and regulate the actin network during cytoskeletal rearrangements (Arbuzova *et al.*, 2002) occurring during infections. *Leishmania* promastigotes down-regulate MRP (Corradin *et al.*, 1999), which may affect vacuolar trafficking or maturation, in order to deactivate the macrophage glycosylinositol phospholipids (GIPLs) found on amastigotes also inhibit PKC, yet this inhibition may involve a different mechanism than that of LPG.

The LPG-associated kinetoplastid membrane protein 11 may also negatively regulate iNOS activity in infected macrophages as it encloses a structural analogue of L-arginine, which is an inhibitor of iNOS as a result diminishing leishmaniacidal activities (Lacerda *et al.*, 2012). One of the protective role of LPG is to attenuate the induction of oxidative burst mediated by PKC. Gp63 has also been associated with suppression of the oxidative burst (Sørensen *et al.*, 1994). Cytokines play a major role in the host defence mechanism. Majorly IL-12, IL-10, TGF-β production are altered during infection of *Leishmania*. Amastigotes are known for impairing the production of agonist induced IL-1 (Reiner, 1987; Reiner *et al.*, 1990b). LPG inhibits IL-1B gene expression by suppressing transcription through a unique sequence of IL-1B promoter.

In one of our study findings reveals a pro-parasitic role of host Bcl-2 and the capacity of host-derived IL-13 to modulate NO levels during infection via Bcl-2. The increased BCL2 results from of *Leishmania* infection and helps the parasite to facilitate their survival in the macrophage (Pandey *et al.*, 2016). Manipulation of cytokine production allows *Leishmania* to survive by deceiving the immune response of the host.

In summary, a large repertoire of defensive mechanisms exists in the *Leishmania* parasite. Some of these are unique to the parasite not being present in the host and therefore, are drug targets. Co-evolution of the host and parasite has enabled both the parasite and the mammalian host to co-exist and at a given time it is the efficiency of the hosts or the parasites defensive arsenals determine the outcome of infection.

Acknowledgements

A part of this work was supported by grants to the Centre for Molecular Medicine, National Institute of Immunology, New Delhi (grant no. BT/PR/14549/MED/14/1291) and the J.C. Bose fellowship to C.S. (award no. SR/S2/JCB-12/2008). S.D. acknowledges the Department of Science and Technology for financial assistance in the form of the DST Inspire Faculty Award fellowship (Award Sanction No.-DST/INSPIRE/04/2015/002785).

References

Aich, A., and Shaha, C. (2013). Novel role of calmodulin in regulating protein transport to mitochondria in a unicellular eukaryote. Mol. Cell. Biol. 33, 4579–4593. https://doi.org/10.1128/MCB.00829-13.

Alphey, M.S., Bond, C.S., Tetaud, E., Fairlamb, A.H., and Hunter, W.N. (2000). The structure of reduced tryparedoxin peroxidase reveals a decamer and insight into reactivity of 2Cys-peroxiredoxins. J. Mol. Biol. *300*, 903–916. https://doi.org/10.1006/jmbi.2000.3881.

Andrade, B.B., de Oliveira, C.I., Brodskyn, C.I., Barral, A., and Barral-Netto, M. (2007). Role of sand fly saliva in human and experimental leishmaniasis: current insights. Scand. J. Immunol. *66*, 122–127. https://doi.org/10.1111/j.1365-3083.2007.01964.x.

Andrade, B.B., Teixeira, C.R., Barral, A., and Barral-Netto, M. (2005). Haematophagous arthropod saliva and host defense system: a tale of tear and blood. An. Acad. Bras. Cienc. *77*, 665–693. https://doi.org//S0001-37652005000400008.

Antoine, J.C., Prina, E., Jouanne, C., and Bongrand, P. (1990). Parasitophorous vacuoles of *Leishmania amazonensis*-infected macrophages maintain an acidic pH. Infect. Immun. *58*, 779–787.

Antoine, J.C., Prina, E., Lang, T., and Courret, N. (1998). The biogenesis and properties of the parasitophorous vacuoles that harbour *Leishmania* in murine macrophages. Trends Microbiol. *6*, 392–401.

Arbuzova, A., Schmitz, A.A., and Vergeres, G. (2002). Cross-talk unfolded: MARCKS proteins. Biochem. J. *362*, 1–12.

Ariyanayagam, M.R., and Fairlamb, A.H. (2001). Ovothiol and trypanothione as antioxidants in trypanosomatids. Mol. Biochem. Parasitol. *115*, 189–198.

Bente, M., Harder, S., Wiesgigl, M., Heukeshoven, J., Gelhaus, C., Krause, E., Clos, J., and Bruchhaus, I. (2003). Developmentally induced changes of the proteome in the protozoan parasite *Leishmania donovani*. Proteomics *3*, 1811–1829. https://doi.org/10.1002/pmic.200300462.

Blanchette, J., Racette, N., Faure, R., Siminovitch, K.A., and Olivier, M. (1999). *Leishmania*-induced increases in activation of macrophage S.H.P.-1 tyrosine phosphatase are associated with impaired I.F.N.-gamma-triggered J.A.K2 activation. Eur. J. Immunol. *29*, 3737–3744. https://doi.org/10.1002/(S.I.C.I)1521-4141(199911)29:11<3737::AID-IMMU3737>3.0.CO;2-S.

Böck, A., Forchhammer, K., Heider, J., and Baron, C. (1991). Selenoprotein synthesis: an expansion of the genetic code. Trends Biochem. Sci. *16*, 463–467.

Brittingham, A., Morrison, C.J., McMaster, W.R., McGwire, B.S., Chang, K.P., and Mosser, D.M. (1995). Role of the *Leishmania* surface protease gp63 in complement fixation, cell adhesion, and resistance to complement-mediated lysis. J. Immunol. *155*, 3102–3111.

Cáceres-Dittmar, G., Sánchez, M.A., Oriol, O., Kraal, G., and Tapia, F.J. (1992). Epidermal compromise in American cutaneous leishmaniasis. J. Invest. Dermatol. *99*, 95S–98S.

Carrera, L., Gazzinelli, R.T., Badolato, R., Hieny, S., Muller, W., Kuhn, R., and Sacks, D.L. (1996). *Leishmania* promastigotes selectively inhibit interleukin 12 induction in bone marrow-derived macrophages from susceptible and resistant mice. J. Exp. Med. *183*, 515–526.

Castro, H., Sousa, C., Santos, M., Cordeiro-da-Silva, A., Flohe, L., and Tomas, A.M. (2002). Complementary antioxidant defense by cytoplasmic and mitochondrial peroxiredoxins in *Leishmania infantum*. Free Radic. Biol. Med. 33, 1552–1562.

Castro, H., Teixeira, F., Romao, S., Santos, M., Cruz, T., Florido, M., Appelberg, R., Oliveira, P., Ferreira-da-Silva, F., and Tomas, A.M. (2011). *Leishmania* mitochondrial peroxiredoxin plays a crucial peroxidase-unrelated role during infection: insight into its novel chaperone activity. PLOS. Pathog. 7, e1002325. https://doi.org/10.1371/journal.ppat.1002325.

Corradin, S., Mauël, J., Ransijn, A., Stürzinger, C., and Vergères, G. (1999). Down-regulation of MARCKS-related protein (MRP) in macrophages infected with *Leishmania*. J. Biol. Chem. *274*, 16782–16787.

Croft, A.M., Taylor, N.A., and Rodenhurst, K.E. (2006). Sandflies and leishmaniasis. Lancet *367*, 112.

Cuervo, P., de Jesus, J.B., Saboia-Vahia, L., Mendonca-Lima, L., Domont, G.B., and Cupolillo, E. (2009). Proteomic characterization of the released/secreted proteins of *Leishmania (Viannia) braziliensis* promastigotes. J. Proteomics *73*, 79–92. https://doi.org/10.1016/j.jprot.2009.08.006.

Culley, F.J., Harris, R.A., Kaye, P.M., McAdam, K.P., and Raynes, J.G. (1996). C-reactive protein binds to a novel ligand on *Leishmania donovani* and increases uptake into human macrophages. J. Immunol. *156*, 4691–4696.

DeLeo, F.R., Allen, L.A., Apicella, M., and Nauseef, W.M. (1999). NADPH oxidase activation and assembly during phagocytosis. J. Immunol. *163*, 6732–6740.

Desjardins, M., and Descoteaux, A. (1997). Inhibition of phagolysosomal biogenesis by the *Leishmania* lipophosphoglycan. J. Exp. Med. *185*, 2061–2068.

Desjeux, P. (2004). Leishmaniasis. Nat. Rev. Microbiol. *2*, 692. https://doi.org/10.1038/nrmicro988.

Dobbin, C.A., Smith, N.C., and Johnson, A.M. (2002). Heat shock protein 70 is a potential virulence factor in murine toxoplasma infection via immunomodulation of host NF-kappa B and nitric oxide. J. Immunol. *169*, 958–965.

Dolai, S., Yadav, R.K., Pal, S., and Adak, S. (2009). Overexpression of mitochondrial *Leishmania major* ascorbate peroxidase enhances tolerance to oxidative stress-induced programmed cell death and protein damage. Eukaryot. Cell *8*, 1721–1731. https://doi.org/10.1128/EC.00198-09.

Dumas, C., Ouellette, M., Tovar, J., Cunningham, M.L., Fairlamb, A.H., Tamar, S., Olivier, M., and Papadopoulou, B. (1997). Disruption of the trypanothione reductase gene of *Leishmania* decreases its ability to survive oxidative stress in macrophages. EMBO J. *16*, 2590–2598. https://doi.org/10.1093/emboj/16.10.2590.

El-Benna, J., Dang, P.M., Gougerot-Pocidalo, M.A., Marie, J.C., and Braut-Boucher, F. (2009). p47phox, the phagocyte NADPH oxidase/NOX2 organizer: structure, phosphorylation and implication in diseases. Exp. Mol. Med. *41*, 217–225. https://doi.org/10.3858/emm.2009.41.4.058.

Evdonin, A.L., Martynova, M.G., Bystrova, O.A., Guzhova, I.V., Margulis, B.A., and Medvedeva, N.D. (2006). The release of Hsp70 from A431 carcinoma cells is mediated by secretory-like granules. Eur. J. Cell Biol. *85*, 443–455. https://doi.org/10.1016/j.ejcb.2006.02.008.

Flohé, L., Hecht, H.J., and Steinert, P. (1999). Glutathione and trypanothione in parasitic hydroperoxide metabolism. Free Radic. Biol. Med. *27*, 966–984.

Forchhammer, K., and Böck, A. (1991). Selenocysteine synthase from *Escherichia coli*. Analysis of the reaction sequence. J. Biol. Chem. *266*, 6324–6328.

Forget, G., Siminovitch, K.A., Brochu, S., Rivest, S., Radzioch, D., and Olivier, M. (2001). Role of host phosphotyrosine phosphatase SHP-1 in the development of murine leishmaniasis. Eur. J. Immunol. *31*, 3185–3196. https://doi.org/10.1128/IAI.00853-05.

Gordon, S. (2003). Alternative activation of macrophages. Nat. Rev. Immunol. *3*, 23–35. https://doi.org/10.1038/nri978.

Handman, E. (1999). Cell biology of *Leishmania*. Adv. Parasitol. *44*, 1–39.

Handman, E. (2001). Leishmaniasis: current status of vaccine development. Clin. Microbiol. Rev. *14*, 229–243. https://doi.org/10.1128/CMR.14.2.229-243.2001.

Harder, S., Bente, M., Isermann, K., and Bruchhaus, I. (2006). Expression of a mitochondrial peroxiredoxin prevents programmed cell death in *Leishmania donovani*. Eukaryotic Cell *5*, 861–870. https://doi.org/10.1128/EC.5.5.861-870.2006.

Hatzigeorgiou, D.E., Geng, J., Zhu, B., Zhang, Y., Liu, K., Rom, W.N., Fenton, M.J., Turco, S.J., and Ho, J.L. (1996). Lipophosphoglycan from *Leishmania* suppresses agonist-induced interleukin 1 beta gene expression in human monocytes via a unique promoter sequence. Proc. Natl. Acad. Sci. U.S.A. *93*, 14708–14713.

Hawn, T.R., Ozinsky, A., Underhill, D.M., Buckner, F.S., Akira, S., and Aderem, A. (2002). *Leishmania* major activates IL-1 alpha expression in macrophages through a MyD88-dependent pathway. Microbes. Infect. *4*, 763–771.

Henderson, G.B., Yamaguchi, M., Novoa, L., Fairlamb, A.H., and Cerami, A. (1990). Biosynthesis of the trypanosomatid metabolite trypanothione: purification and characterization of trypanothione synthetase from *Crithidia fasciculata*. Biochemistry *29*, 3924–3929.

Hermoso, T., Fishelson, Z., Becker, S.I., Hirschberg, K., and Jaffe, C.L. (1991). Leishmanial protein kinases phosphorylate components of the complement system. EMBO J. *10*, 4061–4067.

Holzer, T.R., McMaster, W.R., and Forney, J.D. (2006). Expression profiling by whole-genome interspecies microarray hybridization reveals differential gene expression in procyclic promastigotes, lesion-derived amastigotes, and axenic amastigotes in *Leishmania mexicana*. Mol. Biochem. Parasitol. *146*, 198–218. https://doi.org/10.1016/j.molbiopara.2005.12.009.

Ilg, T., Stierhof, Y.D., McConville, M.J., and Overath, P. (1995). Purification, partial characterization and immunolocalization of a proteophosphoglycan secreted by *Leishmania mexicana* amastigotes. Eur. J. Cell Biol. *66*, 205–215.

Iniesta, V., Gómez-Nieto, L.C., and Corraliza, I. (2001). The inhibition of arginase by N(omega)-hydroxy-L-arginine controls the growth of *Leishmania* inside macrophages. J. Exp. Med. *193*, 777–784.

Iniesta, V., Gomez-Nieto, L.C., Molano, I., Mohedano, A., Carcelen, J., Miron, C., Alonso, C., and Corraliza, I. (2002). Arginase I induction in macrophages, triggered by Th2-type cytokines, supports the growth of intracellular *Leishmania* parasites. Parasite Immunol. 24, 113–118.

Iyer, J.P., Kaprakkaden, A., Choudhary, M.L., and Shaha, C. (2008). Crucial role of cytosolic tryparedoxin peroxidase in *Leishmania donovani* survival, drug response and virulence. Mol. Microbiol. *68*, 372–391. https://doi.org/10.1111/j.1365-2958.2008.06154.x.

Jayakumar, P.C., Musande, V.V., Shouche, Y.S., and Patole, M.S. (2004). The Selenophosphate synthetase gene from *Leishmania major*. DNA Seq. *15*, 66–70.

Kar, S., Ukil, A., Sharma, G., and Das, P.K. (2010). MAPK-directed phosphatases preferentially regulate pro- and anti-inflammatory cytokines in experimental visceral leishmaniasis: involvement of distinct protein kinase C isoforms. J. Leukoc. Biol. *88*, 9–20. https://doi.org/10.1189/jlb.0909644.

Kautz-Neu, K., Noordegraaf, M., Dinges, S., Bennett, C.L., John, D., Clausen, B.E., and von Stebut, E. (2011). Langerhans cells are negative regulators of the anti-*Leishmania* response. J. Exp. Med. *208*, 885–891. https://doi.org/10.1084/jem.20102318.

Khan, M.O. (2007). Trypanothione reductase: a viable chemotherapeutic target for antitrypanosomal and antileishmanial drug design. Drug Target Insights *2*, 129–146.

Kima, P.E., and Soong, L. (2013). Interferon gamma in leishmaniasis. Front Immunol *4*, 156. https://doi.org/10.3389/fimmu.2013.00156.

Krauth-Siegel, R.L., and Comini, M.A. (2008). Redox control in trypanosomatids, parasitic protozoa with trypanothione-based thiol metabolism. Biochim. Biophys. Acta *1780*, 1236–1248. https://doi.org/10.1016/j.bbagen.2008.03.006.

Kropf, P., Fuentes, J.M., Fähnrich, E., Arpa, L., Herath, S., Weber, V., Soler, G., Celada, A., Modolell, M., and Müller, I. (2005). Arginase and polyamine synthesis are key factors in the regulation of experimental leishmaniasis *in vivo*. FASEB J. *19*, 1000–1002. https://doi.org/10.1096/fj.04-3416fje.

Kropf, P., Herath, S., Klemenz, R., and Müller, I. (2003). Signaling through the T1/ST2 molecule is not necessary for Th2 differentiation but is important for the regulation of type 1 responses in nonhealing *Leishmania major* infection. Infect. Immun. *71*, 1961–1971. https://doi.org/10.1128/IAI.71.4.1961-1971.2003.

Lacerda, D.I., Cysne-Finkelstein, L., Nunes, M.P., De-Luca, P.M., Genestra, M.S., Leon, L.L., Berredo-Pinho, M., Mendonca-Lima, L., Matos, D.C., Medeiros, M.A., and Mendonca, S.C. (2012). Kinetoplastid membrane protein-11 exacerbates infection with *Leishmania amazonensis* in murine macrophages. Mem. Inst. Oswaldo Cruz *107*, 238–245.

Lainson, R., Ward, R.D., and Shaw, J.J. (1977). *Leishmania* in phlebotomid sandflies: VI. Importance of hindgut development in distinguishing between parasites of the *Leishmania mexicana* and *L. braziliensis* complexes. Proc. R. Soc. Lond. B Biol. Sci. *199*, 309–320.

Lambertz, U., Silverman, J.M., Nandan, D., McMaster, W.R., Clos, J., Foster, L.J., and Reiner, N.E. (2012). Secreted virulence factors and immune evasion in visceral leishmaniasis. J. Leukoc. Biol. *91*, 887–899. https://doi.org/10.1189/jlb.0611326.

Lancaster, G.I., and Febbraio, M.A. (2005). Exosome-dependent trafficking of HSP70: a novel secretory pathway for cellular stress proteins. J. Biol. Chem. *280*, 23349–23355. https://doi.org/10.1074/jbc.M502017200.

Legare, D., Papadopoulou, B., Roy, G., Mukhopadhyay, R., Haimeur, A., Dey, S., Grondin, K., Brochu, C., Rosen, B.P., and Ouellette, M. (1997). Efflux systems and increased trypanothione levels in arsenite-resistant *Leishmania*. Exp. Parasitol. *87*, 275–282. https://doi.org/10.1006/expr.1997.4222.

Levick, M.P., Tetaud, E., Fairlamb, A.H., and Blackwell, J.M. (1998). Identification and characterisation of a functional peroxidoxin from *Leishmania major*. Mol. Biochem. Parasitol. *96*, 125–137.

Lobanov, A.V., Gromer, S., Salinas, G., and Gladyshev, V.N. (2006). Selenium metabolism in Trypanosoma: characterization of selenoproteomes and identification of a Kinetoplastida-specific selenoprotein. Nucleic Acids Res. *34*, 4012–4024. https://doi.org/10.1093/nar/gkl541.

Lodge, R., and Descoteaux, A. (2006). Phagocytosis of *Leishmania donovani* amastigotes is Rac1 dependent and occurs in the absence of NADPH oxidase activation. Eur. J. Immunol. *36*, 2735–2744. https://doi.org/10.1002/eji.200636089.

Maiorino, M., Roche, C., Kiess, M., Koenig, K., Gawlik, D., Matthes, M., Naldini, E., Pierce, R., and Flohe, L. (1996). A selenium-containing phospholipid-hydroperoxide glutathione peroxidase in *Schistosoma mansoni*. Eur. J. Biochem. *238*, 838–844.

Mantegazza, A.R., Savina, A., Vermeulen, M., Perez, L., Geffner, J., Hermine, O., Rosenzweig, S.D., Faure, F., and Amigorena, S. (2008). NADPH oxidase controls phagosomal pH and antigen cross-presentation in human dendritic cells. Blood *112*, 4712–4722. https://doi.org/10.1182/blood-2008-01-134791.

Matte, C., Maion, G., Mourad, W., and Olivier, M. (2001). *Leishmania donovani*-induced macrophages cyclooxygenase-2 and prostaglandin E2 synthesis. Parasite Immunol. *23*, 177–184.

McNeely, T.B., and Turco, S.J. (1987). Inhibition of protein kinase C activity by the *Leishmania donovani* lipophosphoglycan. Biochem. Biophys. Res. Commun. *148*, 653–657.

Mukhopadhyay, C., Nath, G., Gulati, A.K., and Mohapatra, S.C. (2001). Prevalence of HIV among low and high risk population of eastern part of northern India. J Commun Dis 33, 136–142.

Muleme, H.M., Reguera, R.M., Berard, A., Azinwi, R., Jia, P., Okwor, I.B., Beverley, S., and Uzonna, J.E. (2009). Infection with arginase-deficient *Leishmania major* reveals a parasite number-dependent and cytokine-independent regulation of host cellular arginase activity and disease pathogenesis. J. Immunol. *183*, 8068–8076. https://doi.org/10.4049/jimmunol.0803979.

Murray, H.W. (2001). Clinical and experimental advances in treatment of visceral leishmaniasis. Antimicrob. Agents Chemother. *45*, 2185–2197. https://doi.org/10.1128/AAC.45.8.2185-2197.2001.

Nandan, D., and Reiner, N.E. (1995). Attenuation of gamma interferon-induced tyrosine phosphorylation in mononuclear phagocytes infected with *Leishmania donovani*: selective inhibition of signaling through Janus kinases and Stat1. Infect. Immun. *63*, 4495–4500.

Olivier, M., Gregory, D.J., and Forget, G. (2005). Subversion mechanisms by which *Leishmania* parasites can escape the host immune response: a signaling point of view. Clin. Microbiol. Rev. *18*, 293–305. https://doi.org/10.1128/CMR.18.2.293-305.2005.

Pal, S., Dolai, S., Yadav, R.K., and Adak, S. (2010). Ascorbate peroxidase from *Leishmania major* controls the virulence of infective stage of promastigotes by regulating oxidative stress. PLOS ONE *5*, e11271. https://doi.org/10.1371/journal.pone.0011271.

Pandey, R.K., Mehrotra, S., Sharma, S., Gudde, R.S., Sundar, S., and Shaha, C. (2016). *Leishmania donovani*-Induced Increase in Macrophage Bcl-2 Favors Parasite Survival. Front Immunol *7*, 456. https://doi.org/10.3389/fimmu.2016.00456.

Peters, C., Aebischer, T., Stierhof, Y.D., Fuchs, M., and Overath, P. (1995). The role of macrophage receptors in adhesion and uptake of *Leishmania mexicana* amastigotes. J. Cell. Sci. *108*, 3715–3724.

Peters, C., Kawakami, M., Kaul, M., Ilg, T., Overath, P., and Aebischer, T. (1997). Secreted proteophosphoglycan of *Leishmania mexicana* amastigotes activates complement by triggering the mannan binding lectin pathway. Eur. J. Immunol. *27*, 2666–2672. http://doi.org/10.1002/eji.1830271028.

Peters, W., Evans, D.A., and Lanham, S.M. (1983). Importance of parasite identification in cases of leishmaniasis. J. R. Soc. Med. *76*, 540–542.

Prasad, L.S. (1999). Kala azar. Indian J. Pediatr. *66*, 539–546.

Puentes, S.M., Dwyer, D.M., Bates, P.A., and Joiner, K.A. (1989). Binding and release of C3 from *Leishmania donovani* promastigotes during incubation in normal human serum. J. Immunol. *143*, 3743–3749.

Puentes, S.M., Sacks, D.L., Da Silva, R.P., and Joiner, K.A. (1988). Complement binding by two developmental stages of *Leishmania major* promastigotes varying in expression of a surface lipophosphoglycan. J. Exp. Med. *167*, 887–902.

Reiner, N.E. (1987). Parasite accessory cell interactions in murine leishmaniasis. I. Evasion and stimulus-dependent suppression of the macrophage interleukin 1 response by *Leishmania donovani*. J. Immunol. *138*, 1919–1925.

Reiner, N.E., and Malemud, C.J. (1984). Arachidonic acid metabolism in murine leishmaniasis (*Donovani*): ex-vivo evidence for increased cyclooxygenase and 5-lipoxygenase activity in spleen cells. Cell Immunol. *88*, 501–510.

Reiner, N.E., and Malemud, C.J. (1985). Arachidonic acid metabolism by murine peritoneal macrophages infected with *Leishmania donovani*: *in vitro* evidence for parasite-induced alterations in cyclooxygenase and lipoxygenase pathways. J. Immunol. *134*, 556–563.

Reiner, N.E., Ng, W., Wilson, C.B., McMaster, W.R., and Burchett, S.K. (1990). Modulation of *in vitro* monocyte cytokine responses to *Leishmania donovani*. Interferon-gamma prevents parasite-induced inhibition of interleukin 1 production and primes monocytes to respond to *Leishmania* by producing both tumor necrosis factor-alpha and interleukin 1. J. Clin. Invest *85*, 1914–1924. https://doi.org/10.1172/JCI114654.

Requena, J.M., Montalvo, A.M., and Fraga, J. (2015). Molecular chaperones of *Leishmania*: central players in many stress-related and -unrelated physiological processes. Biomed. Res. Int. *2015*, 301326. https://doi.org/10.1155/2015/301326.

Rhee, S.G., Kang, S.W., Chang, T.S., Jeong, W., and Kim, K. (2001). Peroxiredoxin, a novel family of peroxidases. IUBMB Life *52*, 35–41. https://doi.org/10.1080/15216540252774748.

Rybicka, J.M., Balce, D.R., Chaudhuri, S., Allan, E.R., and Yates, R.M. (2012). Phagosomal proteolysis in dendritic cells is modulated by NADPH oxidase in a pH-independent manner. EMBO J. *31*, 932–944. https://doi.org/10.1038/emboj.2011.440.

Savina, A., Peres, A., Cebrian, I., Carmo, N., Moita, C., Hacohen, N., Moita, L.F., and Amigorena S. (2009). The small GTPase Rac2 controls phagosomal alkalinization and antigen crosspresentation selectively in CD8(+) dendritic cells. Immunity. *30*, 544–555. https://doi.org/10.1016/j.immuni.2009.01.013.

Singh, N.K., Jha, T.K., Singh, I.J., and Jha, S. (1995). Combination therapy in Kala-azar. J. Assoc. Physicians India *43*, 319–320.

Sørensen, A.L., Hey, A.S., and Kharazmi, A. (1994). *Leishmania major* surface protease Gp63 interferes with the function of human monocytes and neutrophils *in vitro*. APMIS *102*, 265–271.

Späth, G.F., Garraway, L.A., Turco, S.J., and Beverley, S.M. (2003). The role(s) of lipophosphoglycan (LPG) in the establishment of *Leishmania major* infections in mammalian hosts. Proc. Natl. Acad. Sci. U.S.A. *100*, 9536–9541. https://doi.org/10.1073/pnas.1530604100.

Sundar, S., Rosenkaimer, F., Lesser, M.L., and Murray, H.W. (1995). Immunochemotherapy for a systemic intracellular infection: accelerated response using interferon-gamma in visceral leishmaniasis. J. Infect. Dis. *171*, 992–996.

Sundar, S., Singh, V.P., Sharma, S., Makharia, M.K., and Murray, H.W. (1997). Response to interferon-gamma plus pentavalent antimony in Indian visceral leishmaniasis. J. Infect. Dis. *176*, 1117–1119.

Tetaud, E., Giroud, C., Prescott, A.R., Parkin, D.W., Baltz, D., Biteau, N., Baltz, T., and Fairlamb, A.H. (2001). Molecular characterisation of mitochondrial and cytosolic trypanothione-dependent tryparedoxin peroxidases in *Trypanosoma brucei*. Mol. Biochem. Parasitol. *116*, 171–183.

Thakur, B.B. (2003). Breakthrough in the management of visceral leishmaniasis. J Assoc Physicians India *51*, 649–651.

Vinet, A.F., Fukuda, M., Turco, S.J., and Descoteaux, A. (2009). The *Leishmania donovani* lipophosphoglycan excludes the vesicular proton-ATPase from phagosomes by impairing the recruitment of synaptotagmin V. PLOS Pathog. *5*, e1000628. https://doi.org/10.1371/journal.ppat.1000628.

Weinheber, N., Wolfram, M., Harbecke, D., and Aebischer, T. (1998). Phagocytosis of *Leishmania mexicana* amastigotes by macrophages leads to a sustained suppression of I.L.-12 production. Eur. J. Immunol. *28*, 2467–2477. https://doi.org/10.1002/

(S.I.C.I)1521-4141(199808)28:08<2467::AID-IMMU2467>3.0.CO;2-1.

Williams, D.L., Pierce, R.J., Cookson, E., and Capron, A. (1992). Molecular cloning and sequencing of glutathione peroxidase from *Schistosoma mansoni*. Mol. Biochem. Parasitol. *52*, 127–130.

Wyllie, S., Mandal, G., Singh, N., Sundar, S., Fairlamb, A.H., and Chatterjee, M. (2010). Elevated levels of tryparedoxin peroxidase in antimony unresponsive *Leishmania donovani* field isolates. Mol. Biochem. Parasitol. *173*, 162–164. https://doi.org/10.1016/j.molbiopara.2010.05.015.

Molecular Regulation of Macrophage Class Switching in Indian Post-kala-azar Dermal *Leishmaniasis* (PKDL)

5

Mitali Chatterjee*, Srija Moulik, Debkanya Dey, Debanjan Mukhopadhyay, Shibabrata Mukherjee and Susmita Roy

Department of Pharmacology, Institute of PG Medical Education and Research, Kolkata, India.

*Correspondence: ilatim@vsnl.net and ilatimc@gmail.com

https://doi.org/10.21775/9781910190715.05

Abstract

Leishmania donovani, the causative parasite responsible for visceral leishmaniasis (VL) and its chronic dermal sequel, post-kala-azar dermal leishmaniasis (PKDL) manipulates host monocytes/macrophages for ensuring its survival. Information regarding macrophage polarization is primarily derived from murine models, but growing evidence is emphasizing the inadequacy of direct inter-species translation. Accordingly, the status of monocytes/macrophages with regard to plasticity and polarization in circulation and dermal lesions of patients with PKDL was characterized. The raised plasma levels of IL-4/IL-13 and IL-10 in patients with PKDL confirmed the presence of a microenvironment conducive for alternative activation of monocytes/macrophages. Furthermore, the mRNA expression of *il-10, ifn-γ, mrc-1, arg-1, vitamin D signalling pathway* (*vdr, cyp27b1, ll-37*), was examined in circulation and lesional sites, wherein there was a consistent increase in expression of M2 markers. Conversely, the classical macrophage activation markers showed a consistent decline in generation of reactive oxygen and nitrogen intermediates, expression of Toll-like receptors 2 and 4 (CD282/284) along with impairment of the MAP kinase pathway. Taken together, impairment of antigen presentation along with an increased presence of alternatively activated monocyte/macrophage subsets sustained parasite survival and facilitated disease persistence. Accordingly, immunomodulators capable of evoking a reduction in M2 monocytes/macrophage population or enhancing their switching from M2 to M1 could be an effective chemotherapeutic strategy against leishmaniasis.

Introduction

*Leishmania*ses comprises a group of heterogeneous parasitic diseases caused by the protozoan parasite of the genus *Leishmania.* Depending on the infecting species, *Leishmania* parasites can give rise to a wide array of clinical manifestations that includes (1) localized cutaneous leishmaniasis (LCL) with single to multiple skin ulcers, satellite lesions or nodular lymphangitis, (2) CL having mucosal involvement, mucocutaneous leishmaniasis (MCL), (3) CL with disseminated lesions, diffuse cutaneous leishmaniasis (DCL) or (4) visceral leishmaniasis (VL), where there is involvement of the liver, spleen and bone marrow, which if not appropriately treated is potentially lethal (Bailey and Lockwood, 2007).

Cunningham and Borovsky first demonstrated *Leishmania* parasites in Oriental sore lesions, but it was the American pathologist James Homer Wright (1869–1928) who is credited with the discovery of *L. tropica.* In 1903, Wright published a detailed description of the organism from the specimen of a sore of an Armenian girl and named the parasite *Helcosoma tropicum* (Steverding, 2017). However, in 1906, the German physician cum zoologist Max Lühe changed the name to *Leishmania tropica*. In 1914, two Russian physicians, Wassily Larionovich Yakimoff and Nathan Isaakovich Schokhor, proposed that based on the size of the parasites identified in skin lesions, *L. tropica* may be subdivided into two subspecies, namely *L. tropica minor* and *L. tropica major* (Steverding, 2017). Later Bray *et al.* (1973) proposed a modification in this classification to *L. tropica* and *L. major* respectively.

In terms of the dermal presentation of leishmaniasis, CL is the most prevalent clinical form and 90% of CL cases occur in seven countries, namely Afghanistan, Algeria, Brazil, Iran, Peru, Saudi Arabia and Syria (Handler *et al.*, 2015). According to the Eurocentric world view, *Leishmania* parasites causing CL can be divided into Old World species, *L. (L.) major*, *L. infantum*, and *L. (L.) tropica* (prevalent around the Mediterranean basin, the Middle East, the horn of Africa, and the Indian subcontinent), and New World species, namely *L. (L.) amazonensis*, *L. (L.) chagasi*, *L. mexicana*, *L. (V.) naiffi*, *L. (V.) braziliensis*, and *L. (V.) guyanensis*, endemic in Middle and South America. Although most Old World species cause self-limiting ulcers, the New World species can cause a syndrome called American tegumentary leishmaniasis that includes CL along with a variety of other manifestations, such as MCL and the much rarer diffuse and disseminated cutaneous leishmaniasis, DCL (Kevric *et al.*, 2015).

In South Asia, a unique form of dermal leishmaniasis appears called post-kala-azar dermal leishmaniasis (PKDL), which is caused by the protozoan parasite *Leishmania donovani*. Amongst the different dermal presentations of leishmaniasis, PKDL is possibly the most intriguing clinically and scientifically, as it generally develops as a sequel after apparent successful cure from VL (Ganguly *et al.*, 2010a; Mukhopadhyay *et al.*, 2014). Unlike the preceding disease of VL where patients suffer from prolonged fever, hepatosplenomegaly, weight loss, and anaemia, the manifestations of PKDL are limited to macular, papular or nodular lesions in the skin (Das *et al.*, 2014).

PKDL is confined to two geographically distinct zones, namely South Asia (India, Nepal and Bangladesh) and East Africa, mainly Sudan (Ziljstra *et al.*, 2003; Mukhopadhyay *et al.*, 2014). Dependent on the geographical region, in *L. donovani* endemic areas, between 5% and 60% of patients develop this dermatosis during or after treatment (Ziljstra *et al.*, 2016). Although the condition has been described for about 80 years, it is only in recent times that its relevance is being fully recognized, as it is considered as the strongest contender for being the disease reservoir, especially during the interepidemic periods of VL (Ganguly *et al.*, 2010a; Ramesh *et al.*, 2015; Bhattacharya *et al.*, 2017).

In South Asia, another area where dermal leishmaniasis occurs is Sri Lanka, where the first locally acquired case of CL was described in 1992, the causative species being *Leishmania donovani* MON-37 (Manamperi *et al.*, 2017; Karunaweera *et al.*, 2003). This is closely related to *Leishmania donovani* MON-2 that causes VL in the Indian subcontinent (Alam *et al.*, 2009). This CL causing zymodeme has also been isolated from cutaneous lesions in a tribal population from Kerala, India (Kumar *et al.*, 2015).

The first evidence for existence of CL in India was based on a clinico-epidemiological analysis of cases during a large-scale outbreak of the disease in Bikaner in 1973. During this period, over 2000 people suffered from this infection and sporadic cases were detected in pockets in the Thar Desert of Rajasthan (Sharma *et al.*, 1973; Kumar *et al.*, 2007). Interestingly, Mohan *et al.* (1975) reported that Indian desert gerbils, *Meriones hurrianae*, and dogs were the reservoirs of this infection. However, information regarding the vector species transmitting CL infection remains open ended, although a few studies have demonstrated that in this region, one or both species of the sandfly, *Phlebotomus paptasi* and *P. sergenti*, are the vectors (Kumar *et al.*, 2007).

Historical background of PKDL

PKDL is a dermatosis generally observed in patients with a previous history of VL (Ganguly *et al.*, 2010a; Zijlstra *et al.*, 2003). It was first described in 1922 by the eminent Indian physician-scientist, Sir U.N. Brahmachari (1873–1946), who, at a meeting of the Asiatic Society of Bengal, presented four cases with unique dermal involvement, each of whom had been successfully treated for kala-azar or VL (Brahmachari, 1922). He proposed the term 'dermal leishmanoid' to describe this condition, as Leishman-Donovan (LD) bodies were observed in slit-skin smear microscopic preparations. Eventually, studies by Shortt and Brahmachari, Acton and Napier, Knowles and Das Gupta, and other workers of the Calcutta School of Tropical Medicine led to this unique variant of dermal leishmaniasis to be renamed as post-kala-azar dermal leishmaniasis (Sen Gupta, 1947 and references therein).

Clinical features of PKDL

Although PKDL is confined to two geographically distinct zones, namely South Asia (India, Nepal, and Bangladesh) and East Africa (Sudan), there are important differences between these two variants, in both their clinical features and immune responses. In the South Asian variant of PKDL, *Leishmania* parasites remain restricted to the skin, and the majority manifest as nodular, papular, hypopigmented macular lesions, erythematous plaques and/or a mixed phenotype, broadly termed as polymorphic (Das *et al.* 2014; Mukhopadhyay *et al.*, 2014). However, in the African variant, they primarily present with a papular rash (Zijlstra *et al.*, 2003). In South Asia, the lag period ranges from 2 to 10 years between cure from VL and onset of PKDL, and it has been suggested that PKDL echoes the epidemic of VL (Stauch *et al.*, 2011), whereas in Africa it often presents concomitantly with VL or within a year of VL (Zijlstra *et al.*, 2016). In the African variant, majority (85%) show spontaneous resolution of PKDL within 12 months, and only a minority need treatment (http://apps.who.int/iris/bitstream/10 665/101164/1/9789 241504102_eng.pdf). However, in South Asia, reports on self-healing are limited and majority need intensive treatment (Ganguly *et al.*, 2010a; Zijlstra *et al.*, 2016).

In a consortium meeting on PKDL held in New Delhi, India, in 2012, it was brought to focus that a deeper understanding of the different forms of PKDL related to VL infection be undertaken, as also the identification of asymptomatic cases of VL and HIV-VL co-infected patients. Another important component that demands attention is xenodiagnosis in the sandfly populations to assess infectiousness, along with validation of potential biomarkers to replace xenodiagnosis, coupled with establishment of when/whether PKDL and VL patients are sterile after treatment (Desjeux *et al.*, 2013).

The current drive by the Government of India in conjunction with neighbouring countries to eliminate leishmaniasis from South Asia has identified PKDL as the strongest contender to be the disease reservoir. This emphasizes the necessity to consider its eradication, which hinges on its early diagnosis and management to achieve the 2017 target of leishmaniasis elimination (Mondal *et al.*, 2009; Bhattacharya *et al.*, 2017). The presence of PKDL in the community could well initiate another outbreak of VL as highlighted in West Bengal, India, where the epidemic of VL in 1980 was traced to a PKDL patient who developed lesions in 1976 (Addy and Nandy, 1992).

In view of the fact that patients with PKDL constitute an important residual reservoir, detecting and treating them are important, especially at times of low prevalence (Ganguly *et al.*, 2015). A major limitation is that patients with PKDL have cosmetic disfigurement, but no major physical impairment. Therefore, they do not seek treatment and even tend to default treatment (Basher *et al.*, 2015). Accordingly, active surveillance drives have been initiated, wherein first-line health workers or kala-azar technical supervisors (KTS) using standard case definitions and well-defined risk factors e.g. living in an endemic area, living in a family in which someone has had or does have VL, having an epidemiological link (past history of VL) refer suspected cases to centres where confirmation is possible (http://nvbdcp.gov.in/Doc/opertional-guideline-KA-2015.pdf). It is expected that this approach of active case detection would substantially diminish disease transmission by shortening the patient's infectious period (Bhattacharya *et al.*, 2017; Ramesh *et al.*, 2015; Ganguly *et al.*, 2010a).

Alternative activation of macrophages and disease etiopathogenesis of PKDL

The etiopathogenesis of PKDL is still unclear and a consensus is yet to emerge regarding possible reasons why the generally viscerotropic *L. donovani* parasite becomes dermatropic and causes PKDL (Mukhopadhyay *et al.*, 2014). It is essential that the *Leishmania* parasite to survive within dermal monocytes/macrophages effectively nullifies their microbicidal effector mechanisms (Kaye and Scott, 2011). Indeed, in PKDL, similar to other leishmaniasis, *Leishmania* have developed several strategies to outmanoeuvre host immunity via subverting and/or suppressing macrophage microbicidal activities (Mukhopadhyay *et al.*, 2014). To achieve this, the robust parasite has developed diverse strategies and elegantly performs a well-coordinated orchestra which directly or indirectly modulates the host–parasite interactions.

The direct component involves attenuation of the oxidative burst within macrophages, thus weakening its microbicidal armamentarium by preventing induction of pro-inflammatory molecules such as nitric oxide (NO), superoxide, $HOCl^-$, etc. (Bogdan, 2001). Collectively, the resultant redox imbalance causes an enhanced vulnerability to intracellular pathogens and disease progression is achieved. In leishmaniasis, macrophages are activated by IFN-γ and control infection through activation of iNOS. However for parasite survival, *Leishmania* inhibit production of IL-12, concomitant with increased release of IL-10 and TGF-β, which translates into a Th2 predominance and susceptibility to disease ensues (Rodrigues *et al.*, 2016).

Osorio *et al.* (2012) established that hamsters infected with *L. (Viannia) panamensis* or *L. (Leishmania) donovani* are capable of switching on the M2 macrophage-like phenotype, secondary to phosphorylation of STAT-6 and induction of arginase-1. The role of IL-4/IL-13-dependent M2 causing susceptibility to *L. major* infection was established by Holscher *et al.* (2006) in a macrophage/neutrophil-specific IL-4Rα-deficient mice model, where despite the sustained increase of Th2 cytokines, mice delayed disease progression due to enhanced induction of a M1 classical macrophage activation phenotype, These IL-4Rα-dependent mechanisms were shown to be dependent on arginase-1, that counter-regulated production of NO by competing for the common substrate L-arginine, as also by inducing the production of polyamines favoured parasite replication (Holscher *et al.*, 2006).

Status of cytokine milieu in dermal leishmaniasis

Cytokines in PKDL

The Th1/Th2 prototype in the adaptive immune response, conferring resistance or susceptibility, respectively, to *Leishmania*, is strictly a feature of experimental cutaneous leishmaniasis caused by *Leishmania major* (Locksley and Scott, 1991), as in murine experimental and clinical VL, a mixed Th1/Th2 response has been documented (Kenney *et al.*, 1998). With regard to PKDL, although the immunological profile is reasonably well characterized in the Sudanese form, information regarding the CMI response in Indian PKDL is relatively limited, primarily due to its low incidence and indeterminate time intervals.

The microenvironment within which the macrophages reside dictates their polarization towards either classically activated macrophages (or M1) or alternatively activated macrophages (M2; Martinez *et al.*, 2008; Biswas *et al.*, 2012). The activation of macrophages towards an alternative phenotype is generally facilitated by T helper 2 (Th2) cytokines interleukin 4 (IL-4) and interleukin-13 (IL-13), which are quite distinct from interferon-γ (IFN-γ). These alternatively activated macrophages participate in a range of physiological and pathological processes like wound healing and tissue repair, enhancement of parasitic infections, hypersensitivity etc. (Gordon and Martinez, 2010).

Pro-inflammatory cytokines namely IL-6, IL-8, IL-1β, TNF-α and IL-12 are the domain of M1 monocytes/macrophages, whereas IL-4, IL-13, IL-10 and TGF-β secreted by M2 monocytes help sustain the immunosuppressive environment (Gordon, 2003). In PKDL, although a dermal disease, monocytes/macrophages in circulation showed an enehanced production of IL-4, IL-10 and IL-13 (Table 5.1). Alongside, circulating monocytes in patients with PKDL generated lower amounts of IL-6, IL-1β and IL-8 (Ganguly *et al.*, 2008). However, a significant population of monocytes expressed the pro-inflammatory IL-12p40 (Table 5.1 see also Mukhopadhyay *et al.* 2015).

At the lesional sites, a raised mRNA expression of both pro- and anti-inflammatory cytokines has

Table 5.1 Status of circulating and lesional cytokines in patients with PKDL at disease presentation

Cytokine	Intramonocytic[a]	Circulation[a]	[a]mRNA expression at lesional sites	References
IFN-γ	ND	Increased	Increased	Ganguly *et al*. (2008, 2010b), Ansari *et al*. (2006)
IFN-γ R1	ND	ND	Decreased	Ansari *et al*. (2006)
TNF α	ND	Increased	Increased	Mukhopadhyay *et al*. (2011), Ansari *et al*. (2008)
IL 6	Decreased	Increased but not significant	Increased	Mukhopadhyay *et al*. (2015), Ansari *et al*. (2006)
IL-8	Decreased	Increased	ND	Mukhopadhyay *et al*. (2011, 2015)
IL-1β	Decreased	ND	ND	Mukhopadhyay *et al*. (2015)
IL-12p40	Increased	ND	Increased	Mukhopadhyay *et al*. (2015)
IL-10	Increased	Increased	Increased	Mukhopadhyay *et al*. (2011, 2015), Ganguly *et al*. (2008, 2010b), Ansari *et al*. (2006, 2008), Katara *et al*. (2011)
IL-4	Increased	Increased	Increased	Mukhopadhyay *et al*. (2011, 2015), Ansari *et al*. (2006)
IL-13	Increased	Increased	ND	Mukhopadhyay *et al*. (2011, 2015)
LAP-TGF-β1	Decreased	ND	Increased	Mukhopadhyay *et al*. (2015)
TGF β	ND	Increased	Increased	Mukhopadhyay *et al*. (2011), Ansari *et al*. (2006)

[a]The frequency and expression of intramonocytic cytokines was measured by flow cytometry, status of circulating cytokines by ELISA while at the lesional site, mRNA expression was examined by PCR.
ND, not done.

been reported (Ansari *et al.*, 2006; Ganguly *et al.*, 2010a; Fig. 5.1). Although the mRNA expression of pro-inflammatory cytokines such as IFN-γ was augmented at disease presentation, it was associated with a significantly lowered expression of its receptor IFN-γ R1, thus accounting for its inability to mediate its anticipated host protective action against leishmaniasis (Ansari *et al.*, 2006). Furthermore, the raised mRNA expression of counter regulatory cytokines such as TGF-β and IL-10 (Ansari *et al.*, 2006; Mukhopadhyay *et al.*, 2011; Ganguly *et al.*, 2010a; Katara *et al.*, 2011; see also Table 5.1 and Fig. 5.1), possibly curtailed the pro-inflammatory cytokines based immune responses, and instead supported an immunosuppressive milieu accounted for parasite persistence. After treatment (sodium stibogluconate: 20 mg/kg b.w./day for 3 months; or miltefosine:100 mg/day p.o. for 2 months), there was a significant decrease in the lesional mRNA expression of IFN-γ and IL-10 and are proposed to be good biomarkers for monitoring patients with PKDL. (Ganguly *et al.*, 2010b; Fig. 5.1).

Cytokine profiles in New World cutaneous leishmaniasis

L. mexicana is the causative of localized (LCL) and diffuse (DCL) leishmaniasis, the former being benign while the latter has a more progressive form (Carrada *et al.*, 2007). This was reflected in their cytokine profile in that patients with LCL following antigenic stimulation of their peripheral blood mononuclear cells demonstrated higher levels of th pro-inflammatory cytokines TNF-α, IL-15, IL-18 but low IL-12 (Carrada *et al.*, 2007). This was corroborated at lesional sites, wherein these patients with LCL also showed a high mRNA expression of IFN-γ, TNF-α, IL-1α, IL-6 and IL-12, but alongside IL-10 and TGF-β too were increased (Melby *et al.*, 1994, 1996; Valencia Pacheo *et al.*, 2014). In contrast, patients with DCL, had lower levels of IL-12 and TNF-α, suggesting a reduced ability of monocytes to produce cell-activating pro-inflammatory cytokines, and this allowed for disease progression (Carrada *et al.*, 2007).

Lesions caused by *L. braziliensis*, generally manifest as LCL, mucocutaneous (MCL) or DCL, with

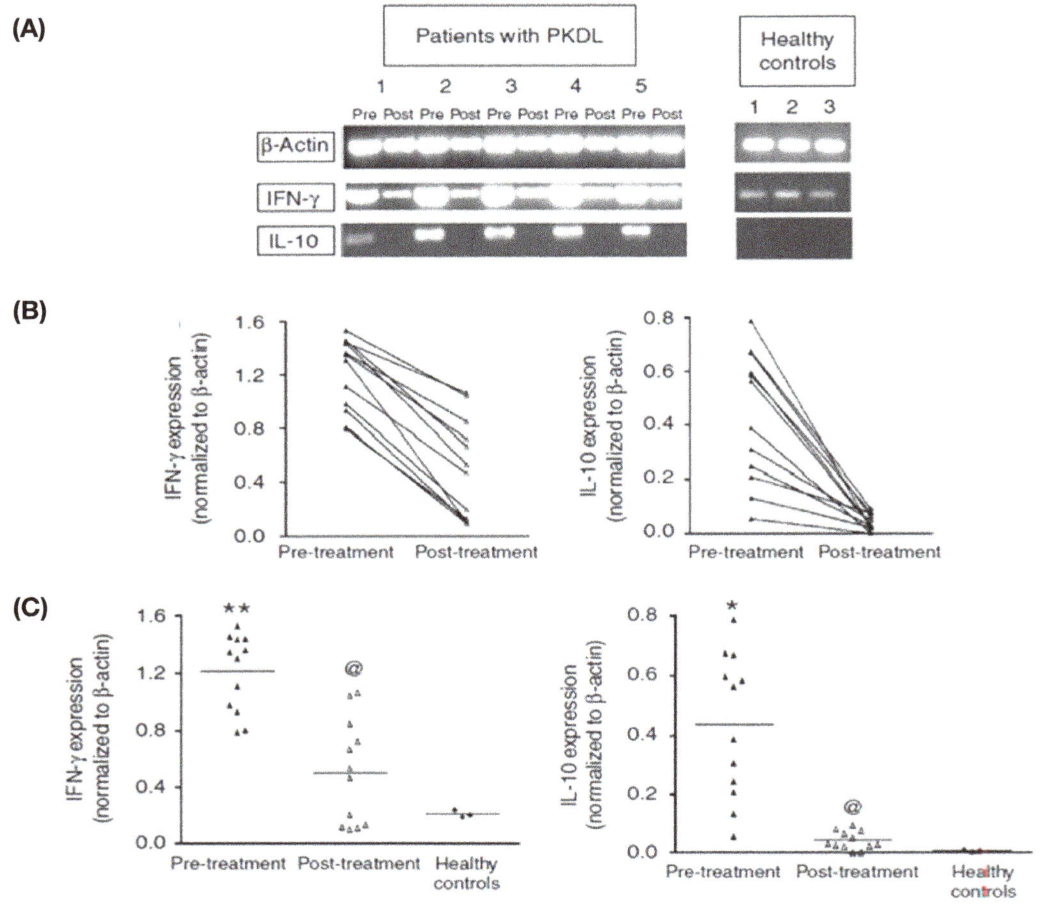

Figure 5.1 Longitudinal monitoring of lesional expression of IFN-g and IL-10 in patients with post-kala-azar dermal leishmaniasis (PKDL) before and after treatment. (A) Representative profile of mRNA expression of IFN-γ and IL-10 in lesional tissue from patients with PKDL (*n*=5) before and after treatment and normal skin tissue from healthy individuals (*n*=3). Isolated RNA was subjected to reverse transcriptase-PCR (RT-PCR); products were resolved using agarose gel electrophoresis with bands specific for β-actin, IFN-γ, and IL-10, visualized using UV transillumination, and quantified using densitometric analysis. (B) Before after plots of IFN-γ and IL-10 expression in patients with PKDL (*n*=12) before (m) and after completion of treatment (n). (C) Scatter plots of expression values for IFN-γ and IL-10 in patients with PKDL (*n*=12) before (m) and after treatment (n) as well as in healthy controls (*n*=3); horizontal lines indicate mean values. *$P<0.01$, **$P<0.001$, significantly different from controls; @$P<0.001$, significantly different from pre-treatment.

a Th1 predominance in LCL, accounting for the disease resolution, whereas in the more severe DCL and MCL, a mixed Th1/Th2 response occurs with a Th2 predominance, leading to disease progression (Cáceres-Dittmar *et al.*, 1993; Pirmez *et al.*, 1993). Akin to *L. mexicana*, LCL lesions showed a higher mRNA expression of IFN-γ, TNF-β, IL-2, IL-8, IL-1β, IL-12, whereas the expression of IL-4, IL-5 and IL-10 was weak (Cáceres-Dittmar *et al.*, 1993; Pirmez *et al.*, 1993). Similarly, patients with DCL showed a high lesional mRNA expression of IFN-γ, TNF-β, IL-1β, IL-8, IL-12; notably absent was IL-6 and there was a weak expression of TNF-α, whereas IL-4 was markedly high (Cáceres-Dittmar *et al.*, 1993). With regard to the MCL variant, lesional mRNA expression of both Th1 (IFN-γ, TNF-β, IL-8, IL-1β, TNF-α, IL-6) and Th2 (IL-5, IL-4 and IL-10) were raised, evidence of disease chronicity (Cáceres-Dittmar *et al.*, 1993; Pirmez *et al.*, 1993). Taken together, irrespective of the causative species, the cytokine profiles in resolving CL, e.g. LCL, show features of Th1 dominance whereas patients with disseminated CL, e.g. DCL and MCL, demonstrate a mixed Th1–Th2 profile, favouring disease exacerbation.

Cytokine status of Old World cutaneous leishmaniasis

In experimental models, the *Leishmania* species *L. major*, causative agent of Old World CL, has been shown to have a classical Th1 dominance, i.e. increased IFN-γ in case of disease resolution, and a Th2 response for disease progression. However, this classical pattern is absent in patients with *L. major* causing CL, as their PBMC when co-cultured with *L. major*, showed high expression of IFN-γ, which was regulated by IL-10 and IL-12 (Rogers *et al.*, 2004). Similarly, the lesional mRNA expression profile indicated a dominance of Th1 cytokines TNF-α, IL-6, IFN-γ in majority of patients, raised IL-12 and IL-10 in 40–70% patients and IL-4 in only 9% patients, suggesting a mixed Th1/Th2 response (Louzir *et al.*, 1998).

In India, CL is mainly caused by *Leishmania tropica* and their lesional mRNA expression indicated an increase in IFN-γ, IL-1β, IL-8, TNF-α, as also IL-10 and IL-4 (Kumar *et al.*, 2010). This was corroborated by a cytokine cDNA array of these lesions, wherein an increase in IFN-γ, TNF-α, IL-1β, IL-10 and IL-13, along with IL-17 and IL-23 was demonstrated (Katara *et al.*, 2013). Another pocket of CL is Sri Lanka, where the mRNA expression of lesional IFN-γ was significantly increased, as also was IL-4 (Manamperi *et al.*, 2017). They reported that the mRNA expression of IFN-γ and TNF-α was especially enhanced in longstanding lesions, and IL-4 mRNA expression correlated better with patients having lesions that did not respond to antimonial therapy. Accordingly, they concluded that a Th1 response supported resolution of disease, whereas a Th2-biased milieu was associated with non-responsiveness to antimony and delayed healing of lesions (Manamperi *et al.*, 2017).

Toll-like receptors in PKDL

In leishmaniasis, the Toll-like receptors TLR2 and TLR4 hold special importance as the parasite possesses molecules which interact with these receptors (Das *et al.*, 2012). These TLRs are established regulators of iNOS and their down-regulation could potentially account for the decreased generation of NO, which would enhance essential for parasite survival. In PKDL, the frequency of $CD14^+$ monocytes expressing TLR-2+ and TLR-4+ was significantly decreased compared with healthy controls (Mukhopadhyay *et al.*, 2015). The expression (Geometric Mean Fluorescence Channel) of TLR-2 was significantly down-regulated in patients compared with healthy controls, but this was not true of TLR-4 (Mukhopadhyay *et al.*, 2015).

Status of reactive oxygen and nitrogen intermediates in PKDL

In leishmaniasis, parasite survival necessitates evasion of its antimicrobial functions, and includes modulation of nitric oxide (NO), the principal antimicrobial molecule generated within monocytes/macrophages from L-arginine via the enzyme, inducible nitric oxide synthase (iNOS). Accordingly, the fate of the intracellular parasite depends on the type of signal the macrophages receive as the type 1 cytokine interferon-γ (IFN-γ) would induce classical activation of macrophages (M1) concomitant with expression of NOS2 that oxidizes L-arginine into nitric oxide (NO), a metabolite responsible for parasite killing. Conversely, the key Th2 cytokine IL-4 results in alternative activation of macrophages (M2) and the induction of arginase (Muller *et al.*, 2008). In monocytes of patients with VL, intracellular levels of NO were significantly decreased as compared to healthy individuals (Roy *et al.*, 2015). Similarly, in PKDL, the *ex vivo* levels of NO in monocytes were significantly attenuated as was generation of ROS. Alongside, the production of superoxide too was significantly decreased, but changed minimally with treatment (Mukhopadhyay *et al.*, 2015).

To establish whether generation of free radicals influenced the chemotherapeutic response, Sarkar *et al.*, 2012 evaluated promastigotes from isolates causing self-healing or delayed/non-self-healing cutaneous leishmaniasis (CL) or visceral leishmaniasis (VL) for their susceptibility to nitric oxide (NO), antimony and miltefosine. In the self-healing CL strain of *Leishmania major* (5ASKH), susceptibility to NO and antimony was higher than other species. On similar lines of thought, the *Leishmania amazonensis* strain, M2269 showed a greater susceptibility to NO and antimony vis-à-vis other species (Sarkar *et al.*, 2012). Collectively, the study suggested that self-healing isolates are more susceptible to oxidative stress. Indeed, NO-generating compounds are also used in the treatment of CL (Buates and Matlashewski, 1999; Arevalo *et al.*, 2001).

Attenuated phosphorylation of p-38 and ERK in PKDL

This remarkable propensity of *Leishmania* to survive within macrophages hinges on their ability to devise strategies to evade or impair host defence mechanisms which includes modulation of the signalling components of the MAPK pathway, which is intimately interconnected with the status of TLRs and intramonocytic generation of NO (Liévin-Le Moal and Loiseau, 2016). In *Leishmania donovani*-infected peritoneal macrophages, a down-regulation of p-p38 MAPK was associated with a decreased mRNA expression of *iNOS*, *IL-12* and *IFN-γ* (Saha *et al.*, 2011). Conversely, a concomitant enhancement in phosphorylation of ERK was associated with increased mRNA expression of *IL-10* and was attributed to modulation of the CD40-CD40L signalling (Mathur *et al.*, 2004). On the contrary, in murine bone marrow-derived macrophages, *L. donovani* caused deactivation of ERK and tumour necrosis factor α (TNF-α) (Privé and Descoteaux, 2000) suggesting diversity in host–pathogen interactions and inadequacy of extrapolation of data.

As TLR signalling occurs through the MAPK pathway, its activation status was examined in patients with PKDL following addition of LPS. In healthy controls, LPS significantly enhanced the phosphorylation of p38 MAPK (7.85 ± 3.04 to 15.18 ± 3.41, $P < 0.01$; Fig. 5.2a and b). However, in patients with PKDL, LPS failed to enhance p-p38 MAPK (7.40 ± 1.8 to 11.40 ± 2.48; Fig. 5.2a and b). Akin to p-38 MAPK, LPS significantly increased

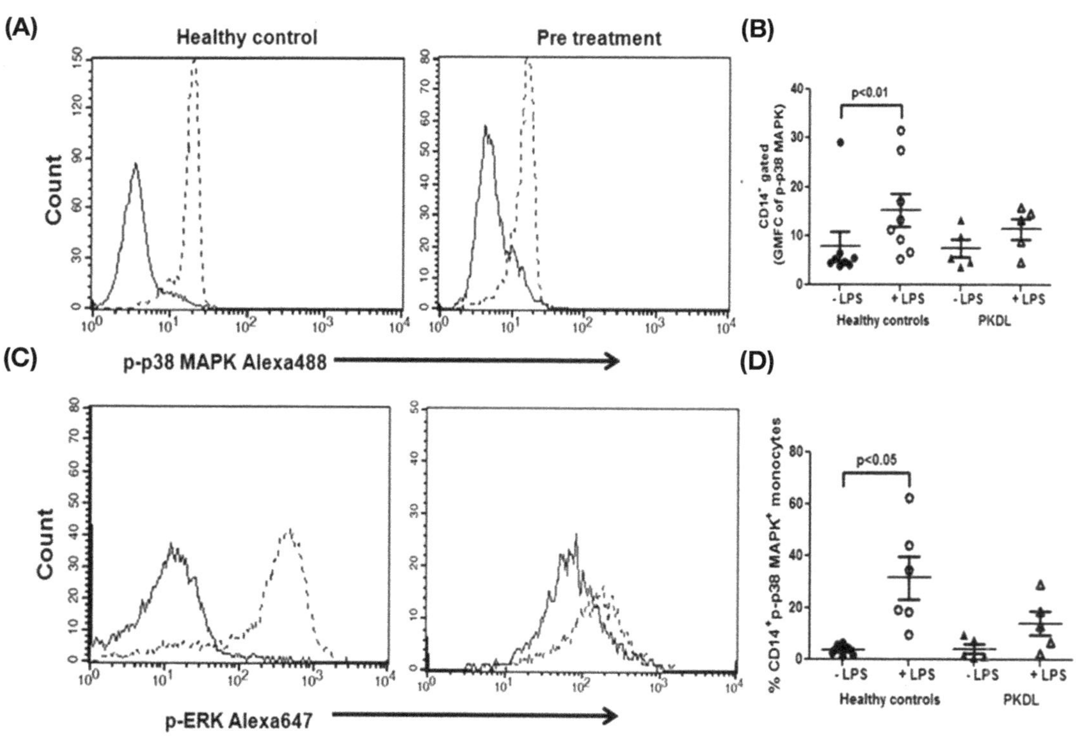

Figure 5.2 Intramonocytic phosphorylation status of p38 MAPK and ERK. (A) and (C) Representative profile showing baseline (—) and LPS-stimulated (---) expression of p-p38 MAPK (A) and p-ERK (C) in monocytes from a healthy control and a patient with PKDL. (B) and (D) Scatterplots showing baseline and LPS-stimulated expression of p-p38 MAPK (B) and p-ERK (D) in healthy controls (●, O) and patients with PKDL (▲, △).

by 5-fold the phosphorylation of ERK in healthy controls (GMFC, 26.43 ± 8.77 to 130.60 ± 59.80, $P < 0.01$; Fig. 5.2c and d), whereas, in patients with PKDL, the phosphorylation of ERK upon stimulation with LPS remained unchanged (24.08 ± 12.01 to 68.27 ± 30.95; Fig. 5.2c and d).

This inability of patients with PKDL to increase the levels of p-p-38MAPK and p-ERK post stimulation with LPS (Fig. 5.2) indicated an impairment of the p-p38MAPK and p-ERK signalling pathway. It may therefore be proposed that in PKDL, the decreased expression of TLR-2 along with impaired phosphorylation of p38 MAPK and ERK, led to a decreased generation of NO, which facilitated the generation of alternatively activated M2 monocytes (Mukhopadhyay *et al.*, 2014, 2015).

Role of host and parasite arginase 1 in PKDL

Four enzymes that play a vital role in the polyamine pathway in *Leishmania* include arginase (*ARG*), ornithine decarboxylase (ODC), spermidine synthase (SPDSYN), and S-adenosylmethionine decarboxylase (ADOMETDC), and these enzymes sequentially generate ornithine, putrescine and spermidine respectively (Boitz *et al.*, 2017). From spermidine, further polyamines such as spermine and its isomer thermospermine are synthesized by spermine synthase. In *Leishmania*, the components of the polyamine pathway are located in the cytosol and glycosome, with *ARG* being located in the glycosome, while ODC, SPDSYN, and ADOMETDC are cytosolic (Roberts *et al.*, 2004). Importantly, spermine, the prevalent polyamine of higher eukaryotes, is not synthesized by *Leishmania* (Jiang *et al.*, 1999), as corroborated by the absence of a spermine synthase homologue in the *Leishmania* genome (El-Sayed *et al.*, 2005). Gene knock out studies in *L. donovani* promastigotes, have validated that ODC, SPDSYN, and ADOMETDC as essential for survival and growth, and their absence leads to polyamine auxotrophy (Jiang *et al.*, 1999, Roberts *et al.*, 2002, Boitz *et al.*, 2017). These polyamines are also critical for infectivity as mouse lesions caused by Δarg mutants were associated with a lowered parasite burden in the liver, but not in the spleen (Boitz *et al.*, 2017). Similarly, null mutants at the *ARG* locus in *L. mexicana*, *L. major*, and *L. amazonensis* (Reguera *et al.*, 2009, de Silva *et al.*, 2012) caused polyamine auxotrophy, as also caused were associated with a significant reduction in their infectivity (Gaur *et al.*, 2007, Muleme *et al.*, 2009). Mou *et al.* (2013) showed that arg^- *L. major* caused a chronic infection, that was attributed to PD-1-mediated clonal exhaustion of T-cells. Taken together, parasite-derived ARG was critical for both its survival and infectivity.

In mammalian macrophages, to promote the growth of intracellular *Leishmania* parasites, the balance must tilt in favour of increased arginase activity so as to initiate degradation of arginine, as also hinder the generation of NO, thus serving two purposes namely (1) increased synthesis of polyamines necessary for parasite growth and (2) decreased availability of the microbicidal molecule NO (França-Costa *et al.*, 2015). In studies regarding human leishmaniasis, raised levels of arginase have been demonstrated in neutrophils and low-density granulocytes (Abebe *et al.*, 2012). Similarly, in diffuse cutaneous leishmaniasis (DCL), a more chronic form of leishmaniasis more akin to PKDL, elevation of the arginase pathway i.e. arginase-1, ornithine decarboxylase and polyamines was demonstrated (França-Costa *et al.*, 2015). In mice, Raes *et al.* (2005) established that the M2 phenotype via IL-4 AND IL-13-dependent STAT 6 activation enhances transcription of genes typical of M2 polarization, notably Mannose receptor (*Mrc1*), Arginase (*Arg1*), PPARγ (*PPARG*) and *Fizz1* among others. However, efforts to simulate this murine scenario by pulsing human monocyte derived macrophages with IL-4 or IL-13 failed to demonstrate arginase-1 expression, based on which it was inferred that *Arg1* is strictly restricted to mouse M2 cells (Raes *et al.*, 2005). However, it is possible that human monocytes require multiple signals to establish the entire spectrum of alternative activation, that may well be missing in an *ex vivo* scenario (Scott, 2009). Indeed, that is the case as Babu *et al.* (2009) reported an increase in expression of arginase-1 in M2 monocytes in patients with filariasis. This was mirrored in the study by Mukhopadhyay *et al.* (2015) who showed that in patients with PKDL, an enhanced expression of arginase-1 is a feature of M2 polarized human monocytes/macrophages (Fig. 5.3). The study reported an increased accumulation of

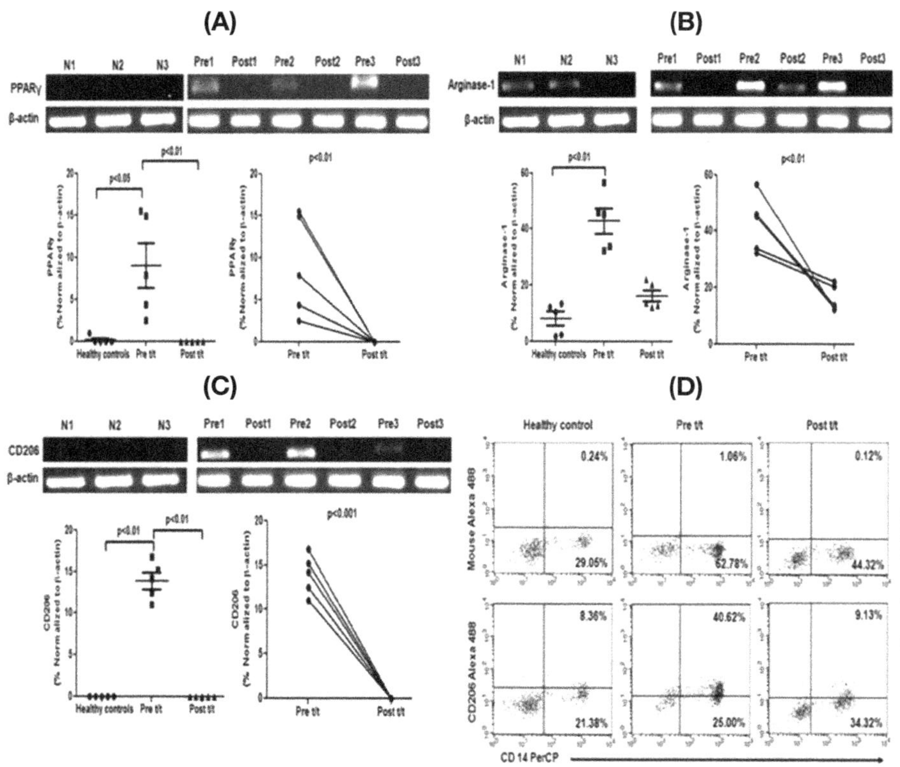

Figure 5.3 Raised expression of PPAR-γ was accompanied by an increased expression of arginase-1 and mannose receptor in circulating monocytes. (A–C) Representative mRNA expression in circulating monocytes of (A) PPAR-γ, (B) arginase-1 and (C) mannose receptor sourced from healthy controls (N1–3), patients with PKDL (Pre 1–3) and post treatment (Post 1–3). Scatterplots indicate RT-PCR products in healthy controls (●), patients with PKDL (Pre t/t □) and after treatment (Post t/t,▲). These RT-PCR products were quantified by densitometric analysis after normalization with β-actin, along with before and after plots of the same. (D) Representative quadrant plot showing frequency of CD206 in a healthy control, a patient with PKDL (Pre t/t) and post treatment (Post t/t). Scatterplots shows frequency of CD206 within CD14 gated monocytes of healthy controls (●), patients with PKDL (Pre t/t, □) and after treatment (Post t/t,▲).

PPARγ mRNA as compared to controls and post treatment and was accompanied by a significant increase in the mRNA of *ARG1* (Fig. 5.3). At disease presentation, the serum arginase activity was significantly higher in patients with PKDL than in healthy controls (mean ± SEM, 559.60 ± 56.45 vs. 365.80 ± 28.07 U/l; $P < 0.05$) which endorsed the role of Arginase in parasite persistence (Mukhopadhyay *et al.*, 2011). To date, studies regarding the status of arginase in *L. donovani* causing PKDL has not been evaluated. In murine CL, Kropf *et al.* (2005) have demonstrated a correlation between parasite load and arginase activity, and showed that inhibition of arginase activity translated into reduced pathology, along with a more efficient control of parasite replication.

Status of co-stimulatory molecules, CD80/CD 86 in PKDL

Monocytes strengthen the interaction between T-cell receptor (TCR) and major histocompatibility complex (MHC) via costimulatory molecules B7.1 (CD80) and B7.2 (CD86). Induction of this secondary signalling is important as it helps in proliferation, differentiation, activation, survival and generation of T and B-cell memory (Abou Fakher *et al.*, 2009; Jain *et al.*, 2013; Maroof *et al.*, 2009). In patients with PKDL, a down-regulation of the costimulatory molecule CD28 that interacts with CD80 and CD86 was shown on $CD8^+$ T-cells (Ganguly *et al.*, 2010b). Additionally, immunophenotypic analysis indicated that expression of CD86 on CD14 monocytes was significantly attenuated at presentation (Mukhopadhyay *et al.*, 2011).

CTLA-4 (CD152) via its ability to transmit inhibitory signal to T-cells, serves as a critical immune checkpoint whose binding to CD80/86 results in down-regulation of immune responses. In fact, its 20-fold higher affinity for binding to CD80/86 than CD28 (Linsley *et al.*, 1994) allows it to outcompete CD28 for binding with co-stimulatory ligands. This was beneficial for parasite survival as demonstrated by Katara *et al.* (2011), who reported that levels of CTLA-4 were enhanced at presentation, indicating a predominance of inhibitory signals in PKDL.

Role of vitamin D in Indian PKDL

Studies on lesional patterns in PKDL have been shown to mirror the clothing habits of individuals, which strongly suggest a link between exposure to UV light and the pathogenicity of PKDL (Das *et al.*, 2014). UV, vitamin D_3, and dihydroxylated vitamin D_3 [1α,25$(OH)_2D_3$] have been proposed to have a potent immunomodulatory role. It has been shown that UV light enhances the synthesis of vitamin D_3 as 7-dehydrocholesterol absorbs UV light and is converted into vitamin D_3 by a series of enzymatic and non-enzymatic reactions. Available data from Indian PKDL showed a down-regulation of co-stimulatory molecules similar to that resulting from UV irradiation, and an increased expression of cytokines TNF-α and IL-10 at the lesional site (Table 5.1). These data endorse a causal role for UV light in both disease pathogenesis and immunosuppression (Mukhopadhyay *et al.*, 2014).

The influence of vitamin D_3 on the immune system appears controversial, as on the one hand, it induces synthesis of potentially host-protective antibacterial peptides from macrophages (Di Rossa *et al.*, 2011), it conversely inhibits the Toll-like receptor (TLR)-induced activation of macrophages, coupled with down-regulation of co-stimulatory molecules and pro-inflammatory cytokines (Di Rossa *et al.*, 2011). In case of PKDL, the ability of 1α,25$(OH)_2D_3$ or vitamin D_3 to up-regulate the arginase pathway is evidence of its ability to facilitate alternative activation of macrophages, and was associated with a decreased expression of TLRs 2/4 (Mukhopadhyay *et al.*, 2014, 2015). This study was in concordance with a *L. major* model wherein VDR knockout mice showed resistance to infection (Ehrchen *et al.*, 2007, Martinez and Gordon, 2015). Additionally, the raised serum 25$(OH)D_3$ was accompanied by an enhanced mRNA expression of *CYP27B1*, *VDR* and *LL-37*, indicating that infection up-regulated the molecular switch needed for monocyte polarization towards a M2 phenotype (Fig. 5.4; Mukhopadhyay *et al.*, 2014, 2015).

The suppressive effects of vitamin D playing a contributory role in the establishment of infection has been reported by Whitcomb *et al.* (2012), who corroborated that enzymatic depletion of the active form of vitamin D mimicked the ablation of VDR, and resulted in an increased resistance to *L. major*. However, in BALB/c mice infected with *L. mexicana*, Ramos *et al.* (2013) examined the effect of supplementation with 1,25 $(OH)_2D_3$ on the development of skin lesions. They determined the *in situ* expression of IL-6, IL-10, IL-12, IFN-γ and TGF-β and found that vitamin D supplementation did not affect parasite survival, suggesting the possibility of inter-species differences.

Status of MRC 1 in Indian PKDL

CD206, also termed MRC1 (C-type mannose receptor 1), is a 175-kDa type I transmembrane glycoprotein which binds and internalizes glycoproteins and collagen ligands. Several types of tissue resident macrophages express CD206 in mice and humans, such as cardiac-resident macrophages, peritoneal macrophages, adipose tissue macrophages and dermal macrophages (Roszer *et al.*, 2015). PPARγ, a transcription factor of the nuclear hormone receptor family acts downstream of STAT6 signalling to regulate macrophage metabolism via the enhanced expression of CD206-mannose receptor (Bouhlel *et al.*, 2007). Indeed, Mukhopadhyay *et al.* (2015) showed that in PKDL, a 14-fold increased expression of PPARγ occurred at disease presentation in lesional macrophages and circulating monocytes, providing strong evidence of M2 monocyte/macrophage polarization (Fig. 5.4).

Taken together, in PKDL the *Leishmania* parasite has deviously evolved a multipronged approach to survive and proliferate within lesional monocytes/macrophages, employing diverse strategies aimed at driving their polarization towards an anti-inflammatory phenotype. Therefore, future

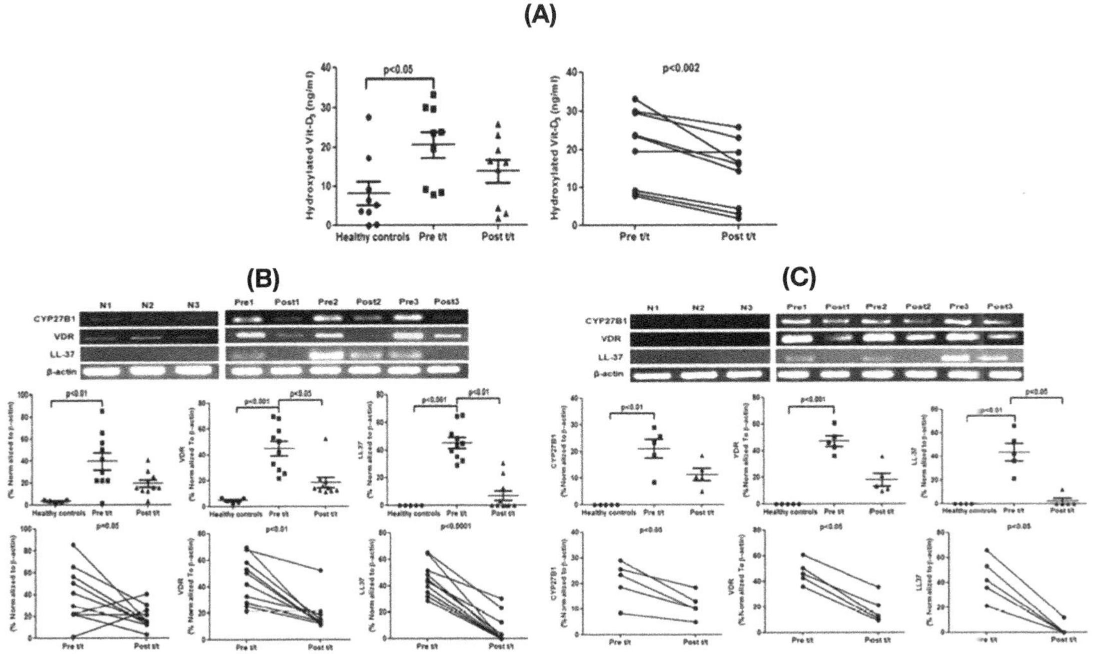

Figure 5.4 Status of vitamin D3 and associated downstream signalling molecules in PKDL. (A) Plasma levels of hydroxylated vitamin D3 in healthy controls (●) and patients with PKDL pre (Pre t/t, □) and post treatment (Post t/t,▲) along with before and after plots of the same. (B) Representative mRNA expression profile of CYP27B1, VDR, LL-37 mRNA in circulating monocytes of healthy controls (N1–3), patients with PKDL (Pre 1–3) and post treatment (Post 1–3). Scatter plots indicate RT-PCR products in healthy controls (●), patients with PKDL (Pre t/t, □) and after treatment (Post t/t,▲). These RT-PCR products were quantified by densitometric analysis after normalization with β-actin, along with before and after plots of the same. (C) Representative mRNA expression of CYP27B1, VDR, LL-37 in skin samples from healthy controls (N 1–3), patients with PKDL (Pre 1–3) and post treatment (Post 1–3). Scatter plot indicates expression as in (B) above.

therapeutic strategies aimed at preventing the parasite from sustaining this phenotype are worthy of future pharmacological consideration.

References

Abebe, T., Hailu, A., Woldeyes, M., Mekonen, W., Bilcha, K., Cloke, T., Fry, L., Seich Al Basatena, N.K., Corware, K., Modolell, M., Munder, M., Tacchini-Cottier, F., Müller, I., and Kropf, P. (2012). Local increase of arginase activity in lesions of patients with cutaneous leishmaniasis in Ethiopia. PLOS Negl. Trop. Dis. *6*, e1684. https://doi.org/10.1371/journal.pntd.0001684.

Abou, Fakher, F.H., Rachinel, N., Klimczak, M., Louis, J., and Doyen, N. (2009). TLR9-dependent activation of dendritic cells by DNA from *Leishmania major* favors Th1 cell development and the resolution of lesions. J. Immunol. *182*, 1386–1396.

Addy, M., and Nandy, A. (1992). Ten years of kala-azar in west Bengal, Part I. Did post-kala-azar dermal leishmaniasis initiate the outbreak in 24-Parganas? Bull. World Health Organ. *70*, 341–346.

Alam, M.Z., Kuhls, K., Schweynoch, C., Sundar, S., Rijal, S., Shamsuzzaman, A.K., Raju, B.V., Salotra, P., Dujardin, J.C., and Schönian, G. (2009). Multilocus microsatellite typing (MLMT) reveals genetic homogeneity of *Leishmania donovani* strains in the Indian subcontinent. Infect. Genet. Evol. *9*, 24–31. https://doi.org/10.1016/j.

Ansari, N.A., Ramesh, V., and Salotra, P. (2006). Interferon (IFN)-gamma, tumor necrosis factor-alpha, interleukin-6, and IFN-gamma receptor 1 are the major immunological determinants associated with post-kala azar dermal leishmaniasis. J. Infect. Dis. *194*, 958–965. https://doi.org/10.1086/506624.

Arevalo, I., Ward, B., Miller, R., Meng, T.C., Najar, E., Alvarez, E., Matlashewski, G., and Llanos-Cuentas, A. (2001). Successful treatment of drug-resistant cutaneous leishmaniasis in humans by use of imiquimod, an immunomodulator. Clin. Infect. Dis. *33*, 1847–1851. https://doi.org/10.1086/324161.

Babu, S., Kumaraswami, V., and Nutman, T.B. (2009). Alternatively activated and immunoregulatory monocytes in human filarial infections. J. Infect. Dis. *199*, 1827–1837. https://doi.org/10.1086/599090.

Bailey, M.S., and Lockwood, D.N. (2007). Cutaneous leishmaniasis. Clin. Dermatol. *25*, 203–211. https://doi.org/10.1016/j.clindermatol.2006.05008.

Basher, A., Nath, P., Nabi, S.G., Selim, S., Rahman, M.F., Sutradhar, S.R., Faiz, A., Bhuiyan, M.R., Ahmed, B.N., and Rahman, R. (2015). A Study on Health Seeking Behaviors of Patients of Post-Kala-Azar Dermal

Leishmaniasis. Biomed Res Int *2015*, 314543. https://doi.org/10.1155/2015/314543.

Bhattacharya, S.K., and Dash, A.P. (2017). Elimination of Kala-Azar from the Southeast Asia Region. Am. J. Trop. Med. Hyg. *96*, 802–804. https://doi.org/10.4269/ajtmh.16-0279.

Biswas, S.K., Chittezhath, M., Shalova, I.N., and Lim, J.Y. (2012). Macrophage polarization and plasticity in health and disease. Immunol. Res. *53*, 11–24. https://doi.org/10.1007/s12026-012-8291-9.

Bogdan, C. (2001). Nitric oxide and the immune response. Nat. Immunol. *2*, 907–916. https://doi.org/10.1038/ni1001-907.

Boitz, J.M., Gilroy, C.A., Olenyik, T.D., Paradis, D., Perdeh, J., Dearman, K., Davis, M.J., Yates, P.A., Li, Y., Riscoe, M.K., Ullman, B., and Roberts, S.C. (2017). Arginase Is essential for survival of *Leishmania donovani* promastigotes but not intracellular amastigotes. Infect. Immun. *85*, e00554-16. https://doi.org/10.1128/IAI.00554-16.

Bouhlel, M.A., Derudas, B., Rigamonti, E., Dièvart, R., Brozek, J., Haulon, S., Zawadzki, C., Jude, B., Torpier, G., Marx, N., *et al.* (2007). PPARgamma activation primes human monocytes into alternative M2 macrophages with anti-inflammatory properties. Cell Metab. *6*, 137–143.

Brahmachari, U.N. (1922). A new form of cuatnaeous Leishmaniasis-dermal leishmanoid. Ind. Med. Gaz. *57*, 125.

Bray, R.S., Ashford, R.W., and Bray, M.A. (1973). The parasite causing cutaneous leishmaniasis in Ethiopia. Trans. R. Soc. Trop. Med. Hyg. *67*, 345–348.

Buates, S., and Matlashewski, G. (1999). Treatment of experimental leishmaniasis with the immunomodulators imiquimod and S-28463: efficacy and mode of action. J. Infect. Dis. *179*, 1485–1494.

Cáceres-Dittmar, G., Tapia, F.J., Sánchez, M.A., Yamamura, M., Uyemura, K., Modlin, R.L., Bloom, B.R., Convit, J.(1993). Determination of the cytokine profile in American cutaneous leishmaniasis using the polymerase chain reaction. Clin. Exp. Immunol. *91*, 500–505. PMID: 8443970.

Carrada, G., Cañeda, C., Salaiza, N., Delgado, J., Ruiz, A., Sanchez, B., Gutiérrez-Kobeh, L., Aguirre, M., and Becker, I. (2007). Monocyte cytokine and costimulatory molecule expression in patients infected with *Leishmania mexicana*. Parasite Immunol. *29*, 117–126. PMID: 17266739. https://doi.org/10.1111/j.1365-3024.2006.00924.x.

da Silva, M.F., Zampieri, R.A., Muxel, S.M., Beverley, S.M., and Floeter-Winter, L.M. (2012). *Leishmania amazonensis* arginase compartmentalization in the glycosome is important for parasite infectivity. PLOS. One *7*, e34022. https://doi.org/10.1371/journal.pone.0034022.

Das, N.K., Dutta, P.K., Mukhopadhyay, D., and Chatterjee, M. (2014). Post Kala-azar Dermal Leishmaniasis: An update. In: Ghosh S, editor. Recent Advances in Dermatology. Jaypee Brothers, New Delhi. pp 155–175.

Das, S., Pandey, K., Kumar, A., Sardar, A.H., Purkait, B., Kumar, M., Kumar, S., Ravidas, V.N., Roy, S., Singh, D., *et al.* (2012). TGF-β1 re-programs TLR4 signaling in *L. donovani* infection: enhancement of SHP-1 and ubiquitin-editing enzyme A20. Immunol. Cell Biol. *90*, 640–654. https://doi.org/10.1038/icb.2011.80.

Desjeux, P., Ghosh, R.S., Dhalaria, P., Strub-Wourgaft, N., and Zijlstra, E.E. (2013). Report of the Post Kala-azar Dermal Leishmaniasis (PKDL) Consortium Meeting, New Delhi, India, 27-29 June 2012. Parasit Vectors *6*, 196. https://doi.org/10.1186/1756-3305-6-196.

Di Rosa, M., Malaguarnera, M., Nicoletti, F., and Malaguarnera, L. (2011). Vitamin D3: a helpful immuno-modulator. Immunology *134*, 123–139. https://doi.org/10.1111/j.1365-2567.2011.03482.x.

Ehrchen, J., Helming, L., Varga, G., Pasche, B., Loser, K., Gunzer, M., Sunderkötter, C., Sorg, C., Roth, J., and Lengeling, A. (2007). Vitamin D receptor signaling contributes to susceptibility to infection with *Leishmania major*. FASEB J. *21*, 3208–3218.

El-Sayed, N.M., Myler, P.J., Blandin, G., Berriman, M., Crabtree, J., Aggarwal, G., Caler, E., Renauld, H., Worthey, E.A., Hertz-Fowler, C., *et al.* (2005). Comparative genomics of trypanosomatid parasitic protozoa. Science *309*, 404–409.

França-Costa, J., Van Weyenbergh, J., Boaventura, V.S., Luz, N.F., Malta-Santos, H., Oliveira, M.C., Santos de Campos, D.C., Saldanha, A.C., dos-Santos, W.L., Bozza, P.T., Barral-Netto, M., Barral, A., Costa, J.M., and Borges, V.M. (2015). Arginase I, polyamine, and prostaglandin E2 pathways suppress the inflammatory response and contribute to diffuse cutaneous leishmaniasis. J. Infect. Dis. *211*, 426–435. https://doi.org/10.1093/infdis/jiu455.

Ganguly, S., Saha, P., Chatterjee, M., Roy, S., Ghosh, T.K., Guha, S.K., Kundu, P.K., Bera, D.K., Basu, N., and Maji, A.K. (2015). PKDL – a silent parasite pool for transmission of leishmaniasis in kala-azar endemic areas of Malda District, West Bengal, India. PLOS Negl. Trop. Dis. *9*, e0004138. https://doi.org/10.1371/journal.pntd.0004138.

Ganguly, S., Das, N.K., Barbhuiya, J.N., and Chatterjee, M. (2010a). Post-kala-azar dermal leishmaniasis – an overview. Int. J. Dermatol. *49*, 921–931. https://doi.org/10.1111/j.1365-4632.2010.04558.x.

Ganguly, S., Mukhopadhyay, D., Das, N.K., Chaduvula, M., Sadhu, S., Chatterjee, U., Rahman, M., Goswami, R.P., Guha, S.K., Modak, D., Mallik, S., Gonju, D., Pramanik, N., Barbhuiya, J.N., Saha, B., and Chatterjee, M. (2010b). Enhanced lesional Foxp3 expression and peripheral anergic lymphocytes indicate a role for regulatory T-cells in Indian PKDL. J. Invest. Dermatol. *130*, 1013–1022. https://doi.org/10.1038/jid.2009.393.

Ganguly, S., Das, N.K., Panja, M., Pal, S., Modak, D., Rahaman, M., Mallik, S., Guha, S.K., Pramanik, N., Goswami, R., *et al.* (2008). Increased levels of interleukin-10 and IgG3 are hallmarks of Indian post-kala-azar dermal leishmaniasis. J. Infect. Dis. *197*, 1762–1771. https://doi.org/10.1086/588387.

Gaur, U., Roberts, S.C., Dalvi, R.P., Corraliza, I., Ullman, B., and Wilson, M.E. (2007). An effect of parasite-encoded arginase on the outcome of murine cutaneous leishmaniasis. J. Immunol. *179*, 8446–8453.

Gordon, S. (2003). Alternative activation of macrophages. Nat. Rev. Immunol. *3*, 23–35. https://doi.org/10.1016/j.immuni.2010.05.007.

Gordon, S., and Martinez, F.O. (2010). Alternative activation of macrophages: mechanism and functions. Immunity 32, 593–604. https://doi.org/10.1016/j.immuni.2010.05.007.

Handler, M.Z., Patel, P.A., Kapila, R., Al-Qubati, Y., and Schwartz, R.A. (2015). Cutaneous and mucocutaneous leishmaniasis: Clinical perspectives. J. Am. Acad. Dermatol. 73, 897–908. https://doi.org/10.1016/j.jaad.2014.08.051.

Hölscher, C., Arendse, B., Schwegmann, A., Myburgh, E., and Brombacher, F. (2006). Impairment of alternative macrophage activation delays cutaneous leishmaniasis in nonhealing BALB/c mice. J. Immunol. *176*, 1115–1121. PMID:16394000.

Jain, S., Chodisetti, S.B., and Agrewala, J.N. (2013). CD40 signaling synergizes with TLR-2 in the BCR independent activation of resting B-cells. PLOS. One. *8*, e54392. https://doi.org/10.1371/journal.pone.0054392.

Jiang, Y., Roberts, S.C., Jardim, A., Carter, N.S., Shih, S., Ariyanayagam, M., Fairlamb, A.H., and Ullman, B. (1999). Ornithine decarboxylase gene deletion mutants of *Leishmania donovani*. J. Biol. Chem. *274*, 3781–3788.

Karunaweera, N.D., Pratlong, F., Siriwardane, H.V., Ihalamulla, R.L., and Dedet, J.P. (2003). Sri Lankan cutaneous leishmaniasis is caused by *Leishmania donovani* zymodeme MON-37. Trans. R. Soc. Trop. Med. Hyg. 97, 380–381.

Katara, G.K., Raj, A., Kumar, R., Avishek, K., Kaushal, H., Ansari, N.A., Bumb, R.A., and Salotra, P. (2013). Analysis of localized immune responses reveals presence of Th17 and Treg cells in cutaneous leishmaniasis due to *Leishmania tropica*. BMC Immunol. *14*, 52. https://doi.org/10.1186/1471-2172-14-52.

Katara, G.K., Ansari, N.A., Verma, S., Ramesh, V., and Salotra, P. (2011). Foxp3 and IL-10 expression correlates with parasite burden in lesional tissues of post kala azar dermal leishmaniasis (PKDL) patients. PLOS Negl. Trop. Dis. *5*, e1171. https://doi.org/10.1371/journal.pntd.0001171.

Kaye, P., and Scott, P. (2011). Leishmaniasis: complexity at the host–pathogen interface. Nat. Rev. Microbiol. *9*, 604–615. https://doi.org/10.1038/nrmicro2608.

Kenney, R.T., Sacks, D.L., Gam, A.A., Murray, H.W., and Sundar, S. (1998). Splenic cytokine responses in Indian kala-azar before and after treatment. J. Infect. Dis. *177*, 815–818.

Kevric, I., Cappel, M.A., and Keeling, J.H. (2015). New World and Old World *Leishmania* Infections: A Practical Review. Dermatol. Clin. *33*, 579–593. https://doi.org/10.1016/j.det.2015.03.018.

Kropf, P., Fuentes, J.M., Fähnrich, E., Arpa, L., Herath, S., Weber, V., Soler, G., Celada, A., Modolell, M., and Müller, I. (2005). Arginase and polyamine synthesis are key factors in the regulation of experimental leishmaniasis *in vivo*. FASEB J. *19*, 1000–1002.

Kumar, R., Bumb, R.A., Ansari, N.A., Mehta, R.D., and Salotra, P. (2007). Cutaneous leishmaniasis caused by *Leishmania tropica* in Bikaner, India: parasite identification and characterization using molecular and immunologic tools. Am. J. Trop. Med. Hyg. *76*, 896–901. PMID: 1748891.

Kumar, R., Bumb, R.A., and Salotra, P. (2010).Evaluation of localized and systemic immune responses in cutaneous leishmaniasis caused by *Leishmania tropica*: interleukin-8, monocyte chemotactic protein-1 and nitric oxide are major regulatory factors. Immunology. *130*, 193–201. https://doi.org/10.1111/j.1365-2567.2009.03223.x.

Kumar, N.P., Srinivasan, R., Anish, T.S., Nandakumar, G., and Jambulingam, P. (2015). Cutaneous leishmaniasis caused by Leishmania donovani in the tribal population of the Agasthyamala biosphere reserve forest, Western Ghats, Kerala, India. J. Med. Microbiol. *64*, 157–163. https://doi.org/10.1099/jmm.0.076695-0.

Liévin-Le Moal, V., and Loiseau, P.M. (2016). *Leishmania* hijacking of the macrophage intracellular compartments. FEBS J. *283*, 598–607. https://doi.org/10.1111/febs.13601.

Linsley, P.S., Greene, J.L., Brady, W., Bajorath, J., Ledbetter, J.A., and Peach, R. (1994). Human B7-1 (CD80) and B7-2 (CD86) bind with similar avidities but distinct kinetics to CD28 and CTLA-4 receptors. Immunity *1*, 793–801.

Locksley, R.M., and Scott, P. (1991). Helper T-cell subsets in mouse leishmaniasis: induction, expansion and effector function. Immunol. Today. *12*, A58–61. PMID: 1829891.

Louzir, H., Melby, P.C., Ben, Salah. A., Marrakchi, H., Aoun, K., Ben, Ismail. R., and Dellagi, K. (1998). Immunologic determinants of disease evolution in localized cutaneous leishmaniasis due to *Leishmania major*. J. Infect. Dis. *177*, 1687–1695. PMID: 9607850.

Manamperi, N.H., Oghumu, S., Pathirana, N., de Silva, V.C., Abeyewickreme, W., Satoskar, A.R., and Karunaweera, N.D. (2017). In situ immunopathological changes in cutaneous leishmaniasis due to *Leishmania donovani*. Parasite. Immunol. *39*. https://doi.org/10.1111/pim.12413.

Maroof, A., Beattie, L., Kirby, A., Coles, M., and Kaye, P.M. (2009). Dendritic cells matured by inflammation induce CD86-dependent priming of naive $CD8^+$ T-cells in the absence of their cognate peptide antigen. J. Immunol. *183*, 7095–7103. https://doi.org/10.4049/jimmunol.0901330.

Martinez, F.O., and Gordon, S. (2015). The evolution of our understanding of macrophages and translation of findings toward the clinic. Expert Rev Clin Immunol *11*, 5–13. https://doi.org/10.1586/1744666X.2015.985658.

Martinez, F.O., Helming, L., and Gordon, S. (2008). Alternative activation of macrophages: an immunologic receptors and leishmaniasis. Infect. Immun. *76*, 866–72. https://doi.org/10.1007/s00430-010-0156-z.

Mathur, R.K., Awasthi, A., Wadhone, P., Ramanamurthy, B., and Saha, B. (2004). Reciprocal CD40 signals through p38MAPK and ERK-1/2 induce counteracting immune responses. Nat. Med. *10*, 540–544. https://doi.org/10.1038/nm1045.

Melby, P.C., Andrade-Narvaez, F., Darnell, B.J., and Valencia-Pacheco, G. (1996). In situ expression of interleukin-10 and interleukin-12 in active human cutaneous leishmaniasis. FEMS Immunol. Med. Microbiol. *15*, 101–107.

Melby, P.C., Andrade-Narvaez, F.J., Darnell, B.J., Valencia-Pacheco, G., Tryon, V.V., and Palomo-Cetina, A. (1994). Increased expression of proinflammatory cytokines in chronic lesions of human cutaneous leishmaniasis. Infect. Immun. *62*, 837–842. PMID: 8112853.

Mohan, K., and Suri, J.C. (1975). Studies on cutaneous leishmaniasis in India V. Isolation of *Leishmania tropica* from gerbils, sandflies and human. J. Com. Dis. *7*, 353–357.

Mondal, D., Singh, S.P., Kumar, N., Joshi, A., Sundar, S., Das, P., Siddhivinayak, H., Kroeger, A., and Boelaert, M. (2009). Visceral leishmaniasis elimination programme in India, Bangladesh, and Nepal: reshaping the case finding/case management strategy. PLOS Negl. Trop. Dis *3*, e355. https://doi.org/10.1371/journal.pntd.0000355.

Mou, Z., Muleme, H.M., Liu, D., Jia, P., Okwor, I.B., Kuriakose, S.M., Beverley, S.M., and Uzonna, J.E. (2013). Parasite-derived arginase influences secondary anti-Leishmania immunity by regulating programmed cell death-1-mediated CD4$^+$ T-cell exhaustion. J. Immunol. *190*, 3380–3389.

Mukhopadhyay, D., Dalton, J.E., Kaye, P.M., and Chatterjee, M. (2014). Post kala-azar dermal leishmaniasis: an unresolved mystery. Trends Parasitol. *30*, 65–74. https://doi.org/10.1016/j.pt.2013.12.004.

Mukhopadhyay, D., Das, N.K., Roy, S., Kundu, S., Barbhuiya, J.N., and Chatterjee, M. (2011). Miltefosine effectively modulates the cytokine milieu in Indian post kala-azar dermal leishmaniasis. J. Infect. Dis. *204*, 1427–1436. https://doi.org/10.1093/infdis/jir551.

Mukhopadhyay, D., Mukherjee, S., Roy, S., Dalton, J.E., Kundu, S., Sarkar, A., Das, N.K., Kaye, P.M., and Chatterjee, M. (2015). M2 Polarization of Monocytes-Macrophages Is a Hallmark of Indian Post Kala-Azar Dermal Leishmaniasis. PLOS Negl. Trop. Dis. *9*, e0004145. https://doi.org/10.1371/journal.pntd.0004145.

Muleme, H.M., Reguera, R.M., Berard, A., Azinwi, R., Jia, P., Okwor, I.B., Beverley, S., and Uzonna, J.E. (2009). Infection with arginase-deficient *Leishmania major* reveals a parasitenumber-dependent and cytokine-independent regulation of host cellular arginaseactivity and disease pathogenesis. J. Immunol. *183*, 8068–8076. https://doi.org/10.4049/jimmunol.0803979.

Müller, I., Hailu, A., Choi, B.S., Abebe, T., Fuentes, J.M., Munder, M., Modolell, M., and Kropf, P. (2008). Age-related alteration of arginase activity impacts on severity of leishmaniasis. PLOS Negl. Trop. Dis *2*, e235. https://doi.org/10.1371/journal.pntd.0000235.

Osorio, E.Y., Zhao, W., Espitia, C., Saldarriaga, O., Hawel, L., Byus, C.V., Travi, B.L., and Melby, P.C. (2012). Progressive visceral leishmaniasis is driven by dominant parasite-induced STAT6 activation and STAT6-dependent host arginase 1 expression. PLOS Pathog. *8*, e1002417. https://doi.org/10.1371/journal.ppat.1002417.

Pirmez, C., Yamamura, M., Uyemura, K., Paes-Oliveira, M., Conceição-Silva, F., and Modlin, R.L. (1993). Cytokine patterns in the pathogenesis of human leishmaniasis. J. Clin. Invest. *91*, 1390–1395. https://doi.org/10.1172/JCI116341.

Privé, C., and Descoteaux, A. (2000). *Leishmania donovani* promastigotes evade the activation of mitogen-activated protein kinases p38, c-Jun N.-terminal kinase, and extracellular signal-regulated kinase-1/2 during infection of naive macrophages. Eur. Immunol, J. *30*, 2235–2244. https://doi.org/10.1002/1521-4141(2000)30:8<2235::AID-IMMU2235>3.0.CO;2-9.

Raes, G., Van den Bergh, R., De Baetselier, P., Ghassabeh, G.H., Scotton, C., Locati, M., Mantovani, A., and Sozzani, S. (2005). Arginase-1 and Ym1 are markers for murine, but not human, alternatively activated myeloid cells. J. Immunol. *174*, 6561.

Ramesh, V., Kaushal, H., Mishra, A.K., Singh, R., and Salotra, P. (2015). Clinico-epidemiological analysis of Post kala-azar dermal leishmaniasis (PKDL) cases in India over last two decades: a hospital based retrospective study. BMC Public Health. *15*, 1092. https://doi.org/10.1186/s12889-015-2424-8.

Ramos-Martínez, E., Villaseñor-Cardoso, M.I., López-Vancell, M.R., García-Vázquez, F.J., Pérez-Torres, A., Salaiza-Suazo, N., and Pérez-Tamayo, R. (2013). Effect of 1,25(OH)2D3 on BALB/c mice infected with *Leishmania mexicana*. Exp. Parasitol. *134*, 413–421. https://doi.org/10.1016/j.exppara.2013.05.009.

Reguera, R.M., Balaña-Fouce, R., Showalter, M., Hickerson, S., and Beverley, S.M. (2009). *Leishmania* major lacking arginase (ARG) are auxotrophic for polyamines but retain infectivity to susceptible BALB/c mice. Mol. Biochem. Parasitol. *165*, 48–56. https://doi.org/10.1016/j.molbiopara.2009.01.001.

Roberts, S.C., Tancer, M.J., Polinsky, M.R., Gibson, K.M., Heby, O., and Ullman, B. (2004). Arginase plays a pivotal role in polyamine precursor metabolism in *Leishmania*. Characterization of gene deletion mutants. J. Biol. Chem. *279*, 23668–23678. https://doi.org/10.1074/jbc.M402042200.

Roberts, S.C., Scott, J., Gasteier, J.E., Jiang, Y., Brooks, B., Jardim, A., Carter, N.S., Heby, O., and Ullman, B. (2002). S adenosylmethionine decarboxylase from *Leishmania donovani*. Molecular, genetic, and biochemical characterization of null mutants and overproducers. J. Biol. Chem. *277*, 5902–59099. https://doi.org/10.1074/jbc.M110118200.

Rodrigues, V., Cordeiro-da-Silva, A., Laforge, M., Silvestre, R., and Estaquier, J. (2016). Regulation of immunity during visceral Leishmania infection. Parasit Vectors *9*, 118. https://doi.org/10.1186/s13071-016-1412-x.

Rogers, K.A., and Titus, R.G. (2004). The human cytokine response to *Leishmania major* early after exposure to the parasite *in vitro*. J. Parasitol. *90*, 557–563. https://doi.org/10.1645/GE-3317.

Rőszer, T. (2015). Understanding the mysterious M2 macrophage through activation markers and effector mechanisms. Mediators. Inflamm. 2015, 816460. https://doi.org/10.1155/2015/816460. Review.

Roy, S., Mukhopadhyay, D., Mukherjee, S., Ghosh, S., Kumar, S., Sarkar, K., Pal, D., Bhowmik, P., Mandal, K., Modak, D., Guha, S.K., Pramanik, N., Goswami, R.P., Saha, B., and Chatterjee, M. (2015). A defective oxidative burst and impaired antigen presentation are hallmarks of human Visceral Leishmaniasis. J Clin. Immunol. *35*, 56–67. https://doi.org/10.1007/s10875-014-0115-3.

Saha, P., Bhattacharjee, S., Sarkar, A., Manna, A., Majumder, S., and Chatterjee, M. (2011). Berberine chloride mediates its anti-leishmanial activity via differential regulation of the mitogen activated protein kinase pathway in macrophages. PLOS ONE. *6*, e18467. https://doi.org/10.1371/journal.pone.0018467.

Sarkar, A., Ghosh, S., Pakrashi, S., Roy, D., Sen, S., and Chatterjee, M. (2012). *Leishmania* strains causing self-healing cutaneous leishmaniasis have greater susceptibility towards oxidative stress. Free Radic. Res. *46*, 665–673. https://doi.org/10.3109/10715762.2012.668186.

Scott, A.L. (2009). The alternatively activated human – redux. J. Infect. Dis. *199*, 1723–1725. https://doi.org/10.1086/599091.

Sengupta, P.C. (1947). History of kala-azar in India. Ind. Med. Gaz. *82*, 283.

Sharma, M.I.D., Suri, J.C., Kalra, N.L., and Mohan, K. (1973). Studies on cutaneous leishmaniasis in India III. Detection of a zoonotic focus of cutaneous leishmaniasis in Rajasthan. J. Com. Dis. *5*, 149–153.

Sharma, U., and Singh, S. (2009). Immunobiology of leishmaniasis. Indian J. Exp. Biol. *47*, 412–423.

Stauch, A., Sarkar, R.R., Picado, A., Ostyn, B., Sundar, S., Rijal, S., Boelaert, M., Dujardin, J.C., and Duerr, H.P. (2011). Visceral leishmaniasis in the Indian subcontinent: modelling epidemiology and control. PLOS Negl. Trop. Dis *5*, e1405. https://doi.org/10.1371/journal.pntd.0001405.

Steverding, D. (2017). The history of leishmaniasis. Parasit Vectors *10*, 82. https://doi.org/10.1186/s13071-017-2028-5.

Valencia-Pacheco, G., Loría-Cervera, E.N., Sosa-Bibiano, E.I., Canché-Pool, E.B., Vargas-Gonzalez, A., Melby, P.C., and Andrade-Narvaez, F.J. (2014). In situ cytokines (IL-4, IL-10, IL-12, IFN-γ) and chemokines (MCP-1, MIP-1α) gene expression in human *Leishmania* (*Leishmania*) *mexicana* infection. Cytokine *69*, 56–61. PMID: 25022962. https://doi.org/10.1016/j.cyto.2014.05.016.

Whitcomb, J.P., Deagostino, M., Ballentine, M., Fu, J., Tenniswood, M., Welsh, J., Cantorna, M., and McDowell, M.A. (2012). The Role of vitamin D and vitamin D receptor in immunity to *Leishmania major* Infection. J. Parasitol. Res. *2012*, 134645. https://doi.org/10.1155/2012/134645.

Zijlstra, E.E. (2016). The immunology of post-kala-azar dermal leishmaniasis (PKDL). Parasit. Vectors *9*, 464. https://doi.org/10.1186/s13071-016-1721-0.

Zijlstra, E.E., Musa, A.M., Khalil, E.A., el-Hassan, I.M., and el-Hassan, A.M. (2003). Post-kala-azar dermal leishmaniasis. Lancet Infect Dis *3*, 87–98.

Leishmania Exploits Host's Defence Machineries for Survival: A Tale of Immune Evasion

6

Amrita Saha[1] and Anindita Ukil[2]*

[1]Infectious Diseases and Immunology Division, CSIR-Indian Institute of Chemical Biology, Kolkata, India.
[2]Department of Biochemistry, University of Calcutta, Kolkata, India.

*Correspondence: u.anindita@gmail.com

https://doi.org/10.21775/9781910190715.06

Abstract

The disease leishmaniasis, caused by protozoan parasite of genus *Leishmania*, is a major health concern as over 12 million people are affected worldwide. The disease comprises a spectrum of clinical manifestations ranging from self-healing cutaneous lesions through life-threatening infection of visceral organs to death. The lifestyle of *Leishmania* is that of an obligate intracellular parasite that resides and multiplies within the phagolysosome of host macrophages. Interestingly, these macrophages are the sentinels of the immune system, that are specialized for the destruction of invading pathogens. Therefore, in order to successfully dwell within its niche, *Leishmania* has had to evolve range of sophisticated mechanisms to subvert normal macrophage functioning. There are multiple ways by which intracellular pathogens such as *Leishmania* make use of host cell's machinery in order to survive and replicate, either by employing strategies to inhibit proteins that play a positive role in immune cell activation or by activating molecules that act as negative regulators of immune cell signalling and function. Current research has focused on parasite induced alteration of host cell signalling pathways, combating microbicidal free radical generation and modulation of cytokine/chemokine profiles that alter immune cell activation and antigen presentation. The aim of this chapter is to highlight the molecular mechanisms underlying immune evasion strategies whereby *Leishmania* can subvert host surveillance, thereby exploiting the macrophage environment in its favour for its persistence. Understanding these mechanisms in detail may open up new perspectives on potential molecular pathways that are prone to pharmacological manipulation and can prove as promising areas for preclinical and clinical research.

Introduction

The complicated interactions between *Leishmania* and the macrophages ultimately decides the final fate of the disease outcome. Macrophages serve as the primary host for *Leishmania*, they phagocytose the parasites and the parasite proliferates within the parasitophorous vacuole of the infected macrophages (Liu and Uzonna 2012). However, these cells are also the major effector cells to contribute in the elimination of infection. *Leishmania* has evolved various strategies to conquer and evade the antimicrobial arsenals, mounted by macrophages (Gupta *et al.*, 2013). Living inside this niche is quite profitable for the parasite if it could escape the immune system that would otherwise be directed against it, obtain nutrients, and multiply within the phagolysosomes of macrophages. Therefore, to live a life within the macrophages, *Leishmania* must evade the host's defensive immune machineries. This includes inhibition of microbicidal activity, prevention of host apoptosis, and impairment of protective cytokine and chemokine responses, etc. (Shio *et al.*, 2012). The strategy to win against *Leishmania* infections therefore depends on a thorough understanding of the immune evasion mechanisms adopted by the parasite within the host macrophages. We will discuss how the current knowledge will allow further refinement of our understanding of the interplay between *Leishmania* and its hosts that leads to either survival or elimination. How *Leishmania* survives and dwells inside this hostile environment as well as sabotage host defence mechanisms in its favour resulting in the establishment of disease will be addressed in this review.

In the present study, we tried to elucidate the mechanism exercised by the parasite to modulate the defensive signalling mechanisms of the host during various stages of infection, i.e. before their entry into macrophages, during their internalization, and sustained suppression of immune response after successful internalization. These subversive activities are dictated by the need to evade microbicidal effector function, as well as to avoid pro-inflammatory pathology that can destabilize the host–parasite interaction.

Early stages of infection

Subversion of macrophage Toll-like receptors

The Toll-like receptors (TLRs), which recognize pathogen-associated molecular patterns (PAMPs), are the means for first line of host defence against *Leishmania*, after invasion (Oda and Kitano 2006). TLRs are transmembrane glycoproteins that confer specificity to the macrophages by recognition of different pathogens that cause human disease. Activation of TLRs converges into series of events leading to leucocyte recruitment, microbicidal activity, and production of inflammatory mediators such as IL-12 and TNF-α, ultimately resulting in clearance of the pathogen (Janssens and Beyaert 2003). *Leishmania* are capable of modulating this host TLR response, and presents unusual challenges to the immune system with their capacity to down-regulate TLR-mediated inflammatory response (Faria *et al.*, 2012). LPS binding to TLR4 triggers the recruitment of the adaptor protein myeloid differentiation primary response gene 88 (MyD88), followed by formation of signalosome complex consisting of TRAF6, cIAP1/2, Ubc13, and TRAF3. Ubiquitination at K63 of TRAF6 enables it to ubiquitinate cIAP1/2, which in turn ubiquitinates TRAF3 at K48, leading to its proteasomal degradation (Akira and Takeda 2004). This is followed by subsequent release and translocation of the entire complex to the cytosol, resulting in activation of MAPK and NF-κB. A study by Mukherjee *et al.*, stated that Sb^RLD infection induced miR-466i, which down-regulates MyD88 expression, results in elevated IL-10 level and aggressive pathology (Mukherjee *et al.*, 2015). Moreover, it was reported that MyD88-dependent pathways are crucial for the development of protective IL-12-mediated Th1 response against *L. major* infection (Debus *et al.*, 2003).

Our study reported that upon *Leishmania donovani* infection, Ubc13 dissociates from the signalosome complex, causing defective K63-linked ubiquitination of TRAF6. This leads to defective K63-linked ubiquitination of cIAP1/2, thereby resulting in insufficient K48-linked ubiquitination of TRAF3. Therefore, in consequence TRAF3 degradation is restricted. TRAF3, as well as the whole signalosome complex, persists in the membrane, resulting in inappropriate TAK1 and subsequent NF-κB activation, ultimately dampening the pro-inflammatory responses (Gupta *et al.*, 2014) (Fig. 6.1).

All the species of *Leishmania* possess a cell surface glycophospholipid lipophosphoglycan (LPG), which is a TLR2 ligand (Tuon *et al.*, 2008). Studies suggested that *Leishmania* LPG capable of activating NK cells and monocytes upon binding TLR2. *Leishmania major* LPG enhances IFN-γ and TNF-α production and nuclear translocation of NF-κB in human NK cells (Becker *et al.*, 2003). LPG serves as a major adhesin facilitating binding and entry of *Leishmania* into macrophages. *Leishmania donovani* LPG has been shown to elicit the production of IL-12 in macrophages (Rodriguez-Sosa *et al.*, 2001; Shweash *et al.*, 2011). The role of LPG in establishing virulence remains restricted to promastigote stage only (Ilg 2000). However, infection by *L. donovani* promastigotes suppresses pro-inflammatory response in naive macrophages in spite of presence of LPG. Although the ability of *Leishmania* to inhibit inflammatory signalling pathways

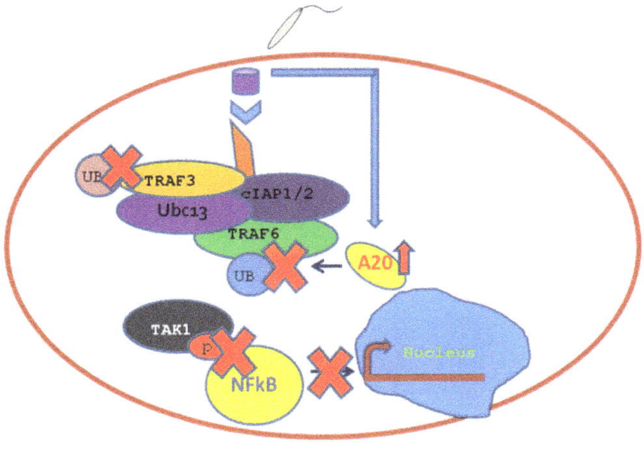

Figure 6.1 Defying recognition of the host TLRS. Attachment of *Leishmania* parasites on macrophage surface results in (1) activation of de-ubiquitinating enzyme A20 which disrupts the upstream TRAF6-TAK-TAB assembly and (2) Ubc13 dissociates from the signalosome complex, causing defective K63-linked ubiquitination of TRAF6 and of cIAP1/2, thereby resulting in insufficient K48-linked ubiquitination of TRAF3. The whole signalosome complex, persists in the membrane. The two processes culminates into inhibition of TAK1 and subsequent NF-kB activation, ultimately dampening the TLR-mediated pro-inflammatory response. This allows the parasite to stealthily enter macrophages without being recognized by the host.

has been nominated as a virulence mechanism, the molecular events underlying this process remained to be explored. Our data demonstrated that *Leishmania* brings about deubiquitination of TRAF6 by inducing the host deubiquitinating enzyme A20, thereby, inhibiting the proper assembly of signalling complexes consisting of TRAF6–TAK1–TAB2. Disorientation of the signalling complexes prevented IKK–NFκB–mediated downstream signalling, thereby leading to the inhibition of host defensive cytokine response (Srivastav *et al.*, 2012) (Fig. 6.1).

Therefore, subverting the TLR signalling cascade allows the parasite to successfully enter and sabotage the host macrophages in order to avoid the otherwise pathogen-recognition-triggered immune response.

Employment of co-stimulatory molecule, CD40, by *Leishmania*

CD40, a transmembrane costimulatory receptor, is expressed on macrophages. CD40 induces IL-12-dependent and interferon-γ (IFN-γ)-dependent host-protective immune responses against *Leishmania* (Tuladhar *et al.*, 2011). Interestingly, IL-10 acts as another CD40-induced cytokine in macrophages and promotes *Leishmania* infection (Srivastava *et al.*, 2011). *Leishmania* skews CD40 signalling towards ERK-1/2, inducing IL-10, which inhibits activation of CD40-induced p38MAPK and expression of inducible nitric oxide synthase-2 (iNOS-2) and IL-12 (Mathur *et al.*, 2004). *Leishmania* infection enhanced CD40-induces Raf-MEK-ERK signalling but inhibits PI3K-MKK-p38MAPK signalling cascade. As Raf and PI3K both possess a Ras-binding domain, they exert opposite effects on *Leishmania* infection (Chakraborty *et al.*, 2015). Report suggests that *L. major* enhances N-Ras and H-Ras expression but inhibits expression of K-Ras in macrophages, leading to inhibition of the p38MAPK pathway (Chakraborty *et al.*, 2015). Therefore, *Leishmania* exploits CD40 to generate cytokine response, favourable for its survival.

Disarming microbicidal arsenals: reactive oxygen species and reactive nitrogen species

After internalization, the major microbicidal arsenal generated by macrophages upon *Leishmania* internalization is reactive oxygen species (ROS), the generation of which is inhibited following infection (Carneiro *et al.*, 2016). They are induced by IFN-γ or Toll-like receptors (Marcato *et al.*, 2008). At early stages, both ROS and RNS produced by macrophages act together to control infection by *L. chagasi* (Gantt *et al.*, 2001) and *L. donovani*

(Bhattacharyya *et al.*, 2002; Murray and Nathan 1999). Therefore, *Leishmania* must have some strategies to circumvent the toxic effects of these reactive species. Some *Leishmania* defensive molecules reported against ROS include intracellular thiols (Channon and Blackwell, 1985), lipophosphoglycan (Chan *et al.*, 1989), iron superoxide dismutase (Paramchuk *et al.*, 1997), HSP70 (Miller *et al.*, 2000), ovothiol A and trypanothione (Ariyanayagam and Fairlamb, 2001), and peroxidoxins (Barr and Gedamu, 2003; Castro *et al.*, 2002). Glutathione has recently been proved beneficial in protecting *L. major* from NO-induced cytotoxicity (Romão *et al.*, 1999). Peroxidoxin1 enhances *Leishmania chagasi* survival in macrophages by providing defence against the most important antimicrobial ROS and RNS molecules (Barr and Gedamu 2003). Down-regulation of iNOS expression in *L. amazonensis*-infected macrophages was reported due to p50/p50 NF-κB activation. Moreover, p50/p50 NF-κB activation induced during *L. amazonensis* infection depends on PI3K/Akt activation (Calegari-Silva *et al.*, 2015). However, no report addressed the possible involvement of host mitochondria, a major ROS generation site. Several studies have emphasized the role of mitochondrial inner membrane UCPs in regulating the mitochondrial ROS generation in diverse cellular and disease context. Report suggested that *L. donovani* infection is associated with strong up-regulation of UCP2 in macrophages, resulting in marked induction of protein tyrosin phosphatases (PTP) activity possibly by inhibition of mitochondrial ROS generation (Basu Ball *et al.*, 2011). This was attested by the fact that silencing of UCP2 resulted in increased ROS generation and inhibition of PTP activity in infected macrophages, thereby promoting positive MAPK signal transduction and induction of proinflammatory cytokine synthesis (Calegari-Silva *et al.*, 2015).

Mid stages of infection

Modulation of the kinase–phosphatase equilibrium

The balance between kinases and phosphatases is prerequisite for mounting appropriate immune responses against pathogens. Any shift in this balance results in dysregulated immune response. Recent studies have demonstrated that host protein phosphatases are exploited in order to evade the host immune response mounted during *Leishmania* infection (Forget *et al.*, 2006; Abu-Dayyeh *et al.*, 2008). A number of reports have demonstrated that SHP-1 plays an indispensable role in establishing and augmenting the disease outcome in the course of *Leishmania* parasite infection (Abu-Dayyeh *et al.*, 2008; Forget *et al.*, 2005). The defensive strategies of macrophages against invading pathogens are dependent on the appropriate phosphorylation of MAP kinases, and there are studies reporting that *Leishmania* interferes with MAPK signalling pathway in macrophages (Nandan and Reiner 1995). MAPK phosphorylation is differentially regulated by dual-specificity MAPK phosphatases (MKPs). The work by Srivastava *et al.*, reported, *L. major*, exploits intracellular phosphatases MKP-1 and MKP-3 reciprocally to modulate CD40 signalling in its favour (Srivastava *et al.*, 2011). *L. donovani* infection was found to significantly up-regulate the expression of MKP-1, MKP-3 and PP2A. Inhibition of phosphatases in *L. donovani*-infected BALB/c mice shifted the cytokine balance in favour of the host by inducing TNF-α and iNOS expression (Kar *et al.*, 2010).

Delaying of phagosome maturation

Metacyclic promastigotes actively modulate normal phagosome progression by delaying fusion with lysosomes, endowing the parasites the time required to mature into the intracellular adapted form, a more antimicrobial-resistant amastigote stage (Polando *et al.*, 2013). Studies documented the necessity for CD11b and FcγR in delaying maturation of phagosome and actin recruitment upon phagocytosis of *Leishmania* species because in the absence of CD11b and FcγR, faster *Leishmania* phagosome maturation takes place in the absence of opsonization (Polando *et al.*, 2013). *Leishmania* LPG has also been stated to inhibit the ability of phagosomes to fuse with late endocytic organelles and lysosomes (Dermine *et al.*, 2000). *Leishmania* promastigotes use LPG directly to alter the formation of phagosome lipid microdomains and/or disorganize these structures after phagocytosis (Dermine *et al.*, 2005). This process is likely to be beneficial for the parasite as phagosome microdomains contain proteins that are involved in different aspects of phagolysosome biogenesis. *L. amazonensis* parasites exploit host CD36 to foster their prolonged survival: amastigotes expose CD36 ligands to stimulate fusion with endolysosomal vesicles and to promote parasitophorous vacuole (PV) enlargement, thereby diluting NO levels in the local environment so as to enhance replication (Okuda *et al.*, 2016). On the other hand, promastigotes do not expose CD36 ligands and therefore PV enlargement is delayed until the promastigotes differentiate into the strong resistant amastigotes.

Alteration of major histocompatibility complex expression and antigen presentation

Blooming of effective host defence against *Leishmania* necessitates effective antigen presentation by the phagocytes to T-lymphocytes (Awasthi *et al.*, 2004). Macrophages when present antigens through class I or II MHC molecules expressed at the plasma membrane, microbicidal activities of macrophages can be enhanced or induced by effector T-lymphocytes. *Leishmania* inhibits the ability of macrophages to display the parasite-specific antigens to other components of the immune system (Saha *et al.*, 1995). Some studies have shown that *L. donovani* infection prevents the induction of MHC class I and II expression by IFN-γ (Reiner *et al.*, 1987). Studies have also proposed that class II MHC molecules may be sequestered in the phagolysosome, and even endocytosed and are degraded by *L. amazonensis* amastigotes (De Souza Leao *et al.*, 1995). Amastigote antigens were found to be sequestered, and therefore were unavailable for presentation by MHC class II molecules (Kima *et al.*, 1996).

Membrane lipid rafts function as important platforms for antigen presentation as it concentrates MHC class II molecules into microdomains, which allow efficient presentation of antigen even at low peptide densities (Kima *et al.*, 1996). To evade effective antigen presentation, *Leishmania donovani* increases the fluidity of lipid rafts in macrophages resulting in reduced T-cells responses (Chakraborty *et al.*, 2005). The protein that is primarily regulates lipid raft formation is sterol regulatory element-binding protein 2 (SREBP2).

SREBP2

Leishmania donovani utilizes host SREBP2, a critical transcription factor, to regulate macrophage plasma membrane cholesterol (Mukherjee *et al.*, 2014). The induced SREBP2 thereafter, transcriptionally up-regulate two distinct genes, (3-hydroxy-3-methyl glutaryl CoA reductase) HMGCR and UCP2, leading to cholesterol biosynthesis. The increased cholesterol thereby ensures formation of plasma membrane (PM) cholesterol-enriched rafts for efficient attachment and entry of *Leishmania* promastigotes Moreover, knockdown of SREBP2 significantly suppressed HMGCR which in turn abrogated parasite internalization. Induction of UCP2 further lead to suppression of mitochondrial-ROS generation to assist parasite survival. SREBP2 induction resulted in generation of anti-inflammatory milieu promoting prolonged parasite burden inside the harsh environment of macrophages.

Late stages of infection

Modulation of various regulatory molecules

Lipid mediators

Lipid metabolites of arachidonic acid (AA), including leukotrienes (LTs) and prostaglandins (PGs), have emerged as important mediators having pathophysiological functions (Ricciotti and FitzGerald 2011). They are synthesized through the metabolism of AA and can be subsequently metabolized by different pathways, involving cyclooxygenase (COX) enzymes and lipoxygenase (LO) generating a range of bioactive eicosanoids, named PGs and LTs, respectively. LTs, particularly LTB_4, play important role in immune response to *Leishmania* infection by promoting leishmanicidal activity and consequently, restriction of infection. A study suggested that the promastigotes down-modulate BLT1 expression decreasing macrophage response to LTB_4, allowing sustained parasite survival (Morato *et al.*, 2014). Our study reported *Leishmania* infection up-regulates the PGE_2 production and resulted in increased intracellular cAMP level in peritoneal macrophages through G protein-coupled E-series prostanoid (EP) 2 receptor. Negative modulation of PGE_2 signalling reduced infection-induced polarization towards the anti-inflammatory cytokines. Moreover, effect of PGE_2 on cytokine and chemokine production was found to be differentially modulated by two cAMP downstream effector molecules: cAMP-dependent protein kinase A (PKA) and exchange protein directly activated by cAMP (EPAC) (Saha *et al.*, 2014) (Fig. 6.2). Administration of indomethacin suppressed PGs, promoting the development of a protective Th1 type response in susceptible mice by enhancement of IL-12, IFN-γ, and NO production (Perez-Santos and Talamas-Rohana, 2001). A study also reported that PGE_2 induction in *L. donovani*-infected macrophages depends upon PKC and cycloxygenase-2 activation (Matte *et al.*, 2001) and increased PGE_2 has been found to be associated with visceralization of *L. donovani* in malnourished mice (Anstead *et al.*, 2001).

PI3K, IRAK-M and SOCS-1

PI3K, IRAK-M and SOCS-1 contribute to negative signalling cascades in TLR signalling that are essential for suppression of excessive inflammation and alteration of the Th1/Th2 balance.

PI3K

Activation of the PI3K/Akt pathway in *L. amazonensis* infection inhibits pro-inflammatory cytokine IL-12 production in macrophages (Ruhland and Kima, 2009). A recent report also stated the use of the PI3Kγ inhibitor for the treatment of experimental L. mexicana infection in mice (Oghumu and Satoskar, 2013). Administration of PI3Kγ inhibitor resulted in significantly lower parasite burdens and reduced lesion sizes than in wild-type untreated infected mice. Studies attested that PI3K/Akt signalling is activated upon *Leishmania* infection and this pathway confers to the infected cell, the capacity to resist against activators induced apoptosis (Ruhland *et al.*, 2007; Vazquez-Lopez *et al.*, 2015).

Suppressor of cytokine signalling 1

Suppressor of cytokine signalling 1 (SOCS1) is a key regulator of cytokine signal transduction, SOCS1 is reported to regulate diverse cytokine signalling responses, including IFN-α, IL-2, IL-6, and TNF-α (Fujimoto and Naka 2010;Alexander 2002). Studies have shown that LPS-induced pro-inflammatory cytokine production and NF-κB

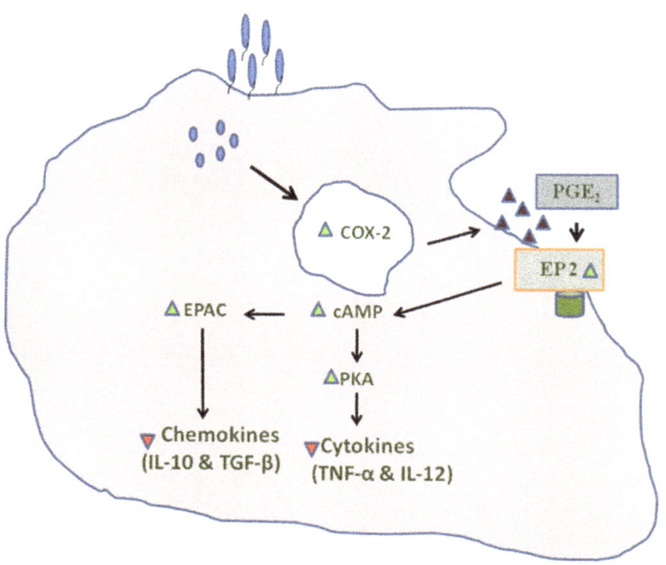

Figure 6.2 Immunomodulatory mechanism of PGE_2. Upon infecting macrophages, *Leishmania donovani* up-regulates COX-2 expression and its product PGE_2. *L. donovani* activates the G-protein coupled receptor EP2. PGE_2 then exerts its action by binding to EP2 leading to up-regulation of intracellular cAMP production. cAMP downstream effector molecules, PKA and EPAC, differentially modulates pro-inflammatory cytokines and chemokines generation respectively, which in turn augments *Leishmania* growth.

activation are prohibited in SOCS1 overexpressing cells (Nakagawa *et al.*, 2002; Kinjyo *et al.*, 2002). *Leishmania major* LPG induces the production of suppressors of the cytokine signalling family proteins, SOCS-1 and SOCS-3, whose activation are associated with diminished cytokine production and TLR4 signalling (de Veer *et al.*, 2003). Our report suggested that *L. donovani* employs host SOCS proteins to subvert macrophage apoptotic machinery and microbicidal responses, thereby creating safe, replicative niche inside the host macrophages (Srivastav *et al.*, 2014).

IL-1 receptor-associated kinase M

IL-1 receptor-associated kinase M (IRAK-M), a negative regulator of TLR signalling, down-regulates macrophage pro-inflammatory response (Kobayashi *et al.*, 2002). Our data revealed a sharp decline in IRAK1 and IRAK4 phosphorylation along with markedly reduced association of IRAK1–TNF receptor-associated factor 6, which is mandatory for TLR activation. IRAK-M induction coincided with increased stimulation of TGF-β, a hallmark cytokine of visceral infection. TGF-β-dependent signalling-mediated induction of SMAD family of proteins, 2, 3, and 4 played important roles in transcriptional up-regulation of IRAK-M. In infected macrophages, silencing of IRAK-M displayed enhanced IRAK1and IRAK4 phosphorylation with a concomitant increase in downstream NF-κB activity and reduced parasite survival (Srivastav *et al.*, 2015).

Polarization of inflammatory response

Cytokines

Pro- and anti-inflammatory cytokines play a major role and their balance decides the fate of the clinical outcome of leishmaniasis (Kar *et al.*, 2010). Pro-inflammatory cytokines are necessary for a protective response, but may also cause excessive inflammation and tissue damage. Anti-inflammatory cytokines counteract the effects of pro-inflammatory cytokines to limit the inflammations. *Leishmania* exploits this inflammatory balance of its host for disease progression.

Pro-inflammatory response

Th1 cytokines

Leishmania infection inhibits the production of pro-inflammatory Th1 cytokines, primarily IL-1β, TNFα, or IL-12. IL-12 enhances the cytotoxic response, a function that remains suppressed in patients with visceral leishmaniasis (Ribeiro-de-Jesus *et al.*, 1998). IL-10 nullifies the effect of IL-12 production in inducing IFN-γ production in *L. chagasi*-stimulated PBMC from visceral leishmaniasis patients (Bacellar *et al.*, 2000). Secretion of IL-1 upon LPS stimulation is inhibited in macrophages following infection with either *L. donovani* or *L. major* (Hawn *et al.*, 2002). TNF-α also gets down-regulated upon *Leishmania* infection but this may be an indirect effect resulting from induction of immunosuppressive cytokine IL-10 (Mosser and Edwards 2008). The expression of anti-inflammatory cytokine IL-10 gets up-regulated upon induction of various negative regulators such as PGE_2, by *Leishmania*. IL-12 plays a crucial role in the regulation of cellular immune responses. A growing body of evidence supports the host-protective role of IL-12-induced differentiation of Th1 cells that produce IFN-γ, thereby triggering NO production in macrophages (Ukil *et al.*, 2005),(Diefenbach *et al.*, 1999). The phosphoglycan moiety of LPG plays an important role in IL-12 repression. Several studies reported the inhibition of IL-12 production by *L. donovani, L. major* or *L. mexicana*. Moreover, phagocytic uptake alone can contribute significantly to suppression of LPS-induced IL-12p70 during *L. major* infection (Lapara and Kelly, 2010).

Th17 cytokines

Recent interest has focused on the inflammatory Th17 cytokine IL-17, a key cytokine reported to be involved in a variety of immune responses, including the induction of other cytokines (Chaudhry *et al.*, 2013). Both IL-17 and IL-22 are reported to be associated with protection against human kala-azar caused by *Leishmania donovani* (Pitta *et al.*, 2009). One of our recent studies documented that experimental treatment of *L. donovani*-infected mice with curdlan strongly induces Th17 and Th1 cytokines production. We hypothesized that IL-17 and IL-23 might function symbiotically in natural protection against visceral leishmaniasis (Ghosh *et al.*, 2013). We also reported that *L. donovani* suppresses pro-inflammatory IL-17 production by exploiting macrophage PGE2 (Saha *et al.*, 2014). A study by Lapara *et al.* showed that in presence of LPS, IL-17 production was suppressed to varying degrees by both *L. major* and *L. mexicana* (Lapara and Kelly 2010).

Anti-inflammatory response

Anti-inflammatory cytokines fosters parasite survival. This is because these cytokines are responsible for down-regulating host mediated pro-inflammatory and oxidative microbicidal molecule generation. *Leishmania* infection up-regulates the production of TGF-β and IL-10 (Bhattacharya *et al.*, 2016). IL-10 promotes human VL, by inhibition of Th1 cell type responses and/or deactivating parasitized tissue macrophages (Murray *et al.*, 2005). IL-10 decreases the production of IL-17 by cells from cutaneous and mucocutaneous leishmaniasis patients (Gonzalez-Lombana *et al.*, 2013). *Leishmania*-induced IL-10 was found to be mediated by cAMP effector molecule, PKA. TGF-β is another anti-inflammatory cytokine that is able to regulate the inflammatory immune response by suppressing the differentiation of $CD4^+$ effector cells, promoting the conversion of naiveT-cells into Treg cells (Wan and Flavell 2007), thereby inhibiting the production of IL-2 and IFN-γ and suppressing the activity of macrophages, dendritic cells and natural killer cells (Yoshimura *et al.*, 2010). Our study reported that *Leishmania* infection up-regulates the phosphorylation of both the TGF-β RI and II subunits and administration of TGF-β-specific antibody abrogated infection induced expression IRAK-M, a negative regulatory molecule (Srivastav *et al.*, 2015).

Chemokines

Cell recruitment to the site of infection is essential for the development of host cellular immune response, the process is controlled by chemotactic cytokines or chemokines. In response to *Leishmania* infection, cytokines and chemokines act synergistically to eliminate parasite infection (Teixeira *et al.*, 2006). Visceral leishmaniasis has been found to be associated with rapid changes in chemokine expression in both humans and mice, several studies suggested that infection with *Leishmania* induces the expression of a number of chemokine genes in the host (Racoosin and Beverley, 1997; Ritter and Körner, 2002) and this could possibly be beneficial to the parasite through recruitment of host cells it can infect, survive in and proliferate. Interestingly, *L. donovani*-infected mice that lack MIP-1α (or CCL3) demonstrated low Ag-specific IFN-γ production during the early phases of infection (Sato *et al.*, 1999). Administration of C57BL/6 mice with Met RANTES or anti-CCL5, rendered mice more susceptible to *Leishmania* by skewing the immune response towards Th2 phenotype (Santiago *et al.*, 2004). However, little is known about the signalling cascades that lead to altered expression of these chemokines upon *Leishmania* infection. CCL2 and MIP-1α can control the intracellular growth and parasite burden of *L. donovani* via a nitric oxide-dependent regulatory mechanism (Bhattacharyya *et al.*, 2002). Our study suggested *L. donovani* reduces the production of inflammatory chemokines, CCL3 and CCL5, and this may possibly be mediated by PGE_2/EP2/cAMP and its effector molecule, exchange protein directly activated by cAMP (EPAC) (Saha *et al.*, 2014) (Fig. 6.2). In *Leishmania*-infected mice, the ability of DCs to transport the parasites to the draining lymph nodes depends on the expression of chemokine receptors, CCR2 and CCR7. *L. donovani* infection resulted in the down-regulation of CCR7 expression, impairing DC migration and thereby contributing to disease progression (Schmidt *et al.*, 2012). Adoptive transfer of CCR7 expressing dendritic cells during treatment of infected mice significantly reduced parasite burdens.

Protecting own niche: prevention of host cell apoptosis

The ultimate way by which the host macrophages tries to defend themselves against invading *Leishmania* pathogens is induction of apoptosis (Donovan *et al.*, 2009). The intelligent parasites, on the other side, relentlessly try to defeat the host defence systems and pursues various mechanisms to inhibit host cell apoptosis in order to proliferate. Apoptosis, induced by cytotoxic immune effector cells serve as a defence mechanism against intracellular pathogens (Jorgensen *et al.*, 2017). A recent study reported the involvement of PI3K/AKT in the prevention of host cell apoptosis during *L. amazonensis* and *L. major* infection (Ruhland *et al.*, 2007). The activation of PI3K by promastigotes may repress an initial apoptotic response triggered by parasite uptake. After establishment of infection and differentiation from promastigote to amastigote form, *Leishmania* evades macrophage apoptosis by inhibiting the release of cytochrome c. A report stated that *Leishmania*-infected cells exhibited increased levels of phospho-Bad, suggesting that release of cytochrome c may be governed by the Bcl-2 family of proteins (Carmen and Sinai 2007). Our study revealed that *L. donovani* up-regulated MCL-1 expression through signalling pathways involving the transcription factor CREB. Infection-induced MCL-1 maintained the mitochondrial membrane integrity by sequestering BAK thereby preventing its oligomerization and subsequent release of apoptogenic proteins. Knockdown of MCL-1 in the spleen of infected mice showed increased induction of splenocyte apoptosis with simultaneous reduction in parasite burden. Therefore, MCl-1 played a crucial function in regulating *L. donovani* mediated inhibition of apoptotic machinery of the host cell (Giri *et al.*, 2016). Infection with *L. donovani* was found to induce SOCS (suppressors of cytokine signalling) proteins, inhibiting ROS-dependent macrophage apoptosis (Srivastav *et al.*, 2014). Our recent report stated that *L. donovani* induces AKT activation resulting in modulation of GSK-3β/β-catenin/FOXO-1 signalling pathway, thereby causing inhibition of both host cell apoptosis and immune response, essential for intra-macrophage survival of the parasite (Gupta *et al.*, 2016).

Conclusion

Leishmania, being an obligate intracellular parasite, has evolved various strategies to successfully survive within the harsh environment of phagocytes and typically produces long-lasting chronic infection. After entering into a well-equipped host cell, *Leishmania* must bypass host defensive arsenals at different stages of infection. In the early stages of infection, these parasites must avoid inappropriate activation of their host cell or microbicidal molecules. Next, parasites have to reduce pro-inflammatory cytokines or chemokines response and at the same time required to generate an anti-inflammatory milieu that is rich with immunologic activity. Lastly, the parasite have to protect its niche for proliferation. This review has addressed how *Leishmania* achieves its goal by exploiting the signalling mechanisms of the host cells in its own favour. Such strategies to manipulate the host immune response may be true for various intracellular pathogens and by no means unique to *Leishmania*. This review has covered some of the evasion mechanisms employed by *Leishmania*, which would boost further studies in knowing the unidentified strategies that any intracellular parasite employs in subverting host immune system. The knowledge of the signalling molecules present in the signalling pathway may open up new perspectives on potential molecular pathways that are prone to pharmacological manipulation not only during leishmaniasis but also for macrophage-associated disease, in general.

References

Abu-Dayyeh, I., Shio, M.T., Sato, S., Akira, S., Cousineau, B., and Olivier, M. (2008). *Leishmania*-induced IRAK-1 inactivation is mediated by SHP-1 interacting with an evolutionarily conserved KTIM motif. PLoS Negl. Trop. Dis. *2*, e305. https://doi.org/10.1371/journal.pntd.0000305.

Akira, S., and Takeda, K. (2004). Toll-like receptor signalling. Nat. Rev. Immunol. *4*, 499–511. https://doi.org/10.1038/nri1391.

Alexander, W.S. (2002). Suppressors of cytokine signalling (SOCS) in the immune system. Nat. Rev. Immunol. *2*, 410–416. https://doi.org/10.1038/nri818.

Anstead, G.M., Chandrasekar, B., Zhao, W., Yang, J., Perez, L.E., and Melby, P.C. (2001). Malnutrition alters the innate immune response and increases early visceralization following *Leishmania donovani* infection. Infect. Immun. *69*, 4709–4718. https://doi.org/10.1128/IAI.69.8.4709-4718.2001.

Ariyanayagam, M.R., and Fairlamb, A.H. (2001). Ovothiol and trypanothione as antioxidants in trypanosomatids. Mol. Biochem. Parasitol. *115*, 189–198. https://doi.org/10.1016/S0166-6851(01)00285-7.

Awasthi, A., Mathur, R.K., and Saha, B. (2004). Immune response to *Leishmania* infection. Indian J. Med. Res. *119*, 238–258.

Bacellar, O., D'Oliveira, A., Jr., Jeronimo, S., and Carvalho, E.M. (2000). IL-10 and IL-12 are the main regulatory cytokines in visceral leishmaniasis. Cytokine*12*, 1228–1231. https://doi.org/10.1006/cyto.2000.0694.

Barr, S.D., and Gedamu, L. (2003). Role of peroxidoxins in *Leishmania chagasi* survival. Evidence of an enzymatic defense against nitrosative stress. J. Biol. Chem. *278*, 10816–10823. https://doi.org/10.1074/jbc.M212990200.

Basu, Ball, W., Kar, S., Mukherjee, M., Chande, A.G., Mukhopadhyaya, R., and Das, P.K. (2011). Uncoupling protein 2 negatively regulates mitochondrial reactive oxygen species generation and induces phosphatase-mediated anti-inflammatory response in experimental visceral leishmaniasis. J. Immunol. *187*, 1322–1332. https://doi.org/10.4049/jimmunol.1004237.

Becker, I., Salaiza, N., Aguirre, M., Delgado, J., Carrillo-Carrasco, N., Kobeh, L.G., Ruiz, A., Cervantes, R., Torres, A.P., Cabrera, N., *et al.* (2003). *Leishmania* lipophosphoglycan (LPG) activates NK cells through toll-like receptor-2. Mol. Biochem. Parasitol. *130*, 65–74. https://doi.org/10.1016/S0166-6851(03)00160-9.

Bhattacharya, P., Ghosh, S., Ejazi, S.A., Rahaman, M., Pandey, K., Ravi Das, V.N., Das, P., Goswami, R.P., Saha, B., and Ali, N. (2016). Induction of IL-10 and TGFbeta from CD4+CD25+FoxP3+ T Cells Correlates with Parasite Load in Indian Kala-azar Patients Infected with *Leishmania donovani*. PLoS Negl. Trop. Dis. *10*, e0004422. https://doi.org/10.1371/journal.pntd.0004422.

Bhattacharyya, S., Ghosh, S., Dasgupta, B., Mazumder, D., Roy, S., and Majumdar, S. (2002). Chemokine-induced leishmanicidal activity in murine macrophages via the generation of nitric oxide. J. Infect. Dis. *185*, 1704–1708. https://doi.org/10.1086/340820.

Calegari-Silva, T.C., Vivarini, A.C., Miqueline, M., Dos Santos, G.R., Teixeira, K.L., Saliba, A.M., Nunes de Carvalho, S., de Carvalho, L., and Lopes, U.G. (2015). The human parasite *Leishmania amazonensis* downregulates iNOS expression via NF-kappaB p50/p50 homodimer: role of the PI3K/Akt pathway. Open Biol. *5*, 150118. https://doi.org/10.1098/rsob.150118.

Carmen, J.C., and Sinai, A.P. (2007). Suicide prevention: disruption of apoptotic pathways by protozoan parasites. Mol. Microbiol. *64*, 904–916. https://doi.org/10.1111/j.1365-2958.2007.05714.x.

Carneiro, P.P., Conceicao, J., Macedo, M., Magalhaes, V., Carvalho, E.M., and Bacellar, O. (2016). The role of nitric oxide and reactive oxygen species in the killing of *Leishmania braziliensis* by monocytes from patients with cutaneous leishmaniasis. PLoS One *11*, e0148084. https://doi.org/10.1371/journal.pone.0148084.

Castro, H., Sousa, C., Santos, M., Cordeiro-da-Silva, A., Flohe, L., and Tomas, A.M. (2002). Complementary antioxidant defense by cytoplasmic and mitochondrial peroxiredoxins in *Leishmania infantum*. Free Radic. Biol. Med. *33*, 1552–1562. https://doi.org/10.1016/S0891-5849(02)01089-4.

Chakraborty, D., Banerjee, S., Sen, A., Banerjee, K.K., Das, P., and Roy, S. (2005). *Leishmania donovani* affects antigen presentation of macrophage by disrupting lipid rafts. J. Immunol. 175, 3214–3224. https://doi.org/10.4049/jimmunol.175.5.3214.

Chakraborty, S., Srivastava, A., Jha, M.K., Nair, A., Pandey, S.P., Srivastava, N., Kumari, S., Singh, S., Krishnasastry, M.V., and Saha, B. (2015). Inhibition of CD40-induced N-Ras activation reduces leishmania *major* infection. J. Immunol. *194*, 3852–3860. https://doi.org/10.4049/jimmunol.1401996.

Chan, J., Fujiwara, T., Brennan, P., McNeil, M., Turco, S.J., Sibille, J.C., Snapper, M., Aisen, P., and Bloom, B.R. (1989). Microbial glycolipids: possible virulence factors that scavenge oxygen radicals. Proc. Natl. Acad. Sci. U.S.A. *86*, 2453–2457.

Channon, J.Y., and Blackwell, J.M. (1985). A study of the sensitivity of *Leishmania donovani* promastigotes and amastigotes to hydrogen peroxide. II. Possible mechanisms involved in protective H2O2 scavenging. Parasitology *91*, 207–217.

Chaudhry, H., Zhou, J., Zhong, Y., Ali, M.M., McGuire, F., Nagarkatti, P.S., and Nagarkatti, M. (2013). Role of cytokines as a double-edged sword in sepsis. *In vivo 27*, 669–684.

De, Souza Leao, S., Lang, T., Prina, E., Hellio, R., and Antoine, J.C. (1995). Intracellular *Leishmania amazonensis* amastigotes internalize and degrade MHC class II molecules of their host cells. J. Cell. Sci. *108*, 3219–3231.

de Veer, M.J., Curtis, J.M., Baldwin, T.M., DiDonato, J.A., Sexton, A., McConville, M.J., Handman, E., and Schofield, L. (2003). MyD88 is essential for clearance of *Leishmania major*: possible role for lipophosphoglycan and Toll-like receptor 2 signaling. Eur. J. Immunol. *33*, 2822–2831. https://doi.org/10.1002/eji.200324128.

Debus, A., Glasner, J., Rollinghoff, M., and Gessner, A. (2003). High levels of susceptibility and T helper 2 response in MyD88-deficient mice infected with *Leishmania major* are interleukin-4 dependent. Infect. Immun. *71*, 7215–7218. https://doi.org/10.1128/IAI.71.12.7215–7218.2003.

Dermine, J.F., Goyette, G., Houde, M., Turco, S.J., and Desjardins, M. (2005). *Leishmania donovani* lipophosphoglycan disrupts phagosome microdomains in J774 macrophages. Cell. Microbiol. *7*, 1263–1270. https://doi.org/10.1111/j.1462-5822.2005.00550.x.

Dermine, J.F., Scianimanico, S., Prive, C., Descoteaux, A., and Desjardins, M. (2000). *Leishmania* promastigotes require lipophosphoglycan to actively modulate the fusion properties of phagosomes at an early step of phagocytosis. Cell. Microbiol. *2*, 115–126. https://doi.org/10.1046/j.1462-5822.2000.00037.x.

Diefenbach, A., Schindler, H., Rollinghoff, M., Yokoyama, W.M., and Bogdan, C. (1999). Requirement for type 2 NO synthase for IL-12 signaling in innate immunity. Science*284*, 951–955. https://doi.org/10.1126/science.284.5416.951.

Donovan, M.J., Maciuba, B.Z., Mahan, C.E., and McDowell, M.A. (2009). *Leishmania* infection inhibits cycloheximide-induced macrophage apoptosis in a strain-dependent manner. Exp. Parasitol. *123*, 58–64. https://doi.org/10.1016/j.exppara.2009.05.012.

Faria, M.S., Reis, F.C., and Lima, A.P. (2012). Toll-like receptors in leishmania infections: guardians or promoters? J. Parasitol Res. *2012*, 930257. https://doi.org/10.1155/2012/930257.

Forget, G., Gregory, D.J., Whitcombe, L.A., and Olivier, M. (2006). Role of host protein tyrosine phosphatase SHP-1 in *Leishmania donovani*-induced inhibition of nitric oxide production. Infect. Immun. *74*, 6272–6279. https://doi.org/10.1128/IAI.00853-05.

Forget, G., Matte, C., Siminovitch, K.A., Rivest, S., Pouliot, P., and Olivier, M. (2005). Regulation of the *Leishmania*-induced innate inflammatory response by the protein tyrosine phosphatase SHP-1. Eur. J. Immunol. 35, 1906–1917. https://doi.org/10.1002/eji.200526037.

Fujimoto, M., and Naka, T. (2010). SOCS1, a negative regulator of cytokine signals and tlr responses, in human liver diseases. Gastroentero.l Res. Pract. *2010*. https://dx.doi.org/10.1155/2010/470468.

Gantt, K.R., Goldman, T.L., McCormick, M.L., Miller, M.A., Jeronimo, S.M., Nascimento, E.T., Britigan, B.E., and Wilson, M.E. (2001). Oxidative responses of human and murine macrophages during phagocytosis of *Leishmania chagasi*. J. Immunol. 167, 893–901. https://doi.org/10.4049/jimmunol.167.2.893.

Ghosh, K., Sharma, G., Saha, A., Kar, S., Das, P.K., and Ukil, A. (2013). Successful therapy of visceral leishmaniasis with curdlan involves T-helper 17 cytokines. J. Infect. Dis. *207*, 1016–1025. https://doi.org/10.1093/infdis/jis771.

Giri, J., Srivastav, S., Basu, M., Palit, S., Gupta, P., and Ukil, A. (2016). *Leishmania donovani* exploits myeloid cell leukemia 1 (MCL-1) protein to prevent mitochondria-dependent host cell apoptosis. J. Biol. Chem. *291*, 3496–3507. https://doi.org/10.1074/jbc.M115.672873.

Gonzalez-Lombana, C., Gimblet, C., Bacellar, O., Oliveira, W.W., Passos, S., Carvalho, L.P., Goldschmidt, M., Carvalho, E.M., and Scott, P. (2013). IL-17 mediates immunopathology in the absence of IL-10 following *Leishmania major* infection. PLoS Pathog. *9*, e1003243. https://doi.org/10.1371/journal.ppat.1003243.

Gupta, G., Oghumu, S., and Satoskar, A.R. (2013). Mechanisms of immune evasion in leishmaniasis. Adv. Appl. Microbiol. *82*, 155–184. https://doi.org/10.1016/B978-0-12-407679-2.00005-3.

Gupta, P., Giri, J., Srivastav, S., Chande, A.G., Mukhopadhyaya, R., Das, P.K., and Ukil, A. (2014). *Leishmania donovani* targets tumor necrosis factor receptor-associated factor (TRAF) 3 for impairing TLR4-mediated host response. FASEB J. *28*, 1756–1768. https://doi.org/10.1096/fj.13-238428.

Gupta, P., Srivastav, S., Saha, S., Das, P.K., and Ukil, A. (2016). *Leishmania donovani* inhibits macrophage apoptosis and pro-inflammatory response through AKT-mediated regulation of beta-catenin and FOXO-1. Cell Death Differ. *23*, 1815–1826. https://doi.org/10.1038/cdd.2016.101.

Hawn, T.R., Ozinsky, A., Underhill, D.M., Buckner, F.S., Akira, S., and Aderem, A. (2002). *Leishmania major* activates IL-1 alpha expression in macrophages through a MyD88-dependent pathway. Microbes Infect. *4*, 763–771. https://doi.org/10.1016/S1286-4579(02)01596-4.

Ilg, T. (2000). Lipophosphoglycan is not required for infection of macrophages or mice by *Leishmania mexicana*. EMBO J. 19, 1953–1962. https://doi.org/10.1093/emboj/19.9.1953.

Janssens, S., and Beyaert, R. (2003). Role of Toll-like receptors in pathogen recognition. Clin. Microbiol. Rev. *16*, 637–646. https://doi.org/10.1128/CMR.16.4.637-646.2003.

Jorgensen, I., Rayamajhi, M., and Miao, E.A. (2017). Programmed cell death as a defence against infection. Nat. Rev. Immunol. https://doi.org/10.1038/nri.2016.147.

Kar, S., Ukil, A., Sharma, G., and Das, P.K. (2010). MAPK-directed phosphatases preferentially regulate pro- and anti-inflammatory cytokines in experimental visceral leishmaniasis: involvement of distinct protein kinase C isoforms. J. Leukoc. Biol. *88*, 9–20. https://doi.org/10.1189/jlb.0909644.

Kima, P.E., Soong, L., Chicharro, C., Ruddle, N.H., and McMahon-Pratt, D. (1996). *Leishmania*-infected macrophages sequester endogenously synthesized parasite antigens from presentation to CD4$^+$ T-cells. Eur. J. Immunol. *26*, 3163–3169. https://doi.org/10.1002/eji.1830261249.

Kinjyo, I., Hanada, T., Inagaki-Ohara, K., Mori, H., Aki, D., Ohishi, M., Yoshida, H., Kubo, M., and Yoshimura, A. (2002). SOCS1/JAB is a negative regulator of LPS-induced macrophage activation. Immunity *17*, 583–591. https://doi.org/10.1016/S1074-7613(02)00446-6.

Kobayashi, K., Hernandez, L.D., Galan, J.E., Janeway, C.A., Jr., Medzhitov, R., and Flavell, R.A. (2002). IRAK-M is a negative regulator of Toll-like receptor signaling. Cell *110*, 191–202. https://doi.org/10.1016/S0092-8674(02)00827-9.

Lapara, N.J., 3rd, and Kelly, B.L. (2010). Suppression of LPS-induced inflammatory responses in macrophages infected with *Leishmania*. J. Inflamm. (Lond) *7*, 8. https://doi.org/10.1186/1476-9255-7-8.

Liu, D., and Uzonna, J.E. (2012). The early interaction of *Leishmania* with macrophages and dendritic cells and its influence on the host immune response. Front Cell Infect Microbiol. *2*, 83. https://doi.org/10.3389/fcimb.2012.00083.

Marcato, L.G., Ferlini, A.P., Bonfim, R.C., Ramos-Jorge, M.L., Ropert, C., Afonso, L.F., Vieira, L.Q., and Sobrinho, A.P. (2008). The role of Toll-like receptors 2 and 4 on reactive oxygen species and nitric oxide production by macrophage cells stimulated with root canal pathogens. Oral Microbiol. Immunol. *23*, 353–359. https://doi.org/10.1111/j.1399-302X.2008.00432.x.

Mathur, R.K., Awasthi, A., Wadhone, P., Ramanamurthy, B., and Saha, B. (2004). Reciprocal CD40 signals through p38MAPK and ERK-1/2 induce counteracting immune responses. Nat. Med. *10*, 540–544. https://doi.org/10.1038/nm1045.

Matte, C., Maion, G., Mourad, W., and Olivier, M. (2001). *Leishmania donovani*-induced macrophages cyclooxygenase-2 and prostaglandin E2 synthesis. Parasite Immunol. *23*, 177–184. https://doi.org/10.1046/j.1365-3024.2001.00372.x.

Miller, M.A., McGowan, S.E., Gantt, K.R., Champion, M., Novick, S.L., Andersen, K.A., Bacchi, C.J., Yarlett, N., Britigan, B.E., and Wilson, M.E. (2000). Inducible resistance to oxidant stress in the protozoan *Leishmania chagasi*. J. Biol. Chem. *275*, 33883–33889. https://doi.org/10.1074/jbc.M003671200.

Morato, C.I., da Silva, I.A., Jr., Borges, A.F., Dorta, M.L., Oliveira, M.A., Jancar, S., Serezani, C.H., and Ribeiro-Dias, F. (2014). Essential role of leukotriene B4 on *Leishmania (Viannia) braziliensis* killing by human macrophages. Microbes Infect. *16*, 945–953. https://doi.org/10.1016/j.micinf.2014.08.015.

Mosser, D.M., and Edwards, J.P. (2008). Exploring the full spectrum of macrophage activation. Nat. Rev. Immunol. *8*, 958–969. https://doi.org/10.1038/nri2448.

Mukherjee, B., Paul, J., Mukherjee, S., Mukhopadhyay, R., Das, S., Naskar, K., Sundar, S., Dujardin, J.C., Saha, B., and Roy, S. (2015). Antimony-Resistant *Leishmania donovani* Exploits miR-466i To Deactivate Host MyD88 for Regulating IL-10/IL-12 Levels during Early Hours of Infection. J. Immunol. *195*, 2731–2742. https://doi.org/10.4049/jimmunol.1402585.

Mukherjee, M., Basu Ball, W., and Das, P.K. (2014). *Leishmania donovani* activates SREBP2 to modulate macrophage membrane cholesterol and mitochondrial oxidants for establishment of infection. Int. J. Biochem. Cell Biol. *55*, 196–208. https://doi.org/10.1016/j.biocel.2014.08.019.

Murray, H.W., Flanders, K.C., Donaldson, D.D., Sypek, J.P., Gotwals, P.J., Liu, J., and Ma, X. (2005). Antagonizing deactivating cytokines to enhance host defense and chemotherapy in experimental visceral leishmaniasis. Infect. Immun. *73*, 3903–3911. https://doi.org/10.1128/IAI.73.7.3903-3911.2005.

Murray, H.W., and Nathan, C.F. (1999). Macrophage microbicidal mechanisms *in vivo*: reactive nitrogen versus oxygen intermediates in the killing of intracellular visceral *Leishmania donovani*. J. Exp. Med. *189*, 741–746.

Nakagawa, R., Naka, T., Tsutsui, H., Fujimoto, M., Kimura, A., Abe, T., Seki, E., Sato, S., Takeuchi, O., Takeda, K., *et al.* (2002). SOCS-1 participates in negative regulation of LPS responses. Immunity*17*, 677–687. https://dx.doi.org/10.1016/S1074-7613(02)00449-1.

Nandan, D., and Reiner, N.E. (1995). Attenuation of gamma interferon-induced tyrosine phosphorylation in mononuclear phagocytes infected with *Leishmania donovani*: selective inhibition of signaling through Janus kinases and Stat1. Infect. Immun. *63*, 4495–4500.

Oda, K., and Kitano, H. (2006). A comprehensive map of the toll-like receptor signaling network. Mol. Syst. Biol. *2*, 2006 0015. https://doi.org/10.1038/msb4100057.

Oghumu, S., and Satoskar, A.R. (2013). PI3K-gamma inhibitors in the therapeutic intervention of diseases caused by obligate intracellular pathogens. Commun. Integr. Biol. *6*, e23360. https://doi.org/10.4161/cib.23360.

Okuda, K., Tong, M., Dempsey, B., Moore, K.J., Gazzinelli, R.T., and Silverman, N. (2016). *Leishmania amazonensis* Engages CD36 to Drive Parasitophorous Vacuole Maturation. PLoS Pathog. *12*, e1005669. https://doi.org/10.1371/journal.ppat.1005669.

Paramchuk, W.J., Ismail, S.O., Bhatia, A., and Gedamu, L. (1997). Cloning, characterization and overexpression of two iron superoxide dismutase cDNAs from *Leishmania chagasi*: role in pathogenesis. Mol. Biochem. Parasitol. *90*, 203–221. https://doi.org/10.1016/S0166-6851(97)00141-2.

Perez-Santos, J.L., and Talamas-Rohana, P. (2001). *In vitro* indomethacin administration upregulates interleukin-12 production and polarizes the immune response towards a Th1 type in susceptible BALB/c mice infected with *Leishmania mexicana*. Parasite Immunol. *23*, 599–606. https://doi.org/10.1046/j.1365-3024.2001.00421.x.

Pitta, M.G., Romano, A., Cabantous, S., Henri, S., Hammad, A., Kouriba, B., Argiro, L., el Kheir, M., Bucheton, B., Mary, C., *et al.* (2009). IL-17 and IL-22 are associated with protection against human kala azar caused by *Leishmania donovani*. J. Clin. Invest. *119*, 2379–2387. https://doi.org/10.1172/JCI38813.

Polando, R., Dixit, U.G., Carter, C.R., Jones, B., Whitcomb, J.P., Ballhorn, W., Harintho, M., Jerde, C.L., Wilson, M.E., and McDowell, M.A. (2013). The roles of complement receptor 3 and Fcgamma receptors during *Leishmania* phagosome maturation. J. Leukoc. Biol. *93*, 921–932. https://doi.org/10.1189/jlb.0212086.

Racoosin, E.L., and Beverley, S.M. (1997). *Leishmania major*: promastigotes induce expression of a subset of chemokine genes in murine macrophages. Exp. Parasitol. *85*, 283–295. https://doi.org/10.1006/expr.1996.4139.

Reiner, N.E., Ng, W., and McMaster, W.R. (1987). Parasite-accessory cell interactions in murine leishmaniasis. II. *Leishmania donovani* suppresses macrophage expression of class I and class II major histocompatibility complex gene products. J. Immunol. *138*, 1926–1932.

Ribeiro-de-Jesus, A., Almeida, R.P., Lessa, H., Bacellar, O., and Carvalho, E.M. (1998). Cytokine profile and pathology in human leishmaniasis. Braz. J. Med. Biol. Res. *31*, 143–148.

Ricciotti, E., and FitzGerald, G.A. (2011). Prostaglandins and inflammation. Arterioscler. Thromb. Vasc. Biol. *31*, 986–1000. https://doi.org/10.1161/ATVBAHA.110.207449.

Ritter, U., and Körner, H. (2002). Divergent expression of inflammatory dermal chemokines in cutaneous leishmaniasis. Parasite Immunol. *24*, 295–301. https://doi.org/10.1046/j.1365-3024.2002.00467.x.

Rodriguez-Sosa, M., Monteforte, G.M., and Satoskar, A.R. (2001). Susceptibility to *Leishmania mexicana* infection is due to the inability to produce IL-12 rather than lack of IL-12 responsiveness. Immunol. Cell Biol. *79*, 320–322. https://doi.org/10.1046/j.1440-1711.2001.01014.x.

Romão, P.R., Fonseca, S.G., Hothersall, J.S., Noronha-Dutra, A.A., Ferreira, S.H., and Cunha, F.Q. (1999). Glutathione protects macrophages and *Leishmania major* against nitric oxide-mediated cytotoxicity. Parasitology *118*, 559–566.

Ruhland, A., and Kima, P.E. (2009). Activation of PI3K/Akt signaling has a dominant negative effect on IL-12 production by macrophages infected with *Leishmania amazonensis* promastigotes. Exp. Parasitol. *122*, 28–36. https://doi.org/10.1016/j.exppara.2008.12.010.

Ruhland, A., Leal, N., and Kima, P.E. (2007). *Leishmania* promastigotes activate PI3K/Akt signalling to confer host cell resistance to apoptosis. Cell. Microbiol. *9*, 84–96. https://doi.org/10.1111/j.1462-5822.2006.00769.x.

Saha, A., Biswas, A., Srivastav, S., Mukherjee, M., Das, P.K., and Ukil, A. (2014). Prostaglandin E2 negatively regulates the production of inflammatory cytokines/chemokines and IL-17 in visceral leishmaniasis. J. Immunol. *193*, 2330–2339. https://doi.org/10.4049/jimmunol.1400399.

Saha, B., Das, G., Vohra, H., Ganguly, N.K., and Mishra, G.C. (1995). Macrophage-T-cell interaction in experimental visceral leishmaniasis: failure to express costimulatory molecules on *Leishmania*-infected macrophages and its implication in the suppression of cell-mediated immunity. Eur. J. Immunol. *25*, 2492–2498. https://doi.org/10.1002/eji.1830250913.

Santiago, H.C., Oliveira, C.F., Santiago, L., Ferraz, F.O., de Souza, D.G., de-Freitas, L.A., Afonso, L.C., Teixeira, M.M., Gazzinelli, R.T., and Vieira, L.Q. (2004). Involvement of the chemokine RANTES (CCL5) in resistance to experimental infection with *Leishmania major*. Infect. Immun. *72*, 4918–4923. https://doi.org/10.1128/IAI.72.8.4918-4923.2004.

Sato, N., Kuziel, W.A., Melby, P.C., Reddick, R.L., Kostecki, V., Zhao, W., Maeda, N., Ahuja, S.K., and Ahuja, S.S.

(1999). Defects in the generation of IFN-gamma are overcome to control infection with *Leishmania donovani* in CC chemokine receptor (CCR) 5-, macrophage inflammatory protein-1 alpha-, or CCR2-deficient mice. J. Immunol. *163*, 5519–5525.

Schmidt, S.V., Nino-Castro, A.C., and Schultze, J.L. (2012). Regulatory dendritic cells: there is more than just immune activation. Front Immunol. 3, 274. https://doi.org/10.3389/fimmu.2012.00274.

Shio, M.T., Hassani, K., Isnard, A., Ralph, B., Contreras, I., Gomez, M.A., Abu-Dayyeh, I., and Olivier, M. (2012). Host cell signalling and leishmania mechanisms of evasion. J. Trop. Med. *2012*, 819512. https://doi.org/10.1155/2012/819512.

Shweash, M., Adrienne McGachy, H., Schroeder, J., Neamatallah, T., Bryant, C.E., Millington, O., Mottram, J.C., Alexander, J., and Plevin, R. (2011). *Leishmania mexicana* promastigotes inhibit macrophage IL-12 production via TLR-4 dependent COX-2, iNOS and arginase-1 expression. Mol. Immunol. *48*, 1800–1808. https://doi.org/10.1016/j.molimm.2011.05.013.

Srivastav, S., Basu Ball, W., Gupta, P., Giri, J., Ukil, A., and Das, P.K. (2014). *Leishmania donovani* prevents oxidative burst-mediated apoptosis of host macrophages through selective induction of suppressors of cytokine signaling (SOCS) proteins. J. Biol. Chem. *289*, 1092–1105. https://doi.org/10.1074/jbc.M113.496323.

Srivastav, S., Kar, S., Chande, A.G., Mukhopadhyaya, R., and Das, P.K. (2012). *Leishmania donovani* exploits host deubiquitinating enzyme A20, a negative regulator of TLR signaling, to subvert host immune response. J. Immunol. 189, 924–934. https://doi.org/10.4049/jimmunol.1102845.

Srivastav, S., Saha, A., Barua, J., Ukil, A., and Das, P.K. (2015). IRAK-M regulates the inhibition of TLR-mediated macrophage immune response during late in vitro *Leishmania donovani* infection. Eur. J. Immunol. 45, 2787–2797. https://doi.org/10.1002/eji.201445336.

Srivastava, N., Sudan, R., and Saha, B. (2011). CD40-modulated dual-specificity phosphatases MAPK phosphatase (MKP)–1 and MKP-3 reciprocally regulate *Leishmania major* infection. J. Immunol. 186, 5863–5872. https://doi.org/10.4049/jimmunol.1003957.

Teixeira, M.J., Teixeira, C.R., Andrade, B.B., Barral-Netto, M., and Barral, A. (2006). Chemokines in host–parasite interactions in leishmaniasis. Trends Parasitol. *22*, 32–40. https://doi.org/10.1016/j.pt.2005.11.010.

Tuladhar, R., Natarajan, G., and Satoskar, A.R. (2011). Role of co-stimulation in Leishmaniasis. Int. J. Biol. Sci. *7*, 1382–1390. https://doi.org/10.7150/ijbs.7.1382.

Tuon, F.F., Amato, V.S., Bacha, H.A., Almusawi, T., Duarte, M.I., and Amato Neto, V. (2008). Toll-like receptors and leishmaniasis. Infect. Immun. *76*, 866–872. https://doi.org/10.1128/IAI.01090-07.

Ukil, A., Biswas, A., Das, T., and Das, P.K. (2005). 18 Beta-glycyrrhetinic acid triggers curative Th1 response and nitric oxide up-regulation in experimental visceral leishmaniasis associated with the activation of NF-kappa B. J. Immunol. *175*, 1161–1169. https://doi.org/10.4049/jimmunol.175.2.1161.

Vazquez-Lopez, R., Argueta-Donohue, J., Wilkins-Rodriguez, A., Escalona-Montano, A., Aguirre-Garcia, M., and Gutierrez-Kobeh, L. (2015). *Leishmania mexicana* amastigotes inhibit p38 and JNK and activate PI3K/AKT: role in the inhibition of apoptosis of dendritic cells. Parasite Immunol. 37, 579–589. https://doi.org/10.1111/pim.12275.

Wan, Y.Y., and Flavell, R.A. (2007). 'Yin–Yang' functions of transforming growth factor-beta and T regulatory cells in immune regulation. Immunol. Rev. *220*, 199–213. https://doi.org/10.1111/j.1600-065X.2007.00565.x.

Yoshimura, A., Wakabayashi, Y., and Mori, T. (2010). Cellular and molecular basis for the regulation of inflammation by TGF-beta. J. Biochem. *147*, 781–792. https://doi.org/10.1093/jb/mvq043.

Ceramide in the Establishment of Visceral *Leishmaniasis*, an Insight into Membrane Architecture and Pathogenicity

7

Junaid Jibran Jawed, Shabina Parveen and Subrata Majumdar*

Division of Molecular Medicine, Bose Institute, Kolkata, India.

*Correspondence: subrata@jcbose.ac.in

https://doi.org/10.21775/9781910190715.07

Abstract

Biological membranes are the permeability barrier for the selective entry of biomolecules but sometimes they are responsible for providing platform for the entry and establishment of infection. Sphingomyelins are the chief component of the most biological membranes, ceramide being one of its important componenst. Generation of ceramide leads to increased susceptibility of the host towards parasite. Increased endogenous ceramide resulted into elevated expression of PKC ζ which causes down-regulation of essential component of MAPK pathways including p-38, JNK and on the other hand increases the phosphorylation of ERK1/2. Alternative activation of different MAPKs ultimately leads to reduced nuclear translocation of NFκB and AP-1 as a result of which the cell show increased synthesis of immunosuppressive cytokines such as TGF-β and IL-10 along with reduced NO generation, leading to severe host immune suppression. On the other hand ceramide also induced by *de novo* pathway which further shows increased membrane fluidity and hampered antigen presentation and reciprocal regulation of different T-cell subsets including T-reg and Th-17 cells. Collectively the roles of ceramide in infectious diseases is of great importance and can be a specific target for designing therapy against infection.

Introduction

Visceral leishmaniasis caused by protozoan parasite *Leishmania donovani* is an obligate intracellular pathogen that infects a large spectrum of mammalian host including human (Jawed *et al.*, 2016). The pathogen resides in the mononuclear phagocytic cell of the infected host and evokes a range of diverse mechanism including altered cytokines signalling by increasing the expression of anti-inflammatory cytokines IL-10, TGF-β and IL-4 and decreasing the expression of Th1 associated cytokines such as IL-12, TNFα and IFN-γ leading to the severe immune suppression of the host. During progression of the disease the antigen presentation ability of the macrophages gets hampered due to increased membrane fluidity, ultimately resulting in altered T-cell response (Hofmeyer *et al.*, 2016). *Leishmania donovani* causes reciprocal regulation of T-cell subsets which include T-reg and Th17 by increasing the proliferation of immunosuppressive T-reg cell whereas the population of Th-17 cell is reduced, resulting in more marked immunosuppression, which is one of the important hallmark of visceral leishmaniasis (Hofmeyer *et al.*, 2016; Jawed *et al.*, 2016). Entry and establishment of the infection are some critical events to get access into the host system bypassing several stages of immune surveillance and leading to the progression of disease (Majumdar *et al.*, 2014). Several component of membrane architecture play important role in providing entry to the invading pathogen making the doorstep for the entry of the parasite into the hostile environment of the macrophages (Ghosh *et al.*, 2002). Ceramide is one of the essential members of such component, which is utilized by the parasite for their entry and exploits the host signalling machinery for their proliferation and infection (Majumder *et al.*, 2012). Studies have revealed that ceramide plays a biphasic role in visceral leishmaniasis during the progression of the disease, which accounts for the increased PKC ζ expression and inhibition of reactive molecules generation such as NO; on the other hand, increased generation of ceramide accounts for membrane cholesterol depletion (Dey *et al.*, 2007; Majumder *et al.*, 2012). Therefore, in this report we have highlighted the importance of ceramide in visceral leishmaniasis and the signalling events which have been regulated during the course of infection. Detailed study of the ceramide exploitation by the parasite apart from being a lipid second messenger for the host cellular environment is necessary in order to understand the subdual of host immunity and to propose a better defence strategy.

Pathways of ceramide biosynthesis

Ceramide is synthesized through three different pathways: (1) the hydrolysis of sphingomyelin through sphingomyelinase, (2) the *de novo* pathway and (3) the salvage pathway for the synthesis of ceramide from constituent molecules. In the sphingomyelinase pathway, the hydrolysis of membrane sphingomyelin through specific sphingomyelinases generates ceramide (Haimovitz-Friedman *et al.*, 1994). The hydrolysis of the sphingomyelin can take place with several isoforms including acidic, neutral and alkaline sphingomyelinases. In case of *de novo* synthesis the simple molecules such as palmitate and serine is utilized in a multiple enzymatic steps of condensation, reduction and acylation which ultimately leads to the generation of ceramide by the involvement of three different enzymes serine-palmitoyl transferase, dihydro ceramide synthase and dihydroceramide desaturase (Hannum and Obeid, 2008). The salvage pathway involves the degradation of sphingolipids and glycosphingolipids in the subcellular acidic compartment, releasing sphingosine, which is ultimately converted into ceramide through the action of ceramide synthase (Kitatani *et al.*, 2008) (Fig. 7.1).

Ceramide an important component of inflammatory response

Sphingolipids are important components of cell membrane which is also involve of diverse cellular functions including cell growth, intracellular trafficking, and adhesion of cell, migration, senescence and inflammation. Among sphingolipids, ceramide is recognized as second messenger that plays an important role in signal transduction (Van Blitterswijk *et al.*, 2003). The relation between ceramide signalling and various inflammatory cytokines has been elucidated in various studies. Ceramide has been found to exert negative effect over TNFα expression whereas the inhibition of ceramide leads to the enhanced TNFα production (Józefowski *et al.*, 2010). In another study it was reported that TNFα and IL-1β treatment induce the generation of ceramide from sphingomyelin through the generation of reactive oxygen species (Singh *et al.*, 1998) suggesting a sequential role of ceramide in inflammation. It has been found that the application of ceramide analogue suppresses the expression of pro-inflammatory cytokines, which include IL-12 and TNFα, and elevates the expression of anti-inflammatory cytokine IL-10 in LPS-challenged mice (Avni *et al.*, 2009). Overall, this highlightis the significant role of ceramide in the regulation of pro-inflammatory cytokines, suppressing inflammation and promoting the expression of Th2-associated response.

Role of ceramide in infectious diseases

Ceramide and sphingomyelinases have found to play critical role in the internalization of various intracellular pathogens. Several viruses, bacteria and even the protozoan parasites infect the mammalian cells by exploiting the acidic and neutral sphingomyelinases to form ceramide which not only help to establish infection but also leads to the release of immunosuppressive cytokines and altered cellular signalling (Castro-Garza *et al.*, 2016; Sanada *et al.*, 2016). Acid sphingomyelinases also known as ASMase, an enzyme required in ceramide generation pathway has been reported to play important role in infectious disease such as *Staphylococcus*

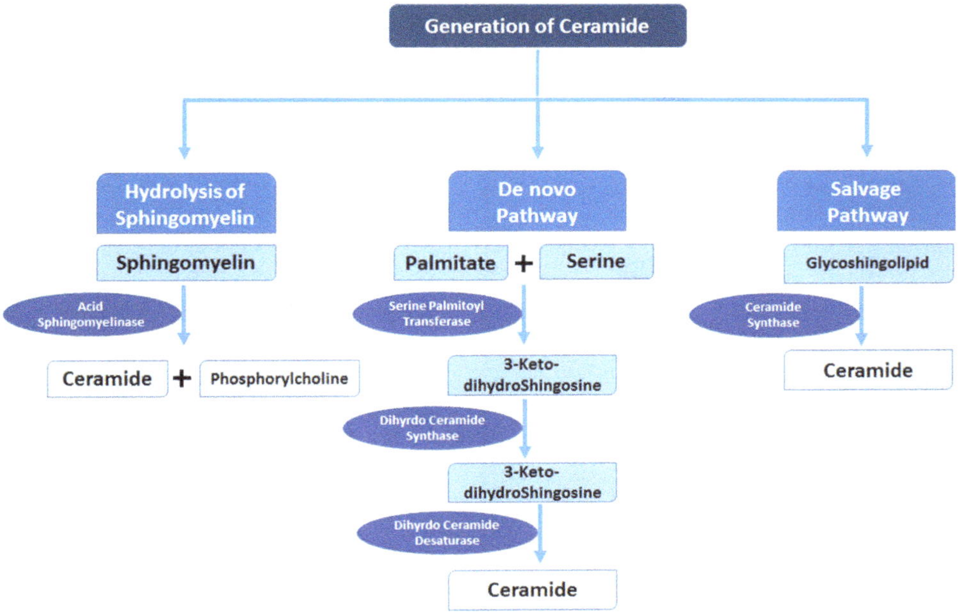

Figure 7.1 Pathways of ceramide generation. Ceramide generated through three different pathways. (1) Hydrolysis of membrane sphingomyelin by acid sphingomyelinase (ASMase) resulting in ceramide generation along with the formation of phosphoryl choline. (2) *De novo* pathway, where ceramide generation takes place through the utilization of palmitic acid and serine with the help of enzyme serine-palmitoyl transferase, dihydro ceramide synthase and dihydro ceramide desaturase which stepwise converts them into ceramide through the intermediates 3-keto-dihydrosphingosine and dihydrosphingosine. (3) Salvage pathway, where membrane glycosphingolipids are transformed to ceramide with the help of enzyme ceramide synthase.

aureus, Neisseria gonorrhoeae, Salmonella spp., *Pseudomonas aeruginosa, Listeria monocytogenes, E. coli* and several viruses including measles virus (Seitz *et al.*, 2015). Infection with *Trypanosoma cruzi*, an example of intracellular protozoan parasite and the causative agent of Chagas' disease have found to increase the expression of ASMase which is required for increased generation of ceramide and the entry of the pathogen, in-contrast the inhibition of ASMase reduces the susceptibility of the host towards invading pathogen by decreased endocytosis and membrane repair. Taken together it was found that ceramide plays a key role in the entry and establishment of various intracellular pathogens, the inhibition of which restores host protective immunity (Fernandes *et al.*, 2011).

Biphasic role of ceramide during visceral leishmaniasis

The interaction between *Leishmania donovani* and macrophages represents a classical example of the parasite-derived intellectual host evasion strategy which masks the host defence mechanism and is required for the initiation and the establishment of successful infection (Colineau *et al.*, 2017). It was reported that *Leishmania donovani* in the early phase of interaction with the host cell causes increase expression of acid sphingomyelinase or ASMase which regulates the generation of ceramide and subsequently causes increased expression of Ca^{2+}-independent PKCζ, ultimately leads to decrease translocation of transcription factors NFκB and AP-1 to the nucleus as a result the expression of pro-inflammatory cytokines such as IL-12, TNFα and IFN-γ decreases (Dey *et al.*, 2007).In the later phase of infection it was found that *Leishmania* activate host *de novo* pathway for the synthesis of ceramide which results into depletion of membrane cholesterol level. Depletion of membrane cholesterol hampers the rigidity of the membrane and lipid rafts thereby increasing the membrane fluidity. Due to increase in membrane fluidity the MHCs does not able to interact with subsequent T-cell; therefore, hampering antigen presentation and ultimately leading to the decreased T-cells mediated adaptive immune response (Majumder *et al.*, 2012) (Fig. 7.2). The inhibition of ceramide or intermediates of ceramide generation pathways attributed to the decrease susceptibility of the host towards pathogen and increased pro-inflammatory response (Majumder *et al.*, 2012), suggesting an alternative strategy to develop an effective therapy against infectious disease.

Ceramide-mediated impairment in antigen presentation

Defective antigen presentation is an important cause for the increased pathology and subversion of innate immunity of the host during visceral leishmaniasis. Parasite-mediated increased ceramide generation not only hampers macrophages' antigen presentation through increased membrane fluidity and cholesterol depletion but also results in the depletion of cd1d surface cluster from the lipid rafts, which is responsible for the activation of NKT-cell mediated activation of host defence mechanism. Impaired NKT-cell response can be corrected if the host cells are immunized with a lipid component of parasite glycosphingophospholipid (GSPL) (Karmakar *et al.*, 2011). GSPL treatment was associated with modification of lipid rafts structure improving the presentation of lipid antigens to the NKT-cells and restoring the subdual mechanism of pathogen through disrupted antigen presentation (Karmakar *et al.*, 2012). However, in another study it was reported that exosomes released from macrophages containing dead cell associated antigen act as a transmitter conveying antigen to the dendritic cell in a ceramide-dependent manner, which results in enhance T-cell response. On the other hand, the T-cell response was hampered by treatment with the sphingomyelinase inhibitor GW4869 (Xu *et al.*, 2016), suggesting a conditional response of ceramide in antigen presentation during apoptosis and infectious disease.

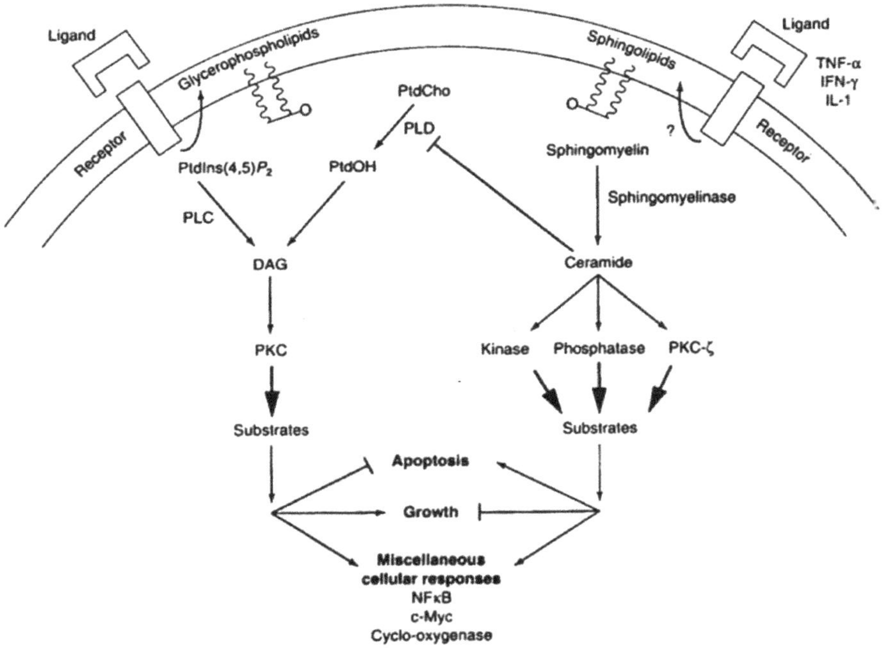

Figure 7.2 Activation of ceramide and its mechanism of action. Ceramide generation can be triggered by the action of various ligands including TNFα, IFN-γ and IL-1 which induced the conversion of membrane sphingomyelin to ceramide with the help of enzyme sphingomyelinase. Increased endogenous ceramide causes diverse cellular functions such as activation of PKCζ, various phosphatases and kinases and simultaneously it blocked the synthesis of di-acyl glycerol (DAG) from phosphatidylcholine (PldCho) by inhibiting the enzyme phospholipase D (PLD). The increased ceramide generation induce apoptosis and subsequently have negative impact on growth.

Ceramide mediated regulation of kinases and phosphatises

MAPKs consists a large family of serine/threonine kinases play an important role during cell signalling, host–pathogen interactions, cytokines response and induction of apoptosis and consist of ERK1/2, p-38 and JNK as important signalling regulators the functions of which are differentially regulated by components of sphingolipids such as ceramide (Chen *et al.*, 2008; Parveen *et al.*, 2016). Activation p-38 and JNK leads to increased expression of calcium-dependent PKC such as PKCα and β which induce nuclear translocation of transcription factors NFκB, resulting into increased expression of Th1 associated cytokines TNFα, IL-12 and IFN-γ, whereas activation of ERK1/2 reverse the Th1cytokines response and in turn leads to increased expression calcium independent PKCζ, restricting the translocation of essential transcription factors for Th1 cytokines such as AP-1 and NFκB and, therefore, resulting in increased anti-inflammatory cytokines synthesis (Parveen *et al.*, 2016). It has been reported that parasite entry results into down-regulation of the p-38 and JNK on the other hand it leads to increased expression of ERK1/2 phosphorylation (Khan *et al.*, 2014; Bandyopadhyay *et al.*, 2015; Parveen *et al.*, 2016) and the modulations of which are correlated with the increased ceramide generation (Gangoiti *et al.*, 2008). *Leishmania* induced generation of ceramide are not limited to MAPKs signalling intermediates only, it was also correlated with deactivation of calcium dependent PKC such as PKCα/β, increased expression of atypical PKC ζ, which ultimately results in decreased IL-12 expression leading to the complete deactivation of macrophages (Fig. 7.3). MAPKs are also implicated in inducing iNOS expression which is responsible for the increased NO generation ultimately reducing the parasite burden, On the other hand it has been shown that increased intracellular ceramide generation suppresses NO production through differential MAPKs pathway regulations (Ghosh *et al.*, 2002). IFN-γ treatment of BALB/c macrophages down-regulated the ceramide and resulted in increased production of NO. Another study showed that endogenous ceramide causes dephosphorylation of protein kinase B (Akt) through activation of protein phosphatase PP2A and simultaneous increase expression of PKC ζ which interact with Akt and inhibit its membrane translocation. Treatment with Fumonisin B1, an inhibitor for endogenous ceramide synthesis

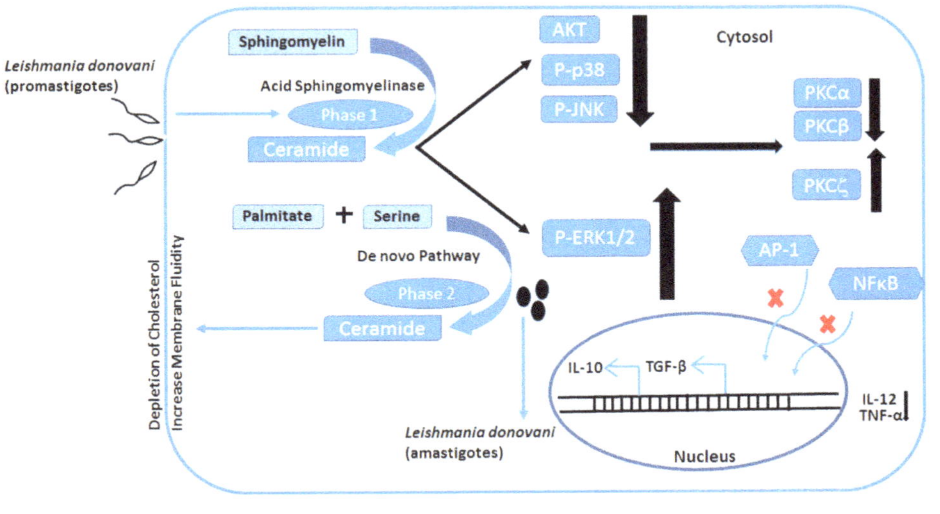

Figure 7.3 Biphasic role of ceramide during *Leishmania donovani* infection. *Leishmania donovani* exploit host ASMase for the generation of ceramide which in turn down-regulate Akt, p-38 and JNK, increased the expression of PKCζ and simultaneously blocked the nuclear translocation of NFκB and AP-1. As a result of inhibition of transcription factors, the expression of Th-1 cytokines such as IL-12 and TNFα is severely = reduced. In the second phase, *Leishmania* amastigotes causes increased endogenous ceramide synthesis through *de novo* pathway which leads to the depletion of membrane cholesterol and resulting into increased membrane fluidity ultimately hampering antigen presentation.

restore Akt membrane translocation through deactivation of protein phosphatase PP2A (Dey *et al.*, 2007), thus highlighting the important aspects of ceramide in perspective of *Leishmania donovani*-induced pathology. (Fig.7.2).

Modulation of T-cell subsets through ceramide generation during visceral leishmaniasis

Leishmania donovani infection playa s critical role in the modulation of T-cell subsets, which includes the population of immune-stimulatory Th-17 and immunosuppressive T-reg cells (Jawed *et al.*, 2016). Studies have reported that *Leishmania donovani* infection shows increased proliferation of T-reg cells on the other hand the population of Th-17 cells were severely reduced (Bhattacharya and Ali, 2013; Jawed *et al.*, 2016). TGF-β is a critical component of immune system which play important role during lineage commitment of T-cell subsets. Increased levels of TGF-β resulted into increased proliferation of T-reg cells whereas significantly low amount of the same is required for Th-17 generation (Zheng, 2013). It was found that treatment with TGF-β leads to increased generation of endogenous ceramide, the elevated production of which in turn resulted into severe down-regulation of pro-inflammatory cytokines and more synthesis of immunosuppressive molecules such as IL-10 and TGF-β (Kue *et al.*, 2013). On the other hand, it was found that C6-ceramide in combination with TGF-β leads to increase differentiation of T-reg cells (Kue *et al.*, 2013). Therefore, from the studies it can be suggested that ceramide plays a pivotal role in the regulation of T-cell subsets during infection and promotes the proliferation and lineage commitment of T-reg cells.

Conclusion

Altogether, ceramide was found to be a critical component of the immune system which has been exploited by the parasite for the entry and the establishment of infection through up-regulation of ASMase, which converts sphingomyelin to ceramide. Subsequently the host MAPKs pathway can be manipulated by increased expression of immunosuppressive cytokines IL-10 and TGF-β, decreased NO generation and altered T-cell response marking the complete deactivation of macrophages (Fig. 7.3). The inhibition of ceramide can be an important strategy to design a therapeutic approach to combat *Leishmania donovani* infection. Therefore, our report has highlighted the significant role of endogenous ceramide and its contribution in the infectious disease taking visceral leishmaniasis as a model and brings about important aspects of ceramide generation pathway which can be utilized to design future drugs targeting endogenous ceramide.

References

Avni, D., Goldsmith, M., Ernst, O., Mashiach, R., Tuntland, T., Meijler, M.M., Gray, N.S., Rosen, H., and Zor, T. (2009). Modulation of TNFalpha, IL-10 and IL-12p40 levels by a ceramide-1-phosphate analog, PCERA-1, *in vivo* and *ex vivo* in primary macrophages. Immunol. Lett. *123*, 1–8. https://doi.org/10.1016/j.imlet.2008.12.011.

Bandyopadhyay, S., Kar Mahapatra, S., Paul Chowdhury, B., Kumar Jha, M., Das, S., Halder, K., Bhattacharyya Majumdar, S., Saha, B., and Majumdar, S. (2015). Toll-like receptor 2 targeted rectification of impaired CD8⁺ T cell functions in experimental *Leishmania donovani* infection reinstates host protection. PLOS. One. 10, e0142800. https://dx.doi.org/10.1371/journal.pone.0142800.

Bhattacharya, P., and Ali, N. (2013). Involvement and interactions of different immune cells and their cytokines in human visceral leishmaniasis. Rev. Soc. Bras. Med. Trop. *46*, 128–134.

Castro-Garza, J., González-Salazar, F., Quinn, F.D., Karls, R.K., De La Garza-Salinas, L.H., Guzmán-de la Garza, F.J., and Vargas-Villarreal, J. (2016). An acidic sphingomyelinase Type C activity from Mycobacterium tuberculosis. Rev. Argent. Microbiol. *48*, 21–26. https://dx.doi.org/10.1016/j.ram.2016.01.001.

Chen, C.L., Lin, C.F., Chang, W.T., Huang, W.C., Teng, C.F., and Lin, Y.S. (2008). Ceramide induces p38 MAPK and JNK activation through a mechanism involving a thioredoxin-interacting protein-mediated pathway. Blood. *111*, 4365–4374. https://dx.doi.org/10.1182/blood-2007-08-106336.

Colineau, L., Clos, J., Moon, K.M., Foster, L.J., and Reiner, N.E. (2017). *Leishmania donovani* chaperonin 10 regulates parasite internalization and intracellular survival in human macrophages. Med. Microbiol. Immunol. *206*, 235–257. https://dx.doi.org/10.1007/s00430-017-0500-7.

Dey, R., Majumder, N., Bhattacharjee, S., Majumdar, S.B., Banerjee, R., Ganguly, S., Das, P., and Majumdar, S. (2007). *Leishmania donovani*-induced ceramide as the key mediator of Akt dephosphorylation in murine macrophages: role of protein kinase CZeta and phosphatase. Infect. Immun. *75*, 2136–2142.

Fernandes, M.C., Cortez, M., Flannery, A.R., Tam, C., Mortara, R.A., and Andrews, N.W. (2011). Trypanosoma cruzi subverts the sphingomyelinase-mediated plasma membrane repair pathway for cell invasion. J. Exp. Med. *208*, 909–921. https://doi.org/10.1084/jem.20102518.

Gangoiti, P., Granado, M.H., Wang, S.W., Kong, J.Y., Steinbrecher, U.P., and Gómez-Muñoz, A. (2008). Ceramide 1-phosphate stimulates macrophage proliferation through activation of the PI3-kinase/PKB, JNK and ERK1/2 pathways. Cell Signal. *20*, 726–736. https://dx.doi.org/10.1016/j.cellsig.2007.12.008.

Ghosh, S., Bhattacharyya, S., Sirkar, M., Sa, G.S., Das, T., Majumdar, D., Roy, S., and Majumdar, S. (2002). *Leishmania donovani* suppresses activated protein 1 and NF-kappa-B activation in host macrophages via ceramide generation: involvement of extracellular signal-regulated kinase. Infect. Immun. *70*, 6828–6838.

Haimovitz-Friedman, A., Kan, C.C., Ehleiter, D., Persaud, R.S., McLoughlin, M., Fuks, Z., and Kolesnick, R.N. (1994). Ionizing radiation acts on cellular membranes to generate ceramide and initiate apoptosis. J. Exp. Med. *180*, 525–535.

Hannun, Y.A., and Obeid, L.M. (2008). Principles of bioactive lipid signalling: lessons from sphingolipids. Nat. Rev. Mol. Cell Biol. *9*, 139–150. https://doi.org/10.1038/nrm2329.

Hofmeyer, K.A., Duthie, M.S., Laurance, J.D., Favila, M.A., Van Hoeven, N., Coler, R.N., and Reed, S.G. (2016). Optimizing Immunization Strategies for the Induction of Antigen-Specific CD4 and CD8 T-cell Responses for Protection against Intracellular Parasites. Clin. Vaccine Immunol. 3,785–794. https://dx.doi.org/10.1128/CVI.00251-16.

Jawed, J.J., Majumder, S., Bandyopadhyay, S., Biswas, S., Parveen, S., and Majumdar, S. (2016). SLA-PGN-primed dendritic cell-based vaccination induces Th17-mediated protective immunity against experimental visceral leishmaniasis: a crucial role of PKCβ. Pathog. Dis. *74*, 1–13. https://dx.doi.org/10.1093/femspd/ftw041.

Józefowski, S., Czerkies, M., Łukasik, A., Bielawska, A., Bielawski, J., Kwiatkowska, K., and Sobota, A. (2010). Ceramide and ceramide 1-phosphate are negative regulators of TNF-α production induced by lipopolysaccharide. J. Immunol. *185*, 6960–6973. https://doi.org/10.4049/jimmunol.0902926.

Karmakar, S., Bhaumik, S.K., Paul, J., and De, T. (2012). TLR4 and NKT-cell synergy in immunotherapy against visceral leishmaniasis. PLOS Pathog. *8*, e1002646. https://doi.org/10.1371/journal.ppat.1002646.

Karmakar, S., Paul, J., and De, T. (2011). *Leishmania donovani* glycosphingolipid facilitates antigen presentation by inducing relocation of CD1d into lipid rafts in infected macrophages. Eur. J. Immunol. *4*, 11376–11387. https://dx.doi.org/10.1002/eji.201040981.

Khan, T.H., Srivastava, N., Srivastava, A., Sareen, A., Mathur, R.K., Chande, A.G., Musti, K.V., Roy, S., Mukhopadhyaya, R., and Saha, B. (2014). SHP-1 plays a crucial role in CD40 signaling reciprocity. J. Immunol. *193*, 3644–3653. https://doi.org/10.4049/jimmunol.1400620.

Kitatani, K., Idkowiak-Baldys, J., and Hannun, Y.A. (2008). The sphingolipid salvage pathway in ceramide metabolism and signaling. Cell. Signal. *20*, 1010–1018. https://doi.org/10.1016/j.cellsig.2007.12.006.

Kue, C.S., Lim, H.X., Jung, M.Y., Hong, H.J., Cho, D., and Kim, T.S. (2013). C6-ceramide in combination with transforming growth factor-β enhances Treg cell differentiation and stable FoxP3 expression *in vitro* and *in vivo*. Immunobiology *218*, 952–959. https://doi.org/10.1016/j.imbio.2012.11.003.

Majumdar, S.B., Bhattacharya, P., Bhattacharjee, S., Majumder, S., Banerjee, S., and Majumdar, S. (2014). Toll like receptor 2 and CC chemokine receptor 5 cluster in the lipid raft enhances the susceptibility of *Leishmania donovani* infection in macrophages. Indian J. Exp. Biol. *52*, 17–29.

Majumder, S., Dey, R., Bhattacharjee, S., Rub, A., Gupta, G., Bhattacharyya Majumdar, S., Saha, B., and Majumdar, S. (2012). *Leishmania*-induced biphasic ceramide generation in macrophages is crucial for uptake and survival of the parasite. J. Infect. Dis. *205*, 1607–1616. https://doi.org/10.1093/infdis/jis229.

Parveen, S., Bandyopadhyay, S., Das, S., Majumdar, S.B., Jawed, J.J., Chowdhury, B.P., Saha, B., and Majumdar, S. (2016). Mycobacterium indicus pranii (Mw)-mediated protection against visceral leishmaniasis by reciprocal regulation of host dual-specificity phosphatases. Int. Immunol. *28*, 585–595. https://dx.doi.org/10.1093/intimm/dxw049.

Sanada, T., Hirata, Y., Naito, Y., Yamamoto, N., Kikkawa, Y., Ishida, Y., Yamasaki, C., Tateno, C., Ochiya, T., and Kohara, M. (2016). Transmission of HBV DNA Mediated by Ceramide-Triggered Extracellular Vesicles. Cell. Mol. Gastroenterol. Hepatol. *3*, 272–283. https://dx.doi.org/10.1016/j.jcmgh.2016.10.003.

Seitz, A.P., Grassmé, H., Edwards, M.J., Pewzner-Jung, Y., and Gulbins, E. (2015). Ceramide and sphingosine in pulmonary infections. Biol. Chem. *396*, 611–620. https://doi.org/10.1515/hsz-2014-0285.

Singh, I., Pahan, K., Khan, M., and Singh, A.K. (1998). Cytokine-mediated induction of ceramide production is redox-sensitive. Implications to proinflammatory cytokine-mediated apoptosis in demyelinating diseases. J. Biol. Chem. *273*, 20354–20362.

Van, Blitterswijk, W.J., van der Luit, A.H., Veldman, R.J., Verheij, M., and Borst, J. (2003). Ceramide: second messenger or modulator of membrane structure and dynamics? Biochem. J. *369*, 199–211. https://dx.doi.org/10.1042/BJ20021528.

Xu, Y., Liu, Y., Yang, C., Kang, L., Wang, M., Hu, J., He, H., Song, W., and Tang, H. (2016). Macrophages transfer antigens to dendritic cells by releasing exosomes containing dead-cell-associated antigens partially through a ceramide-dependent pathway to enhance CD4(+) T-cell responses. Immunology. *149*, 157–171. https://dx.doi.org/10.1111/imm.12630.

Zheng, S.G. (2013). Regulatory T-cells vs Th17: differentiation of Th17 versus Treg, are the mutually exclusive? Am. J. Clin. Exp. Immunol. 2, 94–106.

The Role of Haemproteins in Different Life Cycle Stages of *Leishmania*

8

Subhankar Dolai and Subrata Adak*

Division of Structural Biology and Bio-informatics, CSIR-Indian Institute of Chemical Biology, Kolkata, India.

*Correspondence: adaks@iicb.res.in

https://doi.org/10.21775/9781910190715.08

Abstract

Haemproteins are iron protoporphyrin IX prosthetic group/co-factor containing proteins that are widely distributed among prokaryotes and eukaryotes to carry out vital cellular functions, which are necessary for survival. To perpetuate infection and survive for successful disease establishment, *Leishmania* pathogens adopt different morphology through differentiation, manipulate cellular metabolism, programme cell signalling to thrive within hypoxic arthropod vector gut, employ antioxidant defence systems to evade reactive oxygen species that are generated as a result of cellular metabolism, and reactive nitrogen and oxygen species that are generated within activated vertebrate host macrophages. Genome sequencing reveals the presence of an array of haemproteins in *Leishmania* with hypothetical functions. However, being phylogenetically distant, the *Leishmania* haemproteins are distinct in primary structures and sometimes novel compared to their human host counterparts. In this chapter we have surmised the recent development in understanding of haemprotein functions in *Leishmania* and their physiological roles in parasite life cycle to better understand the pathogenesis, and the novel parasite survival strategies that could be exploited as possible therapeutic targets.

Introduction

Leishmania, the protozoan parasite of the trypanosomatidae family are the causative agents of the disease leishmaniasis. The digenetic life cycle of *Leishmania* alternates between motile flagellated extracellular promastigote in sandfly vector gut and an aflagellated non-motile amastigote within the phagolysosomes of vertebrate host macrophages (Chang, 1979; Chang and Dwyer, 1976; Killick-Kendrick *et al.*, 1977). In sandfly, amastigotes within consumed blood meal differentiate to procyclic promastigotes that undergo rapid multiplication and differentiate to non-dividing infective metacyclic promastigotes which are capable of avoiding host complement system upon infection (Bates, 2008; Cunningham *et al.*, 2001). Infected parasites are phagocytosed by macrophages (Späth *et al.*, 2003), where they further differentiate to amastigotes and replicate within phagolysosomes to establish the disease.

Leishmania encounter distinct environments within sandfly vector and vertebrate hosts. The sandfly gut is hypoxic to anoxic, a condition favourable to generate reactive oxygen species (ROS) in aerobic organisms such as *Leishmania* (Guzy *et al.*, 2007). The vector gut is enriched with proteases and glucose. Infected *Leishmania* in hosts are exposed to elevated temperature. The host macrophage phagolysosome is highly oxidized as a result of a large amount of ROS and reactive nitrogen species (RNS) (Bernstein *et al.*), which are generated by the activated macrophages as microbicidal host defence (Fang, 2004; Nathan and Shiloh, 2000). The energy sources in phagolysosome are amino acids and fatty acids (Burchmore and Barrett, 2001; Naderer and McConville, 2008). Successful completion of parasite life cycle, a prerequisite for mounting disease and its propagation therefore rely on parasite differentiation to suitable morphology (Sacks and Perkins, 1984; Turco and Sacks, 1991), survival through efficient evasion of ROS and RNS (Barr and Gedamu, 2003; Castro *et al.*, 2002), adaptation by adjusting cellular metabolism (McConville and Naderer, 2011; Opperdoes and Coombs, 2007) and regulation of proper protein expression level (Alcolea *et al.*, 2014; De Pablos *et al.*, 2016; Shapira *et al.*, 1988).

In both prokaryotes and eukaryotes, haemproteins are associated with almost all aspects of cell functions that include but are not limited to the storage and transport of oxygen molecule (O_2), mitochondrial respiration, catalysis of peroxides, catalysis of biosynthesis and metabolism of xenobiotics, synthesis and sensing gaseous signalling molecules and regulation of transcription and translation. These diverse functions of haemproteins stem from versatility of the haem groups and the variety of interactions with protein scaffolds that generate unique haem environments that enables haemproteins to interact and transfer electrons from numerous molecules. With the advent of the *Leishmania* (*L. major*, *L. infantum*, and *L. braziliensis*) genome sequencing, a more detailed understanding of its biology, genome organization, and protein inventory has been achieved (Ivens *et al.*, 2005; Peacock *et al.*, 2007). *Leishmania* genome reveals the presence of a large number of haem-containing proteins with hypothetical functions assigned by their domain motif signature. This chapter reviews the understanding of physiological functions of haemproteins in *Leishmania* with an emphasis on post genomic development.

Haem and haem containing proteins

The most abundant form of haem is haem b, which is composed of a porphyrin macrocycle that linked to four methyl groups, two vinyl groups, two propionate groups and an iron ion centrally chelated to four nitrogen atoms each from four pyrrole rings. Haem b serves as immediate template for the synthesis of other types of haems a, c, d and d1 that differ in their side group possession. The hydrophobic porphyrin ring, methyl and vinyl groups; and hydrophilic propionates and iron make haem amphipathic. Further, four coordinated iron in haem leaves two more coordination sites empty. Haem thus can establish a variety of interaction with numerous protein scaffolds (Table 8.1) via axial coordination positions available in iron, polar interactions, as well as hydrophobic interactions to serve as essential component to carry out protein functions in its haem centre or regulate protein functions. The numbers

Table 8.1 Haemproteins in eukaryotes and prokaryotes other than *Leishmania*

General function	Haemprotein	References
Gas carriers	Haemoglobin, myoglobin, neuroglobin, cytoglobin, nitrophorins	Burmester *et al.* (2000), Gros *et al.* (2010),; Montfort *et al.* (2000), Perutz (1979), Trent and Hargrove (2002)
Electron transporters	Cytochrome c, cytochrome c oxidase, cytochrome c oxidoreductase, cytochrome b5	Capaldi (1990), Crofts (2004), Jiang and Wang (2004), Schenkman and Jansson (2003)
Catalysts of biodegradation and biosynthesis	Catalases, peroxidases, cytochrome P450 superfamily, nitric oxide syntheses	Dunford (1999), Guengerich *et al.* (2016), Murad (1999)
Haem-based gas sensors	PAS family proteins (FixL, PAS2), CO-sensing transcription factor (CooA protein), soluble guanylate cyclases, soluble adenylate cyclase, histidine kinase	Derbyshire and Marletta (2012); Dioum *et al.* (2002), Gilles-Gonzalez (1991); Shelver *et al.* (1997)
Haem-regulated proteins	Bach1, δ–aminolevulinate synthase 1, haem-regulated eIF2α kinase, Rev-erb α, Rev-erb β, iron regulatory protein 2α, haem oxygenase-2α and BK channel, Stanniocalcin 1α, R-transferaseα, DGCR8α	Barr *et al.* (2011); Han *et al.* (2001); Hira *et al.* (2007); Hu *et al.* (2008); Kang *et al.* (2003); McCoubrey *et al.* (1997); Raghuram *et al.* (2007); Tang *et al.* (2003); Westberg *et al.* (2011); Yamamoto *et al.* (1988); Yin *et al.* (2007)

of haemproteins are numerous and therefore only explicitly characterized proteins are listed below (Table 8.1)

Leishmania need haem for survival

Eukaryotes synthesize haem through eight step enzymatic reactions distributed in mitochondria and cytosol. The first step, formation of δ-aminolaevulinic acid from glycine and succinyl co-A, and the final three steps involving synthesis of protoporphyrin IX, and insertion of iron atom in protoporphyrin IX take place in mitochondria. Haem is then transported to cytosol and linked to targeted haemproteins. Excess haem is decomposed to biliverdin by haem oxygenases to avoid cellular haem toxicity (Muñoz-Saánchez and Chánez-Cárdenas, 2014). The extracellular dependence on haem for *in vitro* growth of trypanosomatid protozoa are first discovered in *Blastocrithidiaculicis* and *Crithidiaoncopelti* (Chang and Trager, 1974). Later studies, including a study on *L. Mexicana amazonensis*, showed supplementation of haem or haem precursor protoporphyrin IX to support their growth (Chang and Chang, 1985; Chang *et al.*, 1975). Together it is found that the trypanosomatids including *Leishmania* are haem auxotrophs with the lack of complete haem biosynthetic pathway. Consistently *Leishmania* are found to encode only three enzymes that are required for last three steps of haem biosynthesis (Alves *et al.*, 2011). *Leishmania* therefore require to salvage haem or haem precursors from the environment for survival. A large numbers of haemproteins are present in *Leishmania* (Table 8.2) and some of them are characterized.

Source of haem and its acquisition systems in *Leishmania*

In sandfly vector, *Leishmania* promastigotes gain access to haem in the form of haemoglobin (Hb) from sandfly blood meal. In mammalian hosts, macrophage phagolysosomes and macrophage cytosol serve as major pool of haem and iron respectively that generated from Hb metabolism of damaged or dying RBCs, engulfed by erythrophagocytic macrophages (Soares and Hamza, 2016). RBCs are processed by releasing haem from Hb within phagolysosomes. Haem is then translocated to cytosol by the haem responsive gene-1 transporter (Rajagopal *et al.*, 2008). Iron is extracted from haem by haem oxygenase within cytosol (Gozzelino *et al.*, 2010), which is either exported from macrophages or stored intracellularly as inert iron (Fe^{3+}) in ferritin (Harrison and Arosio, 1996). *Leishmania* have been

Table 8.2 Putative haem domain-containing protein in *L. major* was collected from *Leishmania major* Gene DB (geneDB www.genedb.org/Homepage/Lmajor). Cyt, CYP, TB and TC denoted as cytochrome, Cytochrome p450, *Trypanosoma brucei* and *Trypanosoma cruzi*, respectively.

Putative proteins in *L. major*	Systemic ID	General function of haem domain	Orthologues and paralogues in other trypanosomatids	Function of protein in *Leishmania*
Hypothetical Cyt c	LmjF.35.0860	Electron transporters	TC and TB	Unknown
Haem peroxidase	LmjF.35.3460	Catalysis	TC and TB	Unknown
Ascorbate peroxidase	LmjF.34.0070	Catalysis	TC	Scavenge of H_2O_2 (Adak and Datta, 2005; Wilkinson *et al.*, 2002)
CYP	LmjF.34.3330	Catalysis	Unknown	Unknown
Cyt b5	LmjF.36.6670	Electron transporters	TC and TB	Unknown
Hypothetical Cyt b5	LmjF.02.0050	Electron transporters	TC	Unknown
Hypothetical Cyt b5	LmjF.06.1290	Electron transporters	Unknown	Unknown
Flavoprotein with Cyt b5	LmjF.07.0800	Electron transporters	Unknown	Unknown
Cyt b5 with desaturase	LmjF.07.1090	Electron transporters	Unknown	Delta-5 fatty acid desaturase (Tripodi *et al.*, 2006)
Cyt b5	LmjF.09.1490	Electron transporters	TC and TB	Unknown

Table 8.2 Continued

Putative proteins in *L. major*	Systemic ID	General function of haem domain	Orthologues and paralogues in other trypanosomatids	Function of protein in *Leishmania*
Cyt b5 with desaturase	LmjF.14.0510	Electron transporters	Unknown	Putative stearic acid desaturase
Cyt b5 with desaturase	LmjF.14.1340	Electron transporters	LmjF.17.1350	Delta-4 fatty acid desaturase (Tripodi *et al.*, 2006)
Cyt c	LmjF.16.1310	Electron transporters	TC and TB	Electron transporters and substrate for peroxidase (Jasion *et al.*, 2012)
Cyt b5	LmjF.17.1350	Electron transporters	TC	Unknown
Peroxidase	LmjF.21.1567	Catalysis	TC and TB	Pseudoperoxidase (Bose *et al.*, 2012)
Cyt b5 with desaturase	LmjF.24.2250	Electron transporters	TC and TB	Fatty-acid desaturase
Globin domain with adenylate cyclase	LmjF.28.0090	Haem-based gas sensors	Unknown	Oxygen-sensing adenylate cyclase (Sen Satara *et al.*, 2013)
Cyt b5	LmjF.29.0868	Electron transporters	TC and TB	Unknown
Cyt b5, putative	LmjF.31.1170	Electron transporters	Unknown	Unknown
Cyt b5, putative	LmjF.31.1190	Electron transporters	Unknown	Unknown
CYP	LmjF.30.3550	Catalysis	TC	Unknown
Cyt b5 with desaturase	LmjF.36.6950	Electron transporters	Unknown	Delta-6 fatty acid desaturase (Tripodi *et al.*, 2006)
Cyt b5	LmjF.21.1440	Electron transporters	TC and TB	Unknown
CYP	LmjF.27.0090	Catalysis	TC and TB	Unknown
Cyt b5	LmjF.03.0290	Electron transporters	Unknown	Pseudogene
Hypothetical Cyt b5	LmjF.31.1165	Electron transporters	Unknown	Unknown
Hypothetical Cyt b5	LmjF.31.1180	Electron transporters	Unknown	Unknown
Hypothetical Cyt b5	LmjF.31.1200	Electron transporters	Unknown	Unknown
Cyt b5	LmjF.07.0810	Electron transporters	Unknown	Unknown
EIF2AK2	LmjF.34.2150	Haem-regulated proteins	TC and TB	Protein translation (da Silva Augusto *et al.*, 2015)
Cyt c1	LmjF.07.0060	Electron transporters	TC and TB	Unknown
CYP	LmjF.11.1100	Catalysis	TC and TB	Lanosterol 14-demethylase (Chen *et al.*, 2010; Lepesheva *et al.*, 2004; McCall *et al.*, 2015)
Cyt b5	LmjF.15.0190	Electron transporters	TC and TB	Unknown
Cyt c	LmjF.16.1320	Electron transporters	TC and TB	Electron transporters and substrate for peroxidase
Hypothetical Cyt b5	LmjF.26.1430	Electron transporters	TC and TB	Unknown
Ncb5or	LmjF.30.0610	Electron transporters	TC and TB	Electron donor for Δ12 desaturase (Mukherjee *et al.*, 2012)
Hypothetical Cyt b5	LmjF.11.0580	Electron transporters	TC and TB	Unknown
Hypothetical Cyt b5	LmjF.36.4675	Electron transporters	Unknown	Unknown
Hypothetical Cyt b5	LmjF.34.3050	Electron transporters	Unknown	Unknown
Hypothetical Cyt b5	LmjF.22.0015	Electron transporters	TC and TB	Pseudogene

shown to specifically bind haem and haemoglobin (Galbraith and McElrath, 1988; Sengupta *et al.*, 1999). Studies identify presence of Hb receptor in *Leishmania* plasma membrane (Krishnamurthy *et al.*, 2005), and haem transporters in plasma membrane and intracellular compartments (Campos-Salinas *et al.*, 2011; Huynh *et al.*, 2012). Hb receptor in *L. donovani* promastigote has been shown to mediate uptake of Hb by endocytosis. Endocytosed Hb is transported to lysosome by Rab5/7 for degradation and release of haem (Patel *et al.*, 2008). Then the *Leishmania* ATP-Binding Cassette G5 (haem transporter) transfers haem to mitochondria (Campos-Salinas *et al.*, 2011). Haem transporter LHR1 (*Leishmania* Haem Response-1) in plasma membrane and intracellular acidic compartments of *L. amazonensis* allow import of haem to parasite cytosol (Huynh *et al.*, 2012). Acquisition of iron has been shown in *L. amazonensis* through concerted action of transmembrane ferric iron reductase LFR1 (Flannery *et al.*, 2011) and a ferrous iron transporter LIT1 (Jacques *et al.*, 2010).

Role of different types of haemproteins in eukaryotes and prokaryotes

Gas carrier

Globins

Haem-containing globins are widely distributed in all form of living organisms, identified by the presence of conserved 'globin fold'. Globins bind gaseous molecules (O_2, CO, NO and CO_2) reversibly, which facilitates their storage and delivery. Yeast flavin (FAD)-bound haemoglobin (Hb) is implicated in O_2-dependent signalling (Zhao *et al.*, 1996). The chloroplast-localized Hb in *Chlamydomonus* algae is believed to trap O_2 and its delivery to respiratory cytochrome c oxidase (Couture *et al.*, 1994). Leghemoglobin (plant Hb) facilitates diffusion of O_2 to symbiotic bacteria for N_2 fixation (Appleby, 1984). Vertebrate Hb is heterotetrameric that composed of two α- and two β-chains ($\alpha_2\beta_2$), each bound to one haem molecule and localized exclusively in red blood cells, where they transport O_2 from respiratory organs to tissue and CO_2 vice versa (Perutz, 1979). The myoglobin (Mb) in vertebrates is monomeric and distributed in cardiac and skeletal muscle to serve as O_2 store for delivery to myocyte mitochondria (Gros *et al.*, 2010). Neuroglobin (Ng) and cytoglobin (Cg) are two recently discovered globins in vertebrates (Burmester *et al.*, 2000; Trent and Hargrove, 2002). Ng is monomeric with ability to bind O_2 reversibly with an affinity similar to Mb. Ng is distributed in neural cytosol. Cg is dimeric and widely expressed in mammalian tissues. The literature suggests a possible role of Ng and Cg in cellular defence under hypoxia (Schmidt *et al.*, 2004; Sun *et al.*, 2001). However, these types of haemproteins are unknown in *Leishmania* pathogens.

Nitrophorins

Nitrophorins are found in the saliva of blood-feeding insects. Nitrophorins bind and store NO. Localized release of NO by nitrophrins manipulates guanylate cyclase signalling (Montfort *et al.*, 2000).

Electron transporters

Cytochrome c

Cytochrome c (Cyt c) is involved in biological electron transport (ET) processes in both aerobic and anaerobic respiratory chains. In aerobic organisms Cyt c is peripherally associated with mitochondrial inner membrane where they transfer electrons from the transmembrane respiratory complex bc1 complex (complex III) (Crofts, 2004) to Cyt c oxidase (complex IV) to generate ATP (Capaldi, 1990). Cytosolic Cyt c complexed with Apaf 1 to activate caspase protease for initiating programmed cell death or apoptosis (Jiang and Wang, 2004). Apart, Cyt c acts as a redox partner for several redox reactions. In yeast and *L. major* mitochondria Cyt c acts as a redox partner for Cyt c peroxidase (CcP) and *L. major* peroxidase (LMP) respectively to detoxify H_2O_2 (Adak and Pal, 2013). In denitrifying bacteria Cyt c acts as a redox partner for nitrite, nitric oxide, and nitrous oxide reductases. *Leishmania* spp. possess a functional mitochondrial respiratory chain (Guerra *et al.*, 2006) within its single large mitochondrion to carry out oxidative phosphorylation. In *Leishmania* mitochondrial cytochrome c (LmCyt c), haem is covalently linked to the polypeptide chain through single thioether bond to a thiol of cysteine residue of haem-binding motif (F /A)XXCH. Apart from electron transport, LmCyt c functions as a redox partner for mitochondrial peroxidase LMP (Dolai *et al.*, 2008; Jasion *et al.*, 2012). LmCyt c is also implicated in apoptosis (Dolai *et al.*, 2009; Sen *et al.*, 2004). Like mammalian cells, *Leishmania* parasites undergo apoptosis in spite the absence of BH3 domain-containing apoptosis regulatory proteins (Arnoult *et al.*, 2002). Apoptotic *Leishmania* display features of apoptosis such as externalization of phosphatidyl serine and activation of caspase-like proteases under stress that is induced by a variety of pharmacological inhibitors of cellular pathways or by oxidants. The permeabilization of mitochondria, release of LmCyt c to cytosol from mitochondria and subsequent activation of caspase-like proteases are observed during these processes (Das *et al.*, 2001; Dolai *et al.*, 2009, 2011; Sen *et al.*, 2004).

Cytochrome c oxidoreductase (complex III)

Also known as cytochrome bc1 complex, this is an oligomeric electron transfer enzyme complex localized in the inner mitochondrial membrane of eukaryotes and the plasma membrane of respiring bacteria. The active redox centre of bc1 complex contains cytochrome b, cytochrome c1 and 'Rieske' [2Fe–2S]. The bc1 complex transfers electrons from quinol to Cyt c in mitochondrial electron transport chain (ETC) or to bacterial plastocyaninin with the generation of proton gradient across the membrane (Iwata *et al.*, 1998; Trumpower, 1990). Although putative complex III proteins are present in *L. major* gene DB yet the characterization of all proteins in *L. major* mitochondrial complex III are unknown.

Cytochrome c oxidase (complex IV)

Cytochrome c oxidase (CcO) or complex IV is the terminal enzyme in the respiratory ETC in bacterial membrane or mitochondrial inner membrane of eukaryotes. In eukaryotes, CcO is composed of 14 structural subunits, of which the three core components (COX1, COX2 and COX3) that are encoded by mitochondrial DNA are translated within the organelle and are assembled with remaining nuclear encoded subunits (Capaldi, 1990). CcO contains four redox centres designated CuA, haem a, CuB and haem a3. CcO receives four electrons from reduced Cyt c that are used to reduce O_2 to H_2O with the help of four protons ($4H^+ + 4e^- + O_2 = 2H_2O$) from matrix that coupled with translocation of another four protons from matrix to inner mitochondrial space (IMS) to generate electrochemical potential gradient for ATP synthesis.

Trypanosomatids, including *Leishmania*, possess Complex I to IV respiratory complex of ETC like higher eukaryotes to support active oxidative respiration (Santhamma and Bhaduri, 1995). Active respiration is essential for parasite survival (Hart *et al.*, 1981; Van Hellemond and Tielens, 1997). Like higher eukaryotes, *Leishmania* cytochrome c oxidase (Complex IV) is a multiprotein complex composed of 13 subunits (Horvath *et al.*, 2000), where three large subunits are encoded by mitochondrial or kinetoplast genome and remaining 10 are encoded by nuclear genome. Interestingly, *Leishmania* Complex IV is unique having eight out of ten nucleus encoded subunits are novel that display no homology outside trypanosomatids (Speijer *et al.*, 1996). In *Leishmania*, Complex IV is shown to be crucial in parasite survival in macrophages and disease establishment in mice (Cardenas *et al.*, 2015; Dey *et al.*, 2010). Low Complex IV activity with reduced parasite ATP content is associated with low infectivity in macrophage cells and lower virulence in mice (Dey *et al.*, 2010). Depletion of Complex IV in *L. major* is later shown to associate with low macrophage infectivity, which is increased upon restoration of Complex IV by exogenous expression (Cardenas *et al.*, 2015). Apart from respiration, nuclear encoded subunits COXIV and COXVI of complex IV are shown as active partner of *L. tropica* mitochondrial tRNA import complex, thus regulate mitochondrial translation in *Leishmania* (Mukherjee *et al.*, 2007).

Cytochromes b5 (Cyt b5)

Cyt b5 is widely distributed in bacteria, fungi, plants and animals. Cyt b5 is membrane-bound in mitochondria and microsomes, but soluble in bacteria and erythrocytes. Cytb5 possesses two hydrophobic cores, of which the larger hydrophobic core binds haem. Soluble Cyt b5 in erythrocytes reduces methaemoglobin to an active deoxy form that binds oxygen. Microsomal Cyt b5 is a ER transmembrane protein known to function by transferring electrons in fatty acid desaturation, cholesterol biosynthesis, and hydroxylation reactions involving Cyt P450. In addition, *cb*5 domain is found in other proteins such as NAD(P)H cytochrome *b*5 oxidoreductase (Zhu *et al.*, 1999), nitrate reductase (Hyde *et al.*, 1991), sulfite oxidases (Kisker *et al.*, 1997), fatty acid desaturation (Gostincar *et al.*, 2010) and lactate dehydrogenase (Xia *et al.*, 1987).

Fatty acid desaturates are widely distributed in living organisms and ER of eukaryotic cells. Fatty acid desaturates catalyse fatty acid unsaturation by incorporating double bonds to the designated sites of fatty acid carbon backbone using O_2 and reducing partner, usually a cytochrome b (Gostincar *et al.*, 2010). Polyunsaturated fatty acids (PUFAs) contain more than two double bonds, are essential constituents of membrane, where they regulate membrane fluidity and functions of transmembrane proteins. In animals, PUFAs are substrates for eicosanoid synthesis (Funk, 2001). Trypanosomatids contain an elevated level of PUFA that possibly adding membrane fluidity required for transformation to different morphology and adaptation to different environments in vector and host. Fatty acid desaturates act in concert with FA elongases in trypanosomatid ER, to synthesize PUFA (Lee *et al.*, 2006). *Leishmania* genome database reveals the presence of three fatty acid desaturases with cytochrome b5 fused at the N-terminus that is recently characterized (Tripodi *et al.*, 2006). The three fatty acid desaturases (Δ8, Δ5 and Δ6) are shown to catalyse desaturation towards he carboxy terminus of fatty acids (Tripodi *et al.*, 2006).

NAD(P)H cytochrome *b*5 oxidoreductase (Ncb5or) belongs to the cytochrome *b*5 family of flavohaemoprotein that is widely expressed in eukaryotic organisms (Zhu *et al.*, 1999). Ncb5or is an endoplasmic reticulum-bound protein that contains two redox domains; the amino-terminal portion is highly homologous to classical microsomal cytochrome *b*5 (*cb*5), and the carboxyl-terminal portion of the protein is highly similar to cytochrome *b*5 reductase (*cb*5r) domain (Zhu *et al.*, 1999). The *cb*5r domain contains a single FAD prosthetic group and catalyses the transfer of reducing equivalents from NAD(P)H to *cb*5 haem domain. Ncb5or in mammal is implicated in lipid metabolism in hepatocytes and apoptosis in insulin producing β-cells (Xie *et al.*, 2004; Zhu *et al.*, 2004). An accumulation of saturated FA as a result of impaired Δ9 desaturation is observed in Ncb5or deficient mice. Ncb5or deficient β cells display mitochondrial hypertrophy and hyperplasticity, and ER stress associated apoptosis via X-box-binding protein-1 (XBP-1)-C/EBP-homologous protein (CHOP) axis.

Recently Ncb5or in *L. major* (LmNcb5or) has been characterized (Mukherjee *et al.*, 2012), originally identified as putative nitrate reductase (systematic no. LmjF30.0610). The amino acid sequence alignment with *Sus scrofa domestica* cytochrome *b*5 (Ssdcb5, accession number 1106188C) and *Pichia stipitis* NADH-cytochrome *b*5 reductase (Pscb5r, accession no. NP_001153064) attests the 74-residue N-terminal *b*5-type cytochrome domain and a 234-residue C-terminal flavin-containing domain connected with a ≈170 amino acid linker. In *Leishmania*, the LmNcb5or translated to a predicted (≈60 kDa) soluble protein that display UV-visible spectral characteristic of di-histidine-ligated cytochrome *b*5-type haemprotein and catalytic activity towards common electron acceptors like ferrocytochrome *c* in presence of excess electron donor NADH.

To evaluate LmNcb5or physiological functions in *Leishmania*, *L. major* cell lines are generated that are either LmNcb5or null mutant (LmNcb5or-KO) or overexpressing LmNcb5or (LmNcb5or-OE). Analysis of FA composition reveals occurrence of eight predominant FAs in WT *L. major* cells: palmitate (16:0), stearate (18:0), oleate (18:1), linoleate (18:2), linolenate (18:3), eicosatrienate (20:3), arachidonate (20:4) and docasahexaenate (22 :6). Interestingly, LmNcb5or-KO cells show depleted level of linoleate and linolenate, whereas oleate content are found elevated that suggest impaired C18 Δ12 desaturation of oleate by Δ12 FA desaturates. LmNcb5or acts as a redox partner (electron donor) for Δ12 FAD (electron acceptor). LmNcb5or deletion decrease linoleate synthesis, increased oxidative stress, mitochondrial dysfunction and apoptosis. The consequences of LmNcb5or knock out are reversed to WT or LmNcb5or-OE cells by the supplementation of linoleate.

Haemprotein acts as an enzyme

Cytochrome p450 superfamily

The cytochrome P450s (CYPs) constitute a large superfamily that includes about 9000 proteins classified into different families designated by Arabic numbers such as CYP1. CYPs are distributed in all eukaryotes as microsomal membrane bound form and as cytosolic proteins in some prokaryotes. CYPs are classified to more than 800 families due to their extensive sequence variability, but show preservation of oxygen-binding motif: G/AGX(D/E)T and haem-binding motif: FXXGX$_b$XXCXG, where Xb is a basic residue and haem is coordinated to thiolate anion of cysteine residue. CYPs carry out mono-oxygenation reaction in presence of oxygen molecule (O_2) and reducing equivalents NADPH or NADH. One cycle of reaction requires two step reductions. First, the haem iron (Fe^{3+}) of substrate-bound CYP is reduced by one electron (Fe^{2+}) followed by binding to an O_2. The second reduction activates the haem-bound O_2 by splitting the O–O bond. The Fe-ligated oxygen atom (O) is incorporated to substrate to form a hydroxylated molecule whereas the released O atom is reduced to water. Mono-oxygenation of different substrates underlie the diverse biological functions of CYPs.

In human, CYPs carry out mono-oxygenation of cholesterol to synthesize steroid intermediate pregnenolone (Jefcoate *et al.*, 1992) and hydroxylation of several steroid molecules required for biosynthesis of steroid hormones in adrenal cortex and gonuds. CYPs catalyse 14α-demethylation and its oxidative removal from sterol intermediates lanosterol for the biosynthesis of cholesterol in animals (Trzaskos *et al.*, 1986). 7α-hydroxylation of cholesterol by CYPs is the rate limiting step to catabolize cholesterol to bile acids in liver (Danielsson and Sjövall, 1975). CYPs inactivate eicosanoids, arachidonic acid, prostaglandins and leukotrienes by ω-hydroxylation. Kidney CYPs mediate ω, ω-1 hydroxylation and epoxidation of arachidonic acid (Makita *et al.*, 1996). CYPs in liver and lung catalyse ω-hydroxylation of prostaglandins and leukotrienes (Johnson *et al.*, 1996). CYPs in liver and kidney metabolize vitamin D3 to physiologically active 1,25-dihydroxyvitamin D_3 (Takeyama *et al.*, 1997). CYP mediated metabolism of retinoic acid (RA) regulate RA signalling and embryo morphogenesis (Fujii *et al.*, 1997). Synthetic compounds/drugs (xenobiotics) are metabolized by CYPs1–4 family members to less toxic and water-soluble products that facilitate their excretion.

Lanosterol 14α-demethylase (EC 1.14.13.70) is a member of CYP family number 51 (CYP51). CYP51 is considered as an ancestor of all CYP families and is distributed throughout the all living organisms. CYP51 catalyses the oxidative removal of the 14α-methyl group (C32) of lanosterol in fungi and mammals, and obtusifoliol in plants through three successive mono-oxygenation reactions that lead to the synthesis of ergosterol in fungi, phytosterol in plants, and cholesterol in animals with the release of H_2O and formic acid (Debeljak *et al.*, 2003; Lepesheva and Waterman, 2011). Sterols control membrane fluidity and organize membrane domains. Trypanosomatid cell membranes are enriched with ergosterol in comparison to cholesterol in mammals. Inhibitors of CYP51 and fungal sterol biosynthesis are found to be effective against trypanosomatids (Alrajhi *et al.*, 2002; Kauffman *et al.*, 2007). CYP51 is later identified and characterized in *T. cruzi* and *T. brucei* (Chen *et al.*, 2010; Lepesheva *et al.*, 2004, 2006). CYP51 in *L. infantum* (LinJ11.1100) is recently identified and characterized. The *L. infantum* CYP51 display more than 95% amino acid identity with CYP51 from other *Leishmania* spp. The B'helix of *L. infantum* CYP51 contains plant-specific phenylalanine (Phe-104) residue similar to plant CYP51 rather than CYP51 from *T. cruzi* and *T. brucei*. Consistently, *L. infantum* CYP51 displays catalytic preferences towards the C4-monomethylated sterol substrates obtusifoliol and C4-norlanosterol; with lesser activity towards lanosterol (Hargrove *et al.*, 2011). The requirement of CYP51 in *Leishmania* is investigated on *L. donovani* by gene deletion and pharmacological inhibition of *L. donovani* CYP51 (McCall *et al.*, 2015). CYP51 null-mutation in *L. donovani* is feasible only in the presence of episomal expression of *L. donovani* CYP51. The parasites also are found to retain the episomal CYP51 even under negative regulation, suggesting importance of CYP51 for cell survival. The *T. cruzi* CYP51-specific inhibitor shows lower efficacy against *L. donovani*. However, an augmented efficacy is observed in CYP51 depleted parasites. Together these suggest essentiality of CYP51 in *Leishmania*.

Catalase and peroxidases

The 'catalytic' reaction of catalase and peroxidative reactions of peroxidases are carried out within their haem centre through the formation of two oxidizing equivalent higher active species compound I that obtained by reaction with hydrogen peroxide (H_2O_2) (Dunford, 1999). H_2O_2 removes one electron from iron and another from porphyrin ring or an amino acid side chain, usually tryptophan in peroxidase.

Catalatic reaction is completed by subsequent oxidation of a second molecule of H_2O_2, thus producing two molecules of H_2O and one molecule of O_2 by rapid consumption of two molecules of H_2O_2. Catalases are present in almost all aerobic organisms (except trypanosomatids, tetrahymena) where they play an important role in defending cells against oxidative damage by degrading H_2O_2 (Peter Nichols, 2001). The vital physiological roles of catalase include prevention of apoptosis (Sandstrom and Buttke, 1993), nucleic acid damage and mutagenesis (Schraufstätter *et al.*, 1988).

The peroxidative reaction is completed by two-step reduction of compound I by one electron from a molecule of reducing substrate to form compound II which returns to native state by one more electron reduction by another substrate molecule. The diversity of reducing substrate underlies variety of functions of peroxidases in different organisms. Haem peroxidases are widely distributed in plants, animals and micro-organisms. All haem-containing peroxidases are belong to two large superfamilies; peroxidase–cyclooxygenase and peroxidase–catalase superfamily (Za'mocky, 2010). The peroxidase-catalase superfamily contains three classes (Welinder, 1991). Class I peroxidases include intracellular peroxidases of prokaryotic origin-like plant chloroplast and cytosolic APX, yeast CcP, and gene duplicated catalase-peroxidase. Class II peroxidases comprise fungal secretory peroxidases such as lignin and manganese peroxidases from *Phanerochaete chrysosporium* and the ink cap mushroom peroxidase from *Coprinuscinereus*. Examples of class III peroxidases are endoplasmic reticulum mediated classical secretory plant peroxidases such as horseradish peroxidase isoenzyme C. Several haem peroxidases are present in humans like myeloperoxidase, lactoperoxidase, thyroid vw etc. (Kimura and Ikeda-Saito, 1988; Pruitt and Tenovuo, 1985).

Reports on haem enzyme-dependent H_2O_2 consumption in *Leishmania* amastigotes and ascorbate-dependent peroxidase activity in *T. cruzi* epimastigote extracts are dated back to 1980s (Boveris *et al.*, 1980; Channon and Blackwell, 1985). An endoplasmic reticulum-located unusual plant-like ascorbate peroxidase (APX) enzyme is then reported in 2002 (Wilkinson *et al.*, 2002). Adak's laboratory is the first to characterize the sole haem peroxidase from *L. major* (LMP) in 2005 (Adak and Datta, 2005). Comparison of LMP primary sequence shows 35% identity with yeast CcP and 36% identity with cytosolic pea APX (Yadav *et al.*, 2008), the two most well-characterized peroxidases. TargetP V1.0 analysis of LMP primary sequence predicts presence of 12 amino acid N-terminal signal sequence followed by 22-amino-acid-long hydrophobic region, reminiscent of a transmembrane domain. The key proximal and distal site residues to haem of CcP and APX are preserved in LMP. However, the cytochrome c-binding residues (Asp34/Glu35, Tyr39, and Glu290 in CCP) and ascorbate-binding residue (Arg 172 in APX) are absent in LMP. Interestingly, the proximal cation (K^+) binding loop of the LMP is more similar to APX, whereas the c-terminal insertion and proximal Trp radical stabilizing Met residues of LMP are more similar to CcP (Bursey and Poulos, 2000; Pelletier and Kraut, 1992; Sharp *et al.*, 2003). Detailed *in vitro* spectral and biochemical studies confirm LMP as a hybrid of APX and CcP. It shows higher activity towards cytochrome c oxidation (Yadav *et al.*, 2008). Apart, LMP acts as a pseudocatalase in low ascorbate concentration by decomposing H_2O_2 to H_2O and O_2 by monodehydroascorbate, a one-electron oxidation product of ascorbic acid (Dolai *et al.*, 2007). Together, they add plasticity on choice over reducing substrates and more efficient H_2O_2 consumption by LMP.

The intracellular localization of LMP and its regulation under oxidative stress are examined towards the first step to understand the plausible physiological functions of LMP. Combined approaches using Western blot analysis of subcellular fractionations and confocal microscopy show LMP localization in mitochondria (Dolai *et al.*, 2008). The intra-mitochondrial localization is determined through preparation of mitoplast (mitochondria without outer membrane) by digitonin treatment, submitochondrial fractionation, and activity measurement of kynurenine hydroxylase (outer membrane marker), succinate dehydrogenase (inner membrane marker), and malate dehydrogenase (matrix marker) enzymes that reveals mitochondrial inner-membrane localization of LMP with its catalytic domain facing towards IMS (Dolai *et al.*, 2008). This localization is similar to CcP in yeast (Goltz *et al.*, 1982) and consistent with its proficient activity towards LmCyt c over yeast Cyt c and ascorbic acid (Jasion and Poulos, 2012). Therefore, LMP is more likely to be accessible to scavenge endogenous ROS produced in the mitochondrial inner-membrane space. Up-regulation of LMP expression under exogenous H_2O_2 treatment further suggests LMP's role to evade oxidative stress (Dolai *et al.*, 2008).

The physiological functions of LMP are further investigated in detail on LMP overexpressed and LMP knock out (LMP-KO) cell lines using an array of cell biology approach (Dolai *et al.*, 2008, 2009; Pal *et al.*, 2010). Overexpression of LMP prevents oxidant induced mitochondrial cardiolipin and cellular protein oxidation (Dolai *et al.*, 2008, 2009). Cardiolipin plays crucial role in Cyt c functions in mitochondria and Cyt c mediated apoptosis signalling in mammals (McMillin and Dowhan, 2002). LMP overexpressing *L. major* cells show lower accumulation of intracellular H_2O_2 and mitochondrial superoxide, along with resistance towards oxidative stress induced apoptosis (Dolai *et al.*, 2008, 2009). Consistently LMP-KO *L. major* shows heightened accumulation of these ROS molecules and apoptotic cells in culture (Pal *et al.*, 2010). In line with these reports, LMP homologue in *L. donovani* is recently found to prevent ROS mediated apoptosis by antileishmanial drug amphotericin B (Kumar *et al.*, 2014). Interestingly, an elevated percentage of infective metacyclic *L. major* population are observed in LMP-KO culture with concomitant up-regulation of H_2O_2 scavenging non-selenium glutathione peroxidase and mitochondrial superoxide scavenging iron superoxide dismutase; this collectively suggests LMP's role in regulation of intracellular ROS and ROS mediated signalling underlying cellular differentiation (Adak and Pal, 2013; Pal *et al.*, 2010).

Recently a haem containing pseudoperxidase from *L. major* (LmPP; GenBank Accession No. CAJ04972) has been reported. LmPP shows 14–18% sequence identity with class I peroxidase with preservation of key proximal residues: His206 and Glu290 and distal residues: Arg86 and Phe89 residues. However, LmPP is distinct from class I peroxidases by lacking distal site His and proximal site Trp residues; with presence of a Cys205 residue in front of the proximal His206; and the presence of a sequence insertion, of unknown function, near the carboxy terminus. TargetP version 1.0 prediction indicates the presence of an N-terminal signal sequence (1–30 amino acids), followed by a stretch of 22 hydrophobic amino acids, a possible transmembrane domain. UV-visible spectral studies of N-terminal 48 amino acid deleted (Δ48 LmPP) soluble protein reveals presence of Cys-ligated six-coordinated haem. The peroxidase activity is 10^4-fold lower compared to HRP. Interestingly, an efficient peroxynitrite ($ONOO^-$) scavenging property, 600-fold higher compared to iodide oxidation is observed in LmPP.

To gain insight in to LmPP's physiological functions, the distribution of LmPP within *L. major*, the regulation of LmPP expression, and the parasite's ability to tolerate oxidative and nitrosative stress in absence or excess expression of LmPP are studied in detail. Confocal immunofluorescence microscopy detects flagellar pocket colocalization of LmPP with kinetoplastid membrane protein-11 (basal body and flagella-specific protein). Interestingly in amastigotes, LmPP distribution is observed throughout parasite's surface that perfectly colocalize with plasma membrane marker Na^+/K^+ ATPase. An 8-fold expression of LmPP in amastigotes is also observed over promastigotes, indicating LmPP's importance as possible surface scavenger of $ONOO^-$ within host macrophage. LmPP deleted (LmPP-KO) and overexpressed (LmPP-OE) parasite cell lines are generated to validate physiological function of LmPP. All cell lines show similar growth rate. The sensitivity of the cell lines towards exogenous oxidative and nitrosative stress are verified by challenging cell lines to a range of H_2O_2 and SIN1 ($ONOO^-$ donor) and determining the IC_{50} values. LmPP-KO cells display 5-fold sensitivity to SIN 1 compared to LmPP-OE cells, whereas IC_{50} values for H_2O_2 remains similar irrespective of cellular LmPP content, consistent with the higher *in vitro* activity of LmPP towards $ONOO^-$.

Activated macrophage is a major source of $ONOO^-$ (Nathan and Shiloh, 2000). The parasite cell line's ability of survival within activated macrophages is then examined. The LmPP-KO parasites show ≈42-fold more killing compared to LmPP-OE cells within activated macrophage. The killing is significantly ameliorated with the treatment of nitric oxide synthase inhibitor N^{ω}-nitro-L-arginine methyl ester that confirms inability of LmPP-KO parasites to protect against macrophage generated $ONOO^-$. The study is further extended to exam LmPP's role in disease establishment in BALB/c mice model to study *Leishmania* infection. Stationary phase parasites are inoculated in mice footpad. LmPP-KO cells are found unable to develop disease compared to WT-cells. Together, LmPP is a crucial scavenger of $ONOO^-$ required for parasite survival within activated macrophage and disease establishment in host.

Nitric oxide synthase (NOS)

The functionally active NOS is a homodimer, where each monomer contains N-terminal oxidase domain that associated with tetrahydrobiopterin (H4B) and haem reactive centre; and C-terminal reductase domain with binding sites for the redox co-factors NADPH, FMN, FAD and calmodulin. NOS catalyse the two-step, five-electron oxidation of L-arginine (L-Arg) via *N*-hydroxy-L-arginine (NHA) to citrulline and NO. In mammals, NO promotes synthesis of second messenger cyclic GMP by activating guanylate cyclase (Stuehr, 2004); carry out intercellular and intracellular signalling functions including blood-vessel dilatation, neuronal signal transmission, cytotoxicity against pathogens and tumours, coordination of heart rhythm and the regulation of cellular respiration activity (Murad, 1999). The NOS protein is well characterized in human but *Leishmania* lacks an NOS-like gene.

Sensing gaseous molecules

O_2, CO, and NO are physiologically indispensable but sometimes toxic to living organisms (Sasakura *et al.*, 2006). Hence, the machinery for sensing these molecules and responding to them by reprogramming cell signalling pathway is vital for survival. Haem-containing sensors are haemproteins that usually consist of an N-terminal haem-containing sensor domain and a C-terminal effector domain (Hou *et al.*, 2000). These are different from the above-mentioned haemproteins. The structures of haem-based sensor proteins are changed upon binding of O_2, CO, or NO gases with the haem domain. This structural change influences the effector domain for proper functioning. Haem-containing sensor proteins are categorized into four different families, based on haem domain structure (Gilles-Gonzalez and Gonzalez, 2005). The PerARNT-Sim sequence motif (PAS) is present in the regulatory domains of widely distributed PAS family of sensors that linked to effector kinase domains as in FixL that regulate nitrogenase synthesis by phosphorylating FixJ transcription factor in nitrogen fixing bacterium *Rhizobium meliloti* (Monson *et al.*, 1995) and transcription factors of the basic helix–loop–helix and helix–turn–helix classes such as neuronal PAS domain protein 2 that regulate circadian rhythm in mammals (Dioum *et al.*, 2002). The other families include catabolite activator protein (CAP) family of CO sensing CooA transcription factors that activate gene expression upon binding of the CO to the haem site (Shelver *et al.*, 1997). The globin-coupled sensors (GCS) are haem-containing aerotaxis transducers found in bacteria (Gilles-Gonzalez and Gonzalez, 2005; Paoli *et al.*, 2002) and globin-coupled adenylate cyclase in *L. major* (Roy *et al.*, 2014; Sen Santara *et al.*, 2013).The haem NO-binding (HNOB) module containing proteins sense NO to regulate cell signalling that includes eukaryotic soluble guanylate cyclase and some bacterial NO-binding haemproteins (Gilles-Gonzalez and Gonzalez, 2005; Stuehr, 2004).

The adenylate cyclases (ACs) catalyses the cyclization of ATP to cyclic AMP (cAMP) (Sutherland and Rall, 1957) that acts as a vital signalling molecule controlling diverse biological functions in many organisms, including pathogens (Agarwal *et al.*, 2009; Rickenberg, 1974). Recently a soluble globin-coupled AC (HemAC-Lm; systematic name: LmjF.28.0090) is characterized in *L. major* (Roy *et al.*, 2014; Sen Santara *et al.*, 2013) that acts as an oxygen sensor for AC activity. HemAC-Lm contains preserved globin-specific residues: the proximal His in the F helix, Pro in C helix, and Phe in the CD1 region (Bashford *et al.*,

1987) of the N-terminus and signature residues of ACs: lysine (Lys-427) and aspartic acid (Asp-508) residues (Tucker *et al.*, 1998) at the C-terminus. *In vitro* UV-visible spectral analysis and biochemical studies reveal HemAC-Lm's ability to bind O_2 and CO, and AC activity in gaseous molecule-bound state.

The physiological roles of HemAC-Lm are examined on parasites with reduced or elevated level of proteins. *L. major* is found resistant to HemAC-Lm null mutation that suggests its requirement for cell survival. HemAC-Lm depleted (HemAC-Lm-KD) and overexpressed (HemAC-Lm-OE) *L. major* cell lines are obtained by gene knock-down (KD) or overexpression (OE) approach. Under normoxic conditions, the cellular level of HemAC-Lm correlates directly with intracellular concentration of cAMP, downstream activation of protein kinase A (PKA) and cellular oxygen consumption. The HemAC-Lm manipulated cells show slower growth rate as a result of delayed cell cycle progression. The cellular morphology and differentiation occur differently where HemAC-Lm-KD cells show round morphology and HemAC-Lm-OE cells with elevated cAMP content show slender morphology as a result of promoted procyclic to infective metacyclic differentiation. Interestingly the number of apoptotic cells was similar irrespective of HemAC-Lm cellular level. In hypoxic condition, the cellular cAMP content show reduction by ≈50% in all cell types compared to normoxic condition without affecting cellular ATP contents; implicating O_2 dependence on *in vivo* HemAC-Lm activity. However, unlike normoxia, an elevated percentage of apoptotic cells are observed in HemAC-Lm manipulated cells in comparison to control cells. In HemAC-Lm-KD cells, the excess apoptosis are prevented by exogenous supply of cAMP but that worsen HemAC-Lm overexpressed cells; showing tight link between intracellular cAMP content and *L. major* viability. Collectively, HemAC-Lm is shown to regulate O_2-dependent intracellular cAMP level that regulates downstream PKA-mediated antioxidant gene expression and intracellular generation of ROS, thus functions to adapt *Leishmania* parasite in hypoxic environments. Within the sandfly midgut, parasites are densely packed together and likely to create hypoxic or even anoxic environment (Van Hellemond and Tielens, 1997). HemAC-Lm mediated O_2 sensing and adaptation thus could be a survival strategy of *Leishmania* for successful transmission of disease.

Haem-regulated proteins

Haem is not stably associated with haem-regulated proteins; rather it acts as a signalling molecule by binding and regulating their functions (Girvan and Munro, 2013). For example, Bach1, the transcriptional repressor of haem oxygenase-1, contains multiple haem-binding sites (Hira *et al.*, 2007). Binding of haem to Bach1 causes conformational changes that allow dissociation of Bach1 and binding of transcription factors from repressor-binding sites in globins and haem oxygenase 1 promoters. Haem-regulated eIF2α kinase (HRI) regulates globin translation in reticulocytes by regulating eIF2α activity (Han *et al.*, 2001). This is crucial to maintain 2:2:4 ratio of α-globin, β-globin and haem to form stable $\alpha_2\beta_2$ haemoglobin structure and minimize cytotoxic effects by excess production of them. HRI senses intracellular haem concentration through its two haem-binding domains. Under haem deficiency, HRI is activated by multiple autophosphoryation, and phosphorylates α-subunit of eIF2α (P-eIF2α) and subsequent translational inhibition of globins. Rev-erb α, Rev-erb β control ecdysone genes in insect and light/dark cycle genes in vertebrates (Raghuram, 2007; Yin, 2007). Iron regulatory protein2α controls transcription of iron-associated genes (Kang, 2003).

Recently a group of workers have characterized a haem-regulated eIF2α kinase in *Trypanosoma cruzi* (TcK2), the causative agent of Chagas' disease (da Silva Augusto, 2015). TcK2 is a transmembrane protein in parasite's endosomal compartment that also store nutrients in parasite's replicating form. Haem inhibits TcK2 activity by specifically binding to the catalytic domain of kinase that supports parasite growth in nutrient enriched condition. Haem depletion during nutrient deprivation activates TcK2 that leads to cessation of parasite growth and induce cellular differentiation to generate infective trypomastigotes through metacyclogenesis. TcK2 null mutant parasites display growth retardation, impaired cellular differentiation and accumulation of intracellular ROS. These phenotypes are reversed upon episomal expression of native form of TcK2 but not with mutant that unable to bind haem, attesting haem's role in *T. cruzi* growth and differentiation through TcK2 regulation.

Conclusions

The *Leishmania* (*L. major, L. infantum and L. braziliensis*) genome sequence has revealed more than 41 haem-containing protein-coding genes. The biological roles of haemproteins are identified that are crucial for cell survival in different life cycle stages of parasite. *Leishmania* peroxidase is found to maintain cellular redox homeostasis crucial for parasite survival and differentiation. Pseudoperoxidase provides defence against nitrosative stress and promotes amastigote survival within activated macrophage. Fatty acid desaturases synthesize polyunsaturated fatty acids required for normal membrane functions. NAD(P)H cytochrome b5.oxidoreductase serve as a redox partner for Δ12 fatty acid desaturases. Cytochrome c and cytochrome c oxidase carry out mitochondrial respiration. Cytochrome P450 (CYP51) display catalytic preferences towards the C4-monomethylated sterol substrates obtusifoliol and C4-norlanosterol. Haem containing adenylate cyclase regulates cellular cAMP signalling pathway by sensing O_2 level. Haem regulates nutrients accumulation in proliferating parasite forms via eIF2*α* kinase.

Future trends

Identification of enormous number of haem motif containing proteins in *Leishmania* genome sequence are indicative of their important role in parasite survivability and infectivity within the host. However, only a few of them are functionally characterized. Elucidation of the remaining haemprotein's biological functions through the systematic study of haemprotein's biochemical properties, localization within parasites, regulation in different life cycle stages and regulation of parasite physiological functions are prime. It is expected that elucidation of biological functions of the remaining haemproteins in the near future will enrich the current understanding to develop novel haemprotein-based chemotherapies against leishmaniasis.

References

Adak, S., and Datta, A.K. (2005). *Leishmania major* encodes an unusual peroxidase that is a close homologue of plant ascorbate peroxidase: a novel role of the transmembrane domain. Biochem. J. *390*, 465–474.

Adak, S., and Pal, S. (2013). Ascorbate peroxidase acts as a novel determiner of redox homeostasis in *Leishmania*. Antioxid. Redox Signal. *19*, 746–754. https://doi.org/10.1089/ars.2012.4745.

Agarwal, N., Lamichhane, G., Gupta, R., Nolan, S., and Bishai, W.R. (2009). Cyclic AMP intoxication of macrophages by a Mycobacterium tuberculosis adenylate cyclase. Nature *460*, 98–102. https://doi.org/10.1038/nature08123.

Alcolea, P.J., Alonso, A., Gomez, M.J., Postigo, M., Molina, R., Jimenez, M., and Larraga, V. (2014). Stage-specific differential gene expression in *Leishmania infantum*: from the foregut of *Phlebotomus perniciosus* to the human phagocyte. BMC Genomics *15*, 849.

Alrajhi, A.A., Ibrahim, E.A., De Vol, E.B., Khairat, M., Faris, R.M., and Maguire, J.H. (2002). Fluconazole for the treatment of cutaneous leishmaniasis caused by *Leishmania major*. N. Engl. J. Med. *346*, 891–895. https://doi.org/10.1056/NEJMoa011882.

Alves, J.M., Voegtly, L., Matveyev, A.V., Lara, A.M., da Silva, F.M., Serrano, M.G., Buck, G.A., Teixeira, M.M., and Camargo, E.P. (2011). Identification and phylogenetic analysis of heme synthesis genes in trypanosomatids and their bacterial endosymbionts. PLOS ONE *6*, e23518. https://doi.org/10.1371/journal.pone.0023518.

Appleby, C.A. (1984). Leghemoglobin and rhizobium respiration. Ann Rev Plant Physiol 35, 443–478.

Arnoult, D., Akarid, K., Grodet, A., Petit, P.X., Estaquier, J., and Ameisen, J.C. (2002). On the evolution of programmed cell death: apoptosis of the unicellular eukaryote *Leishmania major* involves cysteine proteinase activation and mitochondrion permeabilization. Cell Death Differ 9, 65–81.

Barr, I., Smith, A.T., Senturia, R., Chen, Y., Scheidemantle, B.D., Burstyn, J.N., and Guo, F. (2011). DiGeorge critical region 8 (DGCR8) is a double-cysteine-ligated heme protein. J. Biol. Chem. *286*, 16716–16725. https://doi.org/10.1074/jbc.M110.180844.

Barr, S.D., and Gedamu, L. (2003). Role of peroxidoxins in *Leishmania chagasi* survival. Evidence of an enzymatic defense against nitrosative stress. J. Biol. Chem. *278*, 10816–10823. https://doi.org/10.1074/jbc.M212990200.

Bashford, D., Chothia, C., and Lesk, A.M. (1987). Determinants of a protein fold. Unique features of the globin amino acid sequences. J. Mol. Biol. *196*, 199–216.

Bates, P.A. (2008). *Leishmania* sand fly interaction: progress and challenges. Curr. Opin. Microbiol. *11*, 340–344. https://doi.org/10.1016/j.mib.2008.06.003.

Bernstein, B.E., Michels, P.A., and Hol, W.G. (1997). Synergistic effects of substrate-induced conformational changes in phosphoglycerate kinase activation. Nature *385*, 275–278.

Bose, M., Saha, R., Sen Santara, S., Mukherjee, S., Roy, J., and Adak, S. (2012). Protection against peroxynitrite by pseudoperoxidase from Leishmania major. Free Radic. Biol. Med. 53, 1819-1828.

Boveris, A., Sies, H., Martino, E.E., Docampo, R., Turrens, J.F., and Stoppani, A.O. (1980). Deficient metabolic utilization of hydrogen peroxide in Trypanosoma cruzi. Biochem. J. *188*, 643–648.

Burchmore, R.J., and Barrett, M.P. (2001). Life in vacuoles – nutrient acquisition by *Leishmania* amastigotes. Int. J. Parasitol. *31*, 1311–1320.

Burmester, T., Weich, B., Reinhardt, S., and Hankeln, T. (2000). A vertebrate globin expressed in the brain. Nature *407*, 520–523.

Bursey, E.H., and Poulos, T.L. (2000). Two substrate binding sites in ascorbate peroxidase: the role of arginine 172. Biochemistry *39*, 7374–7379.

Campos-Salinas, J., Cabello-Donayre, M., García-Hernández, R., Pérez-Victoria, I., Castanys, S., Gamarro, F., and Pérez-Victoria, J.M. (2011). A new ATP-binding cassette protein is involved in intracellular haem trafficking in *Leishmania*. Mol. Microbiol. *79*, 1430–1444. https://doi.org/10.1111/j.1365-2958.2010.07531.x.

Capaldi, R.A. (1990). Structure and function of cytochrome c oxidase. Annu. Rev. Biochem. *59*, 569–596. https://doi.org/10.1146/annurev.bi.59.070190.003033.

Cardenas, D., Carter, P.M., Nation, C.S., Pizarro, J.C., Guidry, J., Aiyar, A., and Kelly, B.L. (2015). LACK, a RACK1 ortholog, facilitates cytochrome c oxidase subunit expression to promote *Leishmania major* fitness. Mol Microbiol *96*, 95–109.

Castro, H., Sousa, C., Santos, M., Cordeiro-da-Silva, A., Flohe, L., and Tomas, A.M. (2002). Complementary antioxidant defense by cytoplasmic and mitochondrial peroxiredoxins in *Leishmania infantum*. Free Radic Biol Med *33*, 1552–1562.

Chang, C.S., and Chang, K.P. (1985). Heme requirement and acquisition by extracellular and intracellular stages of *Leishmania mexicana* amazonensis. Mol. Biochem. Parasitol. *16*, 267–276.

Chang, K.P. (1979). *Leishmania donovani*: promastigote – macrophage surface interactions *in vitro*. Exp. Parasitol. *48*, 175–189.

Chang, K.P., and Dwyer, D.M. (1976). Multiplication of a human parasite (*Leishmania donovani*) in phagolysosomes of hamster macrophages *in vitro*. Science *193*, 678–680.

Chang, K.P., and Trager, W. (1974). Nutritional significance of symbiotic bacteria in two species of hemoflagellates. Science *183*, 531–532.

Chang, K.P., Chang, C.S., and Sassa, S. (1975). Heme biosynthesis in bacterium-protozoon symbioses: enzymic defects in host hemoflagellates and complemental role of their intracellular symbiotes. Proc. Natl. Acad. Sci. U.S.A. *72*, 2979–2983.

Channon, J.Y., and Blackwell, J.M. (1985). A study of the sensitivity of *Leishmania donovani* promastigotes and amastigotes to hydrogen peroxide. II. Possible mechanisms involved in protective H2O2 scavenging. Parasitology *91* (*Pt 2*), 207–217.

Chen, C.K., Leung, S.S., Guilbert, C., Jacobson, M.P., McKerrow, J.H., and Podust, L.M. (2010). Structural characterization of CYP51 from Trypanosoma cruzi and Trypanosoma brucei bound to the antifungal drugs posaconazole and fluconazole. PLOS Negl. Trop. Dis. *4*, e651.

Couture, M., Chamberland, H., St-Pierre, B., Lafontaine, J., and Guertin, M. (1994). Nuclear genes encoding chloroplast hemoglobins in the unicellular green alga Chlamydomonas eugametos. Mol. Gen Genet. *243*, 185–197.

Crofts, A.R. (2004). The cytochrome bc1 complex: function in the context of structure. Annu. Rev. Physiol. *66*, 689–733. https://doi.org/10.1146/annurev.physiol.66.032102.150251.

Cunningham, M.L., Titus, R.G., Turco, S.J., and Beverley, S.M. (2001). Regulation of differentiation to the infective stage of the protozoan parasite *Leishmania major* by tetrahydrobiopterin. Science *292*, 285–287.

da Silva Augusto L., N.M., Ramos, T.C., de Jesus T.C., Zhang, M., Castilho, B.A., and Schenkman, S. (2015). A membrane-bound eIF2 Alpha kinase located in endosomes is regulated by heme and controls differentiation and ROS levels in *Trypanosoma cruzi*. PLOS Pathogens. https://doi.org/10.1371/journal.ppat.1004618.

Danielsson, H., and Sjövall, J. (1975). Bile acid metabolism. Annu. Rev. Biochem. *44*, 233–253. https://doi.org/10.1146/annurev.bi.44.070175.001313.

Das, M., Mukherjee, S.B., and Shaha, C. (2001). Hydrogen peroxide induces apoptosis-like death in *Leishmania donovani* promastigotes. J. Cell Sci. *114*, 2461–2469.

De, Pablos, L.M., Ferreira, T.R., and Walrad, P.B. (2016). Developmental differentiation in *Leishmania* lifecycle progression: post-transcriptional control conducts the orchestra. Curr. Opin. Microbiol. *34*, 82–89.

Debeljak, N., Fink, M., and Rozman, D. (2003). Many facets of mammalian lanosterol 14alpha-demethylase from the evolutionarily conserved cytochrome P450 family CYP51. Arch. Biochem. Biophys. *409*, 159–171.

Derbyshire, E.R., and Marletta, M.A. (2012). Structure and regulation of soluble guanylate cyclase. Annu. Rev. Biochem. *81*, 533–559. https://doi.org/10.1146/annurev-biochem-050410-100030.

Dey, R., Meneses, C., Salotra, P., Kamhawi, S., Nakhasi, H.L., and Duncan, R. (2010). Characterization of a *Leishmania* stage-specific mitochondrial membrane protein that enhances the activity of cytochrome c oxidase and its role in virulence. Mol. Microbiol. *77*, 399–414.

Dioum, E.M., Rutter, J., Tuckerman, J.R., Gonzalez, G., Gilles-Gonzalez, M.A., and McKnight, S.L. (2002). NPAS2: a gas-responsive transcription factor. Science *298*, 2385–2387. https://doi.org/10.1126/science.1078456.

Dolai, S., Pal, S., Yadav, R.K., and Adak, S. (2011). Endoplasmic reticulum stress-induced apoptosis in *Leishmania* through Ca2+-dependent and caspase-independent mechanism. J Biol Chem *286*, 13638–13646.

Dolai, S., Yadav, R.K., Datta, A.K., and Adak, S. (2007). Effect of thiocyanate on the peroxidase and pseudocatalase activities of *Leishmania major* ascorbate peroxidase. Biochim. Biophys. Acta *1770*, 247–256.

Dolai, S., Yadav, R.K., Pal, S., and Adak, S. (2008). *Leishmania major* ascorbate peroxidase overexpression protects cells against reactive oxygen species-mediated cardiolipin oxidation. Free Radic. Biol. Med. *45*, 1520–1529.

Dolai, S., Yadav, R.K., Pal, S., and Adak, S. (2009). Overexpression of mitochondrial *Leishmania major* ascorbate peroxidase enhances tolerance to oxidative stress-induced programmed cell death and protein damage. Eukaryot Cell *8*, 1721–1731.

Dunford, H. (1999). Heme Peroxidases (New York, John Wiley & Sons, Inc).

Fang, F.C. (2004). Antimicrobial reactive oxygen and nitrogen species: concepts and controversies. Nat. Rev. Microbiol. *2*, 820–832. https://doi.org/10.1038/nrmicro1004.

Flannery, A.R., Huynh, C., Mittra, B., Mortara, R.A., and Andrews, N.W. (2011). LFR1 ferric iron reductase of *Leishmania amazonensis* is essential for the generation of infective parasite forms. J. Biol. Chem. *286*, 23266–23279. https://doi.org/10.1074/jbc.M111.229674.

Fujii, H., Sato, T., Kaneko, S., Gotoh, O., Fujii-Kuriyama, Y., Osawa, K., Kato, S., and Hamada, H. (1997). Metabolic inactivation of retinoic acid by a novel P450 differentially expressed in developing mouse embryos. EMBO J. *16*, 4163–4173.

Funk, C.D. (2001). Prostaglandins and leukotrienes: advances in eicosanoid biology. Science *294*, 1871–1875. https://doi.org/10.1126/science.294.5548.1871.

Galbraith, R.A., and McElrath, M.J. (1988). Heme binding to *Leishmania mexicana amazonensis*. Mol. Biochem. Parasitol. *29*, 47–53.

Gilles-Gonzalez, M.A., and Gonzalez, G. (2005). Heme-based sensors: defining characteristics, recent developments, and regulatory hypotheses. J. Inorg. Biochem. *99*, 1–22.

Gilles-Gonzalez, M.A., Ditta, G.S., and Helinski, D.R. (1991). A haemoprotein with kinase activity encoded by the oxygen sensor of *Rhizobium meliloti*. Nature *350*, 170–172. https://doi.org/10.1038/350170a0.

Girvan, H.M., and Munro, A.W. (2013). Heme sensor proteins. J. Biol. Chem. *288*, 13194–13203. https://doi.org/10.1074/jbc.R112.422642.

Goltz, S., Kaput, J., and Blobel, G. (1982). Isolation of the yeast nuclear gene encoding the mitochondrial protein, cytochrome c peroxidase. J. Biol. Chem. *257*, 11186–11190.

Gostincar, C., Turk, M., and Gunde-Cimerman, N. (2010). The evolution of fatty acid desaturases and cytochrome b5 in eukaryotes. J. Membr. Biol. *233*, 63–72. https://doi.org/10.1007/s00232-010-9225-x.

Gozzelino, R., Jeney, V., and Soares, M.P. (2010). Mechanisms of cell protection by heme oxygenase-1. Annu. Rev. Pharmacol. Toxicol. *50*, 323–354. https://doi.org/10.1146/annurev.pharmtox.010909.105600.

Gros, G., Wittenberg, B.A., and Jue, T. (2010). Myoglobin's old and new clothes: from molecular structure to function in living cells. J Exp Biol *213*, 2713–2725.

Guengerich, F.P., Waterman, M.R., and Egli, M. (2016). Recent Structural Insights into Cytochrome P450 Function. Trends Pharmacol. Sci. *37*, 625–640. https://doi.org/10.1016/j.tips.2016.05.006.

Guerra, D.G., Decottignies, A., Bakker, B.M., and Michels, P.A. (2006). The mitochondrial FAD-dependent glycerol-3-phosphate dehydrogenase of Trypanosomatidae and the glycosomal redox balance of insect stages of Trypanosoma brucei and *Leishmania* spp. Mol Biochem Parasitol *149*, 155–169.

Guzy, R.D., Mack, M.M., and Schumacker, P.T. (2007). Mitochondrial complex III is required for hypoxia-induced ROS production and gene transcription in yeast. Antioxid. Redox Signal. *9*, 1317–1328. https://doi.org/10.1089/ars.2007.1708.

Han, A.P., Yu, C., Lu, L., Fujiwara, Y., Browne, C., Chin, G., Fleming, M., Leboulch, P., Orkin, S.H., and Chen, J.J. (2001). Heme-regulated eIF2alpha kinase (HRI) is required for translational regulation and survival of erythroid precursors in iron deficiency. EMBO J. *20*, 6909–6918. https://doi.org/10.1093/emboj/20.23.6909.

Hargrove, T.Y., Wawrzak, Z., Liu, J., Nes, W.D., Waterman, M.R., and Lepesheva, G.I. (2011). Substrate preferences and catalytic parameters determined by structural characteristics of sterol 14alpha-demethylase (CYP51) from *Leishmania infantum*. J Biol Chem *286*, 26838–26848.

Harrison, P.M., and Arosio, P. (1996). The ferritins: molecular properties, iron storage function and cellular regulation. Biochim. Biophys. Acta *1275*, 161–203.

Hart, D.T., Vickerman, K., and Coombs, G.H. (1981). Respiration of *Leishmania mexicana* amastigotes and promastigotes. Mol. Biochem. Parasitol. *4*, 39–51.

Hira, S., Tomita, T., Matsui, T., Igarashi, K., and Ikeda-Saito, M. (2007). Bach1, a heme-dependent transcription factor, reveals presence of multiple heme binding sites with distinct coordination structure. IUBMB Life *59*, 542–551.

Horvath, A., Kingan, T.G., and Maslov, D.A. (2000). Detection of the mitochondrially encoded cytochrome c oxidase subunit I in the trypanosomatid protozoan *Leishmania* tarentolae. Evidence for translation of unedited mRNA in the kinetoplast. J Biol Chem *275*, 17160–17165.

Hou, S., Larsen, R.W., Boudko, D., Riley, C.W., Karatan, E., Zimmer, M., Ordal, G.W., and Alam, M. (2000). Myoglobin-like aerotaxis transducers in Archaea and Bacteria. Nature *403*, 540–544. https://doi.org/10.1038/35000570.

Hu, R.G., Wang, H., Xia, Z., and Varshavsky, A. (2008). The N-end rule pathway is a sensor of heme. Proc. Natl. Acad. Sci. U.S.A. *105*, 76–81.

Huynh, C., Yuan, X., Miguel, D.C., Renberg, R.L., Protchenko, O., Philpott, C.C., Hamza, I., and Andrews, N.W. (2012). Heme uptake by *Leishmania amazonensis* is mediated by the transmembrane protein LHR1. PLOS Pathog. *8*, e1002795. https://doi.org/10.1371/journal.ppat.1002795.

Hyde, G.E., Crawford, N.M., and Campbell, W.H. (1991). The sequence of squash NADH:nitrate reductase and its relationship to the sequences of other flavoprotein oxidoreductases. A family of flavoprotein pyridine nucleotide cytochrome reductases. J Biol Chem *266*, 23542–23547.

Ivens, A.C., Peacock, C.S., Worthey, E.A., Murphy, L., Aggarwal, G., Berriman, M., Sisk, E., Rajandream, M.A., Adlem, E., Aert, R., *et al.* (2005). The genome of the kinetoplastid parasite, *Leishmania major*. Science *309*, 436–442.

Iwata, S., Lee, J.W., Okada, K., Lee, J.K., Iwata, M., Rasmussen, B., Link, T.A., Ramaswamy, S., and Jap, B.K. (1998). Complete structure of the 11-subunit bovine mitochondrial cytochrome bc1 complex. Science *281*, 64–71.

Jacques, I., Andrews, N.W., and Huynh, C. (2010). Functional characterization of LIT1, the *Leishmania amazonensis* ferrous iron transporter. Mol. Biochem. Parasitol. *170*, 28–36. https://doi.org/10.1016/j.molbiopara.2009.12.003.

Jasion, V.S., and Poulos, T.L. (2012). *Leishmania major* peroxidase is a cytochrome c peroxidase. Biochemistry *51*, 2453–2460. https://doi.org/10.1021/bi300169x.

Jasion, V.S., Doukov, T., Pineda, S.H., Li, H., and Poulos, T.L. (2012). Crystal structure of the *Leishmania major* peroxidase-cytochrome c complex. Proc. Natl. Acad. Sci. U.S.A. *109*, 18390–18394. https://doi.org/10.1073/pnas.1213295109.

Jefcoate, C.R., McNamara, B.C., Artemenko, I., and Yamazaki, T. (1992). Regulation of cholesterol movement to mitochondrial cytochrome P450scc in steroid hormone synthesis. J. Steroid Biochem. Mol. Biol. *43*, 751–767. https://doi.org/10.1016/0960-0760(92)90305-3.

Jiang, X., and Wang, X. (2004). Cytochrome C-mediated apoptosis. Annu. Rev. Biochem. *73*, 87–106. https://doi.org/10.1146/annurev.biochem.73.011303.073706.

Johnson, E.F., Palmer, C.N., Griffin, K.J., and Hsu, M.H. (1996). Role of the peroxisome proliferator-activated receptor in cytochrome P450 4A gene regulation. FASEB J. *10*, 1241–1248.

Kang, D.K., Jeong, J., Drake, S.K., Wehr, N.B., Rouault, T.A., and Levine, R.L. (2003). Iron regulatory protein 2 as iron sensor. Iron-dependent oxidative modification of cysteine. J. Biol. Chem. *278*, 14857–14864. https://doi.org/10.1074/jbc.M300616200.

Kauffman, C.A., Malani, A.N., Easley, C., and Kirkpatrick, P. (2007). Posaconazole. Nat Rev Drug Discov *6*, 183–184. https://doi.org/10.1038/nrd2270.

Killick-Kendrick, R., Molyneux, D.H., Hommel, M., Leaney, A.J., and Robertson, E.S. (1977). *Leishmania* in phlebotomid sandflies. V. The nature and significance of infections of the pylorus and ileum of the sandfly by leishmaniae of the braziliensis complex. Proc. R. Soc. Lond., B, Biol. Sci. *198*, 191–199.

Kimura, S., and Ikeda-Saito, M. (1988). Human myeloperoxidase and thyroid peroxidase, two enzymes with separate and distinct physiological functions, are evolutionarily related members of the same gene family. Proteins 3, 113–120.

Kisker, C., Schindelin, H., Pacheco, A., Wehbi, W.A., Garrett, R.M., Rajagopalan, K.V., Enemark, J.H., and Rees, D.C. (1997). Molecular basis of sulfite oxidase deficiency from the structure of sulfite oxidase. Cell *91*, 973–983.

Krishnamurthy, G., Vikram, R., Singh, S.B., Patel, N., Agarwal, S., Mukhopadhyay, G., Basu, S.K., and Mukhopadhyay, A. (2005). Hemoglobin receptor in *Leishmania* is a hexokinase located in the flagellar pocket. J. Biol. Chem. *280*, 5884–5891.

Kumar, A., Das, S., Purkait, B., Sardar, A.H., Ghosh, A.K., Dikhit, M.R., Abhishek, K., and Das, P. (2014). Ascorbate peroxidase, a key molecule regulating amphotericin B resistance in clinical isolates of *Leishmania donovani*. Antimicrob. Agents Chemother. *58*, 6172–6184.

Lee, S.H., Stephens, J.L., Paul, K.S., and Englund, P.T. (2006). Fatty acid synthesis by elongases in trypanosomes. Cell *126*, 691–699.

Lepesheva, G.I., and Waterman, M.R. (2011). Structural basis for conservation in the CYP51 family. Biochim. Biophys. Acta *1814*, 88–93. https://doi.org/10.1016/j.bbapap.2010.06.006.

Lepesheva, G.I., Nes, W.D., Zhou, W., Hill, G.C., and Waterman, M.R. (2004). CYP51 from Trypanosoma brucei is obtusifoliol-specific. Biochemistry *43*, 10789–10799. https://doi.org/10.1021/bi048967t.

Lepesheva, G.I., Zaitseva, N.G., Nes, W.D., Zhou, W., Arase, M., Liu, J., Hill, G.C., and Waterman, M.R. (2006). CYP51 from Trypanosoma cruzi: a phyla-specific residue in the B' helix defines substrate preferences of sterol 14alpha-demethylase. J Biol Chem *281*, 3577–3585.

Makita, K., Falck, J.R., and Capdevila, J.H. (1996). Cytochrome P450, the arachidonic acid cascade, and hypertension: new vistas for an old enzyme system. FASEB J. *10*, 1456–1463.

McCall, L.I., El Aroussi, A., Choi, J.Y., Vieira, D.F., De Muylder, G., Johnston, J.B., Chen, S., Kellar, D., Siqueira-Neto, J.L., Roush, W.R., *et al.* (2015). Targeting Ergosterol biosynthesis in *Leishmania donovani*: essentiality of sterol 14 alpha-demethylase. PLOS Negl. Trop. Dis *9*, e0003588.

McConville, M.J., and Naderer, T. (2011). Metabolic pathways required for the intracellular survival of *Leishmania*. Annu. Rev. Microbiol. *65*, 543–561. https://doi.org/10.1146/annurev-micro-090110-102913.

McCoubrey, W.K., Huang, T.J., and Maines, M.D. (1997). Heme oxygenase-2 is a hemoprotein and binds heme through heme regulatory motifs that are not involved in heme catalysis. J. Biol. Chem. *272*, 12568–12574.

McMillin, J.B., and Dowhan, W. (2002). Cardiolipin and apoptosis. Biochim. Biophys. Acta *1585*, 97–107.

Monson, E.K., Ditta, G.S., and Helinski, D.R. (1995). The oxygen sensor protein, FixL, of *Rhizobium meliloti*. Role of histidine residues in heme binding, phosphorylation, and signal transduction. J. Biol. Chem. *270*, 5243–5250.

Montfort, W.R., Weichsel, A., and Andersen, J.F. (2000). Nitrophorins and related antihemostatic lipocalins from *Rhodnius prolixus* and other blood-sucking arthropods. Biochim Biophys Acta *1482*, 110–118.

Mukherjee, S., Basu, S., Home, P., Dhar, G., and Adhya, S. (2007). Necessary and sufficient factors for the import of transfer RNA into the kinetoplast mitochondrion. EMBO Rep. *8*, 589–595.

Mukherjee, S., Sen Santara, S., Das, S., Bose, M., Roy, J., and Adak, S. (2012). NAD(P)H cytochrome b5 oxidoreductase deficiency in *Leishmania major* results in impaired linoleate synthesis followed by increased oxidative stress and cell death. J. Biol. Chem. *287*, 34992–35003.

Muñoz-Sánchez, J., and Chánez-Cárdenas, M.E. (2014). A review on hemeoxygenase-2: focus on cellular protection and oxygen response. Oxid Med Cell Longev *2014*, 604981. https://doi.org/10.1155/2014/604981.

Murad, F. (1999). Discovery of Some of the Biological Effects of Nitric Oxide and Its Role in Cell Signaling (Nobel Lecture). Angew Chem In.t Ed. *38*, 1856–1868.

Naderer, T., and McConville, M.J. (2008). The *Leishmania*–macrophage interaction: a metabolic perspective. Cell. Microbiol. *10*, 301–308.

Nathan, C., and Shiloh, M.U. (2000). Reactive oxygen and nitrogen intermediates in the relationship between mammalian hosts and microbial pathogens. Proc. Natl. Acad. Sci. U.S.A. *97*, 8841–8848.

Opperdoes, F.R., and Coombs, G.H. (2007). Metabolism of *Leishmania*: proven and predicted. Trends Parasitol. *23*, 149–158.

Pal, S., Dolai, S., Yadav, R.K., and Adak, S. (2010). Ascorbate peroxidase from *Leishmania major* controls the virulence of infective stage of promastigotes by regulating oxidative stress. PLOS ONE *5*, e11271. https://doi.org/10.1371/journal.pone.0011271.

Paoli, M., Marles-Wright, J., and Smith, A. (2002). Structure–function relationships in heme-proteins. DNA Cell Biol. *21*, 271–280. https://doi.org/10.1089/104454902753759690.

Patel, N., Singh, S.B., Basu, S.K., and Mukhopadhyay, A. (2008). *Leishmania* requires Rab7-mediated degradation of endocytosed hemoglobin for their growth. Proc. Natl. Acad. Sci. U.S.A. *105*, 3980–3985. https://doi.org/10.1073/pnas.0800404105.

Peacock, C.S., Seeger, K., Harris, D., Murphy, L., Ruiz, J.C., Quail, M.A., Peters, N., Adlem, E., Tivey, A., Aslett, M., *et al.* (2007). Comparative genomic analysis of three *Leishmania* species that cause diverse human disease. Nat. Genet. *39*, 839–847.

Pelletier, H., and Kraut, J. (1992). Crystal structure of a complex between electron transfer partners, cytochrome c peroxidase and cytochrome c. Science *258*, 1748–1755.

Perutz, M.F. (1979). Regulation of oxygen affinity of hemoglobin: influence of structure of the globin on the heme iron. Annu. Rev. Biochem. *48*, 327–386. https://doi.org/10.1146/annurev.bi.48.070179.001551.

Peter, Nichols, I.F., and Peter, Lowen, C. (2001). Enzymology and structure of catalases. *51*, 51–106.

Pruitt, K.M., and Tenuvuo, P.J. (1985) The Lactoperoxidase System: Chemistry and Biological Significance (Immunology) (CRC Press).

Raghuram, S., Stayrook, K.R., Huang, P., Rogers, P.M., Nosie, A.K., McClure, D.B., Burris, L.L., Khorasanizadeh, S., Burris, T.P., and Rastinejad, F. (2007). Identification of heme as the ligand for the orphan nuclear receptors REV-ERBalpha and REV-ERBbeta. Nat. Struct. Mol. Biol. *14*, 1207–1213.

Rajagopal, A., Rao, A.U., Amigo, J., Tian, M., Upadhyay, S.K., Hall, C., Uhm, S., Mathew, M.K., Fleming, M.D., Paw, B.H., *et al.* (2008). Haem homeostasis is regulated by the conserved and concerted functions of HRG-1 proteins. Nature *453*, 1127–1131. https://doi.org/10.1038/nature06934.

Rickenberg, H.V. (1974). Cyclic AMP in prokaryotes. Annu. Rev. Microbiol. *28*, 353–369. https://doi.org/10.1146/annurev.mi.28.100174.002033.

Roy, J., Sen Santara, S., Bose, M., Mukherjee, S., Saha, R., and Adak, S. (2014). The ferrous–dioxy complex of *Leishmania major* globin coupled heme containing adenylate cyclase: the role of proximal histidine on its stability. Biochim. Biophys. Acta *1844*, 615–622.

Sacks, D.L., and Perkins, P.V. (1984). Identification of an infective stage of *Leishmania* promastigotes. Science *223*, 1417–1419.

Sandstrom, P.A., and Buttke, T.M. (1993). Autocrine production of extracellular catalase prevents apoptosis of the human CEM T-cell line in serum-free medium. Proc. Natl. Acad. Sci. U.S.A. *90*, 4708–4712.

Santhamma, K.R., and Bhaduri, A. (1995). Characterization of the respiratory chain of *Leishmania donovani* promastigotes. Mol. Biochem. Parasitol. *75*, 43–53.

Sasakura, Y., Yoshimura-Suzuki, T., Kurokawa, H., and Shimizu, T. (2006). Structure–function relationships of EcDOS, a heme-regulated phosphodiesterase from *Escherichia coli*. Acc. Chem. Res. *39*, 37–43. https://doi.org/10.1021/ar0501525.

Schenkman, J.B., and Jansson, I. (2003). The many roles of cytochrome b5. Pharmacol. Ther. *97*, 139–152.

Schmidt, M., Gerlach, F., Avivi, A., Laufs, T., Wystub, S., Simpson, J.C., Nevo, E., Saaler-Reinhardt, S., Reuss, S., Hankeln, T., *et al.* (2004). Cytoglobin is a respiratory protein in connective tissue and neurons, which is up-regulated by hypoxia. J. Biol. Chem. *279*, 8063–8069. https://doi.org/10.1074/jbc.M310540200.

Schraufstätter, I., Hyslop, P.A., Jackson, J.H., and Cochrane, C.G. (1988). Oxidant-induced DNA damage of target cells. J. Clin. Invest. *82*, 1040–1050. https://doi.org/10.1172/JCI113660.

Sen Santara, S., Roy, J., Mukherjee, S., Bose, M., Saha, R., and Adak, S. (2013). Globin-coupled heme containing oxygen sensor soluble adenylate cyclase in *Leishmania* prevents cell death during hypoxia. Proc. Natl. Acad. Sci. U.S.A. *110*, 16790–16795.

Sen, N., Das, B.B., Ganguly, A., Mukherjee, T., Tripathi, G., Bandyopadhyay, S., Rakshit, S., Sen, T., and Majumder, H.K. (2004). Camptothecin induced mitochondrial dysfunction leading to programmed cell death in unicellular hemoflagellate *Leishmania donovani*. Cell Death Differ. *11*, 924–936.

Sengupta, S., Tripathi, J., Tandon, R., Raje, M., Roy, R.P., Basu, S.K., and Mukhopadhyay, A. (1999). Hemoglobin endocytosis in *Leishmania* is mediated through a 46-kDa protein located in the flagellar pocket. J. Biol. Chem. *274*, 2758–2765.

Shapira, M., McEwen, J.G., and Jaffe, C.L. (1988). Temperature effects on molecular processes which lead to stage differentiation in *Leishmania*. EMBO J. *7*, 2895–2901.

Sharp, K.H., Mewies, M., Moody, P.C., and Raven, E.L. (2003). Crystal structure of the ascorbate peroxidase–ascorbate complex. Nat Struct Biol *10*, 303–307.

Shelver, D., Kerby, R.L., He, Y., and Roberts, G.P. (1997). CooA, a CO-sensing transcription factor from *Rhodospirillum rubrum*, is a CO-binding heme protein. Proc. Natl. Acad. Sci. U.S.A. *94*, 11216–11220.

Soares, M.P., and Hamza, I. (2016). Macrophages and Iron Metabolism. Immunity *44*, 492–504. https://doi.org/10.1016/j.immuni.2016.02.016.

Späth, G.F., Garraway, L.A., Turco, S.J., and Beverley, S.M. (2003). The role(s) of lipophosphoglycan (LPG) in the establishment of *Leishmania major* infections in mammalian hosts. Proc. Natl. Acad. Sci. U.S.A. *100*, 9536–9541. https://doi.org/10.1073/pnas.1530604100.

Speijer, D., Breek, C.K., Muijsers, A.O., Groenevelt, P.X., Dekker, H., de Haan, A., and Benne, R. (1996). The sequence of a small subunit of cytochrome c oxidase from Crithidia fasciculata which is homologous to mammalian subunit IV. FEBS Lett. *381*, 123–126.

Stuehr, D.J., Santolini, J., Wang, Z.Q., Wei, C.C., and Adak, S. (2004). Update on mechanism and catalytic regulation in the NO synthases. J. Biol. Chem. *279*, 36167–36170. https://doi.org/10.1074/jbc.R400017200.

Sun, Y., Jin, K., Mao, X.O., Zhu, Y., and Greenberg, D.A. (2001). Neuroglobin is up-regulated by and protects neurons from hypoxic-ischemic injury. Proc. Natl. Acad. Sci. U.S.A. *98*, 15306–15311. https://doi.org/10.1073/pnas.251466698.

Sutherland, E.W., and Rall, T.W. (1957). The properties of an adenine ribonucleotide produced with cellular particles, ATP, Mg++, and epinephrine or glucagon. J Am Chem Soc 79, 3608.

Takeyama, K., Kitanaka, S., Sato, T., Kobori, M., Yanagisawa, J., and Kato, S. (1997). 25-Hydroxyvitamin D3 1alpha-hydroxylase and vitamin D synthesis. Science *277*, 1827–1830.

Tang, X.D., Xu, R., Reynolds, M.F., Garcia, M.L., Heinemann, S.H., and Hoshi, T. (2003). Haem can bind to and inhibit mammalian calcium-dependent Slo1 BK channels. Nature *425*, 531–535. https://doi.org/10.1038/nature02003.

Trent, J.T., and Hargrove, M.S. (2002). A ubiquitously expressed human hexacoordinate hemoglobin. J. Biol. Chem. *277*, 19538–19545. https://doi.org/10.1074/jbc.M201934200.

Tripodi, K.E., Buttigliero, L.V., Altabe, S.G., and Uttaro, A.D. (2006). Functional characterization of front-end desaturases from trypanosomatids depicts the first polyunsaturated fatty acid biosynthetic pathway from a parasitic protozoan. FEBS J 273, 271–280.

Trumpower, B.L. (1990). Cytochrome bc1 complexes of microorganisms. Microbiol. Rev. *54*, 101–129.

Trzaskos, J., Kawata, S., and Gaylor, J.L. (1986). Microsomal enzymes of cholesterol biosynthesis. Purification of lanosterol 14 alpha-methyl demethylase cytochrome P-450 from hepatic microsomes. J. Biol. Chem. *261*, 14651–14657.

Tucker, C.L., Hurley, J.H., Miller, T.R., and Hurley, J.B. (1998). Two amino acid substitutions convert a guanylyl cyclase, RetGC-1, into an adenylyl cyclase. Proc. Natl. Acad. Sci. U.S.A. *95*, 5993–5997.

Turco, S.J., and Sacks, D.L. (1991). Expression of a stage-specific lipophosphoglycan in *Leishmania major* amastigotes. Mol. Biochem. Parasitol. *45*, 91–99.

Van, Hellemond, J.J., and Tielens, A.G. (1997). Inhibition of the respiratory chain results in a reversible metabolic arrest in *Leishmania* promastigotes. Mol. Biochem. Parasitol. *85*, 135–138.

Welinder, K.G. (1991). Bacterial catalase-peroxidases are gene duplicated members of the plant peroxidase superfamily. Biochim Biophys Acta *1080*, 215–220.

Westberg, J.A., Jiang, J., and Andersson, L.C. (2011). Stanniocalcin 1 binds hemin through a partially conserved heme regulatory motif. Biochem. Biophys. Res. Commun. *409*, 266–269. https://doi.org/10.1016/j.bbrc.2011.05.002.

Wilkinson, S.R., Obado, S.O., Mauricio, I.L., and Kelly, J.M. (2002). Trypanosoma cruzi expresses a plant-like ascorbate-dependent hemoperoxidase localized to the endoplasmic reticulum. Proc. Natl. Acad. Sci. U.S.A. *99*, 13453–13458.

Xia, Z.X., Shamala, N., Bethge, P.H., Lim, L.W., Bellamy, H.D., Xuong, N.H., Lederer, F., and Mathews, F.S. (1987). Three-dimensional structure of flavocytochrome b2 from baker's yeast at 3.0-A resolution. Proc. Natl. Acad. Sci. U.S.A. *84*, 2629–2633.

Xie, J., Zhu, H., Larade, K., Ladoux, A., Seguritan, A., Chu, M., Ito, S., Bronson, R.T., Leiter, E.H., Zhang, C.Y., *et al.* (2004). Absence of a reductase, NCB5OR, causes insulin-deficient diabetes. Proc. Natl. Acad. Sci. U.S.A. *101*, 10750–10755. https://doi.org/10.1073/pnas.0404044101.

Yadav, R.K., Dolai, S., Pal, S., and Adak, S. (2008). Role of tryptophan-208 residue in cytochrome c oxidation by ascorbate peroxidase from *Leishmania major*-kinetic studies on Trp208Phe mutant and wild type enzyme. Biochim. Biophys. Acta *1784*, 863–871. https://doi.org/10.1016/j.bbapap.2008.02.006.

Yamamoto, M., K.S., Engel, J.D., and Hiraga, K. (1988). Structure, turnover, and heme-mediated suppression of the level of mRNA encoding rat liver delta-aminolevulinate synthase. J. Biol. Chem. *263*, 15973–15979.

Yin, L., Wu, N., Curtin, J.C., Qatanani, M., Szwergold, N.R., Reid, R.A., Waitt, G.M., Parks, D.J., Pearce, K.H., Wisely, G.B., *et al.* (2007). Rev-erbalpha, a heme sensor that coordinates metabolic and circadian pathways. Science *318*, 1786–1789.

Za'mocky, M., and Obinger, C. (2010). Molecular Phylogeny Of heme Peroxidases. (Berlin, Springer-Verlag).

Zhao, X.J., Raitt, D., V Burke, P., Clewell, A.S., Kwast, K.E., and Poyton, R.O. (1996). Function and expression of flavohemoglobin in Saccharomyces cerevisiae. Evidence for a role in the oxidative stress response. J. Biol. Chem. *271*, 25131–25138.

Zhu, H., Larade, K., Jackson, T.A., Xie, J., Ladoux, A., Acker, H., Berchner-Pfannschmidt, U., Fandrey, J., Cross, A.R., Lukat-Rodgers, G.S., *et al.* (2004). NCB5OR is a novel soluble NAD(P)H reductase localized in the endoplasmic reticulum. J. Biol. Chem. *279*, 30316–30325. https://doi.org/10.1074/jbc.M402664200.

Zhu, H., Qiu, H., Yoon, H.W., Huang, S., and Bunn, H.F. (1999). Identification of a cytochrome b-type NAD(P)H oxidoreductase ubiquitously expressed in human cells. Proc. Natl. Acad. Sci. U.S.A. *96*, 14742–14747.

Pre-adaptation of *Leishmania* Promastigotes to Intracellular Life: Ensuring a Successful Infection

9

Roma Sinha[1] and Nahid Ali[2*]

[1]Mater Medical Research Institute-University of Queensland, Translational Research Institute, Brisbane, QLD, Australia.
[2]Infectious Diseases and Immunology Division, CSIR-Indian Institute of Chemical Biology, Kolkata, India.

*Correspondence: nali@iicb.res.in

https://doi.org/10.21775/9781910190715.09

Abstract

*Leishmania*sis is a group of diseases with diverse phenotype from self-healing cutaneous lesions to fatal visceral infections. The complex life-cycle of the parasite *Leishmania* in its vector and host as well as the vector and host characteristics determine the severity of infection. *Leishmania* has evolved several strategies to survive as extracellular promastigotes in the sandfly vector and successfully transform to amastigotes sensing an increase in temperature and acidic pH inside the macrophages. In this review we have discussed the physiological and biochemical changes in promastigotes which make them infective and how these infection ready parasites respond to intramacrophagic stress. We have summed up the changes in expression of structural proteins, virulence factors (including species-specific virulence factors), and stress related proteins when the parasites convert to metacyclic promastigotes and then from metacyclics to amastigotes, and how these pathways have been targeted for therapy. The trends in global gene expression of *Leishmania* under various conditions and at various life-stages using modern molecular biology approaches such as genomics, transcriptomics, proteomics and metabolomics have been discussed in conjunction with the theory of pre-adaptation of the parasites for life inside macrophages.

Introduction

*Leishmania*sis is a vector-borne disease ranging from cutaneous to visceral infections caused by different species of *Leishmania*. Visceral leishmaniasis (VL) or kala-azar is a fatal parasitic infection caused by protozoan parasite *Leishmania donovani* and *Leishmania infantum* while self-healing cutaneous or mucocutaneous lesions are caused by *Leishmania major, Leishmania mexicana, Leishmania amazonensis*, etc. Cutaneous ulcers are reported in approximately 700,000–1,200,000 individuals every year (www.cdc.gov/parasites/leishmaniasis/epi.html), while approximately 200,000–400,000 cases of VL occur each year, with majority of cases reported in India, Bangladesh and Nepal (www.cdc.gov/parasites/leishmaniasis/epi.html; Rock *et al.*, 2015). Although host- and vector-driven factors have considerable effect on the infection, parasites are the major determinants of disease outcome. The proteins, lipids, carbohydrates and even nucleic acids of the parasite display differences from species to species as well as between the different stages of the parasite. These differences have evolved from complex genetic adaptations of the parasite that are being extensively investigated by scientists to unravel the disease biology and design clinical interventions.

Promastigotes are the infective stage of *Leishmania* and they are transmitted to the host by bite of a sandfly vector. Metacyclic promastigotes which develop from procyclic forms, both in culture and sandfly gut, can be used as a model to study virulent infection and the factors responsible for virulence can lead to identification of prospective drug targets. Several attempts have been made to detect these virulence factors using genomic, transcriptomic, metabolomic, proteomic, immunologic and bioinformatic approaches coupled with *in vitro* and *in vivo* verification of the identified leads. Although intracellular form of the parasite disseminates the infection inside host, a major factor governing virulence is the preparedness of the promastigote to enter and survive in the intramacrophagic environment which accounts for parasite infectivity and the severity of infection. More interesting is the fact, that during its life cycle *Leishmania* encounters drastically different environmental changes, some of which induce differentiation of the parasite. In the subsequent sections we discuss the important determinants of infectivity in the promastigote form of *Leishmania* spp. which bestow these parasites with invasiveness and preparedness to continue their successful journey as amastigotes. Some proteins which give the visceral species added invasiveness is also discussed. Finally we talk about a global overview of the changes in gene expression of the promastigotes under different conditions of growth and physiological pressure.

Membrane and cytoskeleton-sensing the changes

The amastigotes are ingested by the sandfly during blood meal and get transformed into metacyclic promastigotes after several morphological changes from procyclic to metacyclic form. The metacyclic promastigotes come to the mouth parts of the sandfly from where they are injected into another host during subsequent blood meal. Metacyclic forms have also been isolated from stationary phase cultures (Yao *et al.*, 2008; Saraiva *et al.*, 2005). A plethora of studies described major differences between procyclic and metacyclic promastigotes with special emphasis on virulence factors. As with most pathogens, the initial host–parasite interaction occurs via the cell membrane and thus membrane components are significant virulence factors. It is well known that the major surface glycoprotein gp63 and lipophosphoglycan (LPG) are important virulence factors in *Leishmania*. Infective *Leishmania* exhibit a thickened glycocalyx due to elongation of the phosphoglycan chain of the surface LPG (Bogdan *et al.*, 1996). This feature protects the metacyclics from complement mediated lysis. The initial binding occurs through LPG and gp63 on the parasite surface and their subsequent internalization into phagolysosomes where they convert to amastigotes in approximately 12–24 hours. LPG has been proposed to play many roles in the infectious cycle, including protection against complement and oxidants, serving as the major ligand for macrophage adhesion and as a key factor mitigating host responses by deactivation of macrophage signalling pathways (Spath *et al.*, 2000). Studies comparing infective cultured metacyclic promastigotes with immature procyclic forms also revealed a fall in unsaturated fatty acid levels responsible for the decreased membrane fluidity in the former (Silva *et al.*, 2011). Decrease in membrane fluidity coupled with drastic changes in lipid composition, particularly incorporation of sterol fatty acids in the membrane render them more heat resistant and virulent (Xu *et al.*, 2014).

The zinc metallopeptidase gp63 is another most widely studied virulence factor with role in evasion of complement mediated lysis, parasite phagocytosis, interaction with the host extracellular matrix, host immune suppression, and alteration of host lipid metabolism (Thiakaki *et al.*, 2006; Olivier *et al.*, 2012; Lambertz *et al.*, 2012; Ghosh *et al.*, 2013). By using differential passages of *Leishmania chagasi* Beetham and his group (Beetham *et al.*, 2003) showed decreased expression of promastigote surface antigen (gp46) in late passage (less virulent) promastigotes. Membrane bound acid phosphatases are enzymes implicated in dephosphorylation of organic substrates and are released from promastigotes, as well as amastigotes, from the flagellar pocket into the extracellular space (Lambertz *et al.*, 2012). *Leishmania donovani* virulent and avirulent lines were used to prove both quantitative and qualitative differences in acid phosphatase activity between the two clones and higher production was reported in the virulent phase (Katakura and Kobayashi, 1988; Singla *et al.*, 1992). Although having stage-specific activity in amastigotes, amastins, a group of surface glycoproteins have been implicated in host–parasite interaction as δ-amastin knock-out parasites showed impaired *in vitro* infection and failed to cause infection in mice (de Paiva *et al.*, 2015). Infectivity was restored following re-expression of an RNAi-resistant amastin gene. Thus it is obvious that alterations in parasite membrane have profound effect on the initial interaction with host cells.

As promastigotes are motile, abundance of cytoskeletal and flagellar proteins ensures proper flagellar motility, an important feature of virulent stage. Comparative proteomics revealed higher expression of paraflagellar rod protein 2C and 1D in metacyclic promastigotes (Moore *et al.*, 1996; Mojtahedi *et al.*, 2008). Kinetoplastid membrane protein 11 (KMP11), another cytoskeletal protein displays stage-specific up-regulation in the promastigotes (Lynn *et al.*, 2013). Another stage-specific protein CAP 5.5, indicative of virulence phenotype was reported in *L. chagasi* promastigotes (Yao *et al.*, 2010). Sensing the environmental changes is an important determinant of parasite adaptation to its surroundings. *Leishmania* are equipped with a well-developed sensory system based on mitogen activated protein kinases (MAPKs). These are involved in environmental sensing, virulence and control of flagellar length. The functional significance of MAPKs during developmental changes is beautifully reviewed by Rotureau *et al.* (Rotureau *et al.*, 2009). Various MAPKs have been implicated in parasite survival, virulence and flagellar targeting and put forth as interesting drug targets (Cayla *et al.*, 2014; Wang *et al.*, 2005; Bengs *et al.*, 2005;). In addition, several stress activated protein kinases are produced by the parasite in response to environmental signals which regulates chaperone activity during differentiation and infection (Spath *et al.*, 2015).

Membrane spanning proteins of the ABC transporter family play important roles in substrate transport across membrane and in drug resistance (Linton, 2007). These proteins confer increased fitness to the parasites and are thus important virulence determinants. In *L. major* a mitochondrial ABC transporter LmABCB3 was proposed as a drug target based on its role in the generation of haem and cytosolic iron/sulphur cluster, while its deletion was lethal to the parasite growth and multiplication (Martinez-Garcia *et al.*, 2016). Another haem trafficker of the ABC family (LABCG5) was reported to be crucial for *L. donovani* survival (Campos-Salinas *et al.*, 2011). Experiments with drug sensitive and resistant *L. donovani* parasites showed increased virulence and host modulatory capacity of the latter by virtue of overexpressed multidrug-resistant protein 1, an ABC transporter (Mukherjee *et al.*, 2013). Although debatable, the externalization of phosphatidylserine (PS) has been described as an important determinant of promastigote as well as amastigote virulence (El-Hani *et al.*, 2012). Studies in both cutaneous and visceral species of *Leishmania* displayed increased PS exposure on the outer membrane of metacyclic promastigotes conferring them enhanced virulence and bypass host microbicidal activity (Campos-Salinas *et al.*, 2013). The involvement of ABC transporters in this activity was evidenced by the studies of Campos-Salinas and his group who showed that LABCG2 is responsible for the PS floppase activity (Campos-Salinas *et al.*, 2013). Other transporters having roles in parasite virulence are mostly related to nutrient acquisition like glucose and iron transporters. Omission of glucose transporter gene in *L. mexicana* renders the parasite susceptible to oxidative damage and compromises parasite survival and virulence (Akpunarlieva *et al.*, 2017). The role of iron transporters in growth and virulence of *Leishmania* (Flannery *et al.*, 2011; Jacques *et al.*, 2010; Mittra *et al.*, 2016) has pointed towards iron acquisition as an important determinant of parasite life cycle switch and pathogenicity. Recently, the function of iron-mediated ROS generation as a trigger for differentiation of promastigotes to amastigotes was shown by Mittra *et al.* (2013). Thus, not only the intracellular form but the extracellular promastigotes also sense and utilize iron for successful morphological transformation once inside macrophages.

Parasite exoproteome-modulating the host

Parasites throw out proteins as extracellular signals to their immediate environment. Membrane proteins, secretory proteins as well as proteins expelled due to cell death form part of the exoproteome (Santarem *et al.*, 2013). These proteins have been highly regarded recently as pathological and immunomodulatory signals in *Leishmania* biology (van Zandbergen *et al.*, 2006; Gour *et al.*, 2012). A detailed time kinetics based approach was undertaken to identify stage-specific exoproteome of *L. infantum* which beautifully differentiated the protein components based on their functional role in the parasite at logarithmic and stationary phases (Santarem *et al.*, 2013). The study revealed that stationary promastigotes shed proteins involved in folding, sorting and degradation, proteasome and spliceosome related gene ontology (GO) assignments directly into the medium while transport and catabolism, energy metabolism, replication and repair and chromosome along with many plasma membrane bound proteins are enriched in the vesicles. On the other hand actively dividing logarithmic parasites displayed overrepresentation of gene information processing and metabolic pathway GO terms. Proteins such as gp63, cysteine peptidases (Gerbaba and Gedamu, 2013), Activated Protein Kinase C Receptor (Gomez-Arreaza *et al.*, 2011), KMP11 (Lynn *et al.*, 2013), oligopeptidase B (Swenerton *et al.*, 2011), enolase (Avilan *et al.*, 2011), etc. featured among vesicle bound proteins have known role in virulence in different species of *Leishmania*. Cysteine peptidase b has been proposed as a modulator of gp63 expression in *L. mexicana* (Casgrain *et al.*, 2016). Another protein elongation factor 1 alpha which is secreted out of the promastigotes is known to modulate host macrophage activity by binding Src homology two domain-containing tyrosine phosphatase-1 and deactivating them (Nandan *et al.*, 2002). A lot of other proteins have been known to be secreted out of the cells and are discussed in detail in other sections. Apart from proteins, extracellular vesicles also transmit nucleic acid. The expulsion of small regulatory RNAs via vesicles represents a unique way of modulating the transcriptional machinery during infection (Szempruch *et al.*, 2016).

Cytosolic proteins: a pool of stress busters

The cytosol of the parasite harbours many important virulence factors, mostly enzymes involved in metabolic and signalling pathways. These proteins mostly guard the parasites against oxidative damage and temperature related stress, while some of them act as virulence factors by modulating host cell. The crucial role of cytosolic tryparedoxin peroxidise (TPX) in detoxification of peroxides has been well documented (Iyer *et al.*, 2008). This protein is not only an important virulence factor but has a role in survival and drug response of the parasite (Iyer *et al.*, 2008; Suman *et al.*, 2016). Using TPX overexpressing *L. donovani* promastigotes it was shown that these parasites can withstand peroxide and nitric oxide exposure better along with increased virulence and defended antimonial therapy better than their vector transfected counterparts (Iyer *et al.*, 2008). Although the role of TPX in intracellular amastigotes is more pronounced (Pescher *et al.*, 2011), it is shown that infection-ready promastigotes have an abundance of this protein supporting the pre-adapted hypothesis of *Leishmania* virulence (Alcolea *et al.*, 2016a,c). Interestingly a mitochondrial TPX was identified in *L. infantum* with role in virulence, independent of its peroxidase activity and is thus redundant in the insect stage of the parasite (Castro *et al.*, 2011). The depletion of mitochondrial TPX although did not cause defects in *in vitro* growth of the parasite, significantly reduced its infectivity in BALB/c mice. This property was attributed to a molecular chaperone activity of the protein in combating temperature shift when the parasite is introduced into the host. Careful mining of oxidative stress management in infection ready promastigotes also point towards other peroxidases such as ascorbate peroxidase and glutathione peroxidise which are of more relevance during the extracellular life (Castro and Tomas, 2008). Studied in detail in *L. major*, the ascorbate peroxidise is the only haem peroxidise identified so far in *Leishmania* (Adak and Pal, 2013), with orthologues found in other species. It is co-localized with cytochrome c and defends against oxidative stress while the parasite transforms into infective metacyclics, after which other combating mechanisms take over. Thus, a stage-specific mechanism of tackling oxidative damage is another distinctive strategy of these pathogens. Casein kinase 1, an important enzyme involved in eukaryotic signalling pathways has multiple isoforms in *Leishmania* (Dan-Goor *et al.*, 2013). One of the isoforms Ldck1.4 is unique to the parasite and secreted by the promastigotes. This protein has also been implicated in promastigote survival and virulence (Dan-Goor *et al.*, 2013). Recently casein kinase 2 was also reported to be higher in the secretory proteins of virulent metacyclics of

Leishmania braziliensis where it could modulate parasite–macrophage association (Zylbersztejn *et al.*, 2015)

Molecular chaperones and heat-shock proteins (HSPs) play an important role in parasite growth and survival under stress. Heat-shock proteins are highly conserved among organisms and in a parasite with digenetic life cycle like *Leishmania*, these proteins have crucial roles to play when the parasites encounter sudden changes in temperature or pH. In *L. donovani* HSP100 has been relevant to exosome biogenesis and a proteomic analysis showed that exosomes from the wild-type and a Δhsp100 *L. donovani* mutant have distinct protein cargo, a phenomenon partly attributed to HSP100 (Silverman *et al.*, 2010). Moreover, the same group showed that deletion of HSP100 could reverse the immunosuppressive nature of the exosomes, thus altering the host immune response to a protective phenotype (Silverman *et al.*, 2010). This protein has also been shown to provide thermotolerance (Hubel *et al.*, 1995) and withstand stress by refolding of unfolded proteins (Schaupp *et al.*, 2007) during initial host–parasite interaction and parasite transformation (Hubel *et al.*, 1997). Putative class 2 HSP100s having sequence conservation with *Escherichia coli* ClpY (HslU) and ClpQ (HslV) were annotated in *L. major* with involvement in proteolysis of misfolded proteins (Missiakas *et al.*, 1996). In *L. donovani*, HslU has been linked to reduced promastigote growth although no effect on viability was reported (Chrobak *et al.*, 2012). The HSP90 family is the most abundant protein even under optimal growth conditions and in unstressed *Leishmania* promastigotes, HSP90 make up 2.8% of the total protein (Requena *et al.*, 2015). These proteins are involved in important cellular functions such as homeostasis, proliferation, differentiation and apoptosis and link the cellular compartment to differential environmental impact (Requena *et al.*, 2015). It is found in three forms. The cytoplasmic protein is called HSP83 while the endoplasmic reticulum paralogue is dubbed Grp94 (Requena *et al.*, 2015). Heat shock protein 83 has specific role in differentiation as shown in *L. infantum*, where its level rises during initial phases of promastigote to amastigote transformation (Larreta *et al.*, 2004). In *L. donovani*, Grp94 has crucial involvement in LPG biosynthesis, an important promastigote virulence factor (Descoteaux *et al.*, 2002). The third member, a mitochondrial variant of HSP90 called TRAP-1 or HSP75 has not been fully characterized (Requena *et al.*, 2015). Furthermore, this family of stress proteins works alongside a number of co-chaperones which interact with HSP90 and aid in its chaperonic activity (Requena *et al.*, 2015). Multiple genes encoding for the HSP70 family of proteins exist in *Leishmania* (Requena *et al.*, 2015). They have been extensively studied for their role in host defence mechanism of the parasite particularly when they are introduced into the host macrophages (Folgueira *et al.*, 2005, 2008; Miller *et al.*, 2000). Heat shock protein 70 partners with HSP40 (DNAJ superfamily) for their function. Although not studied in great detail HSP40 expression is shown to be increased during promastigote to amastigote differentiation (Tsigankov *et al.*, 2014). Thus, most of the HSPs have vital contributions in preparing the parasite for an environmental and morphological switch. Apart from usual HSPs, another virulence factor A2 found to be typically expressed in visceral species has a role in heat and oxidative resistance (McCall and Matlashewski, 2010, 2012). This protein is pseudogenized in *L. major* and considered to be expressed specifically in amastigotes of visceral species where high fever is common. Thus, species-specific differences in virulence genes account for the greater invasiveness of visceralizing species over cutaneous ones.

Different classes of proteases play an important role in life cycle and infectivity of *Leishmania* (Besteiro *et al.*, 2007) (Fig. 9.1). Metallopeptidases and cysteine peptidases have direct role in imparting virulent phenotype by modulating host responses. Recently calpain-like peptidases have caught attention of biologists for their role in parasitic protists such as *Leishmania*. Calpains are calcium-dependent cysteine peptidases identified and studied both in mammals as well as lower eukaryotes, and form a large and diverse family of proteins in trypanosomatids (Ersfeld *et al.*, 2005). These proteins perform varied calcium-regulated functions in the parasite, such as signal transduction, cytoskeleton remodelling, cellular proliferation and differentiation, sex determination, membrane fusion, environmentally regulated processes and apoptosis (Branquinha *et al.*, 2013). The abundance of these proteins in *Leishmania* and their life cycle-specific differential expression may be linked to genetic adaptability of the parasites to its immediate environment (Saxena

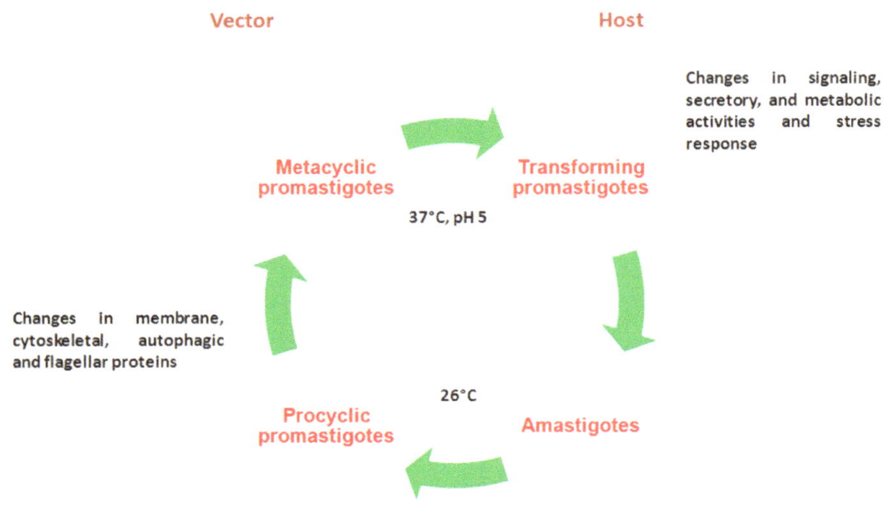

Figure 9.1 Life cycle of *Leishmania. Leishmania* procyclic promastigotes proliferate inside the alimentary tract of phlebotomine sandflies (vector) and mature to virulent metacyclic promastigotes. During a blood meal by the sandfly, metacyclic promastigotes are injected into vertebrate host where it transforms into intracellular amastigotes under the conditions of high temperature and low pH encountered within phagolysosomes of host macrophages. The motile promastigotes require active flagella and well-defined surface components for successful host invasion while once inside the host, these parasites require extensive signalling and secretory mechanisms to respond to temperature and pH induced stress while converting to amastigotes.

et al., 2003). Apparently it appears that calpains up-regulated in the actively dividing forms are involved in morphogenesis and signal transduction activities (Galetovic *et al.*, 2011) while those up-regulated in the quiescent and infective forms mostly possess proteolytic activity and may contribute to virulent phenotype (Ennes-Vidal *et al.*, 2017).

The metabolic adaptations

The whole metabolome of the parasite undergoes changes in order to adapt to the different hosts, namely the insect and the mammal (Rosenzweig *et al.*, 2008b). *Leishmania* promastigotes reside in an extracellular glucose-rich, slightly alkaline environment in the sandfly vector alimentary tract. The initial phase of entry and conversion to amastigote form is very demanding in terms of nutrient acquisition and coping up with the acidic environment. In the macrophage phagolysosome glucose scarcity is supplemented by amino acids. Thus the parasite adapts to use amino acids and fatty acids as the main source of energy and there is rise in gluconeogenesis while glycolysis level drops (Rosenzweig *et al.*, 2008b). At this time the metabolic enzymes for beta-oxidation, amino acid catabolism, tricarboxylic acid cycle, mitochondrial respiratory chain, and oxidative phosphorylation also increase to deal with the changing nutrient availability (Rosenzweig *et al.*, 2008b). Additionally the metabolic pathways directed to tackle oxidative and nitrosative stress are overexpressed including trypanothione and superoxide metabolism (Sardar *et al.*, 2013). Glycosomes are important glycolytic hubs and many of the glycosomal enzymes vary greatly in terms of quantity and quality between the life cycle stages of trypanosomatids. Looking at the metabolic reprogramming of promastigotes inside host cell and reduction in glycolysis (Haanstra *et al.*, 2016), glycosome turnover and enzymes of this pathway have crucial effect on parasite virulence and tropism. An enzyme of the glycolytic pathway present in the glycosomal compartment, glyceraldehyde-3-phosphate dehydrogenase (GAPDH) has been explored as a drug target (de Mesquita Barbosa *et al.*, 2015). More recently it has been shown that the enzyme gives *L. chagasi* and *L. amazonensis* promastigotes a greater tolerance to nitric oxide induced stress (Rios *et al.*, 2015). Another significant finding was that in *L. donovani* and *L. mexicana* a cytosolic variant of the GAPDH (cGApDH) is present. Using cGAPDH null mutant of *L. donovani*, Zhang and his colleagues showed that the gene is not only functional in the species but is important in establishing visceral infection (Zhang *et al.*, 2013). On the other hand, this gene has been pseudogenized in *L. major*, explaining the species-specific invasive capacity of the parasites. Further using ^{1}H NMR spectroscopy in *L. major*, it was shown that metacyclogenesis i.e. conversion of promastigotes to the infective form involves altered valine, leucine and isoleucine biosynthesis, pyruvate metabolism, glutathione metabolism, fatty acid biosynthesis, and sulfur metabolism (Arjmand *et al.*, 2016). As already mentioned in the previous section acquisition and metabolism of iron plays a significant role in both promastigote and amastigote stages. While amastigotes utilize haem from blood, promastigotes mainly take up iron for using iron superoxide dismutase to produce reactive oxygen species which acts as a trigger for cell differentiation (Mittra *et al.*, 2013).

Apart from glucose metabolism, metabolism of phospholipids and sphingolipids has major implications in *Leishmania* pathogenicity. These are abundantly found on the membrane and act as reservoirs of important signalling molecules. Many enzymes involved in sphingolipid and phospholipid metabolism have been studied as drug targets due to their importance in parasite viability (Zhang and Beverley, 2010). Interestingly, in *L. major* it was shown that the cellular level of plasmenyl ethanolamine (a phospholipid) rises significantly at the onset on stationary phase, indicating a critical role in differentiation (Zhang and Beverley, 2010). This lipid possesses rapid membrane fusion capacity (Glaser and Gross, 1994) and may aid in the autophagic process which accompany metacyclogenesis (Besteiro *et al.*, 2006). Authophagic digestion of cellular components is a strategy exploited by the parasites to undergo morphological alteration and this occurs when the actively dividing cells attain stationary phase and convert to infective forms. Failure to do so results in stunted growth and impaired virulence of the promastigotes (Besteiro *et al.*, 2006). It is also noted that autophagic processes are involved in transformation of promastigotes to amastigotes and are thus crucial virulence determinants (Cull *et al.*, 2014). Another related pathway having profound effect on the proliferation and differentiation of virulent parasites is the ubiquitin–proteasome system. Several studies observed stage-specific proteasomal activity in *Leishmania* spp. regulating protein expression. It is very interesting that several important enzymes such as pteridine reductase 1 (PTR1) (Kumar *et al.*, 2007), a kinesin called LmjKIN13-1 (Dubessay *et al.*, 2006), methionine adenosyltransferase (Perez-Pertejo *et al.*, 2011), etc., which are needed at specific growth phases of the promastigotes or amastigotes are degraded or down-regulated by proteasomal system.

Folic acid and pteridines are important cofactors in all organisms. Leishmanial parasites are auxotrophic for both folate and pteridines and acquire these metabolites from their insect and mammalian hosts. Inside the parasite these get reduced by a bifunctional dihydrofolate reductase-thymidylate synthase and the newly identified PTR1 (Nare *et al.*, 2009). Although *Leishmania* has evolved alternative metabolic pathways wherein it does not require many metabolic intermediates of folate pathway and instead salvages them from the host, it was shown that loss of this function attenuates infectivity at least in animals (Vickers and Beverley, 2011). Purines and pyrimidines also serve vital function in all organisms including signalling, biosynthesis and energy storage molecule (Carter *et al.*, 2008). *Leishmania* are auxotrophic for purines and prototrophic for pyrimidines. Interestingly, the pathway involved in assimilation of adenosine nucleotide is uniquely represented in both life forms of *Leishmania*. Adenosine kinase is the first enzyme of this pathway which displays a stage-specific activity pattern having some implication in digenetic life cycle of the parasite (Datta *et al.*, 2008). Other purine salvage enzymes, adenylosuccinate synthetase and adenylosuccinate lyase have been identified in *L. donovani* and their deletion led to attenuated infection *in vitro* and *in vivo* (Boitz *et al.*, 2013). Pyrimidines can be both synthesized *de novo* or salvaged from the host in *Leishmania*. Although the salvage pathway compensates for the *de novo* route for survival, inhibition of the latter has dire effects on the virulence of the parasite as observed in *L. donovani* (Wilson *et al.*, 2012). Dihydroorotase is the third enzyme of the synthetic pathway and it has been studied extensively for development of inhibitors owing to its crucial role in parasite viability (Tiwari *et al.*, 2016).

The global picture of parasite preadaptation

Most of the studies in earlier times were based on identification and characterization of single gene or its product to elucidate its role in the infection process. Several newer techniques in molecular biology together with bioinformatics have now facilitated a greater understanding of different aspects of *Leishmania* adaptation and infectivity and comprehension of the global gene expression under different conditions. Next generation sequencing and high throughput proteomics studies have immensely helped in unravelling the trends in gene expression of both extracellular and intracellular stages of the parasite, distinguish avirulent parasite from virulent ones and identify key differences in axenically grown and intracellular or sandfly gut developed parasites.

The axenically cultured promastigotes have immensely eased the study of different aspects of *Leishmania* biology as they are said to mimic sandfly gut derived parasites to a large extent. An elegant study was taken up by Alcolea *et al.* in *L. infantum* to elucidate the differential gene expression in sandfly-derived promastigotes and amastigotes by Shotgun DNA microarray hybridization analyses (Alcolea *et al.*, 2014). The published data on axenically cultured promastigote gene expression was also compared with promastigotes derived from sandfly foregut to point out the differences in profile. Sandfly derived promastigotes displayed a higher rate of up-regulation as compared to amastigotes, and the up-regulated genes mostly belonged to gene expression regulation and cell signalling (Fig. 9.1). In fact one study proposed that amastigote-like forms are present in the stomodeal valve of sandfly along with the metacyclics owing to the slightly higher temperature and lower pH condition when the sandfly feeds (Anez *et al.*, 2003). This along with transcriptomic profiling of the effect of temperature and pH shift in leishmanial transformation to amastigotes (Alcolea *et al.*, 2010) further strengthens the concept of preadaptation in *Leishmania* promastigotes. In terms of transcript abundance considerable variation was observed in axenic and insect derived promastigotes when compared to amastigotes warranting stringency when cultured parasites are used for studies (Alcolea *et al.*, 2014). Recently, in another study, it was shown that sandfly gut-derived promastigotes are more infective than their cultured counterparts enriched in stationary phase parasites, and this phenomenon was attributed to differential expression of surface molecules such as gp63 and LPG as well as cell signalling molecules (Alcolea *et al.*, 2016a). Global gene expression patterns between insect derived and cultured promastigotes point out differential microenvironmental adaptations which must be taken into consideration while using axenic *Leishmania* for infectivity related studies. A moderately strong positive correlation was reported between naturally developed and culture derived promastigotes (Alcolea *et al.*, 2016b), which authenticates the latter's use but with due caution.

Proteomic comparison of promastigote and amastigotes isolated from THP-1-infected cell lines revealed increased expression of motility and cytoskeletal proteins in promastigotes while cell signalling and vesicular transport are enhanced in the amastigote stage (Biyani and Madhubala, 2012). Nevertheless, we have already discussed about the importance of flagellar and cytoskeletal proteins in macrophage targeting and initial infection in previous section. In general most of the stage-specific proteins belong to five groups, each with functionally related proteins viz. heat and oxidative stress response, cytoskeleton and cell membrane, energy metabolism and phosphorylation, cell cycle and proliferation and amino acid metabolism (Bente *et al.*, 2003), and we have already appreciated the roles of individual proteins under these categories. Major post-translational modifications during differentiation in *L. donovani* included methylation of arginines, aspartic acids, glutamic acids, asparagines, and histidines; acetylation of serines and protein N-terminal acetylations on methionines, serines, alanines, and threonines; and phosphorylations on serines and threonines, but not tyrosines. Novel fucosylation and hexosylations sites were also found using isobaric tagging for relative and absolute quantifications (iTRAQ/LC-MS/MS) (Rosenzweig *et al.*, 2008b). Phosphorylation and dephosphorylation mark one of the crucial signals in *Leishmania* differentiation. Phosphoproteomic data of *L. donovani* promastigote-to-amastigote transformation revealed increased phosphorylation during initial 5 h of interaction and again during onset of amastigote morphogenesis between 10 and

24 hours of infection (Tsigankov *et al.*, 2014). On the other hand there exist intermittent phases when dephosphorylation predominates, i.e. between 5 and 10 hours post infection and again during the maturation phase when parasites are ready for dissemination. This revelation strongly supports the idea that promastigotes are infection ready when they encounter their host for the first time and immediately convey signals that initiate adaptation to higher temperatures and lower pH. A few studies also revealed important differences in early and late passage promastigotes in terms of protein expression. As already known, promastigotes axenically cultured for several passages tend to lose virulence (Moreira *et al.*, 2012). The factors responsible for this loss of virulence were assessed by comparative proteomics which revealed an increased ability of the early passage parasites to combat oxidative and nitrosative stress, as well as increased abundance of virulence factors which enable them to actively interact with and invade the host unlike later passages (Moreira *et al.*, 2012; Magalhaes *et al.*, 2014).

Recently RNASeq technology has been exploited for exploring the changes in *Leishmania* gene expression when it converts from promastigote to amastigote inside macrophages. In *L. mexicana*, RNASeq analysis of promastigote transformation to amastigotes not only revealed differentially expressed genes but also indicated tetraploidy in chromosome 30 which contained most of the genes that were up-regulated in amastigotes (Fiebig *et al.*, 2015), thus reinstating the parasite's adaptation to intracellular life. Many attempts were made to study the transcriptome of different *Leishmania* spp. under differential conditions using RNASeq (Rastrojo *et al.*, 2013; Martin *et al.*, 2014; Haydock *et al.*, 2015; Eliaz *et al.*, 2015; Dillon *et al.*, 2015a,b). Temporal gene expression studies in *L. major* and *L. amazonensis* following infection of macrophages with the promastigotes revealed that major modulation occurs following initial host–parasite interaction or early infection, after which there is a drop in the number of differentially expressed genes both in the parasite and host (Dillon *et al.*, 2015b; Fernandes *et al.*, 2016). These studies provided a comprehensive database for understanding signatures genes of parasite as well as host and how their interaction modulates infection process.

Concluding remarks

Leishmania has a complex life cycle with very contrasting physiological demands. There are two major events in the life cycle of *Leishmania*, first is the generation of metacyclics, which is *ex* host and secondly the transformation to amastigotes which mount the infectivity and affect disease outcome. Promastigotes are preadapted to interact with host cells, thrive inside macrophages, modulate intramacrophagic environment for their survival and finally convert to amastigote form. It is evident from the plethora of studies that cell surface and more importantly cytoskeletal molecules play the most important role in host–parasite interaction, and several proteins are sent out from the parasites as signals to the host cell to facilitate infection. Apart from these, parasite's ability to switch carbon sources and exploit nutrients available in the immediate environment make them fit enough to withstand intracellular stress. Finally, overcoming oxidative and nitrosative stress mark the final steps in establishing infection. A complex yet sequential pattern of changes in the gene expression and appearance of virulence factors ensure safe and successful transformation of extracellular promastigotes to amastigotes. Whereas failure to metacyclogenise can give avirulent phenotype, malfunctioning of promastigote-to-amastigote differentiation can completely jeopardize the intracellular machinery. Careful mining of these pathways and proteins can lead to the identification of many important virulence factors and drug targets. While a lot of these studies are based on axenic promastigote based experiments, it is advisable to standardize culture conditions to ensure relevance, particularly in infection models. High throughput transcriptomic and proteomic analysis have immensely helped in understanding important aspects of gene expression in promastigotes when they are introduced into host cells and we strongly feel that in addition to amastigote or promastigote-specific virulence factors, pathways that show up alterations during morphological transformation are relevant targets for clinical interventions. These gene products act as a trigger for life cycle switch when the parasites undergo a very delicate transition phase and also represent the unique set of proteins required for intracellular adaptation. Although there is still a dearth of understanding in factors governing different forms of leishmaniasis, the information generated from 'omics' studies coupled with bioinformatics can identify global trends in parasitic behaviour and adaptability of intracellular pathogens.

References

Adak, S., and Pal, S. (2013). Ascorbate peroxidase acts as a novel determiner of redox homeostasis in *Leishmania*. Antioxid. Redox Signal. *19*, 746–754. https://doi.org/10.1089/ars.2012.4745.

Akpunarlieva, S., Weidt, S., Lamasudin, D., Naula, C., Henderson, D., Barrett, M., Burgess, K., and Burchmore, R. (2017). Integration of proteomics and metabolomics to elucidate metabolic adaptation in *Leishmania*. J Proteomics *155*, 85–98.

Alcolea, P.J., Alonso, A., Degayon, M.A., Moreno-Paz, M., Jimenez, M., Molina, R., and Larraga, V. (2016a). *In vitro* infectivity and differential gene expression of *Leishmania infantum* metacyclic promastigotes: negative selection with peanut agglutinin in culture versus isolation from the stomodeal valve of *Phlebotomus perniciosus*. BMC Genomics 17, 375.

Alcolea, P.J., Alonso, A., Domínguez, M., Parro, V., Jiménez, M., Molina, R., and Larraga, V. (2016b). Influence of the Microenvironment in the Transcriptome of *Leishmania infantum* Promastigotes: Sand Fly versus Culture. PLOS Negl. Trop. Dis *10*, e0004693. https://doi.org/10.1371/journal.pntd.0004693.

Alcolea, P.J., Alonso, A., Garcia-Tabares, F., Mena, M.C., Ciordia, S., and Larraga, V. (2016c). Increased abundance of proteins involved in resistance to oxidative and nitrosative stress at the last stages of growth and development of *Leishmania amazonensis* promastigotes revealed by proteome analysis. PLOS ONE 11, e0164344.

Alcolea, P.J., Alonso, A., Gomez, M.J., Postigo, M., Molina, R., Jimenez, M., and Larraga, V. (2014). Stage-specific differential gene expression in *Leishmania* infantum: from the foregut of *Phlebotomus perniciosus* to the human phagocyte. BMC Genomics 15, 849.

Alcolea, P.J., Alonso, A., Gomez, M.J., Sanchez-Gorostiaga, A., Moreno-Paz, M., Gonzalez-Pastor, E., Torano, A., Parro, V., and Larraga, V. (2010). Temperature increase prevails over acidification in gene expression modulation of amastigote differentiation in *Leishmania* infantum. BMC Genomics 11, 31.

Añez, N., Tang, Y., Rojas, A., Crisante, G., Killick-Kendrick, M., and Killick-Kendrick, R. (2003). Detection of amastigote-like forms in the valve of Phlebotomus papatasi infected with *Leishmania major*. Mem. Inst. Oswaldo Cruz *98*, 495–498.

Arjmand, M., Madrakian, A., Khalili, G., Najafi Dastnaee, A., Zamani, Z., and Akbari, Z. (2016). Metabolomics-Based Study of Logarithmic and Stationary Phases of Promastigotes in *Leishmania major* by 1H NMR Spectroscopy. Iran. Biomed. J. *20*, 77–83.

Avilan, L., Gualdron-Lopez, M., Quinones, W., Gonzalez-Gonzalez, L., Hannaert, V., Michels, P.A., and Concepcion, J.L. (2011). Enolase: a key player in the metabolism and a probable virulence factor of trypanosomatid parasites-perspectives for its use as a therapeutic target. Enzyme Res 2011, 932549.

Beetham, J.K., Donelson, J.E., and Dahlin, R.R. (2003). Surface glycoprotein PSA (GP46) expression during short- and long-term culture of *Leishmania chagasi*. Mol. Biochem. Parasitol. *131*, 109–117.

Bengs, F., Scholz, A., Kuhn, D., and Wiese, M. (2005). LmxMPK9, a mitogen-activated protein kinase homologue affects flagellar length in *Leishmania mexicana*. Mol Microbiol 55, 1606–1615.

Bente, M., Harder, S., Wiesgigl, M., Heukeshoven, J., Gelhaus, C., Krause, E., Clos, J., and Bruchhaus, I. (2003). Developmentally induced changes of the proteome in the protozoan parasite *Leishmania donovani*. Proteomics 3, 1811–1829.

Besteiro, S., Williams, R.A., Coombs, G.H., and Mottram, J.C. (2007). Protein turnover and differentiation in *Leishmania*. Int. J. Parasitol. 37, 1063–1075.

Besteiro, S., Williams, R.A., Morrison, L.S., Coombs, G.H., and Mottram, J.C. (2006). Endosome sorting and autophagy are essential for differentiation and virulence of *Leishmania major*. J. Biol. Chem. *281*, 11384–11396.

Biyani, N., and Madhubala, R. (2012). Quantitative proteomic profiling of the promastigotes and the intracellular amastigotes of *Leishmania donovani* isolates identifies novel proteins having a role in *Leishmania* differentiation and intracellular survival. Biochim. Biophys. Acta 1824, 1342–1350.

Bogdan, C., Gessner, A., Solbach, W., and Röllinghoff, M. (1996). Invasion, control and persistence of *Leishmania* parasites. Curr. Opin. Immunol. *8*, 517–525.

Boitz, J.M., Strasser, R., Yates, P.A., Jardim, A., and Ullman, B. (2013). Adenylosuccinate synthetase and adenylosuccinate lyase deficiencies trigger growth and infectivity deficits in *Leishmania donovani*. J. Biol. Chem. 288, 8977–8990.

Branquinha, M.H., Marinho, F.A., Sangenito, L.S., Oliveira, S.S., Goncalves, K.C., Ennes-Vidal, V., d'Avila-Levy, C.M., and Santos, A.L. (2013). Calpains: potential targets for alternative chemotherapeutic intervention against human pathogenic trypanosomatids. Curr. Med. Chem. 20, 3174–3185.

Campos-Salinas, J., Cabello-Donayre, M., García-Hernández, R., Pérez-Victoria, I., Castanys, S., Gamarro, F., and Pérez-Victoria, J.M. (2011). A new ATP-binding cassette protein is involved in intracellular haem trafficking in *Leishmania*. Mol. Microbiol. 79, 1430–1444. https://doi.org/10.1111/j.1365-2958.2010.07531.x.

Campos-Salinas, J., León-Guerrero, D., González-Rey, E., Delgado, M., Castanys, S., Pérez-Victoria, J.M., and Gamarro, F. (2013). LABCG2, a new ABC transporter implicated in phosphatidylserine exposure, is involved in the infectivity and pathogenicity of *Leishmania*. PLOS Negl. Trop. Dis. 7, e2179. https://doi.org/10.1371/journal.pntd.0002179.

Carter, N.S., Yates, P., Arendt, C.S., Boitz, J.M., and Ullman, B. (2008). Purine and pyrimidine metabolism in *Leishmania*. Adv. Exp. Med. Biol. *625*, 141–154. https://doi.org/10.1007/978-0-387-77570-8_12.

Casgrain, P.A., Martel, C., McMaster, W.R., Mottram, J.C., Olivier, M., and Descoteaux, A. (2016). Cysteine Peptidase B Regulates *Leishmania mexicana* Virulence through the Modulation of GP63 Expression. PLOS Pathog. *12*, e1005658. https://doi.org/10.1371/journal.ppat.1005658.

Castro, H., and Tomás, A.M. (2008). Peroxidases of trypanosomatids. Antioxid. Redox Signal. *10*, 1593–1606. https://doi.org/10.1089/ars.2008.2050.

Castro, H., Teixeira, F., Romao, S., Santos, M., Cruz, T., Florido, M., Appelberg, R., Oliveira, P., Ferreira-da-Silva, F., and Tomas, A.M. (2011). *Leishmania* mitochondrial peroxiredoxin plays a crucial peroxidase-unrelated role during infection: insight into its novel chaperone activity. PLOS Pathog 7, e1002325.

Cayla, M., Rachidi, N., Leclercq, O., Schmidt-Arras, D., Rosenqvist, H., Wiese, M., and Spath, G.F. (2014). Transgenic analysis of the *Leishmania* MAP kinase MPK10 reveals an auto-inhibitory mechanism crucial for stage-regulated activity and parasite viability. PLOS Pathog 10, e1004347.

Chrobak, M., Förster, S., Meisel, S., Pfefferkorn, R., Förster, F., and Clos, J. (2012). *Leishmania donovani* HslV does not interact stably with HslU proteins. Int. J. Parasitol. *42*, 329–339. https://doi.org/10.1016/j.ijpara.2012.01.008.

Cull, B., Prado Godinho, J.L., Fernandes Rodrigues, J.C., Frank, B., Schurigt, U., Williams, R.A., Coombs, G.H., and Mottram, J.C. (2014). Glycosome turnover in *Leishmania major* is mediated by autophagy. Autophagy *10*, 2143–2157. https://doi.org/10.4161/auto.36438.

Dan-Goor, M., Nasereddin, A., Jaber, H., and Jaffe, C.L. (2013). Identification of a secreted casein kinase 1 in *Leishmania donovani*: effect of protein over expression on parasite growth and virulence. PLOS ONE *8*, e79287. https://doi.org/10.1371/journal.pone.0079287.

Datta, A.K., Datta, R., and Sen, B. (2008). Antiparasitic chemotherapy: tinkering with the purine salvage pathway. Adv. Exp. Med. Biol. *625*, 116–132. https://doi.org/10.1007/978-0-387-77570-8_10.

de Mesquita Barbosa, A., Dos Santos Costa, S., da Rocha, J.R., Montanari, C.A., and Giorgio, S. (2015). Evaluation of the leishmanicidal and cytotoxic effects of inhibitors for microorganism metabolic pathway enzymes. Biomed Pharmacother. *74*, 95–100.

de Paiva, R.M., Grazielle-Silva, V., Cardoso, M.S., Nakagaki, B.N., Mendonca-Neto, R.P., Canavaci, A.M., Souza Melo, N., Martinelli, P.M., Fernandes, A.P., daRocha, W.D., *et al.* (2015). Amastin Knockdown in *Leishmania braziliensis* Affects Parasite-Macrophage Interaction and Results in Impaired Viability of Intracellular Amastigotes. PLOS Pathog 11, e1005296.

Descoteaux, A., Avila, H.A., Zhang, K., Turco, S.J., and Beverley, S.M. (2002). *Leishmania* LPG3 encodes a GRP94 homolog required for phosphoglycan synthesis implicated in parasite virulence but not viability. EMBO J. *21*, 4458–4469.

Dillon, L.A., Okrah, K., Hughitt, V.K., Suresh, R., Li, Y., Fernandes, M.C., Belew, A.T., Corrada Bravo, H., Mosser, D.M., and El-Sayed, N.M. (2015a). Transcriptomic profiling of gene expression and RNA processing during *Leishmania major* differentiation. Nucleic Acids Res. 43, 6799–6813.

Dillon, L.A., Suresh, R., Okrah, K., Corrada Bravo, H., Mosser, D.M., and El-Sayed, N.M. (2015b). Simultaneous transcriptional profiling of *Leishmania major* and its murine macrophage host cell reveals insights into host–pathogen interactions. BMC Genomics 16, 1108.

Dubessay, P., Blaineau, C., Bastien, P., Tasse, L., Van Dijk, J., Crobu, L., and Pages, M. (2006). Cell cycle-dependent expression regulation by the proteasome pathway and characterization of the nuclear targeting signal of a *Leishmania major* Kin-13 kinesin. Mol. Microbiol 59, 1162–1174.

El-Hani, C.N., Borges, V.M., Wanderley, J.L., and Barcinski, M.A. (2012). Apoptosis and apoptotic mimicry in *Leishmania*: an evolutionary perspective. Front Cell. Infect. Microbiol. *2*, 96. https://doi.org/10.3389/fcimb.2012.00096.

Eliaz, D., Doniger, T., Tkacz, I.D., Biswas, V.K., Gupta, S.K., Kolev, N.G., Unger, R., Ullu, E., Tschudi, C., and Michaeli, S. (2015). Genome-wide analysis of small nucleolar RNAs of *Leishmania major* reveals a rich repertoire of RNAs involved in modification and processing of rRNA. RNA Biol. *12*, 1222–1255. https://doi.org/10.1080/15476286.2015.1038019.

Ennes-Vidal, V., Menna-Barreto, R.F., Branquinha, M.H., Dos Santos, A.L., and D'Avila-Levy, C.M. (2017). Why calpain inhibitors are interesting leading compounds to search for new therapeutic options to treat leishmaniasis? Parasitology *144*, 117–123.

Ersfeld, K., Barraclough, H., and Gull, K. (2005). Evolutionary relationships and protein domain architecture in an expanded calpain superfamily in kinetoplastid parasites. J Mol Evol 61, 742–757.

Fernandes, M.C., Dillon, L.A., Belew, A.T., Bravo, H.C., Mosser, D.M., and El-Sayed, N.M. (2016). Dual transcriptome profiling of leishmania-infected human macrophages reveals distinct reprogramming signatures. MBio 7.

Fiebig, M., Kelly, S., and Gluenz, E. (2015). Comparative life cycle transcriptomics revises *Leishmania mexicana* genome annotation and links a chromosome duplication with parasitism of vertebrates. PLOS Pathog 11, e1005186.

Flannery, A.R., Huynh, C., Mittra, B., Mortara, R.A., and Andrews, N.W. (2011). LFR1 ferric iron reductase of *Leishmania amazonensis* is essential for the generation of infective parasite forms. J. Biol. Chem. *286*, 23266–23279. https://doi.org/10.1074/jbc.M111.229674.

Folgueira, C., Carrión, J., Moreno, J., Saugar, J.M., Cañavate, C., and Requena, J.M. (2008). Effects of the disruption of the HSP70-II gene on the growth, morphology, and virulence of *Leishmania infantum* promastigotes. Int. Microbiol. *11*, 81–89.

Folgueira, C., Quijada, L., Soto, M., Abanades, D.R., Alonso, C., and Requena, J.M. (2005). The translational efficiencies of the two *Leishmania infantum* HSP70 mRNAs, differing in their 3′-untranslated regions, are affected by shifts in the temperature of growth through different mechanisms. J Biol Chem 280, 35172–35183.

Galetovic, A., Souza, R.T., Santos, M.R., Cordero, E.M., Bastos, I.M., Santana, J.M., Ruiz, J.C., Lima, F.M., Marini, M.M., Mortara, R.A., *et al.* (2011). The repetitive cytoskeletal protein H49 of *Trypanosoma cruzi* is a calpain-like protein located at the flagellum attachment zone. PLOS ONE 6, e27634.

Gerbaba, T.K., and Gedamu, L. (2013). Cathepsin B gene disruption induced *Leishmania donovani* proteome remodeling implies cathepsin B role in secretome regulation. PLOS ONE 8, e79951.

Ghosh, J., Bose, M., Roy, S., and Bhattacharyya, S.N. (2013). *Leishmania donovani* targets Dicer1 to downregulate miR-122, lower serum cholesterol, and facilitate murine liver infection. Cell Host Microbe *13*, 277–288. https://doi.org/10.1016/j.chom.2013.02.005.

Glaser, P.E., and Gross, R.W. (1994). Plasmenylethanolamine facilitates rapid membrane fusion: a stopped-flow kinetic investigation correlating the propensity of a major plasma membrane constituent to adopt an HII phase with its ability to promote membrane fusion. Biochemistry 33, 5805–5812.

Gomez-Arreaza, A., Acosta, H., Barros-Alvarez, X., Concepcion, J.L., Albericio, F., and Avilan, L. (2011). *Leishmania mexicana*: LACK (*Leishmania* homolog of receptors for activated C-kinase) is a plasminogen binding protein. Exp. Parasitol. 127, 752–761.

Gour, J.K., Kumar, V., Singh, N., Bajpai, S., Pandey, H.P., and Singh, R.K. (2012). Identification of Th1-responsive leishmanial excretory-secretory antigens (LESAs). Exp. Parasitol. *132*, 355–361. https://doi.org/10.1016/j.exppara.2012.04.022.

Haanstra, J.R., González-Marcano, E.B., Gualdrón-López, M., and Michels, P.A. (2016). Biogenesis, maintenance and dynamics of glycosomes in trypanosomatid parasites. Biochim. Biophys. Acta *1863*, 1038–1048. https://doi.org/10.1016/j.bbamcr.2015.09.015.

Haydock, A., Terrao, M., Sekar, A., Ramasamy, G., Baugh, L., and Myler, P.J. (2015). RNA-seq approaches for determining mRNA abundance in *Leishmania*. Methods Mol. Biol. *1201*, 207–219. https://doi.org/10.1007/978-1-4939-1438-8_12.

https://www.cdc.gov/parasites/leishmaniasis/epi.html, Centre for Disease Control and Prevention official website.

Hübel, A., Brandau, S., Dresel, A., and Clos, J. (1995). A member of the ClpB family of stress proteins is expressed during heat shock in *Leishmania* spp. Mol. Biochem. Parasitol. *70*, 107–118.

Hübel, A., Krobitsch, S., Hörauf, A., and Clos, J. (1997). *Leishmania major* Hsp100 is required chiefly in the mammalian stage of the parasite. Mol. Cell. Biol. *17*, 5987–5995.

Iyer, J.P., Kaprakkaden, A., Choudhary, M.L., and Shaha, C. (2008). Crucial role of cytosolic tryparedoxin peroxidase in *Leishmania donovani* survival, drug response and virulence. Mol. Microbiol. *68*, 372–391. https://doi.org/10.1111/j.1365-2958.2008.06154.x.

Jacques, I., Andrews, N.W., and Huynh, C. (2010). Functional characterization of LIT1, the *Leishmania amazonensis* ferrous iron transporter. Mol. Biochem. Parasitol. *170*, 28–36. https://doi.org/10.1016/j.molbiopara.2009.12.003.

Katakura, K., and Kobayashi, A. (1988). Acid phosphatase activity of virulent and avirulent clones of *Leishmania donovani* promastigotes. Infect. Immun. 56, 2856–2860.

Kumar, P., Sundar, S., and Singh, N. (2007). Degradation of pteridine reductase 1 (PTR1) enzyme during growth phase in the protozoan parasite *Leishmania donovani*. Exp. Parasitol. *116*, 182–189.

Lambertz, U., Silverman, J.M., Nandan, D., McMaster, W.R., Clos, J., Foster, L.J., and Reiner, N.E. (2012). Secreted virulence factors and immune evasion in visceral leishmaniasis. J. Leukoc. Biol. *91*, 887–899. https://doi.org/10.1189/jlb.0611326.

Larreta, R., Soto, M., Quijada, L., Folgueira, C., Abanades, D.R., Alonso, C., and Requena, J.M. (2004). The expression of HSP83 genes in *Leishmania infantum* is affected by temperature and by stage-differentiation and is regulated at the levels of mRNA stability and translation. BMC Mol Biol 5, 3.

Linton, K.J. (2007). Structure and function of ABC transporters. Physiology *22*, 122–130.

Lynn, M.A., Marr, A.K., and McMaster, W.R. (2013). Differential quantitative proteomic profiling of *Leishmania infantum* and *Leishmania mexicana* density gradient separated membranous fractions. J. Proteomics 82, 179–192.

Magalhaes, R.D., Duarte, M.C., Mattos, E.C., Martins, V.T., Lage, P.S., Chavez-Fumagalli, M.A., Lage, D.P., Menezes-Souza, D., Regis, W.C., Manso Alves, M.J., *et al.* (2014). Identification of differentially expressed proteins from *Leishmania amazonensis* associated with the loss of virulence of the parasites. PLOS Negl. Trop. Dis. 8, e2764.

Martin, J.L., Yates, P.A., Soysa, R., Alfaro, J.F., Yang, F., Burnum-Johnson, K.E., Petyuk, V.A., Weitz, K.K., Camp, D.G., 2nd, Smith, R.D., *et al.* (2014). Metabolic reprogramming during purine stress in the protozoan pathogen *Leishmania donovani*. PLOS Pathog. 10, e1003938.

Martinez-Garcia, M., Campos-Salinas, J., Cabello-Donayre, M., Pineda-Molina, E., Galvez, F.J., Orrego, L.M., Sanchez-Canete, M.P., Malagarie-Cazenave, S., Koeller, D.M., and Perez-Victoria, J.M. (2016). LmABCB3, an atypical mitochondrial ABC transporter essential for *Leishmania major* virulence, acts in heme and cytosolic iron/sulfur clusters biogenesis. Parasit Vectors 9, 7.

McCall, L.I., and Matlashewski, G. (2010). Localization and induction of the A2 virulence factor in *Leishmania*: evidence that A2 is a stress response protein. Mol. Microbiol. *77*, 518–530. https://doi.org/10.1111/j.1365-2958.2010.07229.x.

McCall, L.I., and Matlashewski, G. (2012). Involvement of the *Leishmania donovani* virulence factor A2 in protection against heat and oxidative stress. Exp. Parasitol. *132*, 109–115. https://doi.org/10.1016/j.exppara.2012.06.001.

Miller, M.A., McGowan, S.E., Gantt, K.R., Champion, M., Novick, S.L., Andersen, K.A., Bacchi, C.J., Yarlett, N., Britigan, B.E., and Wilson, M.E. (2000). Inducible resistance to oxidant stress in the protozoan *Leishmania chagasi*. J. Biol. Chem. *275*, 33883–33889. https://doi.org/10.1074/jbc.M003671200.

Missiakas, D., Schwager, F., Betton, J.M., Georgopoulos, C., and Raina, S. (1996). Identification and characterization of HsIV HsIU (ClpQ ClpY) proteins involved in overall proteolysis of misfolded proteins in *Escherichia coli*. EMBO J. *15*, 6899–6909.

Mittra, B., Cortez, M., Haydock, A., Ramasamy, G., Myler, P.J., and Andrews, N.W. (2013). Iron uptake controls the generation of *Leishmania* infective forms through regulation of ROS levels. J. Exp. Med. *210*, 401–416. https://doi.org/10.1084/jem.20121368.

Mittra, B., Laranjeira-Silva, M.F., Perrone Bezerra de Menezes, J., Jensen, J., Michailowsky, V., and Andrews, N.W. (2016). A trypanosomatid iron transporter that regulates mitochondrial function is required for *Leishmania amazonensis* Virulence. PLOS Pathog 12, e1005340.

Mojtahedi, Z., Clos, J., and Kamali-Sarvestani, E. (2008). *Leishmania major*: identification of developmentally regulated proteins in procyclic and metacyclic promastigotes. Exp. Parasitol 119, 422–429.

Moore, L.L., Santrich, C., and LeBowitz, J.H. (1996). Stage-specific expression of the *Leishmania mexicana* paraflagellar rod protein PFR-2. Mol. Biochem. Parasitol. *80*, 125–135.

Moreira, D., Santarém, N., Loureiro, I., Tavares, J., Silva, A.M., Amorim, A.M., Ouaissi, A., Cordeiro-da-Silva, A., and Silvestre, R. (2012). Impact of continuous axenic cultivation in *Leishmania infantum* virulence. PLOS Negl. Trop. Dis *6*, e1469. https://doi.org/10.1371/journal.pntd.0001469.

Mukherjee, B., Mukhopadhyay, R., Bannerjee, B., Chowdhury, S., Mukherjee, S., Naskar, K., Allam, U.S., Chakravortty, D., Sundar, S., Dujardin, J.C., *et al.* (2013). Antimony-resistant but not antimony-sensitive *Leishmania donovani* up-regulates host IL-10 to overexpress multidrug-resistant protein 1. Proc. Natl. Acad. Sci. U.S.A. *110*, E575–582.

Nandan, D., Yi, T., Lopez, M., Lai, C., and Reiner, N.E. (2002). *Leishmania* EF-1alpha activates the Src homology 2 domain containing tyrosine phosphatase SHP-1 leading to macrophage deactivation. J. Biol. Chem. *277*, 50190–50197. https://doi.org/10.1074/jbc.M209210200.

Nare, B., Garraway, L.A., Vickers, T.J., and Beverley, S.M. (2009). PTR1-dependent synthesis of tetrahydrobiopterin contributes to oxidant susceptibility in the trypanosomatid protozoan parasite *Leishmania major*. Curr. Genet. *55*, 287–299.

Olivier, M., Atayde, V.D., Isnard, A., Hassani, K., and Shio, M.T. (2012). *Leishmania* virulence factors: focus on the metalloprotease GP63. Microbes Infect. *14*, 1377–1389. https://doi.org/10.1016/j.micinf.2012.05.014.

Perez-Pertejo, Y., Alvarez-Velilla, R., Estrada, C.G., Balana-Fouce, R., and Reguera, R.M. (2011). *Leishmania donovani*: proteasome-mediated down-regulation of methionine adenosyltransferase. Parasitology *138*, 1082–1092.

Pescher, P., Blisnick, T., Bastin, P., and Spath, G.F. (2011). Quantitative proteome profiling informs on phenotypic traits that adapt *Leishmania donovani* for axenic and intracellular proliferation. Cell. Microbiol. *13*, 978–991.

Rastrojo, A., Carrasco-Ramiro, F., Martin, D., Crespillo, A., Reguera, R.M., Aguado, B., and Requena, J.M. (2013). The transcriptome of *Leishmania major* in the axenic promastigote stage: transcript annotation and relative expression levels by RNA-seq. BMC Genomics *14*, 223.

Requena, J.M., Montalvo, A.M., and Fraga, J. (2015). Molecular chaperones of leishmania: central players in many stress-related and -unrelated physiological processes. Biomed. Res. Int. 2015, 301326.

Rios, M.C., Silva, W.R., Azevedo, A.F., Santos, P.L., Teixeira, S.A., Muscará, M.N., Thomazzi, S.M., Almeida, R.P., Fernandes, R.P., and Scher, R. (2015). Expression of glyceraldehyde 3-phosphate dehydrogenase is enhanced in *Leishmania* spp naturally resistant to nitric oxide. Genet. Mol. Res. *14*, 7113–7121. https://doi.org/10.4238/2015.June.29.4.

Rock, K.S., le Rutte, E.A., de Vlas, S.J., Adams, E.R., Medley, G.F., and Hollingsworth, T.D. (2015). Uniting mathematics and biology for control of visceral leishmaniasis. Trends Parasitol. *31*, 251–259. https://doi.org/10.1016/j.pt.2015.03.007.

Rosenzweig, D., Smith, D., Myler, P.J., Olafson, R.W., and Zilberstein, D. (2008a). Post-translational modification of cellular proteins during *Leishmania donovani* differentiation. Proteomics 8, 1843–1850.

Rosenzweig, D., Smith, D., Opperdoes, F., Stern, S., Olafson, R.W., and Zilberstein, D. (2008b). Retooling *Leishmania* metabolism: from sand fly gut to human macrophage. FASEB J. *22*, 590–602.

Rotureau, B., Morales, M.A., Bastin, P., and Späth, G.F. (2009). The flagellum-mitogen-activated protein kinase connection in Trypanosomatids: a key sensory role in parasite signalling and development? Cell. Microbiol. *11*, 710–718. https://doi.org/10.1111/j.1462-5822.2009.01295.x.

Santarém, N., Racine, G., Silvestre, R., Cordeiro-da-Silva, A., and Ouellette, M. (2013). Exoproteome dynamics in *Leishmania* infantum. J Proteomics *84*, 106–118. https://doi.org/10.1016/j.jprot.2013.03.012.

Saraiva, E.M., Pinto-da-Silva, L.H., Wanderley, J.L., Bonomo, A.C., Barcinski, M.A., and Moreira, M.E. (2005). Flow cytometric assessment of *Leishmania* spp metacyclic differentiation: validation by morphological features and specific markers. Exp. Parasitol. 110, 39–47.

Sardar, A.H., Kumar, S., Kumar, A., Purkait, B., Das, S., Sen, A., Kumar, M., Sinha, K.K., Singh, D., Equbal, A., *et al.* (2013). Proteome changes associated with *Leishmania donovani* promastigote adaptation to oxidative and nitrosative stresses. J. Proteomics *81*, 185–199.

Saxena, A., Worthey, E.A., Yan, S., Leland, A., Stuart, K.D., and Myler, P.J. (2003). Evaluation of differential gene expression in *Leishmania major* Friedlin procyclics and metacyclics using DNA microarray analysis. Mol. Biochem. Parasitol. *129*, 103–114.

Schaupp, A., Marcinowski, M., Grimminger, V., Bösl, B., and Walter, S. (2007). Processing of proteins by the molecular chaperone Hsp104. J. Mol. Biol. *370*, 674–686.

Silva, A.M., Cordeiro-da-Silva, A., and Coombs, G.H. (2011). Metabolic variation during development in culture of *Leishmania donovani* promastigotes. PLOS Negl. Trop. Dis. *5*, e1451. https://doi.org/10.1371/journal.pntd.0001451.

Silverman, J.M., Clos, J., Horakova, E., Wang, A.Y., Wiesgigl, M., Kelly, I., Lynn, M.A., McMaster, W.R., Foster, L.J., Levings, M.K., *et al.* (2010). *Leishmania* exosomes modulate innate and adaptive immune responses through effects on monocytes and dendritic cells. J. Immunol. *185*, 5011–5022.

Singla, N., Khuller, G.K., and Vinayak, V.K. (1992). Acid phosphatase activity of promastigotes of *Leishmania donovani*: a marker of virulence. FEMS Microbiol. Lett. *73*, 221–225.

Späth, G.F., Drini, S., and Rachidi, N. (2015). A touch of Zen: post-translational regulation of the *Leishmania* stress response. Cell. Microbiol. *17*, 632–638. https://doi.org/10.1111/cmi.12440.

Spath, G.F., Epstein, L., Leader, B., Singer, S.M., Avila, H.A., Turco, S.J., and Beverley, S.M. (2000). Lipophosphoglycan is a virulence factor distinct from related glycoconjugates in the protozoan parasite *Leishmania major*. Proc. Natl. Acad. Sci. U.S.A. 97, 9258–9263.

Suman, S.S., Equbal, A., Zaidi, A., Ansari, M.Y., Singh, K.P., Singh, K., Purkait, B., Sahoo, G.C., Bimal, S., Das, P., *et al.* (2016). Up-regulation of cytosolic tryparedoxin in Amp B resistant isolates of *Leishmania donovani* and its interaction with cytosolic tryparedoxin peroxidase. Biochimie 121, 312–325.

Swenerton, R.K., Zhang, S., Sajid, M., Medzihradszky, K.F., Craik, C.S., Kelly, B.L., and McKerrow, J.H. (2011). The oligopeptidase B of *Leishmania* regulates parasite enolase and immune evasion. J. Biol. Chem. *286*, 429–440. https://doi.org/10.1074/jbc.M110.138313.

Szempruch, A.J., Dennison, L., Kieft, R., Harrington, J.M., and Hajduk, S.L. (2016). Sending a message: extracellular vesicles of pathogenic protozoan parasites. Nat. Rev. Microbiol. *14*, 669–675. https://doi.org/10.1038/nrmicro.2016.110.

Thiakaki, M., Kolli, B., Chang, K.P., and Soteriadou, K. (2006). Down-regulation of gp63 level in *Leishmania amazonensis* promastigotes reduces their infectivity in BALB/c mice. Microbes Infect 8, 1455–1463.

Tiwari, K., Kumar, R., and Dubey, V.K. (2016). Biochemical characterization of dihydroorotase of *Leishmania donovani*: Understanding pyrimidine metabolism through its inhibition. Biochimie 131, 45–53.

Tsigankov, P., Gherardini, P.F., Helmer-Citterich, M., Spath, G.F., Myler, P.J., and Zilberstein, D. (2014). Regulation dynamics of *Leishmania* differentiation: deconvoluting signals and identifying phosphorylation trends. Mol Cell Proteomics 13, 1787–1799.

van Zandbergen, G., Bollinger, A., Wenzel, A., Kamhawi, S., Voll, R., Klinger, M., Muller, A., Holscher, C., Herrmann, M., Sacks, D., *et al.* (2006). *Leishmania* disease development depends on the presence of apoptotic promastigotes in the virulent inoculum. Proc. Natl. Acad. Sci. U.S.A. 103, 13837–13842.

Vickers, T.J., and Beverley, S.M. (2011). Folate metabolic pathways in *Leishmania*. Essays Biochem. *51*, 63–80. https://doi.org/10.1042/bse0510063.

Wang, Q., Melzer, I.M., Kruse, M., Sander-Juelch, C., and Wiese, M. (2005). LmxMPK4, a mitogen-activated protein (MAP) kinase homologue essential for promastigotes and amastigotes of *Leishmania mexicana*. Kinetoplastid Biol Dis 4, 6.

Wilson, Z.N., Gilroy, C.A., Boitz, J.M., Ullman, B., and Yates, P.A. (2012). Genetic dissection of pyrimidine biosynthesis and salvage in *Leishmania donovani*. J. Biol. Chem. *287*, 12759–12770. https://doi.org/10.1074/jbc.M112.346502.

Xu, W., Hsu, F.F., Baykal, E., Huang, J., and Zhang, K. (2014). Sterol biosynthesis is required for heat resistance but not extracellular survival in leishmania. PLOS Pathog. *10*, e1004427. https://doi.org/10.1371/journal.ppat.1004427.

Yao, C., Chen, Y., Sudan, B., Donelson, J.E., and Wilson, M.E. (2008). *Leishmania chagasi*: homogenous metacyclic promastigotes isolated by buoyant density are highly virulent in a mouse model. Exp Parasitol 118, 129–133.

Yao, C., Li, Y., Donelson, J.E., and Wilson, M.E. (2010). Proteomic examination of *Leishmania chagasi* plasma membrane proteins: Contrast between avirulent and virulent (metacyclic) parasite forms. Proteomics Clin. Appl. 4, 4–16.

Zhang, K., and Beverley, S.M. (2010). Phospholipid and sphingolipid metabolism in *Leishmania*. Mol. Biochem. Parasitol. *170*, 55–64. https://doi.org/10.1016/j.molbiopara.2009.12.004.

Zhang, W.W., McCall, L.I., and Matlashewski, G. (2013). Role of cytosolic glyceraldehyde-3-phosphate dehydrogenase in visceral organ infection by *Leishmania donovani*. Eukaryot. Cell 12, 70–77.

Zylbersztejn, A.M., de Morais, C.G., Lima, A.K., Souza, J.E., Lopes, A.H., Da-Silva, S.A., Silva-Neto, M.A., and Dutra, P.M. (2015). CK2 Secreted by *Leishmania braziliensis* mediates macrophage association invasion: a comparative study between virulent and avirulent promastigotes. Biomed. Res. Int. 2015, 167323.

DNA Topoisomerases of Kinetoplastid Parasites: Brief Overview and Recent Perspectives

10

Sourav Saha†, Somenath Roy Chowdhury† and Hemanta K. Majumder*

Infectious Diseases and Immunology Division, CSIR-Indian Institute of Chemical Biology, Kolkata, India.

*Correspondence: hkmajumder@iicb.res.in, hemantamajumder@yahoo.co.in

†Both authors contributed equally.

https://doi.org/10.21775/9781910190715.10

Abstract

Topoisomerases are group of enzymes that resolve DNA topological problems and aid in different DNA transaction processes viz. replication, transcription, recombination, etc. inside cells. These proteins accomplish their feats by steps of DNA strand(s) scission, strand passage or rotation and subsequent rejoining activities. Topoisomerases of kinetoplastid parasites have been extensively studied because of their unusual features. The unique presence of heterodimeric Type IB topoisomerase and prokaryotic 'TopA homologue' Type IA topoisomerase in kinetoplastids still generates immense interest among scientists. Moreover, because of their structural dissimilarity with the host enzymes, topoisomerases of kinetoplastid parasites are attractive targets for chemotherapeutic interventions to kill these deadly parasites. In this chapter, we summarize historical perspectives and recent advances in kinetoplastid topoisomerase research and how these proteins are exploited for drug targeting.

Introduction

DNA topoisomerases are ubiquitous enzymes that help to maintain proper topological states of the double helical DNA molecules and prevent genomic instability inside cell. During replication, transcription, recombination, and other vital nucleic acid metabolic processes, permanent or temporary separations of DNA strands cause different topological constraints. These topological constraints, if not removed, become disastrous and greatly affect cellular health (Chen *et al.*, 2013). DNA topoisomerases are the molecular machines inside all kinds of living cells that resolve these kinds of topological tensions. Not only that, DNA topoisomerases sometimes introduce measured topological tensions in the double helical DNA molecules to aid DNA transaction processes (Champoux, 2001).

The mechanism of action of DNA topoisomerases involves following major steps: (1) binding of topoisomerase to the substrate DNA, (2) cleavage by trans-esterification reaction accompanied by the formation of a transient phosphodiester bond between a tyrosine residue in the protein and one of the ends of the broken strand, (3) strand passage/rotation through/around the break leading to change in the linking number and (4) strand religation and release of the enzyme as the DNA is religated (Stewart *et al.*, 1998; Koster *et al.*, 2005; Champoux, 2001).

Based on structure and mechanism of action, topoisomerases are classified as type I and type II. Those enzymes that cleave a single strand of DNA and pass or rotate another strand through or about the resultant gap are defined as type I topoisomerases and these enzymes change the linking number of the substrate DNA in discrete steps of one or more than one (Champoux, 2001). They are further classified either as type IA or as type IB enzyme. These two subfamilies have no sequence similarities between them and exhibit dissimilar structures and distinct region characteristics. Type IA topoisomerase makes a transient covalent complex with the 5′-end of the broken DNA strand. On the other hand, type IB topoisomerase forms a covalent intermediate with the 3′-end of the DNA (Champoux, 2001). The type II topoisomerases act in a manner fundamentally different from either of the type I enzymes. They require divalent metal ions for their activity and use ATP to actively transport DNA duplex through a transient enzyme-mediated break in the second duplex of the same DNA strand thus changing linking numbers in steps of two. Prokaryotes and lower eukaryotes encode only a single type II topoisomerase whereas vertebrates generally express two discrete forms of the enzyme, topoisomerase IIA and IIB (Champoux, 2001). Both Type IIA and IIB enzymes break the DNA by attacking and bonding to the 5′-phosphate, and have sequence similarities. They are encoded by separate genes, have different molecular masses, show distinct patterns of expression and perform different cellular functions. Topoisomerase IIA helps mainly in the DNA replication and chromosome segregation in proliferating cells and topoisomerase IIβ is mainly required for proper neural development (Deweese and Osheroff, 2009; Yang *et al.*, 2000; Wang, 2002; Champoux, 2001; Velez-Cruz *et al.*, 2004; McClendona and Osheroff, 2007).

Different topoisomerases of kinetoplastid parasites

Topoisomerases of kinetoplastid parasites are of following kinds:

- type IB topoisomerase;
- type IA topoisomerase;
- type II topoisomerase.

Type IB topoisomerase of kinetoplastid parasites

Type IB topoisomerases are generally monomeric enzymes. But in kinetoplastid parasites, type IB topoisomerase is heterodimeric in nature (Das *et al.*, 2004a). The discovery of bi-subunit type IB topoisomerase in kinetoplastid parasites opened up a new avenue in topoisomerase research related to evolution of type IB topoisomerase family and its relevance as a drug target for trypanosomiasis therapeutic interventions (Das *et al.*, 2006).

Type IB DNA topoisomerase activities were first purified from *Leishmania donovani* (Chakraborty and Majumder, 1988; Chakraborty *et al.*, 1993), *Trypanosoma cruzi* (Riou *et al.*, 1983) and *Crithidia fasciculata* (Melendy and Ray, 1987). The purified active enzymes (65–79 kDa) were ATP independent and found to be sensitive to the topoisomerase I-specific inhibitor camptothecin (Melendy and Ray, 1987).

Topoisomerase IB-like gene from the kinetoplastid *L. donovani* was first cloned and reported by Broccoli *et al.* (1999). The amino acid sequence of this gene showed homology with the N-terminal and central core of other eukaryotic type IB topoisomerases, but had a variable C-terminus. Interestingly, this ORF was lacking the SKINYL motif that supplies the catalytic tyrosine (Broccoli *et al.*, 1999). Villa and co-workers first reported the presence of a second small subunit harbouring the SKINYL motif in *Leishmania* parasites (Villa *et al.*, 2003). Similar results were obtained in case of African trypanosomes in the same year (Bodley *et al.*, 2003).

In general, type IB DNA topoisomerases of kinetoplastid parasites are expressed from two different genes and form a heterodimeric active enzyme (Table 10.1). In *L. donovani*, gene for large subunit (LdTOP1L) is present on chromosome 34 encoding a 636 amino acid polypeptide with an estimated molecular weight of 73 kDa and the gene for the small subunit (LdTOP1S) is situated on chromosome 4 encoding a 262 amino acid polypeptide with a molecular weight of 29 kDa (Das *et al.*, 2004a). Bodley *et al.*, 2003 have also identified a heterodimer topoisomerase I in *Trypanosoma brucei* which is comprised of the 90 kDa large subunit and 36 kDa small subunit (Bodley *et al.*, 2003). Large subunit (LdTOP1L) is closely homologous to the core domain of human topoisomerase I and small

Table 10.1 Different topoisomerase genes of kinetoplastid parasites as annotated in GeneDB database (www.genedb.org/Homepage) (Logan-Klumpler *et al.*, 2012)

Protein coding gene	Gene DB systematic IDs	Length
DNA topoisomerase IB large subunit (L)	LmjF.34.3440 (*L. major* Friedlin)	1905 base pairs
	LdBPK_343220.1 (*L. donovani* BPK282A1)	1908 base pairs
	Tb927.4.133 (*T. brucei* 927)	2091 base pairs
	TcCLB.508693.20 (*T. cruzi*)	1995 base pairs
DNA topoisomerase IB small subunit (S)	LmjF.04.0060 (*L. major* Friedlin)	789 base pairs
	LdBPK_040070.1 (*L. donovani* BPK282A1)	789 base pairs
	Tb927.9.6940 (*T. brucei* 927)	822 base pairs
	TcCLB.506625.110 (*T. cruzi*)	819 base pairs
DNA topoisomerase IA	LmjF.21.0125 (*L. major* Friedlin)	2439 base pairs
	LdBPK_210180.1 (*L. donovani* BPK282A1)	2430 base pairs
	Tb927.10.1900 (*T. brucei* 927)	2421 base pairs
	TcCLB.510121.160 (*T. cruzi*)	2493 base pairs
	TcCLB.506493.80 (*T. cruzi*)	2490 base pairs
DNA topoisomerase III	LmjF.28.1780 (*L. major* Friedlin)	2601 base pairs
	LmjF.36.3200 (*L. major* Friedlin)	2844 base pairs
	LdBPK_281900.1 (*L. donovani* BPK282A1)	2604 base pairs
	LdBPK_363350.1 (*L. donovani* BPK282A1)	2844 base pairs
	Tb927.11.9520 (*T. brucei* 927)	2757 base pairs
	Tb927.11.9170 (*T. brucei* 927)	2559 base pairs
	TcCLB.510901.100 (*T. cruzi*)	2766 base pairs
	TcCLB.508851.170 (*T. cruzi*)	2526 base pairs
	TcCLB.511589.120 (*T. cruzi*)	2766 base pairs
DNA topoisomerase II	LmjF.28.2280 (*L. major* Friedlin)	4494 base pairs
	LdBPK_282450.1 (*L. donovani* BPK282A1)	4497 base pairs
	BSAL_28010 (*B. saltans*)	3741 base pairs
	Tb927.11.11550 (*T. brucei* 927)	4368 base pairs
	Tb927.11.11560 (*T. brucei* 927)	4275 base pairs
	TcCLB.508699.10 (*T. cruzi*)	4431 base pairs
	TcCLB.509203.70 (*T. cruzi*)	4455 base pairs
Mitochondrial DNA topoisomerase II	LmjF.15.1290 (*L. major* Friedlin)	3711 base pairs
	LdBPK_151310.1 (*L. donovani* BPK282A1)	3711 base pairs
	Tb927.9.5590 (*T. brucei* 927)	3666 base pairs
	TcCLB.508277.370 (*T. cruzi*)	3696 base pairs
	TcCLB.506445.60 (*T. cruzi*)	3693 base pairs

subunit LdTOP1S contains the phylogenetically conserved 'SKXXY' motif placed at the C-terminal domain of all topoisomerases IB (TOPIB) which harbours a conserved tyrosine residue playing a crucial role in DNA cleavage. Overall, the structure and catalytic machinery of the two enzymes (LdTOPIB and human TOPIB) are highly conserved, despite the fact that LdTOPIB is a heterodimer whereas the human TOPIB is a monomer.

The *in vitro* reconstitution of the two recombinant proteins LdTOP1L and LdTOP1S corresponding to the large and small subunit forms active bi-subunit LdTOP1LS/LdTOPIB. LdTOP1L and LdTOP1S form a direct 1 : 1 heterodimer complex through protein–protein interaction (Das *et al.*, 2004a). Interaction between the two subunits leading to the formation of an active complex could be exploited for development of new therapeutic agents with specific selectivity (Das *et al.*, 2004a). This observation leads to the concept that non-covalent interaction of both subunits is necessary for the activity. Interestingly, deletion of 99 amino acids from the N-terminus of LdTOP1L results in a protein which failed to interact with the smaller subunit. This could be attributed to the presence of many polar residues in this region. Polar interactions are common between the subunits of dimeric proteins. (Das *et al.*, 2005). It was also found that the deletion of the first 39 amino acids from N-terminus of LdTOP1L results in a protein with decreased cleavage activity and sensitivity to CPT (Das *et al.*, 2005). These data argue in favour of the interpretation that N-terminal amino acids of the large subunit regulates DNA dynamics during relaxation by controlling noncovalent DNA-binding or by coordinating DNA contacts by the other parts of the enzyme. Further, it was shown that the amino acids 39–456 of LdTOP1L and 210–262 of LdTOP1S constitute the minimal functionally interacting fragments of LdTOP1LS (Bosedasgupta *et al.*, 2008a) and the LdTOP1L navigates the heterodimeric LdTOP1LS to its cellular DNA targets (Bose Dasgupta *et al.*, 2008b). A recent report also identified the regions of both subunits that interact to each other to form a functional linker in *L. donovani* topoisomerase IB. Amino acids 175 and 180 (RPPVV) of the small subunit interact with a region (amino acids 525–581) of the large subunit (LdTopIL) and have role in sensitivity to camptothecin in *Leishmania* (Prada *et al.*, 2012). Recently it was also shown that LdTOP1LS activity is stimulated by ATP (generally Type IB topoisomerases are ATP independent) in the absence of Mg^{2+} and this stimulation does not need ATP hydrolysis. The ATP binds to a Arg190 residue on LdTOP1L and stimulates the rate of strand rotation by the enzyme (Sengupta *et al.*, 2011).

LdTOP1B localizes in both nucleus and kinetoplast of *L. donovani*. Multiple nuclear localization signals (NLSs) have been mapped in the large subunit of *Leishmania* TOP1B but no NLS has been found in smaller subunits of the enzyme. So it is likely that the subunits interact in the cytosol before nuclear and kinetoplast importation (Das *et al.*, 2004a; Das *et al.*, 2008) and LdTOP1L navigates LdTOP1S by acting as a 'molecular steer' (Bose Dasgupta *et al.*, 2008b). Another recent study reported three nuclear localization signals (NLSs) in LdTOP1B, one in LdTOP1L and two in LdTOP1S and both subunits are transported to the nucleus separately. They also found that LdTOPIB–GFP constructs do not localize to mitochondria (Prada *et al.*, 2013a).

From bioinformatic analysis (multiple sequence alignment) and crystal structure of LdTOP1LS (*L. donovani* topoisomerase IB) captured as a vanadate complex linking the enzyme to a nicked DNA, The crystal structure of the reconstituted *L. donovani* topoisomerase IB (PDB ID 2B9S) reveals that the three-dimensional configuration of the active site is almost similar to that of human topoisomerase I. Five amino acids have been identified as active site residues (Davies *et al.*, 2006). These are Arg314, Lys352, Arg410, His453 and Tyr222. Functions of these amino acids in the *L. donovani* topoisomerase IB have been confirmed by single site directed mutation studies (Díaz González *et al.*, 2007b; Ganguly *et al.*, 2009). Among these, Tyr222 resides in the SKXXY motif in the small subunit and is responsible for the nucleophilic attack on the scissile phosphate group in the DNA backbone and subsequent formation of phosphotyrosine linkage. Arg410 activates Tyr222 for this nucleophilic attack with water acting as a specific base. Mutation of the histidine at position 453 has been shown to alter DNA binding of the enzyme. Of the remaining three, Arg314 and Lys352 stand as potential candidates for acting as the general acid and thereby play a vital role in transesterification reaction (Davies *et al.*, 2006; Ganguly *et al.*, 2009).

Type IB topoisomerases including human topoisomerase I and vaccinia virus topoisomerase I share a common catalytic domain with the tyrosine recombinases that includes bacteriophage HP1 and certain phage integrases and the XerC/D, Cre and Flp recombinases (Corbett and Berger, 2004). In prokaryotic tyrosine recombinases XerD, Cre, HP1 integrase and eukaryotic topoisomerases the active site residues RKHR (H/W)Y and RKRHY resides on a single monomer. In eukaryotic Flp recombinase and *Leishmania* heterodimeric topoisomerase IB, the active site tyrosine is provided by one subunit and the other residues implicated in catalysis comes from another subunit (Chen and Rice, 2003). It may be possible that the active site in case of kinetoplastid topoisomerase I is created by the association of subunits similar to Flp recombinase. The structural diversity of the amino-terminal domain and the linker domain in various eukaryotic topoisomerase IB argues that they have probably evolved by independent gene fusion. Krogh and Shuman described that nuclear topoisomerase I enzymes evolved from the bacterial/poxvirus precursor (Krogh and Shuman, 2002). However, in contrast to vaccinia topoisomerase I, the catalytic tyrosine of human topoisomerase I resides on a small 6 kDa domain and is connected to the central core domain by a 7 kDa coil–coil linker element (Champoux, 2001). It now seems simpler to posit that during evolution of eukaryotic IB topoisomerase, the kinetoplastid parasites (*Leishmania, Trypanosoma*) which arose earlier than multicellular eukaryotes gained some regulatory sequences (nuclear localization signals and sequences for CPT sensitivity) as split domain architecture. Marcotte *et al.* (1999) demonstrated that the fusion of the subunits and insertion of additional domains allow for the rapid evolution of new signalling pathways by the incorporation of a novel interaction domain into a pre-existing peptide. Horizontal gene transfer has been shown to have an important role in the evolution of composite proteins (Simonson *et al.*, 2005). It is possible that horizontal gene transfer did play a key role in propelling the topoisomerase IB protein among the eukaryotes (Krogh and Shuman, 2002). Thus we can suggest that eukaryotic monomeric type IB topoisomerase evolved from the common ancestral bi-subunit enzyme by fusion of the two subunits at genetic level.

Type IA topoisomerases of kinetoplastid parasites

Type IA topoisomerases differ from type IB counterpart in their structures, requirement of Mg^{2+} for activity, substrate specificity (type IA topoisomerases only relax negatively supercoiled DNA) and unlike type IB topoisomerases, type IA topoisomerases form 5′-phosphotyrosine catalytic intermediate (Stewart *et al.*, 1997; Champoux, 2001; Wang, 2002). Apart from type IB topoisomerase, three type IA topoisomerases are there in the parasite genome (Table 10.1), termed as topoisomerase IA, and two topoisomerase III (topoisomerase IIIα and IIIβ). Topoisomerase IA mainly prefers catalysing relaxation reaction whereas topoisomerase III favours catenating/decatenating DNA molecules (Li *et al.*, 2000).

Topoisomerase IA

The presence of topoisomerase IA homologue in the kinetoplastid parasites is very intriguing (Table 10.1). Topoisomerase IA is of prokaryotic origin and is not present in eukaryotes. However, kinetoplastid parasites are exceptions in the eukaryotic world in this issue since they contain topoisomerase IA. This makes topoisomerase IA of these parasites a fantastic drug target due to its absence in the human host and warrants a thorough functional and mechanistic investigation of the type IA topoisomerases of kinetoplastid parasites. Our laboratory is currently involved in characterizing topoisomerase IA of *L. donovani.*

Another group reported topoisomerase IA from *T. brucei*. *T. brucei* topoisomerase IA gene is present on chromosome 10 (NCBI Reference Sequence: XP_822446/Systematic Name: Tb927.10.1900), 2418 bp in size and encodes an 88.9 kDa predicted protein. This enzyme is mitochondrial, essential for late theta structure resolution during kinetoplast DNA (kDNA) replication process and is important for parasite survival (Scocca and Shapiro, 2008).

Topoisomerase IIIα and topoisomerase IIIβ

Like other higher eukaryotes, kinetoplastid parasites also have two putative DNA topoisomerase III (paralogous isoforms of type IA topoisomerase) genes. Very recently, topoisomerase IIIα from *T. brucei* has been shown to play a critical role in antigenic variation by monitoring expression-site-associated VSG switching (Kim and Cross, 2010).

DNA topoisomerase IIIβ from kinetoplastid parasite *L. donovani* (LdTopIIIβ) has been well studied (Banerjee *et al.*, 2011). LdTopIIIβ has high sequence homology with human and drosophila topoisomerase IIIβ. It has features typical for topoisomerase IIIβ proteins including the CXXC type of motifs and a long stretch of G and R residues at its C-terminus. The recombinant protein was purified from *in vitro* transcription-translation reaction and relaxation activity of the enzyme was characterized. GFP-fused LdTopIIIβ localizes both in nucleus and kinetoplast of *L. donovani* parasites indicating the involvement of LdTopIIIβ in DNA processing inside both the parasite organelle.

Wild-type LdTopIIIβ could complement the topoisomerase III mutant yeast with slow-growth phenotype whereas the C-terminal deletion construct of LdTopIIIβ lacking its Zn-binding domain was unable to rescue the mutant yeast revealing that the C-terminal 258 amino acids were indispensable for functional complementation of LdTopIIIβ *in vivo*. Further investigations are required to determine the role of C-terminal end of the LdTopIIIβ protein in DNA binding (Banerjee *et al.*, 2011).

The presence of functionally active topoisomerase III proteins in parasites indicates towards its role in DNA metabolism in the parasites, which requires further studies and might emerge as a new therapeutic target that can be exploited against the deadly parasites (Kim and Cross, 2010; Banerjee *et al.*, 2011).

Type II topoisomerases of kinetoplastid parasites

Eukaryotic type II topoisomerases are mainly involved in chromosome segregation and condensation via their decatenation activity. Eukaryotic topoisomerase II is available in two forms, topo IIα and topo IIβ throughout the phylogeny which exhibits strikingly different distributions among the three domains of life (Gadelle *et al.*, 2003). The distribution of the enzyme among the kinetoplastid parasites is rather trivial. The kinetoplastid parasites harbours two type II topoisomerases viz. topoisomerase II and mitochondrial DNA topoisomerase II (Kulikowicz and Shapiro, 2006) (Table 10.1).

Genes of topoisomerase II have been isolated, cloned and sequenced from *C. fasciculata*, *T. brucei*, *T. cruzi*, *L. donovani*, *L. infantum*, *L. chagasi*, and *B. saltans* and also its activity has been purified from many kinetoplastid parasites (Das *et al.*, 2001, 2004b; Pasion *et al.*, 1992; Strauss and Wang, 1990; Fragoso and Goldenberg, 1992; Gaziová and Lukes, 2003). Compared to other higher eukaryotes topoisomerase II proteins of the parasites were found to be smaller, still they share the same functional domains and are more homologous to eukaryotic than prokaryotic type II enzymes. Unlike *E. coli* DNA gyrase (Wang *et al.*, 1996) none of the parasitic type II enzymes are endowed with supercoiling activity and the enzymatic activities are nearly same as other eukaryotic counterparts (Das *et al.*, 2004b). Though all type IIA topoisomerases are identical in one way that they change the linking number of DNA in an ATP-dependent manner, the eukaryotic type II enzymes are homodimers, while their bacterial counterparts such as gyrase and topo IV are A_2B_2 tetramers; the B and A subunits being the N and C-terminal halves of their eukaryotic counterparts (Lynn *et al.*, 1996).

Topoisomerase II proteins of *L. donovani* and *B. saltans* were found to be localized both in the nucleus and kinetoplast whereas topoisomerase II activity isolated from *C. fasciculata* was shown to be immunolocalized in kinetoplast (Das *et al.*, 2001; Gaziová and Lukes, 2003). Two nuclear topoisomerase II genes from *T. brucei brucei* namely TbTOP2α and TbTOP2β which are separated by 1.7 kb intergenic regions have been reported. TbTOP2α is ATP-dependent and is localized in the nucleus. Silencing of this gene by RNAi leads to growth arrest and causes severe defects in nuclear DNA although mitochondrial DNA remains unaltered. However, silencing of TbTOP2β has no effect on the apparent phenotype of the cell and precise role of TbTOP2β remains unclear (Kulikowicz and Shapiro, 2006). TbTop2α is also essential in bloodstream form *T. brucei* as it is responsible for centromere-specific topoisomerase cleavage activity and has roles in chromosome segregation. TbTop2β gene does not have an obvious role in chromosome segregation (Obado *et al.*, 2010). Mitochondrial topoisomerase II of *Trypanosoma* is involved in replication of the massive kinetoplast DNA network. RNAi of mitochondrial topoisomerase II leads to the progressive degradation of mitochondria (kinetoplast) in *Trypanosoma* (Wang and Englund, 2001). During kDNA replication mini-circle is individually released from the network and this phenomenon creates holes in the kDNA network. After depletion of TbTOP2mt by RNAi these holes continue to exist, indicating that this enzyme is responsible for remodelling the kDNA network during replication to maintain a proper mini-circle density and network structure (Lindsay *et al.*, 2008).

Kinetoplastid parasites diverged early in the eukaryotic evolution at the base of the evolutionary tree well before the emergence of the metazoan kingdom. In spite of having a similarity and identity of 31% and 23% with yeast topoisomerase II, *L. donovani* topoisomerase II (LdTOP2) was found to complement a temperature-sensitive mutant yeast strain (Sengupta *et al.*, 2003).

Just like other eukaryotic topoisomerase II, the *L. donovani* enzyme can also be divided into an N-terminal ATPase, a central DNA-binding domain and an unconserved C-terminal domain (Sengupta *et al.*, 2003, 2005a,b). Though unconserved, the nuclear localization signal and the dimerization domain of this homodimeric enzyme have been mapped in the C-terminus. The C-terminus also contains a stretch of 60 amino acids not present in the human host. Therefore this region can be exploited to develop antileishmanial targets (Sengupta *et al.*, 2003). The parasite enzyme has a greater affinity for DNA and was also stable at a very high salt concentration compared with its human host. These findings were quite consistent with the greater susceptibility of the parasite protein to the anti-topoisomerase II agents. This is because of the fact that an enzyme with more affinity towards DNA would perform more DNA cleavage and thus a greater chance of being trapped in that state by an anti-topoisomerase II drug (Sengupta *et al.*, 2005b). The N-terminal 385 amino acid residues of LdTOP2 were found to possess the ATPase activity. Although the ATPase activity resides in the first 385 amino acid residues, only a larger protein was found to mimic the full-length enzyme kinetics in *in vitro* assay (Sengupta *et al.*, 2005a). This characteristic of the parasite protein was found to be in contrast to the human enzyme where a smaller protein also has ATPase kinetics similar to the full length enzyme (Campbell and Maxwell, 2002) except for the fact that a smaller fragment (1–420 amino acids) fails to be hyper stimulated by DNA. Most interestingly, the N-terminal 385 amino acids of the parasite protein were also found to be inhibited by etoposide which showed that etoposide in addition to being a poison for the parasite enzyme was also a catalytic inhibitor. The study identifies specific amino acids such as Asn^{65}, Asn^{69}, Asn^{96} and Asp^{130} of the parasite protein that are involved in the interaction with ATP and etoposide (Sengupta *et al.*, 2005a). The active site tyrosine implicated in DNA breakage and rejoining for *L. donovani* topoisomerase II has been mapped to be Tyr^{775} (Sengupta *et al.*, 2005b).This tyrosine is the only residue in the parasite protein, which is involved in the trans-esterification reaction and is also homologous to the Tyr^{804} of human (Tsai-Pflugfelder *et al.*, 1988). Surprisingly, the C-terminal truncation mutants of the parasite protein failed to be inhibited by etoposide (Sengupta *et al.*, 2005b) while the full-length protein generates more cleavable complex in the presence of etoposide. Like the human enzyme, the core domain of LdTOP2 contains all the elements essential for sequence preference in protein–DNA interaction but unlike the human enzyme, the C-terminus of the parasite enzyme plays an important role in the protein–drug interaction for the *in vitro* topoisomerase II cleavage reaction.

Exploiting the topoisomerases of kinetoplastid parasites for therapeutic interventions

During its catalytic cycle, topoisomerase generates breaks in the DNA and attaches itself to the newly generated DNA 3′ (type IB topoisomerases) or 5′ termini (all other topoisomerases) via phosphotyrosyl bonds. Under normal physiological conditions, these covalent enzyme–DNA cleavage complexes are short-lived catalytic intermediates and are tolerated by the cell. However, conditions that significantly increase the physiological concentrations or lifetime of these breaks unleash a multitude of deleterious side-effects. Thus, all topoisomerases are fundamentally dualistic in nature. Although they catalyse essential reactions in the cell, they possess an inherent dark side capable of inflicting great harm to the genome of an organism. Topoisomerase-targeting therapeutics currently in use act by stabilizing these covalent topoisomerase–DNA complexes.

The known topoisomerase targeting compounds can be divided into two classes, class I and class II (Liu, 1989; Wang, 1994). The class I inhibitors have been referred to as 'topoisomerase poisons' where as the class II compounds are referred to as 'catalytic inhibitors'. The class I drugs act by stabilizing the covalent topoisomerase-DNA covalent complexes. The class II drugs interfere with catalytic function of DNA topoisomerase without trapping the covalent complexes. A major determinant of cytotoxicity for the class I drug is the conversion of a latent single- or double-stranded break in a drug–topoisomerase-DNA complex into an irreversible double-stranded break. Replication and transcription are the key cellular processes that drive this conversion. However, for class II topoisomerase II drugs, processes other than replication might also be involved. Cell killing by class II topoisomerase II drugs may involve arresting cell cycle progression. Traversing of eukaryotic cells through cell division in the absence of functional DNA topoisomerase II can lead to aneuploidy and chromosomal breakage. For class I drug, cytotoxicity increases with increasing cellular level of target enzyme where as for class II drugs opposite is true. Thus increased levels of topoisomerases render cells hypersensitive to enzyme poisons but resistant to inhibitors. Conversely, decreased enzyme levels render cells resistant to poison but hypersensitive to inhibitors.

Topoisomerase IB inhibitors

Topoisomerase IB of kinetoplastids, being a heterodimeric enzyme, has added a new paradigm in the quest of antitrypanosomatid agents as we have already discussed in the earlier section of this chapter. Continuous effort is being made to identify inhibitors, both from natural and synthetic origin, which selectively target bi-subunit topoisomerase IB of trypanosomatids with negligible or no impact upon host enzyme. The extent of enzyme inhibition varies with structure and nature of the compound. Unfortunately, reports on specific inhibitors of *Trypanosoma* topoisomerase IB remain scarce. 2–6 dimethyl-9-hydroxyellipticinium was shown to inhibit the relaxation activity of *Trypanosoma cruzi* topoisomerase IB (Douc-Rasy *et al.*, 1983). Nonetheless, few studies have been made to observe the effect of different eukaryotic and bacterial topoisomerase I inhibitors upon the growth of *Trypanosoma* species (Zuma *et al.*, 2011; Jobe *et al.*, 2012). Mostly, contribution has been made to develop topoisomerase IB inhibitors targeted to *Leishmania* species. Such inhibitors come from a vast spectrum of chemical identity (Table 10.2). The mode of inhibition primarily follows two distinct mechanisms as conferred in the preceding section. They either hamper the catalytic activity of the enzyme (termed as 'catalytic inhibitors') or stabilize DNA–topoisomerase cleavage complex (termed as 'poisons'). Though such kind of classification is not available for all the inhibitors reported to date, few sporadic theories pertaining to the extent of inhibition have been established. Fatty acids are one of the major groups of compounds studied thoroughly in this regard (Carballeira *et al.*, 2009). Two parameters viz. (1) carbon chain

Table 10.2 Compounds (natural/synthetic) targeting kinetoplastid topoisomerase IB

Compound	Chemical identity	Kinetoplastid	Reference
(2*R*,5*Z*,9*Z*)-2-METHOXY-25-methyl-5,9-hexacosadienoic acid and (2*R*,5*Z*,9*Z*)-2-methoxy-24-methyl-5,9-hexacosadienoic acid	α-Methoxylated Δ5,9 Fatty Acids	*L. infantum*	Carballeira *et al.* (2016)
16α-Hydroxycleroda-3,13(14)Z-dien-15,16-olide	Clerodane diterpenoids	*L. donovani*	Misra *et al.* (2010)
2-methoxy-5,9-eicosadienoic acid, 2-methoxy-5,9-eicosadiynoic acid	α-methoxylated fatty acid and acetylinic analogue	*L. donovani*	Carballeira *et al.* (2013)
2-Octadecynoic acid, 2-hexadecynoic acid, 2-tetradecynoic acid	2-alkylnoic fatty acid	*L. donovani*	Carballeira *et al.* (2012a,b)
3,4-Dihydroxyphenyl, derivative of arylidenefuropyridinediones	Furopyridinediones	*L. donovani*	Mamidala *et al.* (2016)
3–3′-Di-indolylmethane	Acid condensation product of indole-3-carbinol	*L. donovani*	Roy *et al.* (2008)
3β,6β,16β-trihydroxylup-20(29)-ene	Lupane triterpene	*L. amazonensis*	Teles *et al.* (2015)
5-Nitrofuran and 5-nitroimidazole analogues of N-substituted-piperazinyl-1,3,4-thiadiazoles	1,3,4-Thiadiazole derivatives	*Leishmania* spp.	Poorrajab *et al.* (2009)
Anthra[1,2-d]imidazole-6,11-dione derivatives	Imidazole fused anthraquinone derivatives	*L. donovani*	Chaudhuri *et al.* (2007)
Baicalein, luteolin, quercetin	Flavonoids	*L. donovani*	Das *et al.* (2006)
Berberin	Poliheterocyclics	*L. braziliensis* *L. donovani*	Vennerstrom *et al.* (1990), Marquis *et al.* (2003)

Table 10.2 Continued

Compound	Chemical identity	Kinetoplastid	Reference
Bisbenzylisoquinoline alkaloid	bisbenzylisoquinoline alkaloid	*L. donovani*	Kumar *et al.* (2016)
Camptothecin	Quinoline alkaloid	*T. brucei*, *L. donovani*	Bodley and Shapiro (1995)
Copper salisylaldoxime (CuSAL)	Transition metal complex	*L. donovani*	Singh *et al.* (2017)
Diamidine compounds including pentamidine and its analogue DB75	Diamidines	*L. donovani*	Yang *et al.* (2016)
Dichloroacetamido bromobenzyl bromoindolyl bromoquinoline	Indolyl quinolines	*L. donovani*	Ray *et al.* (1997)
Dihydrobetulinic acid	Pentacyclic triterpenoid	*L. donovani*	Chowdhury *et al.* (2003)
Diospyrin, naphthalene derivatives	Naphthoquinones	*L. donovani* *T. brucei* *T. cruzi*	Ray *et al.* (1998); Ganapaty *et al.* (2006)
Disuccinyl betulin, Diglutaryl dihydrobetulin, Disuccinyl dihydrobetulin	Derivatives of betulin, a natural triterpene	*L. donovani*	Chowdhury *et al.* (2011)
Hoechst-33258, Hoechst-33342	Bis-benzimidazoles	*L. donovani*	Walker and Saravia (2004)
Indotecan (LMP400), AM13–55	Indenoisoquinolines	*L. infantum*	Balaña-Fouce *et al.* (2012)
N-6 3-aminopropyl indenoisoquinolines, N-6 3-imidazolylprolyl Indenoisoquinolines, alkylamine repeats N-6	Indenoisoquinolines	*T. brucei*	Bakshi *et al.* (2009)
N-benzyl 2, 2′α3,3′,5′, 6′,7′,7α,α′-octahydro-2methoxycarbonyl-spiro [indole-3, 3′-pyrrolizidine]-2 one	Spirooxindole derivative	*L. donovani*	Saha *et al.* (2016)
Niranthin, Lyoniside, Saracoside	Lignan glycosides	*L. donovani*	Chowdhury *et al.* (2012), Saha *et al.* (2013)
Peganine hydrochloride dihydrate	Peganine hydrochloride dihydrate	*L. donovani*	Misra *et al.* (2008)
Pentostam, glucantime	Pentavalent antimony	*L. donovani*	Chakraborty and Majumder (1988)
Putranoside-D, putranoside-A, putranoside-A methyl ester	Saponin	*L. donovani*	Kumar *et al.* (2014)
Rebeccamycin	Indocarbazoles	*T. brucei*	Deterding *et al.* (2005)
Tetrahydro indeno-1,5-naphthyridines and indeno[1,5]naphthyridines	Indeno-1,5-naphthyridines	*L. infantum*	Tejería *et al.* (2016)
Topotecan, gimatecan, pro-drug irinotecan and active metabolite SN38	Quinoline alkaloid	*L. infantum*	Prada *et al.* (2013b)
Voacamine	Indole alkaloid	*L. donovani* *L. amazonensis* *T. cruzi*	Chowdhury *et al.* (2017)

length and, (2) amount of unsaturation are crucial that govern the DNA relaxation activity by fatty acids. If the nature of unsaturation is kept constant, inhibition capability becomes proportional with the carbon chain length. This correlation is exemplified by 2-alkanoic acids. 2-Octadecynoic acid shows highest inhibition followed by 2-hexacecynoic acid and 2-tetradecynoic acid (Carballeira *et al.*, 2011). Similarly, icosenoic acid is more potent than heptadecenoic acid (Carballeira *et al.*, 2009). On the other hand, if the chain length is kept constant, unsaturated fatty acids show more inhibition than their saturated counterparts. For this reason, 2-methoxy-heptadecynoic acid shows higher inhibition than 2-methoxy-heptadecenoic acid. Their saturated counterpart 2-methoxy-heptadecanoic acid shows no inhibition at all. Probably the weak intermolecular interactions between enzyme active site and unsaturated bonds are responsible for this. In contrast, there are factors that negatively regulate the enzyme inhibition capability of fatty acids. Presence of ciscyclopropane group is one such factor. Another theory suggests that rational modification of core structure of known inhibitors improves inhibition efficacy. Chaudhuri *et al.* (2007) reported that if imidazole group is fused with anthraquinone derivatives, it improves the efficiency. This improved efficacy is probably due to the variation in pK_a value and electronegativity of the side-chain nitrogen. But it is to be remembered all such inhibitors do not necessarily impart cytotoxicity upon parasites. Though good inhibitors generally reflect the same trend with promastigotes and amastigotes, probability still exists that they may not be fruitful in all cases. Normally, fatty acids, being amphipathic, get incorporated into lipid membranes. They perturb the structural integrity of lipid bilayer resulting in cytotoxicity (Carballeira *et al.*, 2016). This argument is corroborated by the finding that presence of triple bond along with α-methoxylation enhances enzyme inhibition. But surprisingly, α-methoxylation decreases the cytotoxicity towards promastigotes. It is reported that despite being potential inhibitors of topoisomerase IB, methoxylated fatty acids are not significant enough as far as parasite viability is concerned. There is no general theory to establish the correlation between fatty acids and their antiproriferative effect upon parasite. Few studies suggest that this effect is probably due to inhibited uptake of glucose or leucine by the protozoa (Chaudhuri *et al.*, 1986). Finally, the most important concern is the host cytotoxicity. Situations often prevail where a particular inhibitor shows significant inhibition of parasite enzyme and induces cell death. But at the same time, the compound exhibits similar or more toxicity to host topoisomerase IB. Carballeira *et al.* (2016) recently reported that few methoxylated fatty acids show more inhibition of human topoisomerase IB than the parasite enzyme. They showed that preference of an inhibitor is shifted from parasite enzyme towards host enzyme due to increase in alkyl chain length. Short chain fatty acids are better inhibitors of *Leishmania* enzyme than human topoisomerase and the case gets reversed for long chain fatty acids. However, more investigation is required to establish these findings in a form of a theory. In the succeeding subsections, we will now deal with two fundamental models of topoisomerase IB inhibition.

Catalytic inhibitors of topoisomerase IB

Catalytic inhibitors show a general trend of interacting with the free enzyme. Their inhibitory effect markedly increases when they are preincubated with the free enzyme prior to DNA substrate addition under *in vitro* conditions. These kinds of inhibitors increase the K_M and follow competitive model of inhibition. However, they cannot bind with DNA–topoisomerase cleavage complex and hence do not alter V_{max} value. Several compounds come under the aegis of catalytic inhibitors (Table 10.2). Derivatives of betulin such as disuccinyl betulin, diglutaryl dihydrobetulin, and disuccinyl dihydrobetulin (Chowdhury *et al.*, 2011), putranoside-D, putranoside-A, putranoside-A methyl ester (Kumar *et al.*, 2014), spirooxiindole derivative (Saha *et al.*, 2016), few bisbenzylisoquinoline alkaloids (Kumar *et al.*, 2016) are some of the examples. All these compounds interact with topoisomerase IB of *Leishmania*. However their binding stoichiometry varies. As an example, three betulin derivatives developed by Chowdhury *et al.* (2011) show 1:1 binding stoichiometry with the enzyme whereas another catalytic inhibitor, *N*-benzyl-octahydro-2-methoxycarbonyl-spiro[indole-3,3′-pyrrolizidine]-2-one, shows a 1:2 stoichiometry (Saha *et al.*, 2016). These catalytic inhibitors reversibly bind with the enzyme and abrogate the DNA-enzyme covalent complex formation.

Topoisomerase IB poisons

Inhibitors falling under this group tend to bind either solely with DNA–enzyme covalent complex or with both free enzyme and DNA–enzyme complex. They may decrease both K_m and V_{max} values acting as uncompetitive inhibitor or may decrease only V_{max} value keeping K_m unaltered acting as non-competitive inhibitor. Many such compounds, including camptothecin, have been established to date as *Leishmania* topoisomerase IB poisons (Table 10.2). Unlike catalytic inhibitors, topoisomerase poisons do not abrogate the DNA-enzyme covalent complex formation, rather stabilize the complex. Such kind of DNA–enzyme adducts ultimately cause DNA single-strand breaks. These breaks are subsequently converted to double-strand breaks upon collision with transcription or replication machineries. Double-strand breaks are detrimental to the parasite. The cell cycle progression is arrested triggering programmed cell death of the parasite. All these poisons can be further classified into two sub-classes depending upon their capability to inhibit the religation reaction. Camptothecin (Das *et al.*, 2006), lyoniside, saracoside (Saha *et al.*, 2013), niranthin (Chowdhury *et al.*, 2012), DIM (Roy *et al.*, 2008) are some of the poisons which inhibit the religation action. In contrast, baicalein, luteolin cannot inhibit this topoisomerase IB-mediated religation step (Das *et al.*, 2006). Moreover, there are few examples of topoisomerase poisons which not only binds with DNA-enzyme covalent complex but also intercalate into the DNA itself at a very high concentration. Quercetin, bicalein, luteolin have been established to possess DNA intercalation property at 300 μM (Das *et al.*, 2006).

Topoisomerase II inhibitors

Topoisomerase II of the kinetoplastid parasites are homodimeric enzymes similar to higher eukaryotes including human. But it is a crucial enzyme for the parasites because of its involvement in replication of kinetoplast DNA network (kDNA) inside mitochondria (Shapiro *et al.*, 1995; Das *et al.*, 2008). Hence, topoisomerase II has also been explored as a promising target for antiparasitic therapeutics (Table 10.3). Interestingly, pentamidine was later found to also be an inhibitor of topoisomerase II (Singh *et al.*, 2007). A distinct aspect of this enzyme in trypanosomatids is the discrete existence of nuclear and kinetoplast topoisomerase II (Kulikowicz and Shapiro, 2006). Some compounds

Table 10.3 Compounds (natural/synthetic) targeting kinetoplastid topoisomerase II

Compound	Chemical identity	Kinetoplastid	Reference
3,5-Bis(4-chlorophenyl)–7-hydroxyisobenzofuran-1(3H)-one and (4-bromo)-30-hydroxy-50-(4-bromophenyl)-benzophenone	Isobenzofuranone derivatives	*L. donovani*	Mishra *et al.* (2014)
5-Nitrofuran and 5-nitroimidazole analogues of N-substituted-piperazinyl-1,3,4-thiadiazoles	1,3,4-Thiadiazole derivatives	*Leishmania* spp.	Poorrajab *et al.* (2009)
Anilinoacridine compounds	Anilinoacridine	*L. chagasi*	Werbovetz *et al.* (1994)
Anilinoacridine compounds	Anilinoacridine	*T. brucei*	Gamage *et al.* (1997)
Dichloroacetamido bromo benzyl bromoindolyl bromo quinoline	Indolyl quinolines	*L. donovani*	Ray *et al.* (1997)
Dihydrobetulinic acid	Pentacyclic triterpenoid	*L. donovani*	Chowdhury *et al.* (2003)
Doxorrubicin	Anthracyclins	*L. donovani*	Singh and Dey (2007)
Enoxacin, Ciprofloxacin	Fluoroquinolones	*L. panamensis*	Romero *et al.* (2005), Cortázar *et al.* (2007)
Etoposide, teniposide	Podophylotoxins	*L. donovani*	Sangeeta *et al.* (2005a,b)
Lute Olin	Flavonoids	*L. donovani*	Muttra *et al.* (2000)
Monoacid and monoacid analogues	4-nitro-benzo-isoquinolinedione and derivatives	*L. chagasi*	Slant *et al.* (1996)
Offloading, ciprofloxacin, offloading	Fluoroquinolones	*T. brucei*	Neonates *et al.* (1999, 2003)
Norfloxacin, ofloxacin	Fluoroquinolones	*T. cruzi*	Gonzales-Perdomo *et al.* (1990)
Novobiocin	Aminocoumarin antibiotic	*L. donovani*	Singh *et al.* (2005)
Novobiocin	Aminocoumarin antibiotic	*T. cruzi*	Gonzales-Perdomo *et al.* (1990)
Pentamidine	Diamidines	*L. donovani*	Singh and Dey (2007)

differentially target nuclear and kinetoplast topoisomerase II whereas some have similar preferences for both. As for example, one subset of mitonafide analogues promotes kDNA linearization in *Leishmania* and another subset affects activity of nuclear topoisomerase II (Slunt *et al.*, 1996). Moving onto anilinoacridines, they have similar effect for both the enzymes. This differential activity is primarily controlled by the presence and position of the nitro groups in the compounds. Now moving onto host toxicity, similar scenario is encountered with topoisomerase II just like we saw with topoisomerase IB. Host toxicity is a critical issue as far as topoisomerase II targeted therapeutic development is concerned. Few researches have shown that parasitic topoisomerase II is more closely related to mammalian counterpart rather than prokaryotic topoisomerase II which has made the situation more complex. This similarity increases the possibility for a protozoan enzyme inhibitor to inhibit host enzyme (Werbovetz *et al.*, 1992). Few compounds tend to inhibit both parasite and host enzyme at comparable range. Compounds such as anilinoacridine, protoberberine coralyne, bisbenzimidazoles, diamidine diminazene, etc. show equal effect on both parasite and human enzymes whereas ellagic acid has preference for macrophage topoisomerase II. Nevertheless second generation fluoroquinolones such as enoxacin and ciprofloxacin, flavonoid quercetin target *L. panamensis* DNA topoisomerase II with substantially less toxicity towards macrophage enzyme (Cortazar *et al.*, 2007). Moving onto *Trypanosoma*, just like topoisomerase IB, reports on *Trypanosoma*-specific topoisomerase II inhibitors are substantially meagre. Effects of few bacterial (ofloxacin, novibiocin, etc.) and eukaryotic (aclarubicin, idarubicin, mitoxantrone, merbarone, etc.) topoisomerase II inhibitors on growth of *Trypanosoma* have been studied. But whether these compounds inhibit *Trypanosoma* topoisomerase II is not known so far. Topoisomerase II inhibitors are also fundamentally subdivided into two subclasses; namely, catalytic inhibitors and poisons. In the succeeding section, we will take a closer view of these two kinds of inhibitors mostly in a *Leishmania* purview.

Catalytic inhibitors of topoisomerase II

Topoisomerase II is fundamentally different from topoisomerase IB in its requirement of ATP to perform its catalytic cycle. Catalytic inhibitors of topoisomerase II have two different mechanisms. They either bind with the free enzyme preventing it to sit on DNA substrate or inhibit the ATPase activity of the enzyme by interacting with ATPase domain. Since they abrogate topoisomerase–DNA interaction, naturally cleavage complex formation is blocked. Mishra *et al.* (2014) reported that two derivatives of isobenzofuranone inhibit *L. donovani* topoisomerase II by interfering at its catalytic centre but neither have any effect in ATPase activity nor do they intercalate within substrate DNA.

Topoisomerase II poisons

Topoisomerase II poisons stabilize enzyme-kinetoplast DNA cleavable complex and disrupt cleavage-religation equilibrium. This irreversible covalent association between topoisomerase II and DNA ultimately cause double-strand breaks which are extremely cytotoxic for the parasites. Luteolin, quercetin (Mittra *et al.*, 2000), some mitonafide analogues fall under this category. They are further classified as intercalating and non-intercalating poisons. Some poisons can intercalate into DNA by virtue of their planar aromatic region (Slunt *et al.*, 1996). During interaction, a local distortion occurs which resembles an intermediate of the topoisomerase II catalytic cycle. If an inhibitor favours that intermediate formation, cleavage complex is stabilized. Amonafide and one mitonafide analogue have been reported to possess this property. Naturally, all poisons do not have this capacity to get intercalated. A classical example of non-intercalating *Leishmania* topoisomerase II poison is etoposide (Sengupta *et al.*, 2005b), which is a well-known anticancer drug. Surprisingly, etoposide can also inhibit the ATPase activity of the recombinant N-terminal domain of *L. donovani* topoisomerase II (Sengupta *et al.*, 2005a). But etoposide has no impact upon parasite viability.

Dual inhibitors of topoisomerase IB and topoisomerase II

There are few compounds which inhibit both type IB and II topoisomerases of *Leishmania*. Nonetheless, there is no theory available to date to functionally connect the structure of an inhibitor with dual inhibition potential. Few indolyl quinoline analogues namely 2-(2′-dichloroacetamidobenzyl)-3-(3′-indolyl)-quinoline,

2-(2′-dichloroacetamido-5′-bromobenzyl)-3′-[3′-(5′-bromoindolyl]-6-bromo quinoline, and 2-(2′-acetamido benzyl)-3-(3′-indolyl)-quinoline (Ray *et al.*, 1997), a pentacyclic triterpenoid viz. dihydrobetulinic acid (DHBA) (Chowdhury *et al.*, 2003) inhibit both the enzyme. DHBA is a catalytic inhibitor of both the enzymes. It is also cytotoxic for the parasites. Though there are no reports on *Trypanosoma*-specific dual inhibitors, antiproliferative effects of several eukaryotic dual inhibitors have been investigated upon *T. cruzi* (Lacombe *et al.*, 2014).

Conclusion and future perspectives

There has been immense progress in the understanding of the biology of kinetoplastid parasites especially of parasitic topoisomerases. The availability of the Tritryp genome database has provided significant impetus towards the study of these enzymes and also provided necessary momentum to the much needed identification of new targets for anti-leishmanial therapeutics. Scientists throughout the world including the author's laboratory have contributed significantly towards the understanding of structures and activities of these parasitic topoisomerases and established *Leishmania* topoisomerases as potential drug targets for leishmaniasis (kala-azar). Topoisomerase genes and proteins characterized from these lower eukaryotes appear to share many characteristics associated with their human homologues. But certain striking differences, including different enzyme activity requirements, variable catalytic sites and different sensitivities to topoisomerase poisons, provide insight for the development of topoisomerase-directed antiparasitic therapeutics. It has been established by several studies that the inhibitors of topoisomerases convert these essential enzymes into intracellular proliferating cell toxins and thereby provide a good tool for preferentially killing of the highly replicative parasite cells within the host. However, future challenges are to overcome the problem related to drug resistance for development of novel anti-leishmanial therapeutics. In this context, the approach of combinatorial therapy becomes very effective and identification of new drugs and new targets is absolutely warranted to combat parasite menace in the foreseeable future.

The search also continues for novel RNA topoisomerase activities in species of *Leishmania* and *Trypanosoma*. This kind of non-canonical function of type IA topoisomerases was first reported for *Escherichia coli* topoisomerase III (Wang *et al.*, 1996) and later in human topoisomerase 3β (Stoll *et al.*, 2013; Xu *et al.*, 2013). An extensive genetic recombination among the 10,000 mini-circles, creating novel guide RNAs followed by RNA editing, might require distinct RNA topoisomerase.

The major obstacle of kinetoplastid topoisomerase research is over-expressing purified recombinant topoisomerases and elucidating their crystal structure. In this way, structural information gained from crystallographic studies can contribute to a better understanding of the molecular mechanisms of enzyme action *in vivo*, including their interaction with inhibitors and poisons screened from natural or synthetic sources. This will cause a paradigm shift in the quest to target the topoisomerases selectively as a means of therapeutic control of the parasites.

Acknowledgements

We thank Dr Samit Chattopadhyay, the Director of our Institute (CSIR-Indian Institute of Chemical Biology), for his interest in this work. This work was supported by Department of Science and Technology (DST) Indo-Brazil project (DST/INT/Brazil/RPO-01/2009/2), Raja Ramanna Fellowship (DAE, Government of India) and NASI-Senior Scientist Platinum Jubilee Fellowship to HKM.

References

Bakshi, R.P., Sang, D., Morrell, A., Cushman, M., and Shapiro, T.A. (2009). Activity of indenoisoquinolines against African trypanosomes. Antimicrob. Agents Chemother. 53, 123–128. https://doi.org/10.1128/AAC.00650-07.

Balaña-Fouce, R., Prada, C.F., Requena, J.M., Cushman, M., Pommier, Y., Álvarez-Velilla, R., Escudero-Martínez, J.M., Calvo-Álvarez, E., Pérez-Pertejo, Y., and Reguera, R.M. (2012). Indotecan (LMP400) and AM13-55: two novel indenoisoquinolines show potential for treating visceral leishmaniasis. Antimicrob. Agents Chemother. 56, 5264–5270. https://doi.org/10.1128/AAC.00499-12.

Banerjee, B., Sen, N., and Majumder, H.K. (2011). Identification of a functional type IA topoisomerase, LdTopIIIβ, from Kinetoplastid parasite *Leishmania donovani*. Enz. Res. 2011 (Article ID 230542). https://doi.org/10.4061/2011/230542.

Berger, J.M., Gamblin, S.J., Harrison, S.C., and Wang, J.C. (1996). Structure and mechanism of DNA topoisomerase II. Nature 379, 225–232. https://doi.org/10.1038/379225a0.

Bodley, A.L., and Shapiro, T.A. (1995). Molecular and cytotoxic effects of camptothecin, a topoisomerase I inhibitor, on trypanosomes and *Leishmania*. Proc. Natl. Acad. Sci. U.S.A. *92*, 3726–3730.

Bodley, A.L., Chakraborty, A.K., Xie, S., Burri, C., and Shapiro, T.A. (2003). An unusual type IB topoisomerase from African trypanosomes. Proc. Natl. Acad. Sci. U.S.A. *100*, 7539–7544. https://doi.org/10.1073/pnas.1330762100.

Bosedasgupta, S., Das, B.B., Sengupta, S., Ganguly, A., Roy, A., Tripathi, G., and Majumder, H.K. (2008a). Amino acids 39–456 of the large subunit and 210–262 of the small subunit constitute the minimal functionally interacting fragments of the unusual heterodimeric topoisomerase IB of *Leishmania*. Biochem. J. 409(2), 481–489. https://doi.org/10.1042/BJ20071358.

BoseDasgupta, S., Ganguly, A., Das, B.B., Roy, A., Khalkho, N.V., and Majumder, H.K. (2008b). The large subunit of *Leishmania* topoisomerase I functions as the 'molecular steer'in type IB topoisomerase. Mol. Microbiol. *67*, 31–46. https://doi.org/10.1111/j.1365-2958.2007.06002.x.

Broccoli, S., Marquis, J.F., Papadopoulou, B., Olivier, M., and Drolet, M. (1999). Characterization of a *Leishmania donovani* gene encoding a protein that closely resembles a type IB topoisomerase. Nucleic Acids Res. *27*, 2745–2752.

Campbell, S., and Maxwell, A. (2002). The ATP-operated clamp of human DNA topoisomerase IIalpha: hyperstimulation of ATPase by 'piggy-back' binding. J. Mol. Biol. *320*, 171–188. https://doi.org/10.1016/S0022-2836(02)00461-8.

Carballeira, N.M., Cartagena, M., Li, F., Chen, Z., Prada, C.F., Calvo-Alvarez, E., Reguera, R.M., and Balaña-Fouce, R. (2012a). First total synthesis of the (±)-2-methoxy-6-heptadecynoic acid and related 2-methoxylated analogs as effectiveinhibitors of the *Leishmania* topoisomerase IB enzyme. Pure Appl. Chem., *84*, 1867–7185. https://doi.org/10.1351/PAC-CON-11-10-21.

Carballeira, N.M., Cartagena, M., Sanabria, D., Tasdemir, D., Prada, C.F., Reguera, R.M., and Balaña-Fouce R. (2012b). 2-Alkynoic fatty acids inhibit topoisomerase IB from *Leishmania donovani*. Bioorg. Med. Chem. Lett. *22*, 6185–6189. https://doi.org/10.1016/j.bmcl.2012.08.019.

Carballeira, N.M., Cartagena, M.M., Prada, C.F., Rubio, C.F., and Balaña-Fouce, R. (2009). Total synthesis and antileishmanial activity of the natural occurring acetylenic fatty acids 6-heptadecynoic acid and 6-icosynoic acid. Lipids *44*, 953–961. https://doi.org/10.1007/s11745-009-3345-z.

Carballeira, N.M., Montano, N., Alvarez-Velilla, R., Prada, C.F., Reguera, R.M., and Balaña-Fouce, R. (2013). Synthesis of marine α-methoxylated fatty acid analogs that effectively inhibit the topoisomerase IB from *Leishmania donovani* with a mechanism different from that of camptothecin. Marine Drugs, 11, 3661–75. https://doi.org/10.3390/md11103661.

Carballeira, N.M., Montano, N., Amador, L.A., Rodríguez, A.D., Golovko, M.Y., Golovko, S.A., Reguera, R.M., Álvarez-Velilla, R., and Balaña-Fouce, R. (2016). Novel very long-chain α-methoxylated Δ5, 9 fatty acids from the sponge *Asteropusniger* are effective inhibitors of topoisomerases IB. Lipids, 51(2), 245–256. https://doi.org/10.1007/s11745-015-4114-9.

Carballeira, N.M., Montano, N., Cintrón, G.A., Márquez, C., Rubio, C.F., Prada, C.F., and Balaña-Fouce, R. (2011). First total synthesis and antileishmanial activity of (Z)-16-methyl-11-heptadecenoic acid, a new marine fatty acid from the sponge Dragmaxia undata. Chem. Phys. Lipids, *164*, 113–117. https://doi.org/10.1016/j.chemphyslip.2010.11.006.

Chakraborty, A.K., and Majumder, H.K. (1988). Mode of action of pentavalent antimonials: Specific inhibition of type I DNA topoisomerase of *Leishmania donovani*. Biochem. Biophys. Res. Commun. *152*, 605–611.

Chakraborty, A.K., Gupta, A., and Majumder, H.K. (1993). A type 1 DNA topoisomerase from the kinetoplast hemoflagellate *Leishmania donovani*. Ind. J. Biochem. Biophys. 30(5), 257–263.

Champoux, J.J. (2001). DNA topoisomerases: structure, function, and mechanism. Annu. Rev. Biochem. *70*, 369–413. https://doi.org/10.1146/annurev.biochem.70.1.369.

Chaudhuri, G., Ghoshal, K., and Banerjee, A.B. (1986). Toxic effects of fatty acids on *Leishmania donovani* promastigotes. Ind. J. Med. Res. *84*, 361–365.

Chaudhuri, P., Majumder, H.K., and Bhattacharya, S. (2007). Synthesis, DNA binding, and leishmania topoisomerase inhibition activities of a novel series of anthra [1, 2-d] imidazole-6, 11-dione derivatives. J. Med. Chem. *50*, 2536–2540. https://doi.org/10.1021/jm0610604.

Chen, S.H., Chan, N.L., and Hsieh, T.S. (2013). New mechanistic and functional insights into DNA topoisomerases. Annu. Rev. Biochem. *82*, 139–170. https://doi.org/10.1146/annurev-biochem-061809-100002.

Chen, Y., and Rice, P.A. (2003). New insight into site-specific recombination from Flp recombinase-DNA structures. Annu. Rev. Biophys. Biomol. Struct. *32*, 135–159. https://doi.org/10.1146/annurev.biophys.32.110601.141732.

Chowdhury, A.R., Mandal, S., Goswami, A., Ghosh, M., Mandal, L., Chakraborty, D., Ganguly, A., Tripathi, G., Mukhopadhyay, S., Bandyopadhyay, S., and Majumder, H.K. (2003). Dihydrobetulinic acid induces apoptosis in *Leishmania donovani* by targeting DNA topoisomerase I and II: implications in antileishmanial therapy. Mol. Med. 9(1–2), 26–36.

Chowdhury, S., Mukherjee, T., Mukhopadhyay, R., Mukherjee, B., Sengupta, S., Chattopadhyay, S., Jaisankar, P., Roy, S., and Majumder, H.K. (2012). The lignan niranthin poisons *Leishmania donovani* topoisomerase IB and favours a Th1 immune response in mice. EMBO Mol. Med. 4(10), 1126–1143. https://doi.org/10.1002/emmm.201201316.

Chowdhury, S., Mukherjee, T., Sengupta, S., Chowdhury, S.R., Mukhopadhyay, S., and Majumder, H.K. (2011). Novel betulin derivatives as antileishmanial agents with mode of action targeting type IB DNA topoisomerase. Mol. Pharmacol. *80*, 694–703. https://doi.org/10.1124/mol.111.072785.

Chowdhury, S.R., Kumar, A., Godinho, J.L.P., Silva, S.T.D.M., Zuma, A.A., Saha, S., Kumari, N., Rodrigues, J.C.F., Sundar, S., Dujardin, J.C. *et al.* (2017). Voacamine alters *Leishmania* ultrastructure and kills parasite by poisoning unususal bi-subunit topoisomerase 1B. Biochem. Pharmacol. *138*, 19-30. https://doi.org/10.1016/j.bcp.2017.05.002.

Corbett, K.D., and Berger, J.M. (2004). Structure, molecular mechanisms, and evolutionary relationships in DNA topoisomerases. Annu. Rev. Biophys. Biomol. Struct. *33*, 95–118. https://doi.org/10.1146/annurev.biophys.33.110502.140357.

Cortázar, T.M., Coombs, G.H., and Walker, J. (2007). *Leishmania panamensis*: comparative inhibition of nuclear DNA topoisomerase II enzymes from promastigotes and human macrophages reveals anti-parasite selectivity of fluoroquinolones, flavonoids and pentamidine. Exp. Parasitol. 116(4), 475–482. https://doi.org/10.1016/j.exppara.2007.02.018.

Das, A., Dasgupta, A., Sengupta, T., and Majumder, H.K. (2004). Topoisomerases of kinetoplastid parasites as potential chemotherapeutic targets. Trends Parasitol. *20*, 381–387. https://doi.org/10.1016/j.pt.2004.06.005.

Das, A., Dasgupta, A., Sharma, S., Ghosh, M., Sengupta, T., Bandopadhyay, S., and Majumder, H.K. (2001). Characterisation of the gene encoding type II DNA topoisomerase from *Leishmania donovani*: a key molecular target in antileishmanial therapy. Nucleic Acids Res. 29(9), 1844–1851.

Das, B.B., Ganguly, A., and Majumder, H.K. (2008). DNA topoisomerases of *Leishmania*: the potential targets for anti-leishmanial therapy. Adv. Exp. Med. Biol. *625*, 103–115. https://doi.org/10.1007/978-0-387-77570-8_9.

Das, B.B., Sen, N., Dasgupta, S.B., Ganguly, A., and Majumder, H.K. (2005). N-terminal region of the large subunit of *Leishmania donovani* bisubunit topoisomerase I is involved in DNA relaxation and interaction with the smaller subunit. J. Biol. Chem. *280*, 16335–16344. https://doi.org/10.1074/jbc.M412417200.

Das, B.B., Sen, N., Roy, A., Dasgupta, S.B., Ganguly, A., Mohanta, B.C., Dinda, B., and Majumder, H.K. (2006). Differential induction of *Leishmania donovani* bi-subunit topoisomerase I–DNA cleavage complex by selected flavones and camptothecin: activity of flavones against camptothecin-resistant topoisomerase I. Nucleic Acids Res. 34(4), 1121–1132. https://doi.org/10.1093/nar/gkj502.

Das, B.B., Sengupta, T., Ganguly, A., and Majumder, H.K. (2006). Topoisomerases of kinetoplastid parasites: why so fascinating? Mol. Microbiol. *62*, 917–927. https://doi.org/10.1111/j.1365-2958.2006.05428.x.

Das, B.B., Sen, N., Ganguly, A., and Majumder, H.K. (2004). Reconstitution and functional characterization of the unusual bi-subunit type I DNA topoisomerase from *Leishmania donovani*. FEBS Lett. *565*, 81–88. https://doi.org/10.1016/j.febslet.2004.03.078.

Davies, D.R., Mushtaq, A., Interthal, H., Champoux, J.J., and Hol, W.G. (2006). The structure of the transition state of the heterodimeric topoisomerase I of *Leishmania donovani* as a vanadate complex with nicked DNA. J. Mol. Biol. *357*, 1202–1210. https://doi.org/10.1016/j.jmb.2006.01.022.

Deterding, A., Dungey, F.A., Thompson, K.A., and Steverding, D. (2005). Anti-trypanosomal activities of DNA topoisomerase inhibitors. Acta Trop. 93,311–316. https://doi.org/10.1016/j.actatropica.2005.01.005.

Deweese, J.E., and Osheroff, N. (2009). The DNA cleavage reaction of topoisomerase II: wolf in sheep's clothing. Nucleic Acids Res. 37,738–748. https://doi.org/10.1093/nar/gkn937.

Díaz, González, R., Pérez Pertejo, Y., Redondo, C.M., Pommier, Y., Balaña-Fouce, R., and Reguera, R.M. (2007). Structural insights on the small subunit of DNA topoisomerase I from the unicellular parasite *Leishmania donovani*. Biochimie, 89, 1517–1527. https://doi.org/10.1016/j.biochi.2007.07.015.

Douc-Rasy, S., Kayser, A., and Riou, G. (1983). A specific inhibitor of type I DNA-topoisomerase of *Trypanosoma cruzi*: dimethyl-hydroxy-ellipticinium. Biochem. Biophys. Res. Commun. 117(1), 1–5.

Fragoso, S.P., and Goldenberg, S. (1992). Cloning and characterization of the gene encoding *Trypanosoma cruzi* DNA topoisomerase II. Mol. Biochem. Parasitol. *55*, 127–134. https://doi.org/10.1016/0166-6851(92)90133-5.

Gadelle, D., Filée, J., Buhler, C., and Forterre, P. (2003). Phylogenomics of type II DNA topoisomerases. Bioessays. *25*, 232–242. https://doi.org/10.1002/bies.10245.

Gamage, S.A., Figgitt, D.P., Wojcik, S.J., Ralph, R.K., Ransijn, A., Mauel, J., Yardley, V., Snowdon, D., Croft, S.L., and Denny, W.A. (1997). Structure–activity relationships for the antileishmanial and antitrypanosomal activities of 1′-substituted 9-anilinoacridines. J. Med. Chem. *40*, 2634–2642. https://doi.org/10.1021/jm970232h.

Ganapaty, S., Steve Thomas, P., Karagianis, G., Waterman, P.G., and Brun, R. (2006). Antiprotozoal and cytotoxic naphthalene derivatives from *Diospyros assimilis*. Phytochem. *67*, 1950–1956. https://doi.org/10.1016/j.phytochem.2006.05.039.

Ganguly, A., Sengupta, S., BoseDasgupta, S., Roy, A., and Majumder, H.K. (2009). Mutational studies reveal lysine 352 on the large subunit is indispensable for catalytic activity of bi-subunit topoisomerase I from *Leishmania donovani*. Mol. Biochem. Parasitol. 165(1), 57–66. https://doi.org/10.1016/j.molbiopara.2009.01.002.

Gaziová, I., and Lukes, J. (2003). Mitochondrial and nuclear localization of topoisomerase II in the flagellate *Bodo saltans* (Kinetoplastida), a species with non-catenated kinetoplast DNA. J. Biol. Chem. *278*, 10900–10907. https://doi.org/10.1074/jbc.M202347200.

Gonzales-Perdomo, M., de Castro, S.L., Meirelles, M.N., and Goldenberg, S. (1990). *Trypanosoma cruzi* proliferation and differentiation are blocked by topoisomerase II inhibitors. Antimicrob. Agents Chemother. *34*, 1707–1714.

Jobe, M., Anwuzia-Iwegbu, C., Banful, A., Bosier, E., Iqbal, M., Jones, K., Lecutier, S.J., Lepper, K., Redmond, M., Ross-Parker, A., and Ward, E. (2012). Differential *in vitro* activity of the DNA topoisomerase inhibitor idarubicin against *Trypanosoma rangeli* and *Trypanosoma cruzi*. Memórias do Instituto Oswaldo Cruz, *107*, 946–950.

Kim, H.S., and Cross, G.A. (2010). TOPO3alpha influences antigenic variation by monitoring expression-site-associated VSG switching in *Trypanosoma brucei*. PLOS

Pathog. 6(7), e1000992. https://doi.org/10.1371/journal.ppat.1000992.

Koster, D.A., Croquette, V., Dekker, C., Shuman, S., and Dekker, N.H. (2005). Friction and torque govern the relaxation of DNA supercoils by eukaryotic topoisomerase IB. Nature 434, 671–674. https://doi.org/10.1038/nature03395.

Krogh, B.O., and Shuman, S. (2002). A poxvirus-like type IB topoisomerase family in bacteria. Proc. Natl. Acad. Sci. U.S.A. 99,1853–1858. https://doi.org/10.1073/pnas.032613199.

Kulikowicz, T., and Shapiro, T.A. (2006). Distinct genes encode type II topoisomerases for the nucleus and mitochondrion in the protozoan parasite *Trypanosoma brucei*. J. Biol. Chem. *281*, 3048–3056. https://doi.org/10.1074/jbc.M505977200.

Kumar, A., Chowdhury, S.R., Chakrabarti, T., Majumdar, H.K., Jha, T., and Mukhopadhyay, S. (2014). A new ellagic acid glycoside and DNA topoisomerase IB inhibitory activity of saponins from *Putranjiva roxburghii*. Nat. Prod. Commun. *9*, 675–677.

Kumar, A., Chowdhury, S.R., Sarkar, T., Chakrabarti, T., Majumder, H.K., Jha, T., and Mukhopadhyay, S. (2016). A new bisbenzylisoquinoline alkaloid isolated from *Thalictrum foliolosum*, as a potent inhibitor of DNA topoisomerase IB of *Leishmania donovani*. Fitoterapia,= *109*, 25–30. https://doi.org/10.1016/j.fitote.2015.11.021.

Lacombe, O.K., Zuma, A.A., da Silva, C.C., de Souza, W., and Motta, M.C.M. (2014). Effects of camptothecin derivatives and topoisomerase dual inhibitors on *Trypanosoma cruzi* growth and ultrastructure. J. Negat. Results Biomed. 13(1), 11. https://doi.org/10.1186/1477-5751-13-11.

Li, Z., Mondragón, A., Hiasa, H., Marians, K.J., and DiGate, R.J. (2000). Identification of a unique domain essential for *Escherichia coli* DNA topoisomerase III-catalysed decatenation of replication intermediates. Mol. Microbiol. 35, 4888–4895.

Lindsay, M.E., Gluenz, E., Gull, K., and Englund, P.T. (2008). A new function of *Trypanosoma brucei* mitochondrial topoisomerase II is to maintain kinetoplast DNA network topology. Mol. Microbiol. *70*, 1465–1476. https://doi.org/10.1111/j.1365-2958.2008.06493.x.

Liu, L.F. (1989). DNA topoisomerase poisons as antitumor drugs. Annu. Rev. Biochem. *58*, 351–375. https://doi.org/10.1146/annurev.bi.58.070189.002031.

Logan-Klumpler, F.J., De Silva, N., Boehme, U., Rogers, M.B., Velarde, G., McQuillan, J.A., Carver, T., Aslett, M., Olsen, C., Subramanian, S. *et al.* (2011) GeneDB—an annotation database for pathogens. Nucl. Acids Res. *40*, 98-108. https://doi.org/10.1093/nar/gkr1032.

Lynn, R., Giaever, G., Swanberg, S.L., and Wang, J.C. (1986). Tandem regions of yeast DNA topoisomerase II share homology with different subunits of bacterial gyrase. Science, *233*, 647–649. http://www.jstor.org/stable/1698013.

Mamidala, R., Majumdar, P., Jha, K.K., Bathula, C., Agarwal, R., Chary, M.T., Majumder, H.K., Munshi, P., and Sen, S. (2016). Identification of *Leishmania donovani* Topoisomerase 1 inhibitors via intuitive scaffold hopping and bioisosteric modification of known Top 1 inhibitors. Sci. Rep., 6. https://doi.org/10.1038/srep26603.

Marcotte, E.M., Pellegrini, M., Ng, H.L., Rice, D.W., Yeates, T.O., and Eisenberg, D. (1999). Detecting protein function and protein–protein interactions from genome sequences. Science 285, 751–753. https://doi.org/10.1126/science.285.5428.751.

Marquis, J.F., Makhey, D., LaVoie, E.J., and Olivier, M. (2003). Effects of topoisomerases inhibitors protoberberine on *Leishmania donovani* growth, macrophage function, and infection. J. Parasitol. 89(5), 1048–1052. https://doi.org/10.1645/GE-3161.

McClendon, A.K., and Osheroff, N. (2007). DNA topoisomerase II, genotoxicity, and cancer. Mutat. Res. *623*, 83–97. https://doi.org/10.1016/j.mrfmmm.2007.06.009.

Melendy, T., and Ray, D.S. (1987). Purification and nuclear localization of a type I topoisomerase from *Crithidia fasciculata*. Mol. Biochem. Parasitol. *24*, 215–225.

Mishra, A., Vinayagam, J., Saha, S., Chowdhury, S., Roychowdhury, S., Jaisankar, P., and Majumder, H.K. (2014). Isobenzofuranone derivatives exhibit antileishmanial effect by inhibiting type II DNA topoisomerase and inducing host response. Pharmacol. Res. Perspect. 2(6). e00070. https://doi.org/10.1002/prp2.70.

Misra, P., Khaliq, T., Dixit, A., SenGupta, S., Samant, M., Kumari, S., Kumar, A., Kushawaha, P.K., Majumder, H.K., Saxena, A.K., and Narender, T. (2008). Antileishmanial activity mediated by apoptosis and structure-based target study of peganine hydrochloride dihydrate: an approach for rational drug design. J. Antimicrob. Chemother. 62(5), 998–1002. https://doi.org/10.1093/jac/dkn319.

Misra, P., Sashidhara, K.V., Singh, S.P., Kumar, A., Gupta, R., Chaudhaery, S.S., Gupta, S.S., Majumder, H.K., Saxena, A.K., and Dube, A. (2010). 16alpha-Hydroxycleroda-3,13 (14)Z-dien-15,16-olide from *Polyalthia longifolia*: a safe and orally active antileishmanial agent. Br. J. Pharmacol. *159*, 1143–1150. https://doi.org/10.1111/j.1476-5381.2009.00609.x.

Mittra, B., Saha, A., Chowdhury, A.R., Pal, C., Mandal, S., Mukhopadhyay, S., Bandyopadhyay, S., and Majumder, H.K. (2000). Luteolin, an abundant dietary component is a potent anti-leishmanial agent that acts by inducing topoisomerase II-mediated kinetoplast DNA cleavage leading to apoptosis. Mol. Med. *6*, 527–541.

Nenortas, E., Burri, C., and Shapiro, T.A. (1999). Antitrypanosomal activity of fluoroquinolones. Antimicrob. Agents Chemother. *43*, 2066–2068.

Nenortas, E., Kulikowicz, T., Burri, C., and Shapiro, T.A. (2003). Antitrypanosomal activities of fluoroquinolones with pyrrolidinyl substitutions. Antimicrob. Agents Chemother. 47(9), 3015-3017. https://doi.org/10.1128/AAC.47.9.3015-3017.2003.

Obado, S.O., Bot, C., Echeverry, M.C., Bayona, J.C., Alvarez, V.E., Taylor, M.C., and Kelly, J.M. (2010). Centromere-associated topoisomerase activity in bloodstream form *Trypanosoma brucei*. Nucleic Acids Res. 39, 1023–1033. https://doi.org/10.1093/nar/gkq839.

Pasion, S.G., Hines, J.C., Aebersold, R., and Ray, D.S. (1992). Molecular cloning and expression of the gene encoding the kinetoplast-associated type II DNA topoisomerase of *Crithidia fasciculata*. Mol. Biochem. Parasitol. 50(1), 57–67.

Poorrajab, F., Ardestani, S.K., Foroumadi, A., Emami, S., Kariminia, A., Behrouzi-Fardmoghadam, M., and Shafiee, A. (2009). Selective leishmanicidal effect of 1, 3, 4-thiadiazole derivatives and possible mechanism of action against *Leishmania* species. Exp. Parasitol. *121,* 323–330. https://doi.org/10.1016/j.exppara.2008.12.004.

Prada, C.F., Álvarez-Velilla, R., Balaña-Fouce, R., Prieto, C., Calvo-Álvarez, E., Escudero-Martínez, J.M., Requena, J.M., Ordóñez, C., Desideri, A., Pérez-Pertejo, Y., and Reguera, R.M. (2013). Gimatecan and other camptothecin derivatives poison *Leishmania* DNA-topoisomerase IB leading to a strong leishmanicidal effect. Biochem. Pharmacol. *85,* 1433–1440. https://doi.org/10.1016/j.bcp.2013.02.024.

Prada, C.F., Álvarez-Velilla, R., Diaz-González, R., Prieto, C., Pérez-Pertejo, Y., Balaña-Fouce, R., and Reguera, R.M. (2012). A pentapeptide signature motif plays a pivotal role in *Leishmania* DNA topoisomerase IB activity and camptothecin sensitivity. Biochim. et Biophys. Acta, *1820,* 2062–2071. https://doi.org/10.1016/j.bbagen.2012.09.005.

Prada, C.F., Álvarez-Velilla, R., Díaz-Gozález, R., Pérez-Pertejo, Y., Balaña-Fouce, R., and Reguera, R.M. (2013). Identification and characterization of the regions involved in the nuclear translocation of the heterodimeric leishmanial DNA topoisomerase IB. PloS One, 8(9), e73565. https://doi.org/10.1371/journal.pone.0073565.

Ray, S., Hazra, B., Mittra, B., Das, A., and Majumder, H.K. (1998). Diospyrin, a bisnaphthoquinone: a novel inhibitor of type I DNA topoisomerase of *Leishmania donovani.* Mol. Pharmacol. *54,* 994–999. https://doi.org/10.1124/mol.54.6.994.

Ray, S., Sadhukhan, P.K., Mandal, N.B., Mahato, S.B., and Majumder, H.K. (1997). Dual inhibition of DNA topoisomerases of *Leishmania donovani* by novel indolyl quinolines. Biochem. Biophys. Res. Commun. *230,* 171–175. https://doi.org/10.1006/bbrc.1996.5874.

Riou, G.F., Gabillot, M., Douc-Rasy, S., Kayser, A., and Barrois, M. (1983). A type I DNA topoisomerase from *Trypanosoma cruzi.* Eur. J. Biochem. *134,* 479–484. https://doi.org/10.1111/j.1432-1033.1983.tb07592.x.

Romero, I.C., Saravia, N.G., and Walker, J. (2005). Selective action of fluoroquinolones against intracellular amastigotes of *Leishmania* (Viannia) *panamensis in vitro.* J. Parasitol. *91,* 1474–1479. https://doi.org/10.1645/GE-3489.1.

Roy, A., Das, B., Ganguly, A., Bose Dasgupta, S., Khalkho, N.V., Pal, C., Dey, S., Giri, V., Jaisankar, P., Dey, S., and Majumder, H.K. (2008). An insight into the mechanism of inhibition of unusual bi-subunit topoisomerase I from *Leishmania donovani* by 3, 3′-di-indolylmethane, a novel DNA topoisomerase I poison with a strong binding affinity to the enzyme. Biochem. J. *409,* 611–622. https://doi.org/10.1042/BJ20071286.

Saha, S., Acharya, C., Pal, U., Chowdhury, S.R., Sarkar, K., Maiti, N.C., Jaisankar, P., and Majumder, H.K. (2016). A novel spirooxindole derivative inhibits the Growth of *Leishmania donovani* parasites both *in vitro* and *in vivo* by targeting type IB Topoisomerase. Antimicrob. Agents Chemother. *60,* 6281–6293. https://doi.org/10.1128/AAC.00352-16.

Saha, S., Mukherjee, T., Chowdhury, S., Mishra, A., Chowdhury, S.R., Jaisankar, P., Mukhopadhyay, S., and Majumder, H.K. (2013). The lignan glycosides lyoniside and saracoside poison the unusual type IB topoisomerase of *Leishmania donovani* and kill the parasite both *in vitro* and *in vivo.* Biochem. Pharmacol. *86,* 1673–1687. https://doi.org/10.1016/j.bcp.2013.10.004.

Scocca, J.R., and Shapiro, T.A. (2008). A mitochondrial topoisomerase IA essential for late theta structure resolution in African trypanosomes. Mol. Microbiol. *67,* 820–829. https://doi.org/10.1111/j.1365-2958.2007.06087.x.

Sengupta, S., Ganguly, A., Roy, A., Bosedasgupta, S., D'Annessa, I., Desideri, A., and Majumder, H.K. (2011). ATP independent type IB topoisomerase of *Leishmania donovani* is stimulated by ATP: an insight into the functional mechanism. Nucleic Acids Res. *39,* 3295–3309. https://doi.org/10.1093/nar/gkq1284.

Sengupta, T., Mukherjee, M., Das, A., Mandal, C., Das, R., Mukherjee, T., and Majumder, H.K. (2005a). Characterization of the ATPase activity of topoisomerase II from *Leishmania donovani* and identification of residues conferring resistance to etoposide. Biochem. J. *390,* 419–426. https://doi.org/10.1042/BJ20042128.

Sengupta, T., Mukherjee, M., Mandal, C., Das, A., and Majumder, H.K. (2003). Functional dissection of the C-terminal domain of type II DNA topoisomerase from the kinetoplastid hemoflagellate *Leishmania donovani.* Nucleic Acids Res. *31,* 5305–5316. https://doi.org/10.1093/nar/gkg727.

Sengupta, T., Mukherjee, M., Das, R., Das, A., and Majumder, H.K. (2005b). Characterization of the DNA-binding domain and identification of the active site residue in the 'Gyr A' half of *Leishmania donovani* topoisomerase II. Nucleic Acids Res. *33,* 2364–2373. https://doi.org/10.1093/nar/gki527.

Shapiro, T.A., and Englund, P.T. (1995). The structure and replication of kinetoplast DNA. Annu. Rev. Microbiol. *49,* 117–143. https://doi.org/10.1146/annurev.mi.49.100195.001001.

Simonson, A.B., Servin, J.A., Skophammer, R.G., Herbold, C.W., Rivera, M.C., and Lake, J.A. (2005). Decoding the genomic tree of life. Proc. Natl. Acad. Sci. U.S.A. *102* (Suppl. 1), 6608–6613.

Singh, G., and Dey, C.S. (2007). Induction of apoptosis-like cell death by pentamidine and doxorubicin through differential inhibition of topoisomerase II in arsenite-resistant *L. donovani.* Acta Trop.*103,* 172–185. https://doi.org/10.1016/j.actatropica.2007.06.004.

Singh, G., Jayanarayan, K.G., and Dey, C.S. (2005). Novobiocin induces apoptosis-like cell death in topoisomerase II over-expressing arsenite resistant *Leishmania donovani.* Mol. Biochem. Parasitol. *141,* 57–69. https://doi.org/10.1016/j.molbiopara.2005.01.014.

Singh, M.K., Bhaumik, S.K., Karmakar, S., Paul, J., Sawoo, S., Majumder, H.K., and Roy, A. (2017). Copper salisylaldoxime (CuSAL) imparts protective efficacy against visceral leishmaniasis by targeting *Leishmania donovani* topoisomerase IB. Exp. Parasitol. *175,* 8–20. https://doi.org/10.1016/j.exppara.2017.02.010.

Slunt, K.M., Grace, J.M., Macdonald, T.L., and Pearson, R.D. (1996). Effect of mitonafide analogs on topoisomerase II of *Leishmania chagasi*. Antimicrob. Agents Chemother. *40*, 706–709.

Stewart, L., Ireton, G.C., and Champoux, J.J. (1997). Reconstitution of human topoisomerase I by fragment complementation. J. Mol. Biol. *269*, 355–372. https://doi.org/10.1006/jmbi.1997.1056.

Stewart, L., Redinbo, M.R., Qiu, X., Hol, W.G., and Champoux, J.J. (1998). A model for the mechanism of human topoisomerase I. Science *279*, 1534–1541. https://doi.org/10.1126/science.279.5356.1534.

Stoll, G., Pietiläinen, O.P., Linder, B., Suvisaari, J., Brosi, C., Hennah, W., Leppä, V., Torniainen, M., Ripatti, S., Ala-Mello, S. *et al.* (2013). Deletion of TOP3β, a component of FMRP-containing mRNPs, contributes to neurodevelopmental disorders. Nature Neurosci. *16*(9), 1228–1237. https://doi.org/10.1038/nn.3484.

Strauss, P.R., and Wang, J.C. (1990). The TOP2 gene of *Trypanosoma brucei*: a single copy gene that shares extensive homology with other TOP2 genes encoding eukaryotic DNA topoisomerase II. Mol. Biochem. Parasitol. *38*, 141–150.

Tejería, A., Pérez-Pertejo, Y., Reguera, R.M., Balaña-Fouce, R., Alonso, C., Fuertes, M., González, M., Rubiales, G., and Palacios, F. (2016). Antileishmanial effect of new indeno-1, 5-naphthyridines, selective inhibitors of *Leishmania infantum* type IB DNA topoisomerase. Eur. J. Med. Chem. *124*, 740–749. https://doi.org/10.1016/j.ejmech.2016.09.017.

Teles, C.B.G., Moreira-Dill, L.S., de Almeida Silva, A., Facundo, V.A., de Azevedo, W.F., da Silva, L.H.P., Motta, M.C.M., Stábeli, R.G., and Silva-Jardim, I. (2015). A lupane-triterpene isolated from *Combretum leprosum* Mart. fruit extracts that interferes with the intracellular development of *Leishmania (L.) amazonensis in vitro*. BMC Complement. Alt. Med. *15*, 165. https://doi.org/10.1186/s12906-015-0681-9.

Tsai-Pflugfelder, M., Liu, L.F., Liu, A.A., Tewey, K.M., Whang-Peng, J., Knutsen, T., Huebner, K., Croce, C.M., and Wang, J.C. (1988). Cloning and sequencing of cDNA encoding human DNA topoisomerase II and localization of the gene to chromosome region 17q21–22. Proc. Natl. Acad. Sci. U.S.A. *85*, 7177–7181.

Velez-Cruz, R., and Osheroff, N.D. (2004). DNA topoisomerases: type II. Encyclopedia of Biological Chemistry, 1, 806À811.

Vennerstrom, J.L., Lovelace, J.K., Waits, V.B., Hanson, W.L., and Klayman, D.L. (1990). Berberine derivatives as antileishmanial drugs. Antimicrob. Agents Chemother. *34*, 918–921.

Villa, H., Otero Marcos, A.R., Reguera, R.M., Balaña-Fouce, R., García-Estrada, C., Pérez-Pertejo, Y., Tekwani, B.L., Myler, P.J., Stuart, K.D., Bjornsti, M.A., and Ordóñez, D. (2003). A novel active DNA topoisomerase I in *Leishmania donovani*. J Biol. Chem. *278*, 3521–6. https://doi.org/10.1074/jbc.M203991200.

Walker, J., and Saravia, N.G. (2004). Inhibition of *Leishmania donovani* promastigote DNA topoisomerase I and human monocyte DNA topoisomerases I and II by antimonial drugs and classical antitopoisomerase agents. J. Parasitol. *90*, 1155–1162. https://doi.org/10.1645/GE-3347.

Wang, H., Di Gate, R.J., and Seeman, N.C. (1996). An RNA topoisomerase. Proc. Natl. Acad. Sci. U.S.A. *93*, 9477–9482.

Wang, J.C. (1994). DNA topoisomerases as targets of therapeutics: an overview. Adv. Pharmacol. 29A, 1–19.

Wang, J.C. (2002). Cellular roles of DNA topoisomerases: a molecular perspective. Nat. Rev. Mol. Cell Biol. *3*, 430–440. https://doi.org/10.1038/nrm831.

Wang, Z., and Englund, P.T. (2001). RNA interference of a trypanosome topoisomerase II causes progressive loss of mitochondrial DNA. EMBO J. *20*, 4674–4683. https://doi.org/10.1093/emboj/20.17.4674.

Werbovetz, K.A., Lehnert, E.K., Macdonald, T.L., and Pearson, R.D. (1992). Cytotoxicity of acridine compounds for *Leishmania* promastigotes *in vitro*. Antimicrob. Agents Chemother. *36*, 495–497.

Werbovetz, K.A., Spoors, P.G., Pearson, R.D., and Macdonald, T.L. (1994). Cleavable complex formation in *Leishmania chagasi* treated with anilinoacridines. Mol. Biochem. Parasitol. 65(1), 1–10.

Xu, D., Shen, W., Guo, R., Xue, Y., Peng, W., Sima, J., Yang, J., Sharov, A., Srikantan, S., Yang, J., *et al.* (2013). Top3β is an RNA topoisomerase that works with fragile X syndrome protein to promote synapse formation. Nat. Neurosci. *16*, 1238–1247. https://doi.org/10.1038/nn.3479.

Yang, G., Choi, G., and No, J.H. (2016). Antileishmanial Mechanism of Diamidines Involves Targeting Kinetoplasts. Antimicrob. Agents Chemother. 60(11), 6828–6836. https://doi.org/10.1128/AAC.01129-16.

Yang, X., Li, W., Prescott, E.D., Burden, S.J., and Wang, J.C. (2000). DNA topoisomerase IIbeta and neural development. Science *287*, 131–134. https://doi.org/10.1126/science.287.5450.131.

Zuma, A.A., Cavalcanti, D.P., Maia, M.C., De Souza, W., and Motta, M.C. (2011). Effect of topoisomerase inhibitors and DNA-binding drugs on the cell proliferation and ultrastructure of *Trypanosoma cruzi*. Int. J. Antimicrob. Agents, 37, 449–56, https://doi.org/10.1016/j.ijantimicag.2010.11.031.

Host–Kinetoplastid Parasite Interaction at the Immune System Interface: Immune Evasion and Immunotherapy

11

Arathi Nair*, Sunil Kumar, Bhaskar Saha and Divanshu Shukla*

National Centre for Cell Science, Ganeshkhind, Pune, India.

*Correspondence: nayerarathi@gmail.com and divanshukla@gmail.com

https://doi.org/10.21775/9781910190715.11

Abstract

Parasites co-evolved with their respective hosts over millions of years. During the process of co-evolution, host and parasite bequeath each other with a striking legacy of immune surveillance and immune evasion, respectively. The host–parasite interaction is thus dynamic and complete with cascades of temporally separated responses that characterize different phases of infection; the phases remain connected to each other as if one response leads to the next. In the first phase, the parasite's cell surface molecules interact with the pattern recognition receptors (PRR) on the host cell surface, followed by internalization and recognition of the parasite's intracellular molecules by the PRRs within the host cells. The results of these interactions include, but are not limited to, parasite survival or degradation, moderation of the antigen-presenting functions of the host cells and elicitation of the T-cell responses that further control parasite survival in a host. The parasite in each phase modulates the host immune responses to engineer its own survival. Therefore, in this chapter, molecular interactions between the hosts and the parasites – *Leishmania* and *Trypanosoma* – will be described emphasizing the anti-parasite immune mechanisms and parasite-devised immune evasion. How this knowledge helps strategize anti-parasite therapy is also discussed.

Introduction

Kinetoplastids or Kinetoplastea form a group of protozoan parasites with a highly flexible unicellular body, ranging in size from 10–100 microns and uniflagellate (trypanosomatida) or biflagellate (bodonina), and derive its name due to the presence of 'kinetoplast', a condensed structure of mitochondrial DNA. Kinetoplast contains interlocked circular DNA as maxi- or mini-circles called as the kinetoplast DNA (kDNA) and some exclusive structural proteins and polymerases. These are missing from the host. The nuclear genome is streamlined with high gene density – with little or no RNA editing – and the transcription of nuclear genes take place as long multicistronic RNA molecules (Myler *et al.*, 1991; Liniger *et al.*, 2001). Early in evolution, these parasites underwent divergence from other eukaryotes and, under strong selection pressure, acquired beneficial metabolic traits as evident from the presence of certain specialized cellular peroxisomes called 'glycosomes', a rich reservoir of many anabolic and catabolic enzymes that enable the parasites to adapt to different environmental conditions; a prerequisite for parasitism and successful survival in the host.

When seen under an electron microscope the complex structural organization of a kinetoplastid shows an outer pellicular membrane underneath which are subpellicular microtubules. Typical eukaryotic cellular structures such as a nucleus with distinct nucleolus, Golgi apparatus and endoplasmic reticulum are seen but mitochondrial structures are stage-dependent. A flagellum with a distinct paraflagellar rod is characteristic of the motile form (Fig. 11.1).

Parasites are critical 'moderators' of the evolution, diversity and ecological status of free-living organisms which play hosts to these parasites (Dunne *et al.*, 2013). The host–parasite interaction undergoes an antagonistic or reciprocal evolution: while the parasites evolve virulence mechanisms, the host reciprocates by evolving anti-parasite immune mechanisms (Dawkins *et al.*, 1979). The evolution of parasitism is accompanied by obsolescence of non-parasitic genes and acquisition of the genes that related to metabolic adaptability and infectivity. Transporters across vacuolar membranes diminished in number with a concomitant increase in the number of membrane transporters for scavenging nutrients and metabolites from the host cells. As parasites evolve expeditiously owing to shorter generation time and higher genetic variation, the diversity in parasites also increases to counter the evolution of the host's immune system.

The evolution of the pathogen and its fitness in its niche, i.e. the host, is dependent on multitude of factors that confer complexity to interaction of the pathogen with its host. As a result of its interaction with the host's immune system, a parasite thus develops an 'immunophenotype' via the process of positive selection of favourable traits. This host–parasite interaction in terms of strength and duration could be short-lived so as to generate an innate cytokine response or long enough to generate memory against the same or closely related parasite. Upon interaction with the host, the immune-phenotype of a pathogen acquires a spatio-temporal attribute that depends on time, strength and number of interacting immune components (Cobey *et al.*, 2014). The parasite influences the evolution of host by putting selection pressure against deleterious mutations in the host, selecting resistant over susceptible host (Kraaijeveld *et al.*, 1997; Webster *et al.*, 1999; Bohannan *et al.*, 2000), thus leading to the evolution of a complex immune system. Not only does the parasite co-evolve with the host, they avoid antigenic niche overlaps and co-occurrence with the parasites sharing common alleles (Chesson *et al.*, 2000), essentially minimizing the intra-species competition.

Complex interactions occur between host and its parasite, as miniaturized in host–virus interaction wherein the virus utilizes the host's cellular machinery for its own replication. In a kinetoplastid–virus or a bacteria–bacteriophage interaction, the genome-integrated pathogens may modify the host's genetic make-up while separating from the host. The evolution of viral sensing receptor and effectors in the host cell depends on the immune suppression by virus. The integration of virus into the host's genome followed by a lytic cycle led to the transfer of genetic material between hosts, which resulted in evolution of the host's immune system (Fig. 11.2).

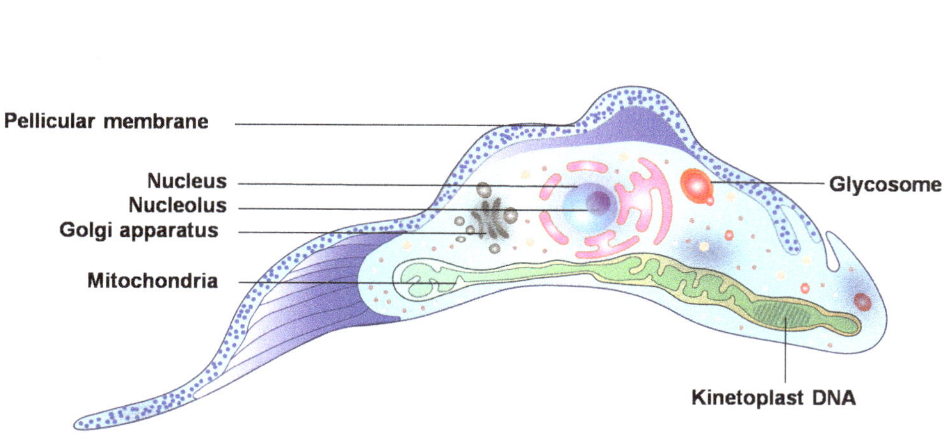

Figure 11.1 General organization of the sub-cellular organelle in a kinetoplastid is shown. The presence of the 'kinetoplast' is the characteristic feature of the kinetoplastid parasites.

Molecular host–parasite interaction and strategies of immune evasion by the parasite

Kinetoplastids are causatives of various diseases in humans, animals and plants. The term 'TriTryps' encompasses the three prevalent kinetoplastid parasitic diseases, leishmaniasis caused by *Leishmania* spp., African sleeping sickness caused by *Trypanosoma brucei* and Chagas' disease by *Trypanosoma cruzi* (Fig. 11.3). This spectrum of diseases mainly affects poor and marginalized tropical and subtropical population. The diseases are vector borne and the parasite is propagated by the different genus of sandfly in leishmaniasis (*Phlebotomus* in Africa, Europe and Asia and *Lutzomya* in the Eastern hemisphere) and Tsetse fly (*Glossina palpalis*) in trypanosomiasis. Nearly 60 million people are affected by sleeping sickness in sub-Saharan regions and it causes greater morbidity than HIV/AIDS, while *T. cruzi* infection is most prevalent

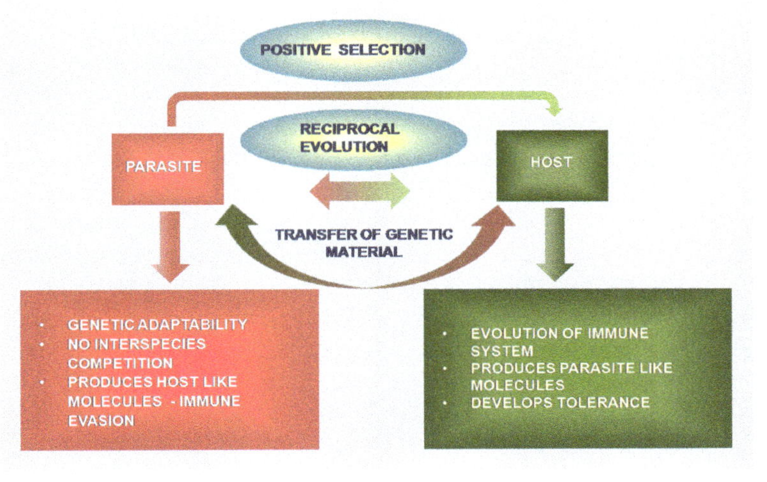

Figure 11.2 Co-evolution of the host and the parasite: the parasite puts positive selection pressure on the host leading to the evolution of Host's immune system, while the host enables evolution of the genetic adaptability in the parasite.

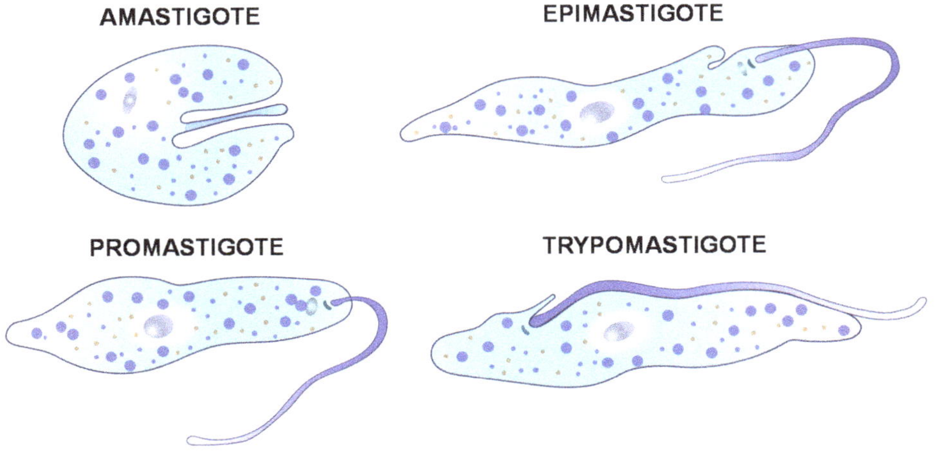

Figure 11.3 Morphological characterization of Tri-tryps based on the absence (amastigote), presence (promastigote) and position of the nucleus with respect to the basal body: epimastigote (anterior nucleus), trypomastigote (posterior nucleus).

in Latin America. 350 million people are affected worldwide by leishmaniases, and together these parasitic diseases not only affect human health but also agriculture and livestock (Dávila *et al.*, 2002). Kinetoplastid caused diseases present an ideal infection model from which much information can be generated on host–parasite interplays and parasitic immune evasion strategies. As the molecular host–parasite interaction is a multidimensional and multifactorial process, deciphering the exact mechanism underlying the pathogenicity of parasites remains a challenge.

Morphological characterization of kinetoplastids

The kinetoplastid parasite has complex multihost life cycle wherein it morphologically transforms into promastigotes, amastigotes, trypomastigotes, epimastigotes, etc. characterized by the presence, absence of the flagella and the position of the kinetoplast with respect to the nucleus. The genome analysis of Tri-Tryps reveals highly conserved cytosolic proteins and highly evolved species-specific cell-surface proteins such as promastigote surface antigen (PSA) in *Leishmania* species, variant surface glycoproteins (VSGs) in *T. brucei* and trans-sialidase (TS) family of proteins in *T. cruzi* (Jackson, 2015). The *Trypanosoma* is further divided into Stercoraria and Salivaria (Hoare, 1966) the former parasite develops in the intestinal track while the latter in the stomach of the invertebrate vector, hence the mode of infection is through the faeces and saliva of the *Glossina* spp., respectively. Kinetoplastea infection begins with initials recognition and interaction with the host cells, followed by penetration of host cell or phagocytosis that leads to an immune response culminating in either the survival or elimination of the parasite. As the parasite infects its mammalian host, the phagocytic cells of the immune system internalize it and results in the self-confining cutaneous lesions to deforming and potentially fatal mucocutaneous and visceral infection in leishmaniasis or neuro- and cardiomyopathies in trypanosomiasis. A better understanding of the molecular interactions of the host–parasite and the immune evasion strategies devised by the parasite would enable generation of safer and more effective immunotherapeutics.

Leishmania

Initial interaction with the host cell

Leishmania parasites secure their survival and propagation within their hosts by an array of manoeuvres that compromises ability of the host macrophages to kill the pathogens or to engage the adaptive immune system. On entering a mammalian host during the blood meal of a female sandfly, the promastigotes devise various strategies to evade or subvert host immune mechanisms. This involves preferential recruitment of cells of the immune system, majorly neutrophils and macrophages to the site of inoculation in order to infect them and establish disease. Neutrophils actively engulf the promastigotes and secrete anti-leishmanial factors such as neutrophil elastase (Ribeiro-Gomes *et al.*, 2007). The short-lived neutrophils serve as an intermediate host that enables the parasite to stealthy gain entry into the host's immune system. *L. major* enhances the expression of chemokines and MIP-1β secretion by polymorphonuclear neutrophils chemoattract macrophages to infected tissues and help establish the infection. Natural killer cells too are recruited at the site of infection that lyse the parasites and generate IFN-γ (Scharton *et al.*, 1993) and iNOS (Prajeeth *et al.*, 2011), which helps to curb the early infection. During early stages of infection NKT-cells also contribute to parasite clearance but its role is strain-specific and organ-specific, whereby it controls infection in skin lesions and the spleen but not in the lymph nodes (Mattner *et al.*, 2006). Once in vicinity of the macrophages the apoptosised PMNs are phagocytosed (van Zandbergen *et al.*, 2004). In chronic infection, chemokine types, amount, and the duration of chemotactic effect have been implicated in parasite clearance or persistence.

The complement receptors (CR1, CR2), mannose-fucose receptor and the fibronectin receptor on the surface of the macrophages play a crucial role in interaction with the promastigote parasite (Kane *et al.*, 2000), while parasite surface molecules LPG and GP63 play pivotal role in the course of infection endowing the parasite with various survival capabilities. The promastigote expressed LPG – an ensemble of repetitive units of disaccharide linked to a phosphate and anchored to membrane by Glycosylphosphatidylinositol (GPI) moiety – play a deciding role in immune evasion during early interaction and phagocytosis by macrophages. The structure of LPG moiety differs between different species and between various stages of the same parasite (McConville *et al.*, 1992, 1995). Owing to a significantly longer LPG, *L. major* metacyclic promastigotes avoid the complement-mediated lysis by preventing attachment of C5b–C9 subunits of complement complex. The *L. donovani* LPG prevents C5 convertase formation by fixing the inactive C3bi subunit on its surface (Puentes *et al.*, 1989). High number of LPG masks GP63 – a zinc-dependent metalloprotease on promastigote surface – thus the immune system fails to recognize this antigen. As the promastigotes transform to amastigote, they not only lose flagella but also substantially reduce the expression of LPG. As a result, GP63 is unmasked (Schneider *et al.*, 1992) and converts C3b to C3bi enabling the opsonization and internalization of the parasite (Brittingham *et al.*, 1995). The initial opsonization of the parasite is by C3b and C3bi that allows its interaction with the complement receptor CR1 (transient) and CR3 of the macrophages. The parasite skews this interaction predominantly towards CR3 and subverts the oxidative burst during phagocytosis (Mosser *et al.*, 1987). The interaction also occurs via the LPG of the parasite and the mannose-fucose receptor on the macrophages (Blackwell *et al.*, 1985). The parasite's protein kinases phosphorylate the components of the complement system to inactivate them (Hermoso *et al.*, 1991). Another receptor that plays a crucial role directly in the attachment of the parasite to its host cell and indirectly in increasing the number of mononuclear phagocyte complement receptors is the fibronectin receptor. *Leishmania* amastigotes rely particularly on proteophosphoglycan (PPG) for invasion of macrophages (Desjardins *et al.*, 1997; Descoteaux *et al.*, 2002).

Thus, a multitude of factors decides the outcome of infection and the host's immunological status, i.e. susceptibility, resistance or tolerance. The resistant host may reduce the parasite load completely and clear the infection, whereas a tolerant host may just limit the fitness effect of the parasite. Parasites manoeuvre the evolution of the host towards a most resistant profile; genetic or phenotypic heterogeneity of the parasite and the host further contributes to a shift in the dynamics of this co-evolution. In the kinetics of infection or the growth of the parasite, a repertoire of interacting factors with their respective spatio-temporal attributes may act in a synergistic or antagonistic manner, and the cumulative effect is seen as the susceptibility or resistance to an infection. An intracellular parasite like *Leishmania* replicating within a heterogeneous host may evolve to specialize, so as to ensure its survival in one cell type while losing its ability to infect or propagate in another cell type. The first set of such interaction occurs between the parasite's ligand and its complementary receptors on the surface of the host cells, which facilitates the entry of the parasite into its niche, i.e. the host. This is followed by attempts from the host cell to regulate the growth of the parasite and clear the invading parasites, and counteracting tactics from the parasite to subvert such attempts.

Internalization and inhibition of host cell functions

Microbicidal activity and oxidative burst

Heat-shock proteins provide thermotolerance to the parasite within the host. Acidic environment along with elevated temperature induces promastigote to amastigotes transformation (Zilberstein *et al.*, 1994). Heat-shock proteins such as HSP90, HSP70, STI1/HOP (stress inducible protein 1/HSP organizer protein) are involved in the regulation of promastigote to amastigote transformation in response to elevated temperature (33–37°C) in the mammalian host (Morales *et al.*, 2010). During promastigote to amastigotes transformation the expression of certain mitochondrial (HSP-100) and endoplasmic reticulum (A2 stress protein) stress-related proteins is up-regulated that is essential for the survival of the parasite in mammalian host (Hubel *et al.*, 1997).

Generation of reactive oxygen species (ROS) and reactive nitrogen species (RNS) and subsequent oxidative burst is an important process targeted at clearance of the parasite. NADPH oxidase or phagocyte oxidase (phox) and inducible nitric oxide synthase iNOS are other enzymes contributing to oxidative burst and parasite clearance. Studies in *L. donovani*-infected macrophages show that during early infection the parasite burden increased in both phox$^{-/-}$ and in iNOS $^{-/-}$ mice whereas during late stages on infection the parasite burden reduced in $phox^{-/-}$ while increased in $iNOS^{-/-}$ mice, underlining the fact that both NADPH oxidase and iNOS are necessary to control the early infection, while iNOS is sufficient to control infection in later stages (Murray *et al.*, 2006). The parasite counteracts this by detoxifying the reactive radicals. The highly conserved enzymes produced by the parasite called Peroxidoxins can detoxify ROS (McGonigle *et al.*, 1998). The amastigote-derived peroxidoxins namely LcPxn1, LcPxn2, superoxide dismutase and LPG deplete nitrite and ROI levels (Chan *et al.*, 1989; Ghosh *et al.*, 2003). The reactive superoxide radicals produced by the activated macrophages are neutralized by promastigote's Superoxide dismutase enzyme, a metalloenzyme that dissociates superoxide to oxygen and hydrogen peroxide that is then eliminated by peroxidases and catalases. SOD is present in resident as well as activated macrophages (Mukherjee *et al.*, 1988). The inhibition of ROIs in *L. donovani*-infected macrophages is dependent on LPG and GP63. Previous studies showed that upon incubation with leishmanial LPG or glycoinositolphospholipids (GIPL), the infected macrophages lose their ability to induce iNOS and thereby to produce NO in response to IFN-γ or LPS. This inability to induce oxidative burst or nitrite production could in effect be due to production of IL-10, TGF-β or inactivation of JAK-STAT pathway (Proudfoot *et al.*, 1995, 1996). The presence of a thiol enzyme in *Leishmania* called the trypanothione enables the parasite to metabolize peroxide, detoxify xenobiotics (Dumas *et al.*, 1997) and hence subvert the oxidative stress.

The competition between the enzymes iNOS and arginase for a common substrate arginine plays a deciding role in disease outcome. The enzyme arginase has been attributed with inflammatory and immunosuppressive functions. T_H2 cytokines (IL-1, IL-10, IL-13) and LPS induce the generation of the enzyme arginase that converts L-arginine to L-ornithine and urea, whereas T_H1 cytokines such as IFN-γ induce iNOS which is utilized by activated macrophages to convert L-arginine to citrulline and NO and other reactive nitrogen intermediates that enable parasite killing (Macmicking *et al.*, 1997). The parasite sabotages the host's arginase pathway by increasing Arginase-I leading to polyamine production (Iniesta *et al.*, 2005) and thus auxotrophic parasite growth.

As internalized promastigotes are susceptible to acidic and hydrolytic lysis, the internalized parasites prevent phagosome–endosome fusion, which is an LPG-dependent process. This provides the parasite with enough time to transform to a resistant amastigote form (Desjardins *et al.*, 1997). Once inside the host cell *Leishmania* resides within membraneous parasitophorous vacuole that has lysosomal membrane-like proteins and MHC-II molecules. *Leishmania* modifies the characteristics of this vacuole from a lysosomal to endosomal compartment (Russell *et al.*, 1992). Owing to the presence of vacuolar H+-ATPase (Antoine *et al.*, 1990) the vacuole is acidic in pH (4.7–5.2), which is the optimal respiratory and metabolic condition for amastigotes (Mukkada *et al.*, 1985). LPG further destabilizes the lipid microdomains of the phagosomal membrane preventing the formation of microbicidal compartments (Dermine *et al.*, 2005). Many metabolic changes also take place as promastigotes differentiate into amastigotes so as to ensure survival within the parasitophorous vacuole. Studies with *L. mexicana* amastigotes revealed that glucose uptake is essential for parasite's virulence (Burchmore *et al.*, 2003). Amino acids obtained through degradation of proteins within the phagolysosomes are the main carbon source, which are taken up by the permease present in amastigote's plasma membrane. An alternative way to obtain amino acids is through uptake and degradation of host's protein in enlarged vacuoles called 'Megasomes' (Fig. 11.4) (Besteiro *et al.*, 2008).

The chemical and enzymatic processed involved in oxidative burst not only limits the growth and number of the parasite but also lead to the influx of extracellular calcium (Ca^{2+}) into the cytosol, and subsequently into the nuclei and the mitochondria. Not only does this calcium signalling crosstalks with ROS signalling and control apoptosis but also regulates the activation of various proteins involved in signal transduction pathway (Ermak *et al.*, 2002). This interaction between Ca^{2+} and ROS is bidirectional, as ROS and Ca^{2+} can regulate mutual signalling: Ca^{2+} up-regulates the free radical generating enzymes (Gordeeva *et al.*, 2003) and ROS regulates the functioning of several calcium transporters such as the redox-sensitive voltage gated calcium channels of the plasma membrane (Mikami *et al.*, 1989).

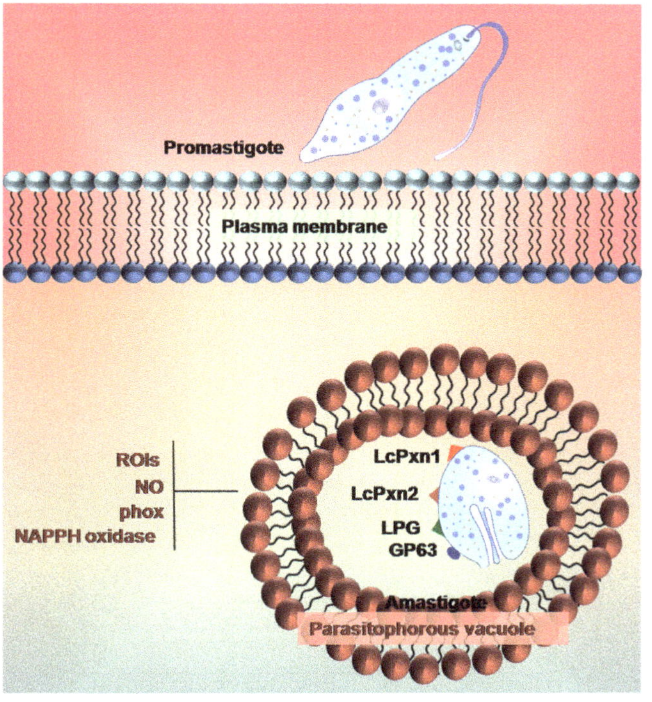

Figure 11.4 Oxidative burst-antimicrobicidal process against Leishmania: reactive oxygen and nitrogen species, NADPH oxidase and phox generate an oxidative stress. Oxidative burst is detoxified by leishmanial GP63, LPG, peridoxins etc.

Host cell signalling pathways

Binding of the ligand to its corresponding receptors on the cell surface, is the first step in the process of signal transduction. The chemical message encoded within the ligand, as it binds to its receptor is transduced into the cell. Receptor-ligand-binding changes the polarization of the plasma membrane and thus generates an electrical potential. Hence the transduction of one form of energy to another (chemical to electrical energy) takes place. Further the enzyme linked, activation–deactivation dependent signalling components relays this message across the cytosol. The activation of transcription factors takes this message into the nucleus, which in turn regulates the gene expression; thus decoding the message.

Pathogen recognition receptors

TLRs play crucial role in recognizing pathogen associated molecular patterns and generating an innate immune response against it. *Leishmania* parasite is known to interfere with the TLR–PAMP recognition. The GPI anchor of *Leishmania* namely LPG binds to the TLR2 of macrophages and NK cells (de Veer *et al.*, 2003; Becker *et al.*, 2003) causes the deactivation of macrophages due to the inhibition of oxidative microbicidal action and the effect of interleukines. TLRs can signal via MyD88-dependent or independent pathway leading to the activation of MAPKs. A study shows that *L. donovani* is recognized through both TLR2 and TLR3, where TLR3 promotes parasite clearance through generation of NO and by pro-inflammatory cytokines and silencing of TLR3 lead to an increased parasite survival and impaired NO and TNF-α production in IFN-γ-primed macrophages (Flandin *et al.*, 2006). The parasite also targets the TLR2 mediated activation of p38MAPK, and hence the production of pro-inflammatory cytokine IL-12 (Chandra and Naik 2008). Moreover, mice deficient in an adaptor protein in TLR signalling: MyD88 are found to be more susceptible to *L. major* infection due to impaired production of T_H1 cytokines (Muraille *et al.*, 2003). As mentioned earlier the competition between enzymes iNOS and arginase for a common substrate arginine plays deciding role in disease outcome. TLR4 is a strong regulator of the enzyme iNOS that aids in parasite clearance. Parasite survival was found to be higher in TLR4 knockouts due to a higher activity of the enzyme arginase and due to the abrogation of IL-12 receptor activity (Fig. 11.5) (Kropf *et al.*, 2004a,b).

The influx of calcium into the cell activates cysteine proteases called Calpain (Calcium-activated papain-like proteases), which plays a major role in cytoskeletal remodelling, signal transduction and regulating programmed cell death and virulence of the parasite (Arnoult *et al.*, 2002; Mottram *et al.*, 2004). The genome of *L. major* has 27 calpain-like sequences (Ersfeld *et al.*, 2005), and the enzyme modulates the susceptibility of strains to antimonial compounds and miltefosine (Vergnes *et al.*, 2007).

Leishmania targets various signalling components of the host cell to ensure its survival and sustenance. The host cell's JAK pathway gets activated when the receptor dimerization occurs following stimulation with a cytokine or a growth factor. The activation of JAK via transphosphorylation leads to the phosphoactivation of Signal Transducer and Activator of Transcription (STAT), which in turn translocates to nucleus and regulates a number of genes responsible for immune activation (Rawlings *et al.*, 2004; Woldman *et al.*, 2001). *Leishmania* cleaves and activates host protein tyrosine phosphatase SHP-1 containing phosphatase-1 (PTP1C), causing dephosphorylation and inactivation of a wide array of proteins. SHP-I co-localizes with Janus Kinase 2 and inactivates it. It negatively regulates several transcription factors involved in macrophage activation leading to inactivation of key molecules such as NO, IL-12, and TNF-α (Forget *et al.*, 2006). SHP-1 is activated through *L. donovani* SHP-1-binding protein called Elongation factor – 1α (EF-1 α). EF-1α is produced by *Leishmania* in the phagosome is exported to the cytosol where it acts on SHP-1. The parasites EF-1α but not the host cells EF-1α binds to SHP-1. Studies using purified *Leishmania* EF-1α blocked the IFN-γ induces iNOS production, thus *Leishmania* EF-1α contributes to disease progression. The mode of activation of SHP-1 is possibly through binding of tyrosine phosphorylated ITIM (immunoreceptor tyrosine-based inhibitory motifs) of EF-1α to the Src homology 2 domain of SHP-1. Additionally, the binding of EF-1α to SHP-1 may affect the distribution pattern as well as kinases, thus activating SHP-1 (Nandan *et al.*, 2002). *L. donovani* also has the ability to block the JAK/STAT signalling pathway in response to IFN-γ stimulation in monocytes and to subvert the NO response of the host cell (Nandan *et al.*, 1995). JAK 2 is affected by *L donovani* infection as the promastigotes rapidly activate host SHP-1 leading to inhibition of IFN-λ-induced JAK2 phosphorylation, therefore avoiding the induction of NO. It also interferes with IFN-γ

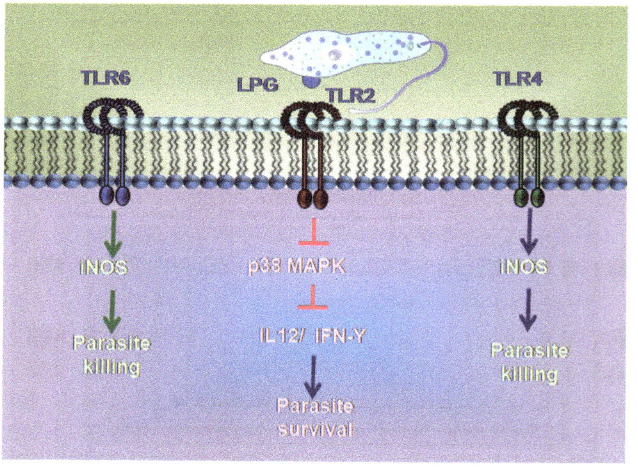

Figure 11.5 TLR mediated innate immune response against *Leishmania*. LPG-TLR 2 interaction enables parasite survival by activating ERK1/2 and generation of IL-10, while signalling through TLR4 and TLR6 leads to parasite killing.

receptor signalling in macrophages by interfering with STATs. This is in part due to the proteosomal degradation of STAT-1 (Fig. 11.6).

Protein kinase C (PKC) a family of serine/threonine kinases, comprises 11 structurally homologous isoforms namely; classical (cPKCs-α, βI, βII, γ), novel (nPKCs-δ, ε, η, θ), atypical (aPKCs-ζ, ι/λ), and PKCμ. The calcium-dependent isoforms have a conserved C2 domain that is absent in other isoforms. The cPKCs additionally require an acidic phopholipid such as phosphatidylserine (PS) and DAG enhances Ca^{2+}-dependent activation (Luo *et al.*, 1993). These enzymes have a regulatory and a highly conserved catalytic domain and phophoactivation leads to its membrane translocation and hence activation (Mukherjee *et al.*, 2016). PKCs are known to play a crucial role in the regulation of macrophage activation via production of IFN-γ and TNF-α that affect NO and ROS generation that leads to parasite clearance (Severn *et al.*, 1992; Warzocha *et al.*, 1995; Turco *et al.*, 1992). The LPG of *Leishmania* binds to the regulatory unit of PKCs to block its activation (Descoteaux *et al.*, 1999). As *Leishmania* enters the host macrophages, it selectively prevents or enhances the translocation and thus the activation of PKC isoforms. *Leishmania* prevents DAG mediated activation of cPKCs α and β while the activation of PKCδ- and ζ is enhanced hence leading to production of anti-inflammatory cytokines and thus the parasite ensures its survival (Sudan *et al.*, 2012). *L. major* infection reduces PKCα, βI, βII, and ε isoforms that activate p38MAPK and production of IL-12 but enhances PKCδ, ζ, and λ isoforms that phophoactivate ERK1/2 and produce IL-10. Lentiviral blockade of PKCδ, ζ, and λ leads to enhanced iNOS expression and restoration of IFN-γ mediated T_H1 response. As calcium acts as a cofactor for PKCα, βI, βII isoforms and in *Leishmania* infection the intracellular Ca^{2+} increases (Olivier *et al.*, 1992), one might expect that the activation of these calcium dependent isoforms should increase in disease condition, but the dependency of these isoforms on not only calcium but DAG and other cofactors for its activation might explain its decreased activation in *Leishmania* infection. Moreover, the activation of Calcineurin, a serine-theronine phosphatase that differentially regulate the isoform-specific activation might be a plausible cause. CD40 a costimulatory molecule expressed on the

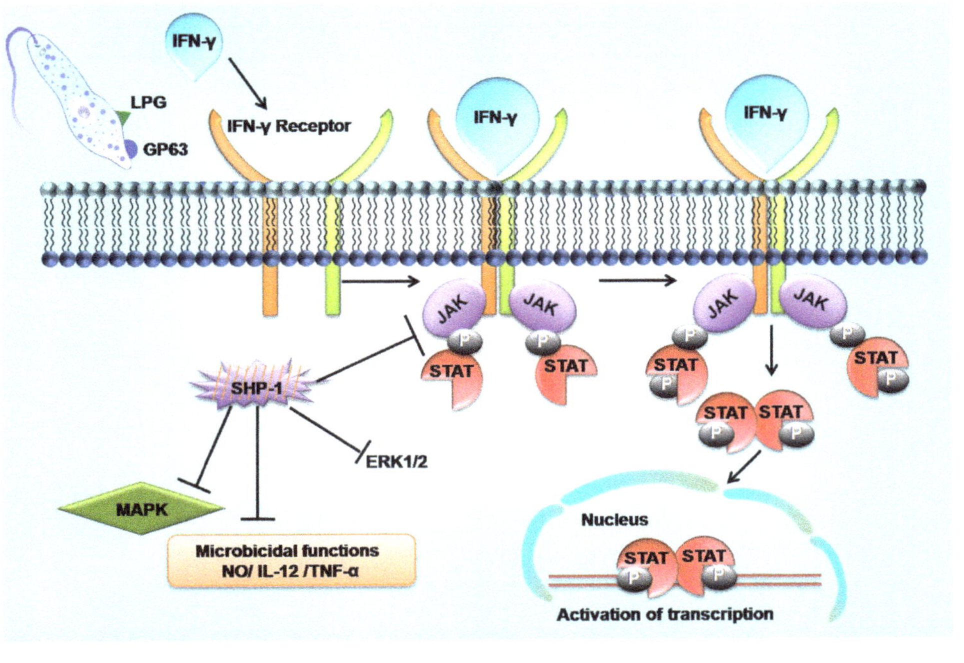

Figure 11.6 *Leishmania* cleaves and activates host protein tyrosine phosphatase SHP-1, which directly inactivates Janus Kinase 2 and ERK-1/2 and negatively regulates several transcription factors involved in macrophage activation leading to inactivation of key molecules such as NO, IL-12, and TNF-α.

surface of macrophages plays an important role in *L. major* infection. The signalling of CD40 from a cholesterol rich lipid rafts may lead to activation of PKC isoforms PKCα, βI, and βII while CD40 signalling from a cholesterol poor non-raft as in *L. major*-infected macrophage activates PKCδ and ζ/λ isoforms (Sudan *et al.*, 2012). Oxidative stress not only leads to calcium influx but also produces Ceramide a pro-apoptotic lipid metabolite that selectively recruits and activates PKC-ζ (Lozano *et al.*, 1994). Moreover, the infection-induced generation of ceramide by the macrophages plays a role in NF-κB inhibition (Ghosh *et al.*, 2002) (Fig. 11.7).

CD40–CD154 interaction is crucial for both humoral and cell mediated responses. The CD40–CD154 interaction recruits the adaptor proteins TRAFs (TNFR-associated factors) to the cytoplasmic domain (Bishop *et al.*, 2007). The development of ulcerating cutaneous lesion on *L. major* infection in resistant C57/BL6 mice deficient for CD40 or CD154 underlines the importance of this receptor–ligand interaction (Campbell *et al.*, 1996). CD154$^{-/-}$ mice subjected to *L. amazonensis* infection showed impaired NO generation. Hence, CD40-CD154 pathway not only confers protection by promoting T_H1 type immune response, it also aids in activating the macrophage for parasite killing (Soong *et al.*, 1996). As CD40 partitions between a cholesterol-rich domain and cholesterol-poor domain, different signalosome formations lead to different immune responses. *Leishmania* depletes the membrane cholesterol affecting CD40 signalosome formation and inhibits the formation of signalosome that produces IL-12, through activation of p38MAPK, involving selective adaptor proteins (TRAF2/3/5) and upstream kinases Lyn. Instead it promotes CD40 signalosome that signals through adaptor protein TRAF6 and upstream kinase Syk activating ERK1/2, producing parasite protective IL-10 (Mathur *et al.*, 2004; Rub *et al.*, 2009).

Leishmania differentially targets the Ras isoforms: small cellular GTPases that regulates cellular responsiveness. When compared to the wild-type c57BL/6 mice, H-Ras or N-Ras deficient mice had a higher *L. major* infection, indicating their role in a T_H1 response (Iborra *et al.*, 2011). Isoform-specific modulation of Ras isoforms

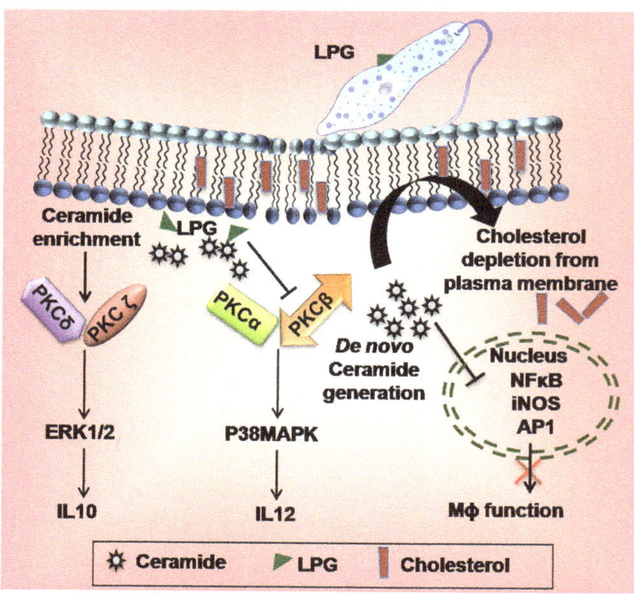

Figure 11.7 Role of PKC isoforms in *Leishmania* infection, *Leishmania* prevents activation of cPKCs α and β while the activation of PKCδ- and ζ is enhanced hence leading to production of anti-inflammtory cytokines and thus the parasite ensures its survival. Ceramide production by *L. donovani*-infected macrophages can also reduce PKC and ERK-1/2 activation. The infection-induced generation of ceramide by the macrophages plays a role in NF-κB inhibition.

by the parasite was further elucidated, wherein it was found that the expression of H-Ras and N-Ras increased during *L. major* infection while that of K-Ras decreased. Blockade of the LPG–TLR2 interaction reversed these *L. major* induced changes in the expression profile. Thus the parasite skews the initial signalling towards Ras-Raf-MEK-ERK, through TLR2–LPG interaction, which provides a cumulative effect to signalling towards ERK1/2 module via CD40. Moreover, the signalling towards p38MAPK module is inhibited by a feedback loop between ERK and upstream kinase Lyn, thus the generation of pro-inflammatory cytokines is abrogated, causing disease progression (Chakraborty *et al.*, 2015).

PI3K is a known effector of Ras GTPase. Studies have revealed the involvement of PI3Kδ/PI3Kγ in *Leishmania* disease progression. PI3K on activation phosphorylates (phosphatidylinositol 3,4-bisphosphate) converting it into PIP3 (phosphatidylinositol 3,4,5-triphosphate), which in turn activates Akt, which acts as a regulator of various cellular processes such as apoptosis, growth and survival. *L. mexicana* phosphoactivated PI3K and hence Akt thus preventing apoptosis, ensuring its survival in the host (Vázquez-López *et al.*, 2015).

Leishmania also renders host MAPKs inactive. Phagocytosis of *L. donovani* promastigotes by macrophages does not lead to the activation of any of the three MAPKs (ERK-1/2, JNK, p38) (Privé *et al.*, 2000). Furthermore, activation of several MAPKs in response to LPS was inhibited in infected cells. For instance, *L. amazonensis* blocks LPS-mediated ERK-1 phosphorylation in infected macrophages (Martiny *et al.*, 1999), and *L. donovani* amastigotes block PMA-induced ERK-1/2 phosphorylation in RAW264 monocytic cell line leading to the inhibition of Elk-1 and c-fos expression (Nandan *et al.*, 1999). Parasite cysteine protease can degrade the host MAPKs and NF-κB, as in the case of *L. mexicana*.

Hence the message decoding in the cell occurs via change in the activation status of a cell on receptor ligand interaction. The parasite devises survival strategies by modulating the innate immune receptors: the TLRs followed by the adaptive immune response of the host cell. The receptor–ligand interaction like CD40-CD154 is important to antigen presentation.

Antigen presentation in host

The *Leishmania* resident within the phagolysosomal compartment is processed into peptide fragments and of specific length (dependent on the type of MHC) and is presented to the lymphocytes. The process of leishmanial antigen presentation through MHC-II is a tightly regulated process. In infected macrophages the amastigotes reside and multiplies within the parasitophorous vacuole, the membrane of which only contains class II MHC (Lang *et al.*, 1994). The MHC-II molecules are synthesized in the E.R and stabilized by association with calnexin. A chaperone called invariant chain binds to the MHC-II α and β chains thus preventing the binding of endogenous antigens to it and guides the complex to endosomal-lysosomal compartment. The invariant chain is proteolytically degraded to form a CLIP (class-II associated invariant chain). Once the endolysosomal compartment is acidified and the processing of the antigenic peptide fragment is accomplished it is loaded on to the MHC-II. In case of intracellular antigen presentation, transporter proteins called TAP transport the antigenic peptides from the cytosol to the lumen of E.R where they are loaded on to MHC-II. The parasite's amastigotes cysteine proteases degrade structures called 'megasomes' that contain MHC-II (De Souza Leao *et al.*, 1995). Macrophages processes and presents peptides derived from *Leishmania* onto MHC-II and present it to $CD4^+$ T-cells.

The parasite not only targets and suppresses the microbicidal activity of the macrophages it also alters the antigen presentation process to immune system. The expression of both class I MHC and class II MHC goes down with *Leishmania* infection. The *Leishmania* parasite (*L. chagasi* and *L. donovani*) inhibits the IFN-γ-induced MHC class II expression in infected macrophages (De Almeida *et al.*, 2003; Reiner *et al.*, 1987). IFN-γ not only up-regulates the expression of MHC-II but also of HLA-DM and a lysosomal thioreductase called GILT (gamma interferon-inducible lysosomal thiol-reductase This enzyme is constitutively expressed in APCs and it catalysis the reduction of internalized proteins and hence facilitates in antigen presentation (Arunachalam *et al.*, 2000). *Leishmania* also interferes with the loading of antigens onto MHC class II molecules aborting the antigen presentation to T-cells. *Leishmania* also interferes with costimulatory molecules required for the activation of T-cells, for example, selective interference with ICAM-1 and CD80 expression (Saha *et al.*, 1995, 1998) or with CD40 signalling in infected macrophages in a p38MAPK-dependent manner (Awasthi *et al.*, 2003). On LPS stimulation the expression of the co-stimulatory molecule B7–1 was reduced plausibly by the action of parasite's prostaglandins (Saha *et al.*, 1995). *L. donovani* blocks LPS-induced expression of CD80 in infected macrophages (Kaye *et al.*, 1994). *Leishmania* also alters the cholesterol-rich lipid rafts within the plasma membrane important for the clustering of MHCs on the surface of APCs, thus compromising the rigidity of the plasma membrane and hence hampering antigen presentation to T-cells (Chakraborty *et al.*, 2005) (Fig. 11.8).

Cytokines are the major deterministic factor in deciding the type of immune response and the parasite targets both innate and adaptive anti-leishmanial immune responses of the host to alter the effector functions and skew it towards an anti-inflammatory response.

Cytokine production

Leishmania suppresses the immune system by disrupting the balance of cytokines and chemokines produced during infection. Cytokines are generated in response to receptor signalling and the response is short-lived owing to its unstable mRNA. The interplay between these immune mediators plays a decisive role in the disease outcome. The susceptibility or resistance in a host is dependent to a large extent on specific pattern of cytokine production during the course of infection. The generation of T_H1 cytokines (IL-12, IL-1, IFN-γ) leads to cell-mediated immune response that culminates in the elimination of the pathogen and hence affords resistance to the host, whereas T_H2 cytokines (IL-10, IL-4, IL-5, TGF-β) leads to humoral response resulting in disseminating infection and hence susceptibility (Pirmez *et al.*, 1993). The disease progression or control is dependent on an early induction of T_H1 response by IL-12 and the generation of IFN-γ $CD4^+$ T-cells (Rosas *et al.*, 2005). The pro-inflammatory cytokines IFN-γ activated mononuclear phagocytes kill the parasites within the tissue-granulomas (Murray, 2000). IFN-γ has a strong role to play in maintenance of inflammation, neutralization of which leads to a decrease in TNF-α production and an increase in the amount of IL-10. IL-10 an anti-inflammatory cytokine has an inhibitory effect on IL-17; a pro-inflammatory cytokine responsible for recruitment of neutrophils at the site of infection. TGF-β suppresses the activity of macrophages and NK cells and the differentiation of $CD4^+$ cells, and enhances the conversion of naive T-cells into Treg cells. It inhibits the production of IL-2 and IFN-γ.

Leishmania interferes with transcription factors (TFs) like NF-κB, STAT-1α, and AP-1 in order to inhibit gene expression of pro-inflammatory cytokines and microbicidal molecules. The

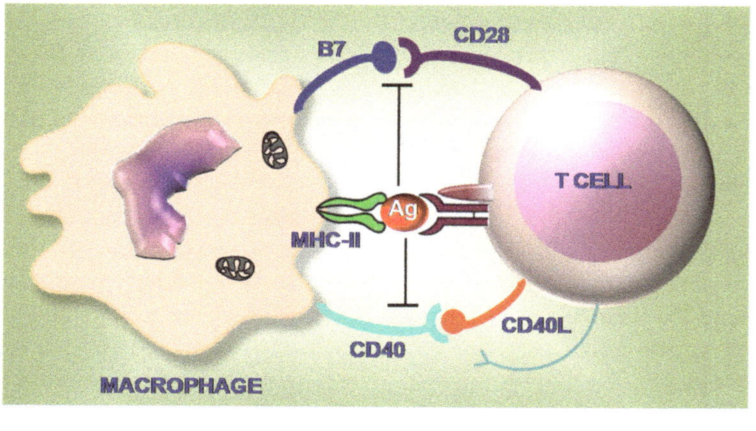

Figure 11.8 *Leishmania* interferes with the antigen presentation in the host by lowering the expression of MHC-I and MHC-II. It also suppresses co-stimulatory molecules such as CD40.

parasite amastigotes selectively favour the nuclear translocation of complexes that play a role in the generation of immunosuppressive cytokines such as IL-10 (Reiner *et al.*, 1988). The IL-10 produced from the infected macrophages suppresses the production of pro-inflammatory cytokine IL-12 and TNF-α (Kane *et al.*, 2001). Not only the cytokines produced but their respective receptor's expression level also determines the cellular responsiveness.

Chemokines and its G-protein coupled receptors play a pivotal role in determining the disease outcome. *Leishmania* induces the expression of a number of chemokine genes in the host, specifically of those responsible for recruitment of phagocytes to the site of infection enabling its installation and infection (Katzman *et al.*, 2008). The chemokines generated at the site of *Leishmania* infection enable infiltration of distinct cell population, which in turn has implications on disease outcome. For instance, chemokines produced at the site of the self healing cutaneous lesions are – CCL2, CXCL9, CXCL10, that chemoattracts macrophages and $CD4^+$ T-cells, lesions of chronic diffuse cutaneous leishmaniasis predominantly produce CCL3 that does not garner many $CD4^+$ T-cells (Ritter *et al.*, 2002). Through the saliva of the sandfly CCL2 is induced at the site of infection that recruits macrophages (Zer *et al.*, 2001). As neutrophils serve as a transitory host to the parasites, *Leishmania* induces CXCL8 that attracts neutrophils (Venuprasad *et al.*, 2002), a T_H2 attractant chemokine CCL7 (Katzman *et al.*, 2008), at the same time it inhibits CXCL10 generated by neutrophils to avoid NK-cell activation and evade killing (van Zanderbergen *et al.*, 2002). Leishmanial LPG inhibits the expression of CCL2 in endothelial cells thus preventing trans-endothelial migration (Lo *et al.*, 1998). The expression of important macrophage chemotactic factor CCL2 and CCR1 was reduced in *Leishmania* infection, thus restricting the number of macrophages at the site of infection leading to disease progression (Panaro *et al.*, 2004).

The distinct contrast in T_H1 versus T_H2 response seen in murine models of leishmaniasis is not so evident in human infection as the dynamics of disease and the immune response is more complex (Castellano *et al.*, 2009).

Trypanosoma

Trypanosome infective forms of different strains have varied characteristics and modes of infection (Table 11.1). The parasite can exist as extracellular (*T. brucei*) or reside and multiply within the macrophage (*T. cruzi*).

Initial interaction with the host cell

On bite of a tsetse fly, the metacyclic *T. brucei* trypomastigotes are deposited in the dermal connective tissue of the mammalian host that leads to chancre, a localized inflammatory response characterized by accumulation of macrophages, lymphocytes and WBCs at the site of inflammation. The parasite enters the bloodstream and lymph and is exposed to the immune surveillance mechanisms of the host. The parasite is covered with a thick coat (about 12–15 nm) of polypeptide shield of variable surface glycoprotein (VSG). An antibody-mediated response is mounted against the parasite that leads to its partial clearance. As parasitaemia continues the parasite exhibits antigenic variation with the surface expression of VSGs. As the infection progresses *in situ*, activation of alternative VSG genes takes place one at a time. The genome of *T. brucei* encodes more than 1000 VSG genes; moreover, pseudogenes of VSGs are also found in the genome containing frameshift and frame stop errors that can undergo recombination forming functional novel VSGs and thus adding further diversity to parasite's surface epitopes (Taylor *et al.*, 2006). This VSG switching also provides the parasite a window to multiply until the immune response is initiated against a new VSG. The parasite divides its resources in maintaining a balance between survival in the vertebrate by asexual replication into slender forms and simultaneously reproducing to generate transmissible stumpy forms for further infectivity (Vassella *et al.*, 1997; Roditi *et al.*, 2008). Moreover, as the process of replication continues a small proportion of parasite has a VSG coat that is not yet recognized by the immune system that results in rapid parasitaemia (McCulloch *et al.*, 2009). This balance has significant effects on virulence and infectiousness of the parasite (Alizon *et al.*, 2009). (Fig. 11.9)

Table 11.1 *Trypanosoma*, differences and similarities between *T. cruzi* and *T. brucei*[a]

Attributes	Trypanosoma brucei	Trypanosoma cruzi
Size	18–42 μm	15–24 μm
Taxonomic position	Salivaria-transmitted through saliva of insect vector	Stercoraria-transmitted through the faeces of insects vector
Nature	Extracellular	Intracellular
Morphology	Trypomastigote and epimastigote in vector	Epimastigotes in vector
		Trypomastigote in blood
		Amastogite inside cells
Vector	Tsetse fly	Reduviid bugs
Multiplication	Parasite multiply in the host's blood and extracellular fluids such as spinal fluid	Parasite multiplies in host macrophages
Immune target	Antibodies, neutrophils, eosinophils	Macrophage activation, T_H1 cells

[a]http://parasite.org.au/para-site/text/cruzi-text.html.

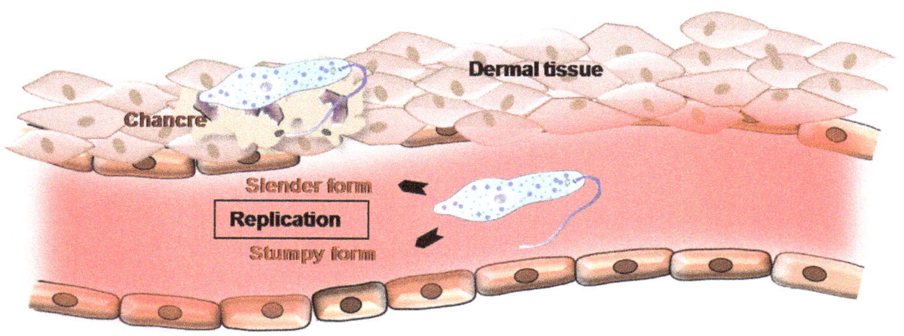

Figure 11.9 *Trypanosoma brucei* deposits in the dermal connective tissue to form chancre. The parasite enters the blood flow and replicates into the asexual slender forms and transmissible stumpy forms.

Following infection, macrophage serves either as host or as effector cells to *Trypanosoma cruzi*. Macrophages are among the first cells to be parasitized by *T. cruzi* and multiplication occurs in resident macrophages, leading to destruction of the host cell releasing infective trypomastigote forms. Various parasites' surface molecules have its complementary receptive structures on the host cell that enables initial attachment and penetration of the host cell. The causative of American trypanosomiasis, *T. cruzi*, has a surface glycoprotein gp83 that binds to the p74 receptor on host cell and plays a pivotal role in the infection of both phagocytic and non-phagocytic cells (Villalta *et al.*, 2001). In a study, the glycoprotein encoded by gp85/trans-sialidase gene superfamily was modelled and cloned (Tc85–11) that was found to be interacting with laminin of the host cell (Quelopana *et al.*, 2004). Fibronectin is also thought to participate in the attachment of *T. cruzi* to mammalian host cell (Ouaissi *et al.*, 1986). Study has shown that antibody generated against attachment domain of fibronectin or parasite receptor (p85) or synthetic peptides can block the cell penetration and decrease infection (Alves *et al.*, 1986; Ouaissi *et al.*, 1986).

Under natural or artificial selection pressure, host population subsets are selected with associated changes in characteristics strains of parasites. In order to determine the extent of heterogeneity, single-cell clones were prepared from various sources and 40% difference was observed in the amount of total DNA/parasite between clones (Dvorak *et al.*, 1984). The clones with faster multiplication rates predominated over other closely related strains under similar culture conditions (Finley *et al.*, 1987). Heterogeneity is also found in surface proteins such as the mucin-associated surface protein (MASP) owing to transcriptional differences in clonal isogenic populations. Other parameters such as growth, spontaneous metacyclogenesis or excretion of different catabolites also contribute to additional variability and a phenotypically heterogenous *T. cruzi*. This heterogeneity along with a multitude of other factors leads to variable immune responses.

Internalization and inhibition of host cell functions

Microbicidal activity and oxidative burst

Once *Trypanosoma* invades the macrophages, it down-regulates the phagocytic activity of the host macrophages, but instead of impeding the maturation of a lysosome-like parasitophorous vacuole called a phagolysosome, it transiently associates with it and disintegrates it through deacylation of internal phagolysosomal membrane proteins by trans-sialidase (Hall *et al.*, 1992). The parasite escapes into the cytoplasm transforms into amastigotes and multiplies by binary fission (Bogdan *et al.*, 1999). The amastigotes differentiate into the trypomastigotes that are the infective blood form, which are released into the bloodstream on the rupture of host cells. *T. cruzi* blood forms can also survive complement activation, as they express glycoproteins such as gp160, gp58/68, calreticulin and T-DAF. These proteins can bind to C3b and C4b, which allow evasion of complement (Norris *et al.*, 1991; Ferreira *et al.*, 2004).

To establish a productive infection inside the macrophage, *T. cruzi* must tolerate the extremely oxidative environment inside the phagolysosome (Piacenza *et al.*, 2013). *T. cruzi* devises a complex network of antioxidant enzymes, such as peroxidases and SODs, that protect the parasite against macrophage-released reactive oxygen (ROs) and nitrogen species (NOs) (Piacenza *et al.*, 2009). During phagocytosis, the trypomastigotes trigger activation of macrophage membrane-associated NADPH oxidase, that produces superoxide radical anions ($O_2\cdot^-$), which can be converted to H_2O_2 by SOD (Piacenza *et al.*, 2013; Fridovich *et al.*, 1995; Radi *et al.*, 2013). Following *T. cruzi* infection and disease progression, reactive oxygen species (ROS; e.g. $O^{2\cdot-}$, H_2O_2 and $\cdot OH$) can be produced as a consequence of immune-mediated cytotoxic reactions, secondary damage to mitochondrion, and tissue destruction caused by the parasite. ROs can oxidize DNA, proteins, and lipids resulting in elimination of the parasite. (Gupta *et al.*, 2009). Following *T. cruzi* acute infection, pro-inflammatory cytokines (IFN-γ and TNF) get triggered that stimulate infected macrophages to produce large amounts of nitric oxide (NO) by the enzymatic activity of inducible nitric oxide synthase (iNOS), which oxidizes L-arginine (Gupta *et al.*, 2009; Munoz-Fernandez *et al.*, 1992; Gutierrez *et al* 2009). NO affects the survival of the parasite in the macrophage by chemically modifying cysteine-containing proteins, inhibiting the catalytic activity of cruzipain, and by binding to parasite metalloproteins (Gutierrez *et al.*, 2009; Venturini *et al.*, 2000). Once generated, NO can react with $O_2\cdot^-$ to produce peroxynitrite (ONOO–), a potent oxidant and cytotoxic molecule that is highly effective against *T. cruzi* (Radi *et al.*, 2013; Alvarez *et al.*, 2011). Moreover, secondary intermediate free radicals produced from $ONOO^-$, such as hydroxyl (OH), nitrogen dioxide (NO_2), and carbonate ($CO_3\cdot^-$) radicals, can participate in the oxidation and nitration of proteins, lipids, and DNA, leading to mutations and transcription errors (Piacenza *et al.*, 2013; Radi *et al.*, 2013). Due to its high tissue-damaging potential (Gutierrez *et al.*, 2009), the NO-induced oxidative stress can be detrimental to the host.

NO has important roles in immunity, including the induction and suppression of apoptosis, and microbicidal activity. *Trypanosoma brucei*-infected mice express iNOS, produce NO, and kill parasites in the presence of L-arginine in peritoneal macrophages. However, parasites proliferate in the vicinity of these macrophages *in vivo*. L-arginine availability also modulates NO production. Trypanosomes use L-arginine for polyamine synthesis, required for DNA and trypanothione synthesis. Additionally, arginase activity is up-regulated in macrophages from infected mice from the first days of infection. Arginase competes with iNOS for their common substrate, L-arginine. Arginase inhibitors decreased urea production, increased macrophage nitrite production, and restored trypanosome killing *in vitro*. *In vivo* study showed dramatic decrease in plasma L-arginine in infected mice (Gobert *et al.* 2000).

Several reports have shown that products of IFN-γ regulated genes inhibit pathogen survival. Indoleamine 2,3-dioxygenase (IDO), iNOS and phagocyte oxidase (Phox) inhibit the survival of intracellular bacteria and protozoa. iNOS is located in the cytosol, where it catalyses the production of NO, which rapidly diffuses to the phagosome to clear the pathogen. Phox is found in plasma membrane, cytosol and intracellular granules, but translocates to the phagosome, where it catalyses the conversion of O_2 to superoxide, which gives rise to molecules that are toxic to phagosome-bound pathogens. It has been suggested that NO combines with superoxide to generate peroxynitrite, which is more toxic to *T. cruzi* than NO (Peluffo *et al.*, 2004). IFN-γ production also depends on the susceptibility and resistance of mice during *T. cruzi* infection. Lymphocytes from C57BL6 mice produced 10-fold more IFN-γ than BALB/c lymphocytes (Hoft *et al.*, 1993). In contrast to classically activated macrophages, alternatively activated macrophages fail to generate NO from L-arginine. Macrophages enhance the activity of arginase following interaction with T_H2 cytokines IL-4 and IL-10 (Munder *et al.*, 1999). *T. cruzi* infection in IFN-γ and IFN-γ receptor deficient mice leads to enhanced susceptibility and diminished NO production. Administration of iNOS inhibitors in *T. cruzi*-infected mice shows higher parasite level and greater mortality than untreated mice (Vespa *et al.*, 1994). Cruzipain – a cysteine proteinase produced by *T. cruzi* – increases arginase activity in macrophages upon activation of tyrosine kinase, protein kinase A (PKA), and p38 mitogen activated protein kinase (MAPK) (Stempin *et al.*, 2004). It also induces an alternatively activated phenotype in macrophages even in the absence of T-cells, due to increased secretion of TGF-β and IL-10 by macrophages (Fig. 11.10) (Stempin *et al.*, 2002)

Trypanosomatids have conserved mitochondrial Ca^{2+} transport mechanism known as MCU (mitochondrial calcium uniporter). In trypanosomes, MCU is important for buffering cytosolic Ca^{2+} changes and for activation of the bioenergetics of the cells (Docampo *et al.*, 2014a). In trypanosomatids, calcium ion (Ca^{2+}) is an important second

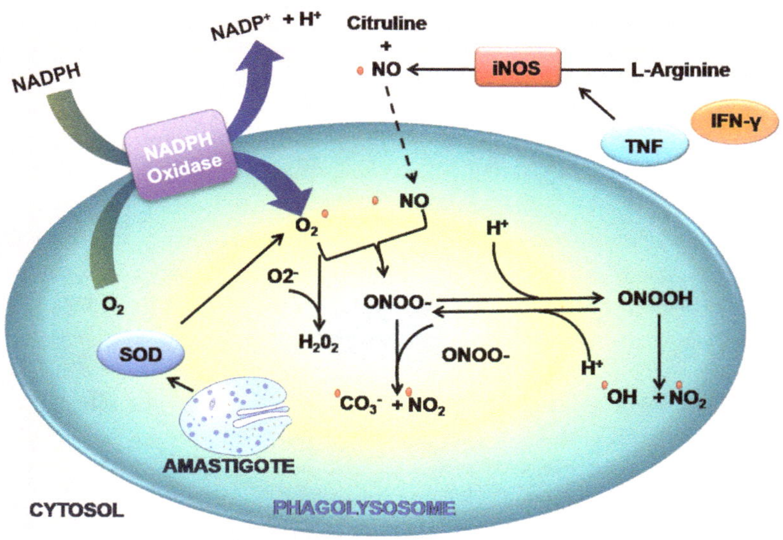

Figure 11.10 Host induced oxidative stress: Post phagocytosis of the parasite, macrophage membrane-associated NADPH oxidase is activated, producing the superoxide radical $O_2^{\bullet-}$ that can be converted into H_2O_2 inside the lumen of the phagolysosome. Macrophages stimulated with pro-inflammatory cytokines (IFN-γ and TNF) induce the expression of nitric oxide synthase (iNOS), by the oxidation of l-arginine. •NO then diffuse into the phagolysosome and reacts with $O_2^{\bullet-}$ to form peroxynitrite ($ONOO^-$). Secondary free radicals, such as carbonate $CO_3^{\bullet-}$, nitrogen dioxide ($\bullet NO_2$), and hydroxyl (•OH) radicals, are produced from $ONOO^-$. To survive in this highly oxidative environment, the parasite has a complex network of antioxidant enzymes such as superoxide dismutase (SOD), which detoxifies ROS.

messenger and is essential for its survival, although prolonged high intracellular Ca^{2+} levels lead to cell death. Like other eukaryotic cells, trypanosomes also use two sources of Ca^{2+} for generating signals, first by releasing Ca^{2+} from intracellular stores and second by Ca^{2+} influx across the plasma membrane. Ca^{2+} released from intracellular stores is regulated by the inositol 1,4,5-trisphosphate receptor (IP_3R) which is located in acidocalcisomes. A number of ion-pumps and exchangers are responsible for Ca^{2+} uptake and release from intracellular compartments. The *Trypanosoma* contains a repertoire of signalling and regulatory proteins that binds to Ca^{2+} as well as many Ca^{2+}-binding proteins that are not well characterized. Mitochondrial Ca^{2+} transporters of trypanosomatids have an important role in the regulation of cell bioenergetics and flagellar Ca^{2+} appears to have roles in sensing the environment (Docampo *et al.*, 2014b).

Trypanosoma invasion leads to transient changes in Ca^{2+} in mammalian cells (Moreno *et al.*, 1994; Tardieux *et al.*, 1994; Caler *et al.*, 2000; Scharfstein *et al.*, 2000). Due to this change in intracellular Ca^{2+} level, it regulates the parasite invasion by promoting transient actin rearrangements and lysosome-plasma membrane fusion (Rodríguez *et al.*, 1995). The Ca^{2+} transients are initiated relatively slowly (1–2 min post addition) in response to live trypomastigotes (Moreno *et al.*, 1994; Tardieux *et al.*, 1994; Caler *et al.*, 2000). It suggests that Ca^{2+} proximity to the host cell is a necessary for induction of Ca^{2+} signalling by this pathogen (Caler *et al.*, 2000; Scharfstein *et al.*, 2000). Following an initial increase in Ca^{2+} in infected and neighbouring uninfected cells in culture, Ca^{2+} levels decline at later times of infection (3–5 days) (Low *et al.*, 1992) indicating that cellular Ca^{2+} homeostasis is disrupted in *T. cruzi*-infected mammalian cells for the duration of the intracellular infection cycle.

Host cell signalling pathways

Innate immunity and pathogen recognition receptors

The innate immune response against African trypanosomes in humans is by a subclass of high-density lipoproteins (HDL) known as the Trypanosome Lytic Factor (TLF-1), which comprises a pore-forming protein namely Apolipoprotein L-1 circulating in the form of haptoglobin-related protein (Hpr). To neutralize this toxin *T. b. rhodensiense* has evolved an inhibitory protein called the serum resistance associated (SRA) protein whereas *T. b. gambiense* develops mutations in the genes encoding SRA, forming a receptor with low binding affinity to TLF-1, resulting in its reduced uptake. *T. cruzi* possesses several molecules capable of stimulating TLRs. The *T. cruzi* trypomastigote mucin GPI anchors are potent stimulators of the extracellular heterodimer TLR2/6. *In vitro* stimulation of TLR2/6 by *T. cruzi* GPI-mucins leads to the production of pro-inflammatory cytokines, such as IL-12 and TNF, as well as nitric oxide attributed T_H1-type immune response responsible for control of parasitaemia and tissue parasitism (Gazzinelli *et al* 2006; Rodrigues *et al.*, 2012; Coelho *et al.*, 2002) With the activation of innate immune receptors – TLR7 by parasite RNA and TLR9 by parasite CpG DNA motifs – *T. cruzi* establishes a T_H1 immune response (Latz *et al.*, 2004; Junqueira *et al.*, 2010). This leads to a *T. cruzi*-specific $CD8^+$ T-cell response that is crucial for regulating parasitaemia (Rosenberg *et al.*, 2010). GIPL, the virulent factor of the parasite, activates TLR4 leading to pro-inflammatory effects (Oliveira *et al.*, 2004). It also links the innate immunity to adaptive immunity by enhancing the potential of immune response mediated by T-cells and B-cells. In the absence of TLRs, the $CD4^+$ and $CD8^+$ T-cell-mediated immune response against *T. cruzi* is limited (Oliveira *et al.*, 2010). Besides TLR, other innate immune receptors important in controlling *T. cruzi* infection are the nucleotide-binding oligomerization domain (Nod)-like receptors (NLR). NLRs are localized in the cytoplasm or are associated with the plasma membrane of mammalian cells. After the parasite escapes from the phagolysosome, it can activate the cytoplasmic NOD1 receptor. Although NOD1 receptor is important for controlling the *T. cruzi* infection, its mechanism of action is still unknown. (Fig. 11.11) (Silva *et al.*, 2010).

Once stable inside the host cell, the parasite targets various crucial signalling components of the host cell. Macrophages and dendritic cells (DCs) from $MyD88^{-/-}TRIF^{-/-}$ mice, in which TLR-dependent activation of innate immunity was abolished, were unable to clear *T. cruzi* and showed impaired induction of IFN-β during *T. cruzi* infection. *T. cruzi* growth was enhanced with

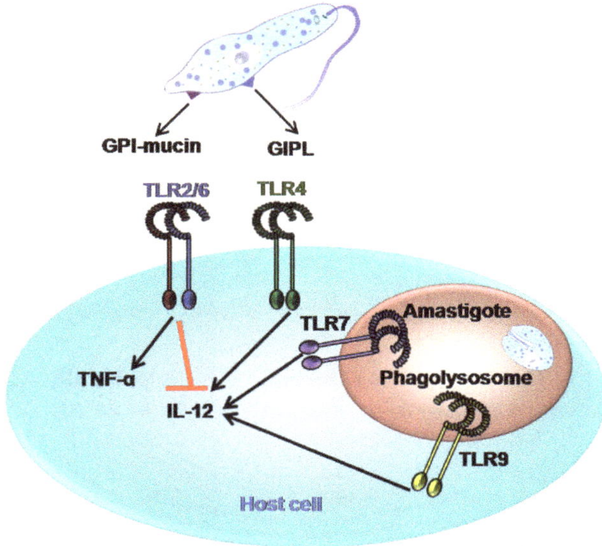

Figure 11.11 TLR activation with *T. cruzi* infection. The activation of the heterodimer TLR2/6 by parasite GPI-mucins leads to TNF production in macrophages or to the inhibition of IL-12 in dendritic cells (in red bar). By contrast, the activation of TLR4 by parasite GIPLs, TLR9 by parasite CpG DNA and TLR7 by parasite RNA result in the production of pro-inflammatory cytokines, such as IL-12.

IFN-β neutralization in MyD88$^{-/-}$ macrophages (Koga *et al.*, 2006). The importance of MyD88 and TLR2 in MAPK activation and cytokine synthesis in *T. cruzi*-infected macrophages has been shown and compared with that induced by tGPIm (glycosylphosphatidylinositol (GPI)-anchored mucin-like glycoproteins of *Trypanosoma cruzi* trypomastigotes). The absence of MAPK phosphorylation was observed in TLR2 and MyD88-deficient macrophages when exposed to tGPIm correlated with the inability to induce cytokine release from these cells. In contrast, activation of MAPK and synthesis of pro-inflammatory cytokines were not abrogated in *T. cruzi*-infected TLR2-deficient macrophages (Ropert *et al.*, 2004). It was recently shown that host cell PI-3 kinase and Akt (a serine/threonine kinase) plays an important in the regulation of *T. cruzi* invasion of non-professional phagocytes (Wilkowsky *et al.*, 2001) as well as macrophages (Todorov *et al.*, 2000). Because PI-3 kinases regulate actin-dependent processes as well as regulated exocytosis (Yang *et al.*, 1996), this signalling pathway may function as a modulator of lysosome-mediated invasion or potentially regulate an actin-associated entry process (Procopio *et al.*, 1999). *T. cruzi*-infected neuronal cells show pro-survival pathways by activation of host cell PI-3 kinase/Akt signalling (Chuenkova *et al.*, 2001). Protein tyrosine phosphorylation plays an important role PI-3-kinase activation (Storz and Toker, 2002). *T. cruzi* stimulates tyrosine phosphorylation of a number of proteins in professional phagocytes (Ruta *et al.*, 1996) whereas dephosphorylation of tyrosine has been correlated with infectivity of rat myoblasts, where inhibitors of tyrosine phosphate hydrolysis significantly reduce efficiency of host cell invasion.

T. cruzi mediated NF-κB activation and induction of pro-inflammatory cytokine synthesis have been extensively studied in macrophages (Ropert *et al.*, 2002), where signalling through TLR2 by tGPIm derived from infective trypomastigotes initiates the response (Campos *et al.*, 2001). In non-professional phagocytic cells, an inverse correlation between transcription factor NF-κB activation in mammalian cells and susceptibility to *T. cruzi* infection has been reported (Hall *et al.*, 2000). The parasite cysteine proteases block the activation of NF-κB p65 and expression of pro-inflammatory cytokine IL-12 but promote the secretion of anti-inflammatory IL-10 and TGF-β supporting the parasite's survival (Doyle *et al.*, 2011; Hunter *et al.*, 1997; Silva *et al.*, 1991).

Role of lymphocytes and cytokine production

With the onset of infection, the microbicidal action of the macrophages and the repertoire of pro-inflammatory cytokines come into action to kill the parasite, followed by the acquired immune response in the form of activation of the subsets of CD4$^+$ T-cells, CD8$^+$ T-cells, and B-cells. A T_H1 response is crucial for parasite clearance while a T_H2 response corroborates with the parasite persistence (DosReis *et al.*, 1997). *T. cruzi* can invade non-professional phagocytes either through fusion of host cell lysosome at the site of parasite entry (Tardieux *et al.*, 1994) or by forming an invagination in the plasma membrane that fuses with the lysosome (Woolsey *et al.*, 2003). PI3-kinase inhibition helps in the blockade of lysosome-mediated *T. cruzi* invasion. A significant fraction of the internalized parasites is not subsequently retained inside host cells for a productive infection. A direct correlation between the lysosomal fusion rates after invasion and the intracellular retention of trypomastigotes was observed. Therefore, formation of a parasitophorous vacuole with lysosomal properties is essential for preventing these highly motile parasites from exiting host cells and for allowing completion of the intracellular life cycle (Andrade *et al.*, 2004).

The trypanosome parasite inhibits the activation of the lymphocytes (Maleckar *et al.*, 1983) and specifically depletes the number of T-cells during the acute stage of infection when the density of the parasite is very high (Chaussabel *et al.*, 2003). This leads to an immune suppression and delayed response of cytotoxic T-cells, resulting in widespread infection (Martin *et al.*, 2006). GPIs and GPI-anchored mucins play an important role in invasion and subversion of host immune scrutiny and play a decisive role in both pro- and anti-inflammatory responses evoked by the parasite. GPI-mucins consist of a large array of proteins that confer variability to the parasite surface leading to variable immune evasion strategies (Buscaglia *et al.*, 2006). GPI-mucins have an immunosuppressive and pro-parasitic role during adaptive immune response. T-cells exposed to GPI-mucin become anergic and fail to produce IL-2 (Argibay *et al.*, 2002) while epimastigote derived GPI-mucins inhibits T-cell activation and IL-2 secretion. *T. cruzi* derived mucins inhibit T-cells by binding to L-selectin on its surface (Alcaide *et al.*, 2004). The antigenic determinants are shielded from the action of complement factor and anti-galactosyl antibodies by sialylation (Kipnis *et al.*, 1981; Joiner *et al.*, 1986; Gazzinelli *et al.*, 1991; Pereira-Chioccola *et al.*, 2000). During the course of infection, B-cells produce abundant non-specific and low affinity antibodies (Ouaissi *et al.*, 2001). Some of these may cross-react against neural and cardiac tissues aggravating secondary pathogenesis (Acosta *et al.*, 1985; Kierszenbaum *et al.*, 1999; Engman *et al.*, 2002).

Not only the lymphocytes but also the DCs are affected by GPI-mucin, as it renders them dysfunctional. The trans-sialidase (TS) enzyme expressed by *T. cruzi* transfers sialic acid from host glycoproteins to parasite GPI-mucin (Buscaglia *et al.*, 2006) and blocks the DC-mediated induction of T-cells activation and IL-12 production (Erdmann *et al.*, 2009). The immunodominant epitope for cytotoxic T-cells is encoded by the trans-sialidase family of genes, which differs according to the infective strain, conferring further variability to the host immune response against the parasite (Martin *et al.*, 2006). The TS interacts with the lymphocytes to stimulate CD4$^+$ T-cell proliferation (Todeschini *et al.*, 2002) but also results in lymphocyte apoptosis by cell resialyation (Mucci *et al.*, 2006). The resialylation of cytotoxic T-cells by TS hampers its ability to kill *T. cruzi*-infected cells (Freire-de-Lima *et al.*, 2010) (Fig. 11.12). Another immune-modulatory molecule of *T. cruzi* is glycoinositolphospholipids (GIPLs), which suppress CD4$^+$ T-cell activation (Gomes *et al.*, 1996). It also skews the immune response towards an anti-inflammatory mode by suppressing the expression of IL-2 and enhancing that of IL-4. Furthermore, GIPL down-regulates the expression of costimulatory molecules and pro-inflammatory cytokines in macrophages and DCs on LPS stimulation (Brodskyn *et al.*, 2002). Activation of B-cells is also attributed to the glycan moiety of GIPLs (Bento *et al.*, 1996).

Cruzipan, a parasite expressed Cysteine proteinase, and Tc52, a glutathione thioltransferase, are essential for the replication of the parasite in the host (Garzon *et al.*, 2003). Cruzipan induces the secretion of IL-10, TGF-β and arginase leading to rapid intracellular replication of the parasite (Stempin *et al.*, 2002). During the stage of acute

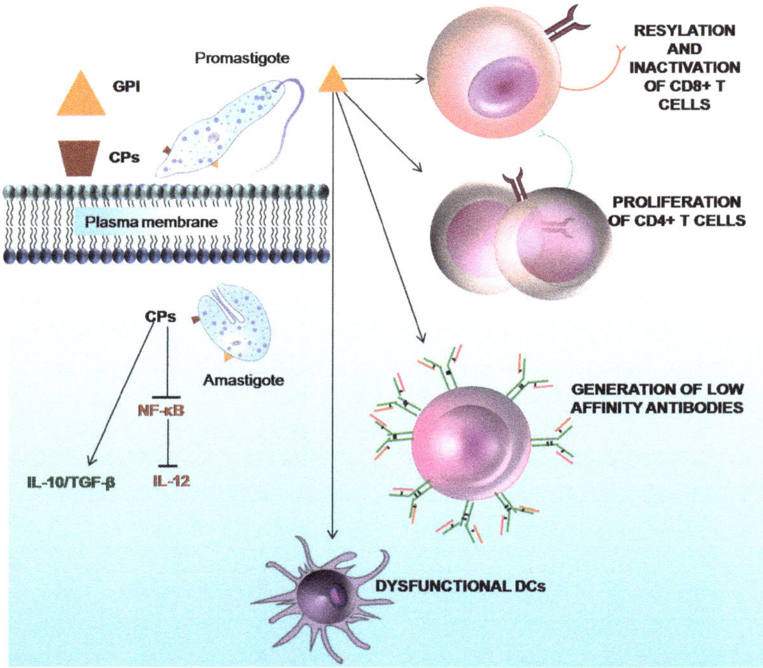

Figure 11.12 *Trypanosoma* affects the lymphocytes activation. B-cells generate low affinity antibodies while the number of T-cells deplete during infection. The generation of pro-inflammatory cytokines is suppressed while that of anti-inflammatory cytokine is enhanced.

infection, a general state of immunosuppression occurs. T-cells from infected mice have been found to respond poorly to mitogens (Kierszenbaum *et al.*, 1999, 2002; Alcaide *et al.*, 2004) and upon activation through TCRs undergo rapid apoptosis (Nunes *et al.*, 1998).

The balance between pro- and anti-inflammatory cytokines is another deciding factor in disease outcome. IL-10 and TGF-β play a major role in inducing an anti-inflammatory response promoting parasitaemia (Hunter *et al.*, 1997; Silva *et al.*, 1991). The acute phase of Chagas' disease is characterized by high parasitaemia, prolonged infection leads to a chronic phase where the parasites reside in the cardiac smooth muscles and muscles of the digestive system (Fig. 11.12).

Antigen presentation in host

As *T. cruzi* resides in the host cytoplasm, it is not surprising that proteins released during normal cellular infection gain direct access to the classical cytosolic MHC class I pathway. This mechanism is supported by the studies showing that the TAP-1-deficient mice are highly susceptible to *T. cruzi* infection (Kumar *et al.*, 1998). *T. cruzi* interferes with processing and presentation of antigenic peptides in the MHC class II pathway. Following *T. cruzi* infection, surface expression of MHC class-I molecules was found to be unaffected or up-regulated *in vitro*. A model system utilized a P-galactosidase (P-gal)-specific murine cytotoxic T lymphocyte (CTL) line (0805B) and showed that *in vitro* infection with *T. cruzi* in mouse peritoneal macrophages or J774 cells did not inhibit MHC class I presentation of exogenous peptide (Buckner *et al.*, 1997).

Development of immunotherapeuetics

Common drug targets in trypanosomatids

VL and CL are commonly treated by the toxoid amphotericin B, which makes the parasitic ergosterol-rich membrane porous (Croft *et al.*, 2006), resulting in altered osmotic pressure and death of the parasite (Ramos *et al.*, 1996).

In Africa, the use of pentavalent antimonial compounds such as sodium stibogluconate to treat leishmaniasis is common (Copeland *et al.*, 2015). Pentavalent antimonials act upon multiple targets and interfere with the bioenergetics of *Leishmania*. It inhibits glycolysis (Mottram *et al.*, 1998; Berman *et al.*, 1987), β-oxidation of fatty acids (Berman *et al.*, 1989), and ADP phosphorylation (Berman *et al.*, 1987). Moreover, it blocks the thiol group of amastigotes parasite's proteins and inhibits DNA topoisomerase-I, thus leading to DNA fragmentation and apoptosis. It also inhibits an enzyme trypanthion reductase responsible for protection against host ROs and nitrogen species (Wyllie *et al.*, 2004).

Miltefosine was initially developed as an anti-cancer drug (Croft *et al.*, 2003). It is the first oral drug used for the treatment of VL, proving to be a major revolution in anti-leishmanial chemotherapy (Jha *et al.*, 1999; Sundar *et al.*, 2006) and reached successfully up to phase IV clinical trial against VL in endemic areas (Bhattacharya *et al.*, 2007). Its mode of action is similar to induction of apoptosis but its teratogenic nature restricts its use in pregnancy. Paromomycin has shown its potential as both anti-leishmanial and antibacterial drug and cures CL and VL effectively (Thakur *et al.*, 2000). In *L. donovani* paramomycin affects its protein synthesis and recycling (Maarouf *et al.*, 1995). Not only has the parasite developed resistance to these non-selective drugs, the use is accompanied with serious side-effects (Fig. 11.13).

The drugs currently used to treat Chagas' disease are nifurtimox (4[(5-nitrofurfurylidene) amino]-3-methylthiomorpholine-1,1-dioxide), derived from nitrofuran, and benznidazole (N-benzyl-2-nitroimidazole-1-acetamide), a nitroimidazole derivative. Nifurtimox and benznidazole acts through the formation of free radicals and/or electrophilic metabolites. The nitro group of the drugs is reduced to an amino group by the action of nitroreductases, with the formation of various free radical intermediates and electrophilic metabolites. This process begins with a reaction catalysed by NADPH-cytochrome P-450 reductase, that acts on the nitro group of R-NO_2-type molecules, producing an intermediary nitro anion radical (R-NO_2^-) (Moreno *et al.*, 1982). Nifurtimox and benznidazole drugs are effective against the circulating form of the parasite (trypomastigotes) during the acute phase of the disease, but not during the chronic stage.

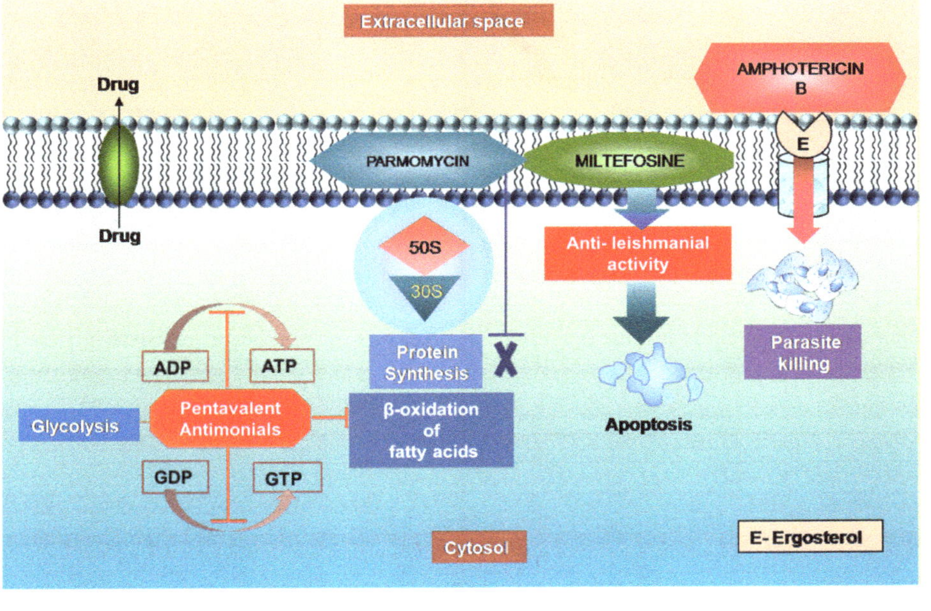

Figure 11.13 An overview of molecular mechanism of drug resistance in *Leishmania* spp.

Drugs used to treat African trypanosomiasis

Eflornithine (α-difluoromethylornithine, DFMO) is a specific and irreversible suicide inhibitor of ornithine decarboxylase, the first enzyme involved in polyamine biosynthesis. Inhibition of ornithine decarboxylase results in the depletion of the polyamines putrescine and spermidine, with the consequence of a slowdown in cell proliferation.

Suramin is effective for the treatment of only the early stages of human trypanosomiasis due to *T. b. gambiense* and *T. rhodesiense*, as it does not cross the Blood Brain Barrier (Hawking *et al.*, 1940) but currently pentamidine is preferred to cure haemolymphatic infections due to *T. b. gambiense*. Suramin has inhibitory effect on wide spectrum of enzymes: dihydrofolate reductase, fumarase, glycerol-3-phosphate dehydrogenase, hexokinase, L-α-glycerophosphate oxidase, reverse transcriptase, receptor-mediated uptake of low-density lipoprotein, RNA polymerase and kinases, thymidine kinase, trypsin, etc. (Pépin *et al.*, 1994; Wang, 1995).

Melarsoprol (MelB) was used for treating late stage sleeping sickness, as a replacement to earlier organo-arsenic compounds, particularly tryparsamide, which had become ineffective due to drug resistance and were ineffective in the treatment of *T. rhodesiense* sleeping sickness. Melarsoprol is the only arsenic-based compound currently used against HAT, but the related product melarsenoxide cysteamine is used to treat trypanosomiasis in domestic animals, especially *T. evansi* infection (Raynaud *et al.*, 1989; Musa *et al.*, 1994).

Diminazene aceturate (Berenil) has been widely used as a chemotherapeutic agent for trypanosomiasis in livestock since 1955. The main biochemical mechanism of Berenil's trypanocidal actions is thought to be by binding to kinetoplast DNA. (Brack *et al.*, 1972). Berenil is not licensed for use in humans because of its serious side-effects observed in animals, which include tremors, itching, sweating, convulsions, dyspnoea, recumbency and vomiting in camels (Riou *et al.*, 1980) and decreased blood pressure (Joubert *et al.*, 2003) and diarrhoea in dogs.

Isometamidium and homidium: the amphiphilic cationic phenanthridine, isometamidium chloride (ISM; Samorin®, Trypamidium®, Veridium®, 8-[(m-amidinophenyl-azo) amino]-3-amino-5-ethyl-6-phenylphenanthridinium chloride hydrochloride ($C_{28}H_{25}C_{l}N_{7}HCl$; MW 531.5)), has been used prophylactically or therapeutically for livestock suffering from trypanosomiasis due to infection with *Trypanosoma congolense* and other *Trypanosoma* spp. (Leach and Roberts, 1981). ISM is mainly concentrated in the kinetoplast, whereas homidium is spread much more diffuse throughout the trypanosome parasite (Boibessot *et al.*, 2002). The mode of action of ISM chloride is thought to be the cleavage of kDNA–topoisomerase complexes, causing the desegregation of the mini-circle network within the kinetoplast (Shapiro and Englund, 1990).

Many sequences are conserved in trypanosomatids that are absent in their mammalian hosts. For example, the sequence that encodes the leishmanial and Trypanosomal receptor for activated C-kinase is highly conserved making it a prominent drug target (McCahill *et al.*, 2002). The peptide inhibitors against these parasitic targets show anti-parasitic activity in both cell culture and in mice models (Qvit *et al.*, 2016). This new approach of targeting specific parasitic proteins also reduces the probability of side-effects, as it is targeted and specific. Moreover, the possibility of parasite developing a resistance against it is very low, as these proteins are evolutionarily conserved and the probability of substitution mutations in genes encoding for these proteins is very low (Thomas *et al.*, 2003). Emergence of resistance against drugs opens a window to shift towards means of devising effective vector eradication programme by using pesticides as an alternative. In 1960 under the National malaria control programme, DDT was used to control the sandfly population and VL cases were nearly eliminated in India (Thakur *et al.*, 1992), but the use of such chemicals pose serious environment related issues, toxicity and related health concerns that led to the development of a novel para-transgenic approach for controlling the infectious disease (Durvasula *et al.*, 1997; Beard *et al.*, 1998; Hillesland *et al.*, 2008). In this strategy, native bacterial flora to disease transmitting vectors are isolated and genetically transformed *in vitro* to export molecules to interfere with pathogen transmission. The genetically altered symbionts are then reintroduced into the host vector where expression of the engineered molecules affects the host's ability to transmit the pathogen. Initially, this approach was implemented in *T. cruzi* (Baines, 1956) as well as in sandfly due to its short life cycle (Durvasula *et al.*, 1997) but the approach failed, as the symbionts were not identified in sandfly. Advancement in basic understanding of immunoparasitology is a prerequisite for the generation of effective vaccines.

Vaccination approaches

The recent trend in vaccine research has shifted towards targeting the stages in the life cycle of the parasite. Many attempts to make a potent vaccine have met with failures and the toxicity, cost, antigenic variation and mixed T_H1/T_H2 response further deters it. To date, only two FDA-approved canine vaccines have been developed in Brazil under the trademarks Leishmune® and Leishtec®. Recently, numerous antigens such as surface molecules, intracellular molecules, stress related molecules, salivary molecules, metabolic and biochemical pathway targets have been selected as vaccine candidates but most of them have failed. Although the availability of complete genome sequence makes it easier to select a novel vaccine candidate, success remains unmet.

Immunization using salivary components of sandfly

In sandfly saliva, a vasodilator molecule called Maxadilan enhances the blood flow following bite and remains in the form of suspension. In immunization studies using Maxadilan, it was found that the salivary proteins triggered an antibody-mediated immune response where immunodominant antigen PpSP30 elicited all IgGs (de Moura *et al.*, 2013). Immunization with fly's salivary gland extract (SGE) was found to be dose dependent in the type of immune response produced; for example, single immunization showed an augmentation in pro-parasitic immune response via generation of IL-10 while immunizing the mice three times with the same SGE showed an increase in the IFN-γ produced in the draining lymph nodes (Carregaro *et al.*, 2013).

Attenuated and mutant parasites as vaccines

Previous exposure to a pathogen provides protection from subsequent infection by the same or related pathogens is the basis for designing of vaccines. 'Leishmanization' was the first adopted vaccination approach against *Leishmania* infection, wherein individuals were inoculated with low number of virulent parasites. Earlier, native tribal population was known to expose their newborns to sandfly bites on their rears in order to prevent deforming facial lesions (Handman *et al.*, 2001). The possibility of combining the attenuated parasite with an appropriate adjuvant was assessed by co-administration of autoclaved *L. major* (ALM) with BCG rectally, that induced type 1 immune response against subsequent *L. major* infection, when compared to ALM alone (Soudi *et al.*, 2011). Another randomized double-blind trial was carried out in Sudan to test the efficacy of vaccine against visceral leishmaniasis. These methods were inconsistent due to a number of variables that decide the efficacy such as the dosage, use of adjuvants, and the route of administration. Due to the limitation of such attenuation methods, in 1952, Pizzi and Prager derived cultured parasites in attenuated form to be used as a vaccine against *T. cruzi* infection (Pizzi *et al.*, 1952), but subsequent studies revealed that these attenuated forms obtained from culture exacerbated in presence of corticosteroids. Since then, several methods were derived by either heat, chemical, pressure treatment or by serial culturing to generate a reliable attenuation protocol. Later in 1968, Menezes *et al.* (1968) postulated that parasites cultured for 15 years lose its pathogenicity due to certain alterations in its nucleotide sequence. The use of live but attenuated parasites was preferred over killed parasite, as it mimics the natural course of infection, helps in representation of antigen to the immune cells and moreover, controlled persistence of parasite evokes memory response which is a prerequisite for successful vaccination. Similar attempts were done to generate a vaccine against *Trypanosoma* by chemically treating epimastigote of *T. rangeli* with glutaraldehyde using saponin as an adjuvant. Breganó and colleagues used plant-infecting *trypanosomatid* parasite *Phytomonas serpens*, which shares antigens with *T. cruzi*, for initial immunization in mice. On further challenge with *T. cruzi*, lower parasitaemia and higher survivability were observed (Breganó *et al.*, 2003).

Pathogenic stages are the most common target sites for attenuation due to limited antigenic diversity, although there are chances of reversion into normal wild-type form. Immunization of animals with pelleted down culture and the culture supernatant exhibited positive xenodiagnosis, myocarditis and myositis. In addition to attenuated parasites, mutant forms were also assessed for immunization in animal models. Monoallelic dhfr-ts mutant generated showed less pathogenicity in mice, by conferring protection against new challenge of *T. cruzi* (Perez *et al.*, 2011).

Vaccination using parasite derived immunogenic components

Immunization using parasite cell fraction

In addition to using whole attenuated, whole live and mutant parasites for vaccination, the kinetoplastid parasite's immunogenic components are being considered for generation of an efficient vaccine. Subcellular fraction obtained through differential centrifugation of *T. cruzi* epimastigotes was used to immunize mice that were subsequently challenged with 25 bloodstream trypomastigotes and it provided protective immunity (Ruiz *et al.*, 1985). Immunization with flagellar portion of the parasite alone resulted in infection symptoms indicating that the fraction alone is responsible for inducing an autoimmune response. Wrightsman and colleagues purified paraflagellar rod protein (PAR) from *T. cruzi* epimastigotes and immunization with PAR protein via s.c. route exhibited 100% protection upon infection whereas the i.p. route exhibited pathogenicity (Wrightsman *et al.*, 1995). A modification of the former study was tried by administration of PAR protein through s.c. route in combination with Freund's adjuvant or alum that provided 100% and 83% protection, respectively, by evoking T_H1 type immune response (Miller *et al.*, 1996).

Immunization using parasite-purified proteins and recombinant protein

Since kinetoplastid proteins are known to play a multifaceted role in establishing infection in the host, various approaches were used to test the efficacy of these proteins as vaccine candidates. Purification of protein from protozoan parasite is tedious process and it requires further validation by challenging to see the augmentation in the immunization profile. Snary *et al.* compared two different stages of *T. cruzi* surface glycoprotein in which 90 kDa found in all stages of *T. cruzi* provided protection against both bloodstream and metacyclic trypomastigote whereas 72 kDa (found only in the insect stage) provided protection against the metacyclic trypomastigote (Snary *et al.*, 1983). Araujo and Morein in 1991 purified 45 and 68 kDa antigenic protein through affinity chromatography from the cell membrane and used it for immunization in combination with QuilA a saponin derivative which elicited 100% protection by generating humoral and cellular immune response when challenged against blood trypomastigotes (Araujo *et al.*, 1991).

Immunization of C57BL/10 mice with TcY72, a lysate of *T. cruzi* Y strain epimastigotes, leads to generation of DTH response, increased IgG antibodies, decreased parasitic burden and, CD4/CD8 ratio upon challenge with epimastigotes, thus demonstrating its immunogenic potential (Gomes *et al.*, 1999). Immunization with trypomastigote's excretory-secretory antigen (ESA) provided 60% protection in BALB/c mice and total protection in Fisher Rats against lethal infection (Taibi *et al.*, 1993). Exoantigens of *T. cruzi* that are released into the circulation were used for immunization experiments and it was found to confer partial protection, whereas upon transfer of lymph node from immunized mice to normal recipient after challenge lead to decrease in parasitic levels (Gruppi *et al.*, 1995). Garcia *et al* administered *T. cruzi* soluble extract (TCSE) in CBA/J mice and challenged it with *T. cruzi* that lead to rise in the production of IFN-γ as well as cellular proliferative response upon stimulation with concanavalin A (Garcia *et al.*, 2000).

The parasite's surface proteins play a crucial role from initial interaction with the host cell to skewing it towards a pro-parasitic immune response. Leishmaniolysin GP63, native or recombinant form has been extensively tested in various models of leishmaniasis. The rGP63 elicited IFN-γ mediated proliferation of T-cell response while the native form failed to show the same in patients with active or cured visceral, mucocutaneous and cutaneous leishmaniasis (Basyoni, 2011). As purified antigens are weak immunogens and need adjuvants to confer immunity, the efficacy of Liposome and MPL-A were tested in various host–parasite pathogenesis models either alone or with different suitable adjuvants. *L. donovani* surface antigen (LAG) in combination with various adjuvants such as cationic liposomes, BCG and MPL-A were tested in VL models and the liposomal–leishmanial antigen immunization showed the highest level of protection in comparison to other LAG–adjuvant combinations (Ravindran *et al.*, 2010). Recombinant GP63 with liposomal adjuvant have been tested in mouse model with promising results in clinical trial and has been approved by FDA. GP63 derived from *L. donovani* used with distearoylphophatidylcholine (DSPC) lead to IFN-γ augmentation and decrease of IL-4 and hence reduced the disease progression in mice (Bhowmick *et al.*, 2008) and with MPL-A resulted in production of IFN-γ producing effector T-cells (Mazumder *et al.*, 2011). rGP63 co-administered with CpG-ODN; a TLR9 ligand in cationic liposome provided protection in mice. Vaccination with GP63 in cationic distearoyl phosphatidylcholine (DPSC) liposomes also provided protection (Bhowmick *et al.*, 2008).

Fucose mannose ligand (FML) is a surface glycoprotein of *Leishmania* with GP36 as its major immunogenic component and is present in all stages of the parasite. Vaccination with FML with QuilA saponin as an adjuvant provided protection in mice and Hamster model against *L. donovani* infection. Phase III human and canine clinical trials are undergoing in endemic regions of Brazil. Vaccination with FML-QuilA formulation also provided long-lasting robust protective efficacy against canine visceral leishmaniasis during phase III in an endemic area of Brazil. These successes lead researchers to register this vaccine under the trademark Leishmune® in Brazil. Recently, purified soluble *L. major* antigen (SLA) was used as a model of first generation vaccine with the nuclease-resistant phosphorothioate CpG oligodeoxynucleotides (PS CpG) or nuclease-sensitive phosphodiester CpG ODNs (PO CpG) as adjuvants in murine CL. Immunogenic peptide derived from GP63 and restricted to HLA-DR1 were effective in T_H1 immune response induction (Rezvan *et al.*, 2013). As mentioned earlier LACK are highly conserved in *Leishmania* and are promising drug targets, in a different approach rLACK antigen administered in the presence of rIL-12 induced protection against *L. major* infection in mice by evoking $CD4^+$ T-cells.

Repetitive amino acid units of Hydrophilic acylated proteins (HASP) are found on the plasma membrane of *Leishmania* parasite, the exact function of this protein is yet to be deciphered. Immunization using HASPB1 induced protection against infection even in the absence of an adjuvant; its mode of action could be similar to that of DNA vaccine (Depledge *et al.*, 2010).

With the advent of genetic engineering it is much easier to use protein expression vectors to generate a large amount of protein in a short period of time. Immunogenic recombinant protein derived from *T. cruzi* cytoplasmic repetitive antigen (CRA) and flagellar repetitive antigen (FRA) was tested. Immunization results showed an induction of IgG3 and IgG1 response by CRA whereas induction of IgG1 response by FRA (Pereira *et al.*, 2003). Luhrs *et al.* (2003) immunized recombinant paraflagellar rod proteins (PFR) and found that it generated a T_H1 type immune response against *T. cruzi* infection. An equimolar combination of various paraflagellar rod proteins, e.g. PFR-1, PFR-2 and PFR-3, leads to a reduction in parasitaemia.

Trypanosoma trans-sialidase is an important enzyme that regulates host lymphocytes. Plasmid p154/13 encoding *T. cruzi* trans-sialidase when used for immunization generated T_H1 response against infection in susceptible mouse strain. A recombinant trans-sialidase protein alone generated T_H2 type response. Whereas simultaneous administration of single dose of both plasmid p154/13 and recombinant induces T_H2 response, a double dose of plasmid followed by booster recombinant protein induces T_H1 type response in mice. This indicated an interaction of DNA with protein possibly through TLR receptor that augments IL-12 generation and T_H1 type response. (Vasconcelos *et al.*, 2003). Mice immunized with mutant inactive trans-sialidase (mTS) with adjuvant IMX (ISCOMATRIX) showed a TS-specific IgG response, increased T_H1/T_H2 ratio, DTH response, as well as balanced production of IFN-γ and IL-10, reduced parasitaemia, parasite load and conferred protection through both cell-mediated and humoral responses (Bontempi *et al.*, 2015).

Cruzipain, a parasite cysteine protease that is essential for its replication, was found to be antigenic in both mice and humans and show differential immune response upon immunization with N or C-terminus proteins. N-terminal domain (catalytic domain) is associated with protective response whereas C-terminal domain is associated with IgG2a antibody response (Cazorla *et al.*, 2010).

Multiantigenic *Leishmania* antigens such as Leish-111f, which includes TSA (thiol-specific antioxidant), LmSTI1 (*Leishmania major* stress-inducible protein1) and LeIF (*Leishmania* elongation initiation factor), having ORF linked in a tandem manner, are also considered as vaccine candidates. rLeish-111f with adjuvant MPL (monophosphoryl lipid A) plus squalene (MPL-SE) shows defensive ability against *L. major* and *L. amazonensis* infection and also showed efficacy against cutaneous, mucocutaneous leishmaniasis in mice, non-human primates and humans (Skeiky *et al.*, 2002). Leish-111f+MPL-SE is the first successful vaccines to reach up to phase-I and II human clinical trials (Coler *et al.*, 2007). As it failed to provide protection in Canine model, currently Leish-110f +MPL-SE in combination with N-methyl meglumine antimoniate (Glucantime) is being assessed as a vaccine against canine visceral leishmaniasis (Miret *et al.*, 2008).

Immunization using DNA

Genetic immunization imitates the effect of a live attenuated vaccine, has less safety concerns, is stable and needs no refrigeration for storage. Efforts have been taken to optimize genetic vaccination by introducing various modifications in vector designing, using viral vectors, prime-boost regimens, formulation with adjuvant or cytokines and improvements in DNA delivery systems. Immunization with TcSP (belonging to trans-sialidase superfamily) gene and their recombinant protein against *T. cruzi* showed mixed T_H1/T_H2 response. However, mice immunized with DNA alone showed protective response (Arce *et al.*, 2011; Salgado *et al.*, 2013). YF-17D that expresses an immunogenic *T. cruzi* amastigote Surface Protein-2 (ASP-2) was used for s.c. immunization in A/J mice. This was followed with two dose of recombinant virus and subsequent challenge with *T. cruzi* exhibited reduced parasitaemia, and higher survivability (Nogueira *et al.*, 2013).

Cazorla *et al.* (2015) administered multiantigen *Salmonella* strains carrying plasmids Cruzipan (cz), thiol-transferase (Tc52) and a 24 kDa flagellar calcium-binding protein (FCaBP or Tc24) to C3H/HeN mice and multiantigen conferred higher protection against *T. cruzi* in comparison with a single antigen.

Currently a therapeutic Chagas' vaccine is under development by a consortium of Mexican (including the Carlos Slim Health Institute) and Texan scientific institutions; based on the evidence of therapeutic efficacy of Tc24 and TSA1 vaccines in reducing *T. cruzi*-induced cardiac disease in mice and dogs. As *Leishmania* is an intracellular parasite, cell mediated immunity is crucial in affording protection to the host. Earlier, DNA vaccination of gp63, PSA-2 etc. have shown promising results against *L. major* infection (Sjolander *et al.*, 1997; Gurunathan *et al.*, 1997; Melby *et al.*, 2001; Marques-da-Silva *et al.*, 2005). Recently, a DNA vaccine comprising minimalistic immunogenically defined gene expression (MIDGE) vector developed by Mologen AG was used to create LEISHDNAVAX, which consists of immunogenic epitopes from pan-species proteins of *Leishmania* encoded in the vector. This vaccine provided protection against *L. donovani* in rodent and elicited $CD4^+$ and $CD8^+$ T-cell responses in cured patients of leishmaniasis (Fig. 11.14) (Das *et al.*, 2014).

Conclusion

As they co-evolve with their hosts, parasites change their genetic make-up by different horizontal gene transfer mechanisms with consequential functional alterations in infectivity, adaptability, drug resistance, virulence and immune evasion. Another level of complexity arises as the parasite cycles through an intermediate host or a vector. Molecular interaction between host and parasite depends on genetic nature of the host and different parasite expressed molecular patterns such as gp63, LPG, VSG, gp85 etc. Interaction of *Leishmania* or *Trypanosoma* with host targets various signalling components, for example, JAK/STAT and MAPKs, which play pivotal roles in microbicidal function by generating NO, IL-12 and TNF-α. These parasites block the activation of NF-κB p65 and expression of pro-inflammatory cytokine IL-12 but promote the secretion of anti-inflammatory cytokines – IL-10 and TGF-β – to engineer their own survival.

Based on the different modes of interactions between the host and the parasites, the targets are defined. In general, activators are designed for the regulators that are less expressed or inhibited during the parasitic infection. Reciprocally inhibitors for

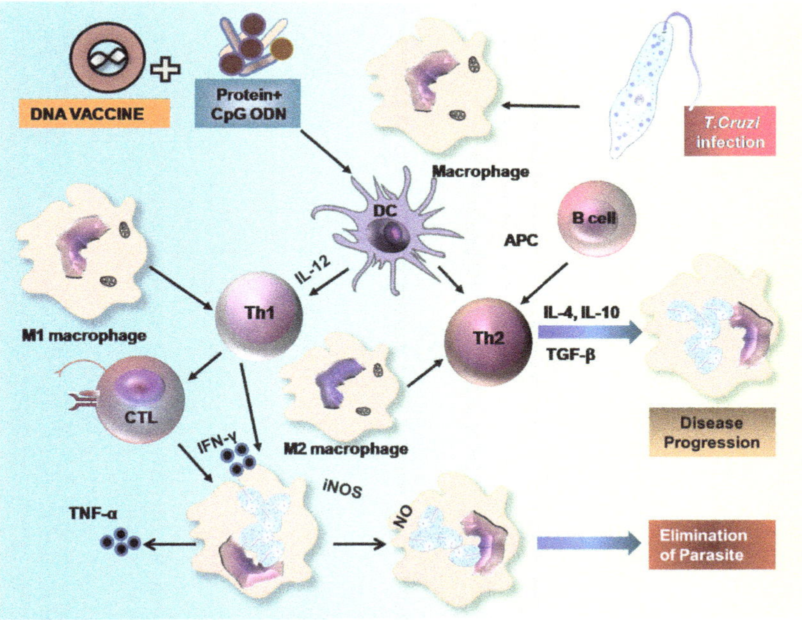

Figure 11.14 A model demonstrating protective and pathogenic ability of *Leishmania* and *Trypanosoma*. During infection both intracellular and extracellular antigens are presented to the immune system. T_H1 subset is associated with host-protective, while T_H2 with parasitic survival.

the regulators that are activated or enhanced by the parasites are designed for therapy. However, the parasite-targeted drugs are also discovered and formulated but that comes with a rider: under the selection pressure exerted by the parasite-targeted drug, the drug-resistant parasites emerge. Therefore, to control the parasitic diseases, novel methods are required. The possible targets are biological control of the vector restricting the transmission of the parasite and develop new vaccine. So, although amphotericin B, pentavalent antimonial, miltefosine and paromomycin were used as anti-parasite drugs, the emergence of resistance, relapse and re-infection compels the development of a vaccine. Many approaches of vaccination were put on trials but all of them failed justifying further fundamental research to understand the mechanism of prophylaxis.

References

Acosta, A.M., and Santos-Buch, C.A. (1985). Autoimmune myocarditis induced by *Trypanosoma cruzi*. Circulation *71*, 1255–1261.

Alcaide, P., and Fresno, M. (2004). The *Trypanosoma cruzi* membrane mucin AgC10 inhibits T-cell activation and IL-2 transcription through L-selectin. Int. Immunol. *16*, 1365–1375.

Alizon, S., Hurford, A., Mideo, N., and Van Baalen, M. (2009). Virulence evolution and the trade-off hypothesis: history, current state of affairs and the future. J. Evol. Biol. *22*, 245–259.

Alvarez, M.N., Peluffo, G., Piacenza, L., and Radi, R. (2011). Intraphagosomal peroxynitrite as a macrophage-derived cytotoxin against internalized *Trypanosoma cruzi*: consequences for oxidative killing and role of microbial peroxiredoxins in infectivity. J. Biol. Chem. *286*, 6627–6640.

Alves, M.J., Abuin, G., Kuwajima, V.Y., and Colli, W. (1986). Partial inhibition of trypomastigote entry into cultured mammalian cells by monoclonal antibodies against a surface glycoprotein of *Trypanosoma cruzi*. Mol. Biochem. Parasitol. *21*, 75–82.

Andrade, L.O., and Andrews, N.W. (2004). Lysosomal fusion is essential for the retention of *Trypanosoma cruzi* inside host cells. J. Exp. Med. *200*, 1135–1143.

Antoine, J.C., Prina, E., Jouanne, C., and Bongrand, P. (1990). Parasitophorous vacuoles of *Leishmania amazonensis*-infected macrophages maintain an acidic pH. Infect. Immun. *58*, 779–787.

Arnoult, D., Akarid, K., Grodet, A., Petit, P.X., Estaquier, J., and Ameisen, J.C. (2002). On the evolution of programmed cell death: apoptosis of the unicellular eukaryote *Leishmania major* involves cysteine proteinase activation and mitochondrion permeabilization. Cell Death Differ. *9*, 65–81.

Araujo, F.G., and Morein, B. (1991). Immunization with *Trypanosoma cruzi* epimastigote antigens incorporated into iscoms protects against letal challenge in mice. Infect Immun. *59*, 2909–2914.

Arce-Fonseca, M., Ramos-Ligonio, A., López-Monteón, A., Salgado-Jiménez, B., Talamás Rohana, P., and Rosales-Encina, J.L. (2011). A DNA vaccine encoding for TcSSP4 induces protection against acute and chronic infection in experimental Chagas disease. Int. J. Biol. Sci. *7*, 1230–1238.

Argibay, P.F., Di Noia, J.M., Hidalgo, A., Mocetti, E., Barbich, M., Lorenti, A.S., Bustos, D., Tambutti, M., Hyon, S.H., Frasch, A.C., and Sánchez, D.O. (2002). *Trypanosoma cruzi* surface mucin TcMuc-e2 expressed on higher eukaryotic cells induces human T-cell anergy, which is reversible. Glycobiology *12*, 25–29.

Arunachalam, B., Phan, U.T., Geuze, H.J., and Cresswell, P. (2000). Enzymatic reduction of disulfide bonds in lysosomes: Characterization of a Gamma-interferon-inducible lysosomal thiol reductase (GILT). Proc. Natl. Acad. Sci. U. S. A. *97*, 745–750.

Awasthi, A., Mathur, R., Khan, A., Joshi, B.N., Jain, N., Sawant, S., Boppana, R., Mitra, D., and Saha, B. (2003). CD40 signaling is impaired in *L. major*-infected macrophages and is rescued by a p38MAPK activator establishing a host-protective memory T-cell response. J. Exp. Med. *197*, 1037–1043.

Baines, S. (1956). The role of the symbiotic bacteria in the nutrition of *Rhodnius prolixus* (Hemiptera). J. Exp Biol. *33*, 533–541.

Basyoni, M.M. (2011). *Leishmania* vaccines updates. Parasitol. United J. *5*, 1–10.

Beard, C.B., Durvasula, R.V., and Richards, F.F. (1998). Bacterial symbiosis in arthropods and the control of disease transmission. Emerging Infect. Dis. *4*, 581–591.

Becker, I., Salaiza, N., Aguirre, M., Delgado, J., Carrillo-Carrasco, N., Kobeh, L.G., Ruiz, A., Cervantes, R., Torres, A.P., Cabrera, N., Gonzalez, A., Maldonado, C., and Isibasi, A. (2003). *Leishmania* lipophosphoglycan (LPG) activates NK cells through toll-like receptor-2. Mol. Biochem. Parasitol. *130*, 65–74.

Bento, C.A., Melo, M.B., Previato, J.O., Mendonça-Previato, L., and Peçanha, L.M. (1996). Glycoinositolphospholipids purified from *Trypanosoma cruzi* stimulate Ig production *in vitro*. J. Immunol. *157*, 4996–5001.

Berman, J.D., and Gallalee, J.V. (1987). *In vitro* antileishmanial activity of inhibitors of steroid biosynthesis and combinations of antileishmanial agents. J. Parasitol. *73*, 671–673.

Berman, J.D., Edwards, N., King, M., and Grogl, M. (1989). Biochemistry of Pentostam resistant *Leishmania*. Am. J. Trop. Med. Hyg. *40*, 159–164.

Besteiro, S., Tonn, D., Tetley, L., Coombs, G.H., and Mottram, J.C. (2008). The AP3 adaptor is involved in the transport of membrane proteins to acidocalcisomes of *Leishmania*. J. Cell Sci. *121*, 561–570.

Bhattacharya, S.K., Sinha, P.K., Sundar, S., Thakur, C.P., Jha, T.K., Pandey, K., Das, V.R., Kumar, N., Lal, C., Verma, N., Singh, V.P., Ranjan, A., Verma, R.B., Anders, G., Sindermann, H., and Ganguly, N.K. (2007). Phase 4 trial of miltefosine for the treatment of Indian visceral leishmaniasis. J. Infect. Dis. *196*, 591–598.

Bhowmick, S., Ravindran, R., and Ali, N. (2008). Gp63 in stable cationic liposomes confers sustained vaccine immunity to susceptible Balb/c mice infected with *Leishmania donovani*. Infect Immun.*76*, 1003–1015.

Bishop, G.A., Moore, C.R., Xie, P., Stunz, L.L., and Kraus, Z.J. (2007). TRAF proteins in CD40 signaling. Adv. Exp. Med. Biol. *597*, 131–151.

Blackwell, J.M., Ezekowitz, R.A., Roberts, M.B., Channon, J.Y., Sim, R.B., and Gordon, S. (1985). Macrophage complement and lectin-like receptors bind *Leishmania* in the absence of serum. J. Exp. Med. *162*, 324–331.

Bohannan, B.J.M., and Lenski, R.E. (2000). Linking genetic change to community evolution: insights from studies of bacteria and bacteriophage. Ecol. Lett. *3*, 362–377.

Boibessot, I., Turner, C.M., Watson, D.G., Goldie, E., Connel, G., McIntosh, A., Grant, M.H., and Skellern, G.G. (2002). Metabolism and distribution of phenanthridine trypanocides in *Trypanosoma brucei*. Acta Trop. *84*, 219–228.

Bontempi, I.A., Vicco, M.H., Cabrera, G., Villar, S.R., González, F.B., Roggero, E.A., Ameloot, P., Callewaert, N., Pérez, A.R., and Marcipar, I.S. (2015). Efficacy of a trans-sialidase-ISCOMATRIX subunit vaccine candidate to protect against experimental Chagas disease. Vaccine *33*, 1274–1283.

Bogdan, C., and Röllinghoff, M. (1999). How do protozoan parasites survive inside macrophages? Parasitol. Today *15*, 22–28.

Breganó, J.W., Picão, R.C., Graça, V.K., Menolli, R.A., Itow Jankevicius, S., Pinge Filho, P., and Jankevicius, J.V. (2003). Phytomonas serpens, a tomato parasite, shares antigens with *Trypanosoma cruzi* that are recognized by human sera and induce protective immunity in mice. FEMS Immunol. Med. Microbiol. *39*, 257–264.

Brittingham, A., Morrison, C.J., McMaster, W.R., McGwire, B.S., Chang, K.P., and Mosser, D.M. (1995). Role of the *Leishmania* surface protease gp63 in complement fixation, cell adhesion, and resistance to complement-mediated lysis. J. Immunol. *155*, 3102–3111.

Brodskyn, C., Patricio, J., Oliveira, R., Lobo, L., Arnholdt, A., Mendonça-Previato, L., Barral, A., and Barral-Netto, M. (2002). Glycoinositolphospholipids from *Trypanosoma cruzi* interfere with macrophages and dendritic cell responses. Infect. Immun. *70*, 3736–3743.

Brack, C., Delain, E., Riou, G., and Festy, B. (1972). Molecular organization of the kinetoplast DNA of *Trypanosoma cruzi* treated with berenil, a DNA interacting drug. J. Ultrastruct. Res. *39*, 568–579.

Buckner, F.S., Wipke, B.T., and Van Voorhis, W.C. (1997). *Trypanosoma cruzi* infection does not impair major histocompatibility complex class I presentation of antigen to cytotoxic T-lymphocytes. Eur. J. Immunol. *27*, 2541–2548.

Burchmore, R.J., Rodriguez-Contreras, D., McBride, K., Merkel, P., Barrett, M.P., Modi, G., Sacks, D., and Landfear, S.M. (2003). Genetic characterization of glucose transporter function in *Leishmania mexicana*. Proc. Natl. Acad. Sci. U.S.A. *100*, 3901–3906.

Buscaglia, C.A., Campo, V.A., Frasch, A.C., and Di Noia, J.M. (2006). *Trypanosoma cruzi* surface mucins: host-dependent coat diversity. Nat. Rev. Microbiol. *4*, 229–236.

Caler, E.V., Morty, R.E., Burleigh, B.A., and Andrews, N.W. (2000). Dual role of signaling pathways leading to Ca(2+) and cyclic AMP elevation in host cell invasion by *Trypanosoma cruzi*. Infect. Immun. *68*, 6602–6610.

Carregaro, V., Costa, D.L., Brodskyn, C., Barral, A.M., Barral-Netto, M., Cunha, F.Q., and Silva, J.S. (2013). Dual effect of Lutzomyia longipalpis saliva on *Leishmania braziliensis* infection is mediated by distinct saliva-induced cellular recruitment into Balb/c mice ear. BMC Microbiol. *13*, 102–112.

Campbell, K.A., Ovendale, P.J., Kennedy, M.K., Fanslow, W.C., Reed, S.G., and Maliszewski, C.R. (1996). CD40 ligand is required for protective cell-mediated immunity to *Leishmania major*. Immunity *4*, 283–289.

Campos, M.A., Almeida, I.C., Takeuchi, O., Akira, S., Valente E.P., Procópio, D.O., Travassos, L.R., Smith, J.A., Golenbock, D.T., and Gazzinelli, R.T. (2001). Activation of Toll-like receptor-2 by glycosylphosphatidylinositol anchors from a protozoan parasite. J. Immunol. *167*, 416–423.

Castellano, L.R., Filho, D.C., Argiro, L., Dessein, H., Prata, A., Dessein, A., and Rodrigues, V. (2009). Th1/Th2 immune responses are associated with active cutaneous leishmaniasis and clinical cure is associated with strong interferon-γ production. Hum. Immunol. *70*, 383– 390.

Cazorla, S.I., Frank, F.M., Becker, P.D., Arnaiz, M., Mirkin, G.A., Corral, R.S., Guzmán, C.A., and Malchiodi, E.L. (2010). Redirection of the immune response to the functional catalytic domain of the cystein proteinase cruzipain improves protective immunity against *Trypanosoma cruzi* infection. J. Infect. Dis. *202*, 136–144.

Cazorla, S.I., Matos, M.N., Cerny, N., Ramirez, C., Alberti, A.S., Bivona, A.E., Morales, C., Guzmán, C.A., and Malchiodi, E.L. (2014). Oral multicomponent DNA vaccine delivered by attenuated *Salmonella* elicited immunoprotection against American trypanosomiasis. J. Infect. Dis. *211*, 698–707.

Chakraborty, D., Banerjee, S., Sen, A., Banerjee, K.K., Das, P., and Roy, S. (2005). *Leishmania donovani* affects antigen presentation of macrophage by disrupting lipid rafts. J. Immunol. *175*, 3214–3224.

Chakraborty, S., Srivastava, A., Jha, M.K., Nair, A., Pandey, S.P., Srivastava, N., Kumari, S., Singh, S., Krishnasastry, M.V., and Saha, B. (2015). Inhibition of CD40-induced N-Ras activation reduces *Leishmania major* infection. J. Immunol. *194*, 3852–3860.

Chan, J., Fujiwara, T., Brennan, P., McNeil, M., Turco, S.J., Sibille, J.C., Snapper, M., Aisen, P., and Bloom, B.R. (1989). Microbial glycolipids: possible virulence factors that scavenge oxygen radicals. Proc. Natl. Acad. Sci. U.S.A. *86*, 2453–2457.

Chandra, D., and Naik, S. (2008). *Leishmania donovani* infection down-regulates TLR2-stimulated IL-12p40 and activates IL-10 in cells of macrophage/monocytic lineage by modulating MAPK pathways through a contact-dependent mechanism. Clin. Exp. Immunol. *154*, 224–234.

Chaussabel, D., Pajak, B., Vercruysse, V., Bisseyé, C., Garzé, V., Habib, M., Goldman, M., Moser, M., and Vray, B. (2003). Alteration of migration and maturation of dendritic cells and T-cell depletion in the course of experimental *Trypanosoma cruzi* infection. Lab. Invest. *83*, 1373–1382.

Chesson, P. (2000). Mechanism of maintenance of species diversity. Annu. Rev. Ecol. Syst. *31*, 343–366.

Chuenkova, M.V., Furnari, F.B., Cavenee, W.K., and Pereira, M.A. (2001). *Trypanosoma cruzi* trans-sialidase: a potent and specific survival factor for human Schwann cells by means of phosphatidylinositol 3-kinase/Akt signaling. Proc. Natl. Acad. Sci. U.S.A. *98*, 9936–9941.

Coelho, P.S., Klein, A., Talvani, A., Coutinho, S.F., Takeuchi, O., Akira, S., Silva, J.S., Canizzaro, H., Gazzinelli, R.T., and Teixeira, M.M. (2002). Glycosylphosphatidylinositol-anchored mucin-like glycoproteins isolated from *Trypanosoma cruzi* trypomastigotes induce *in vivo* leukocyte recruitment dependent on MCP-1 production by IFN-gamma-primed-macrophages. J. Leukoc. Biol. *71*, 837–844.

Cobey, S. (2014). Pathogen evolution and the immunological niche. Ann. N. Y. Acad. Sci. *1320*, 1–15.

Coler, R.N., Goto, Y., Bogatzki, L., Raman, V., and Reed, S.G. (2007). Leish-111f, a recombinant polyprotein vaccine that protects against visceral leishmaniasis by elicitation of CD4$^+$ T-cells. Infect Immun. *75*, 4648–4654.

Copeland, N.K., and Aronson, N.E. (2015). Leishmaniasis: treatment updates and clinical practice guidelines review. Curr. Opin. Infect. Dis. *28*, 426–437.

Croft, S.L., and Coombs, G.H. (2003). Leishmaniasis – current chemotherapy and recent advances in the search for novel drugs. Trends Parasitol. *19*, 502–508.

Croft, S.L., Sundar, S., and Fairlamb, A.H. (2006). Drug resistance in leishmaniasis. Clin. Microbiol. Rev. *19*, 111–126.

Das, S., Freier, A., Boussoffara, T., Das, S., Oswald, D., Losch, F.O., Selka, M., Sacerdoti-Sierra, N., Schönian, G., Wiesmüller, K.H., Seifert, K., Schroff, M., Juhls, C., Jaffe, C.L., Roy, S., Das, P., Louzir, H., Croft, S.L., Modabber, F., and Walden, P. (2014). Modular multiantigen T-cell epitope-enriched DNA vaccine against human leishmaniasis. Sci. Transl. Med. *6*, 234–256.

Dávila, A.M., and Tyler, K.M. (2002). Combating Kinetoplastid diseases. Kinetoplastid Biol. Dis. *1*, 6–7.

Dawkins, R., and Krebs, J.R. (1979). Arms races between and within species. Proc. R. Soc. Lond. B. Biol. Sci. *205*, 489–511.

De Almeida, M.C., Cardoso, S.A., and Barral-Netto, M. (2003). *Leishmania* (*Leishmania*) *chagasi* infection alters the expression of cell adhesion and costimulatory molecules on human monocyte and macrophage. Int. J Parasitol. *33*, 153–162.

De Moura, T.R., Oliveira, F., Carneiro, M.W., Miranda, J.C., Clarêncio, J., Barral-Netto, M., Brodskyn, C., Barral, A., Ribeiro, J.M., Valenzuela, J.G., and de Oliveira, C.I. (2013). Functional transcriptomics of wild-caught Lutzomyia intermedia salivary glands: Identification of a protective salivary protein against *Leishmania braziliensis* infection. PLOS Negl. Trop. Dis. 7, e2242.

Depledge, D.P., MacLean, L.M., Hodgkinson, M.R., Smith, B.A., Jackson, A.P., Ma, S., Uliana, S.R., and Smith, D.F. (2010). *Leishmania*-specific surface antigens show sub-genus sequence variation and immune recognition. PLOS Negl. Trop. Dis. *4*, e829.

Dermine, J.F., Goyette, G., Houde, M., Turco, S.J., and Desjardins, M. (2005). *Leishmania donovani* lipophosphoglycan disrupts phagosome microdomains in J774 macrophages. Cell. Microbiol. 7, 1263–1270.

Descoteaux, A., and Turco, S.J. (1999). Glycoconjugates in *Leishmania* infectivity. Biochim. Biophys. Acta *1455*, 341–352.

Descoteaux, A., and. Turco, S.J. (2002). Functional aspects of the *Leishmania donovani* lipophosphoglycan during macrophage infection. Microbes and Infect. *4*, 975–981.

Desjardins, M., and Descoteaux, A. (1997). Inhibition of phagolysosomal biogenesis by the *Leishmania* lipophosphoglycan. J. Exp. Med. *185*, 2061–2068.

De Souza Leao, S., Lang, T., Prina, E., Hellio, R., and Antoine J.C. (1995). Intracellular Leishmania amazonensis amastigotes internalize and degrade MHC class II molecules of their host cells. J. Cell Sci. *108*, 3219–3231.

de Veer, M.J., Curtis, J.M., Baldwin, T.M., DiDonato, J.A., Sexton, A., McConville, M.J., Handman, E., and Schofield, L. (2003). MyD88 is essential for clearance of *Leishmania major*: possible role for lipophosphoglycan and Toll-like receptor 2 signaling. Eur. J. Immunol. *33*, 2822–2831.

Docampo, R., Vercesi, A.E., and Huang, G. (2014). Mitochondrial calcium transport in trypanosomes. Mol. Biochem. Parasitol. *196*, 108–116.

Docampo, R., and Huang, G. (2015). Calcium signaling in trypanosomatid parasites. Cell Calcium *57*, 194–202.

Doyle, P.S., Zhou, Y.M., Hsieh, I., Greenbaum, D.C., McKerrow, J.H., and Engel, J.C. (2011). The *Trypanosoma cruzi* protease cruzain mediates immune evasion. PLOS Pathog. *7*, e1002139.

DosReis, G.A. (1997). Cell-mediated immunity in experimental Trypanosoma *cruzi* infection. Parasitol. Today *13*, 335–342.

Dumas, C., Ouellette, M., Tovar, J., Cunningham, M.L., Fairlamb, A.H., Tamar, S., Olivier, M., and Papadopoulou, B. (1997). Disruption of the trypanothione reductase gene of *Leishmania* decreases its ability to survive oxidative stress in macrophages. EMBO J. *16*, 2590–2598.

Dunne, J.A., Lafferty, K.D., Dobson, A.P., Hechinger, R.F., Kuris, A.M., Martinez, N.D., McLaughlin, J.P., Mouritsen, K.N., Poulin, R., Reise, K., Stouffer, D.B., Thieltges, D.W., Williams, R.J., and Zander, C.D. (2013). Parasites affect food web structure primarily through increased diversity and complexity. PLOS Biol. 11, e1001579.

Durvasula, R.V., Gumbs, A., Panackal, A., Kruglov, O., Aksoy, S., Merrifield, R.B., Richards, F.F., and Beard, C.B. (1997). Prevention of insect-borne disease: an approach using transgenic symbiotic bacteria. Proc. Natl. Acad. Sci. U.S.A. *94*, 3274–3278.

Dvorak, J.A. (1984). The natural heterogeneity of *Trypanosoma cruzi*: biological and medical implications. J. Cell. Biochem. *24*, 357–371.

Engman, D.M., and Leon, J.S. (2002). Pathogenesis of Chagas heart disease: role of autoimmunity. Acta Trop. *81*, 123–132.

Erdmann, H., Steeg, C., Koch-Nolte, F., Fleischer, B., and Jacobs, T. (2009). Sialylated ligands on pathogenic *Trypanosoma cruzi* interact with Siglec-E (sialic acid-binding Ig-like lectin-E). Cell. Microbiol. *11*, 1600–1611.

Ermak, G., and Davies, K.J. (2002). Calcium and oxidative stress: from cell signaling to cell death. Mol. Immunol. *38*, 713–721.

Ersfeld, K., Barraclough, H., and Gull, K. (2005). Evolutionary relationships and protein domain architecture in an expanded calpain superfamily in kinetoplastid parasites. J. Mol. Evol. *61*, 742–757.

Ferreira, V., Valck, C., Sanchez, G., Gingras, A., Tzima, S., Molina, M.C., Sim, R., Schwaeble, W., and Ferreira, A. (2004). The classical activation pathway of the human complement system is specifically inhibited by calreticulin from *Trypanosoma cruzi*. J. Immunol. *172*, 3042–3050.

Finley, R.W., and Dvorak, J.A. (1987). *Trypanosoma cruzi*: analysis of the population dynamics of heterogeneous mixtures. J. Protozool. *34*, 409–415.

Flandin, J.F., Chano, F., and Descoteaux, A. (2006). RNA interference reveals a role for TLR2 and TLR3 in the recognition of *Leishmania donovani* promastigotes by interferon-gamma-primed macrophages. Eur. J. Immunol. *36*, 411–420.

Forget, G., Gregory, D.J., Whitcombe, L.A., and Olivier, M. (2006). Role of host protein tyrosine phosphatase SHP-1 in *Leishmania donovani*-induced inhibition of nitric oxide production. Infect Immun. *74*, 6272–6279.

Freire-de-Lima, L., Alisson-Silva, F., Carvalho, S.T., Takiya, C.M., Rodrigues, M.M., DosReis, G.A., Mendonça-Previato, L., Previato, J.O., and Todeschini, A.R. (2010). *Trypanosoma cruzi* subverts host cell sialylation and may compromise antigen-specific CD8$^+$ T-cell responses. J. Biol. Chem. *285*, 13388–13396.

Fridovich, I. (1995). Superoxide radical and superoxide dismutases. Annu. Rev. Biochem. *64*, 97–112.

Garcia, C.A., Oliveira, E.C., Sakurada. J.K., and Santos. L.M.B. (2000). Protective immunity induced by a *Trypanosoma cruzi* soluble extract antigen in experimental Chagas' disease. Role of interferon γ. Immunol Invest. *29*, 1–12.

Garzon, E., Borges, M.C., Cordeiro-Da-Silva, A., Nacife, V., Meirelles, M.N., Guilvard, E., Bosseno, M.F., Guevara, A.G., Brenière, S.F., and Ouaissi, A. (2003). *Trypanosoma cruzi* carrying a targeted deletion of a Tc52 protein-encoding allele elicits attenuated Chagas' disease in mice. Immunol Lett. *89*, 67–80.

Gazzinelli, R.T., Pereira, M.E., Romanha, A., Gazzinelli, G., and Brener, Z. (1991). Direct lysis of *Trypanosoma cruzi*: a novel effector mechanism of protection mediated by human anti-gal antibodies. Parasite Immunol. *13*, 345–356.

Gazzinelli, R.T., and Denkers, E.Y. (2006). Protozoan encounters with Toll-like receptor signalling pathways: implications for host parasitism. Nat. Rev. Immunol. *6*, 895–906.

Ghosh, S., Bhattacharyya, S., Sirkar, M., Sa, G.S., Das, T., Majumdar, D., Roy, S., and Majumdar, S. (2002). *Leishmania donovani* suppresses activated protein 1 and NF-κB activation in host macrophages via ceramide generation: involvement of extracellular signal-regulated kinase. Infect. Immun. *70*, 6828–6838.

Ghosh, S., Goswami, S., and Adhya, S. (2003). Role of superoxide dismutase in survival of *Leishmania* within the macrophage. Biochem. J. *369*, 447–452.

Gomes, N.A., Previato, J.O., Zingales, B., Mendonca-Previato, L., and DosReis, G.A. (1996). Down-regulation of T lymphocyte activation *in vitro* and *in vivo* induced by glycoinositolphospholipids from *Trypanosoma cruzi*. Assignment of the T-cell-suppressive determinant to the ceramide domain. J Immunol. *156*, 628–635.

Gomes, Y.M., Abath, F.G.C., Nakazawa, M., Minoprio, P., Vouldoukis, I., and Monjour, L. (1999). Partial protection of mice against *Trypanosoma cruzi* after immunizing with the TcY 72 antigenic preparation. Mem Inst Oswaldo Cruz. *94*, 167–172.

Gordeeva, A.V., Zvyagilskaya, R.A., and Labas, Y.A. (2003). Cross-talk between reactive oxygen species and calcium in living cells. Biochemistry Mosc. *68*, 1077–1080.

Gobert, A.P., Daulouede, S., Lepoivre, M., Boucher, J.L., Bouteille, B., Buguet, A., Cespuglio, R., Veyret, B., and Vincendeau, P. (2000). L-Arginine availability modulates local nitric oxide production and parasite killing in experimental trypanosomiasis. Infect. Immun. *68*, 4653–4657.

Gruppi, A., Cerban, F., Pistoresi-Palencia, M.C., and Vottero-Cima, E. (1995). *Trypanosoma cruzi*: transfer of protection by lymph node cells obtained from mice immunized with exoantigens of pI 4.5. Exp. Parasitol. *80*, 382–389.

Gupta, S., Wen, J.J., and Garg, N.J. (2009). Oxidative Stress in Chagas Disease. Interdiscip Perspect Infect Dis *2009*, 190354.

Gurunathan, S., Sacks, D.L., Brown, D.R., Reiner, S.L., Charest, H., Glaichenhaus, N., and Seder, R.A. (1997). Vaccination with DNA encoding the immunodominant LACK parasite antigen confers protective immunity to mice infected with *Leishmania major*. J. Exp. Med. *186*, 1137–1147.

Gutierrez, F.R., Mineo, T.W., Pavanelli, W.R., Guedes, P.M., and Silva, J.S. (2009). The effects of nitric oxide on the immune system during *Trypanosoma cruzi* infection. Mem. Inst. Oswaldo Cruz. *104* (Suppl. 1), 236–245.

Hall, B.F., Webster, P., Ma, A.K., Joiner, K.A., and Andrews, N.W. (1992). Desialylation of lyso-somal membrane glycoproteins by *Trypanosoma cruzi*: a role for the surface neuraminidase in facilitating parasite entry into the host cell cytoplasm. J. Exp. Med. *176*, 313–325.

Hall, B.S., Tam, W., Sen, R., and Pereira, M.E. (2000). Cell-specific activation of nuclear factor-kappaB by the parasite *Trypanosoma cruzi* promotes resistance to intracellular infection. Mol. Biol. Cell. *11*, 153–160.

Handman, E. (2001). Leishmaniasis: current status of vaccine development. Clin. Microbiol. Rev. *14*, 229–243.

Hawking, F. (1940). Concentration of Bayer 205 (Germanin) in human blood and cerebrospinal fluid after treatment. Trans. R. Soc. Trop. Med. Hyg. *34*, 37–52.

Hermoso, T., Fishelson, Z., Becker, S.I., Hirschberg, K., and Jaffe, C.L. (1991). Leishmanial protein kinases phosphorylate components of the complement system. EMBO J. *10*, 4061–4067.

Hillesland, H., Read, A., Subhadra, B., Hurwitz, I., McKelvey, R., Ghosh, K., Das, P., and Durvasula, R. (2008). Identification of aerobic gut bacteria from the kala azar vector, Phlebotomus argentipes: a platform for potential paratransgenic manipulation of sand flies. Am. J. Trop. Med. Hyg. *79*, 881–886.

Hoare, C.A. (1966). The classification of mammalian trypanosomes. Ergeb.Mikrobiol. Immunitatsforsch. Exp. Ther. *39*, 43–57.

Hoft, D.F., Lynch, R.G., and Kirchhoff, L.V. (1993). Kinetic Analysis of Antigen-Specific Immune Responses in Resistant and Susceptible Mice During Infection with *Trypanosoma cruzi*. J. Immunol. *151*,7038–47.

Hübel, A., Krobitsch, S., Hörauf, A., and Clos, J. (1997). *Leishmania major* Hsp100 is required chiefly in the mammalian stage of the parasite. Mol. Cell. Biol. *17*, 5987–5995.

Hunter, C.A., Ellis-Neyes, L.A., Slifer, T., Kanaly, S., Grünig, G., Fort, M., Rennick, D., and Araujo, F.G. (1997). IL-10 is required to prevent immune hyperactivity during infection with *Trypanosoma cruzi*. J. Immunol. *158*, 3311–3316.

Iborra, S., Soto, M., Stark-Aroeira, L., Castellano, E., Alarcón, B., Alonso, C., Santos, E., and Fernández-Malavé, E. (2011). H-ras and N-ras are dispensable for T-cell development and activation but critical for protective Th1 immunity. Blood *117*, 5102–5111.

Iniesta, V., Carcelen, J., Molano, I., Peixoto, P.M., Redondo, E., Pilar, P., Marina, M., Isabel, M., Maria, C., Carlos, G N., and Inés, C. (2005). Arginase I induction during *Leishmania major* infection mediates the development of disease. Infect. Immun. *73*, 6085–6090.

Jackson, A.P. (2015). Genome evolution in trypanosomatid parasites. Parasitology *142* (Suppl. 1), 40–56.

Jha, T.K., Sundar, S., Thakur, C.P., Bachmann, P., Karbwang, J., Fischer, C., Voss, A., and Berman, J. (1999). Miltefosine, an oral agent, for the treatment of Indian visceral leishmaniasis. N. Engl. J. Med. *341*, 1795–1800.

Joiner, K., Sher, A., Gaither, T., and Hammer, C. (1986). Evasion of alternative complement pathway by *Trypanosomacruzi* results from inefficient binding of factor B. Proc. Natl. Acad. Sci. U.S.A. *83*, 6593–6597.

Joubert, K.E., Kettner, F., Lobetti, R.G., and Miller, D.M. (2003). The effects of diminazene aceturate on systemic blood pressure in clinically healthy adult dogs. J S Afr Vet Assoc *74*, 69–71.

Junqueira, C., Caetano, B., Bartholomeu, D.C., Melo, M.B., Ropert, C., Rodrigues, M.M., and Gazzinelli, R.T. (2010). The endless race between *Trypanosoma cruzi* and host immunity: lessons for and beyond Chagas disease. Expert Rev Mol Med *12*, e29.

Kane, M.M., and Mosser, D.M. (2000). *Leishmania* parasites and their ploys to disrupt macrophage activation. Curr. Opin. Hematol. *7*, 26–31.

Kane, M.M., and Mosser, D.M. (2001). The role of IL-10 in promoting disease progression in leishmaniasis. J. Immunol. *166*, 1141–1147.

Katzman, S.D., and Fowell, D.J. (2008). Pathogen-imposed skewing of mouse chemokine and cytokine expression at the infected tissue site. J. Clin. Invest. *118*, 801–811.

Kaye, P.M., Rogers, N.J., Curry, A.J., and Scott, J.C. (1994). Deficient expression of co-stimulatory molecules on *Leishmania*-infected macrophages. Eur. J. Immunol. *24*, 2850–2854.

Kierszenbaum, F. (1999). Chagas' disease and the autoimmunity hypothesis. Clin. Microbiol. Rev. *12*, 210–223.

Kierszenbaum, F., de Diego, J.L., Fresno, M., and Sztein, M.B. (1999). Inhibitory effects of the *Trypanosoma cruzi* membrane glyco-protein AGC10 on the expression of IL-2 receptor chains and secretion of cytokines by subpopulations of activated human T-lymphocytes. Eur. J. Immunol. *29*, 1684–1691.

Kierszenbaum, F., Fresno, M., and Sztein, M.B. (2002).The *Trypanosoma cruzi* membrane glycoprotein AGC10 inhibits human lymphocyte activation by a mechanism preceding translation of both, interleukin-2 and its high-affinity receptor subunits,' Mol Biochem Parasitol. *125*, 91–101.

Kipnis, T.L., David, J.R., Alper, C.A., Sher, A., and da Silva, W.D. (1981). Enzymatic treatment transforms trypomastigotes of *Trypanosoma cruzi* into activators of alternative complement pathway and potentiates their uptake by macrophages. Proc Natl Acad Sci U S A. *78*, 602–605.

Kraaijeveld, A.R., and Godfray, H.C.J. (1997). Trade-off between parasitoid resistance and larval competitive ability in Drosophila melanogaster. Nature.*389*, 278–280.

Kropf, P., Freudenberg, N., Kalis, C., Modolell, M., Herath, S., Galanos, C., Freudenberg, M., and Müller, I. (2004a). Infection of C57BL/10ScCr and C57BL/10ScNCr mice with *Leishmania major* reveals a role for Toll-like receptor 4 in the control of parasite replication. J. Leukoc. Biol. *76*, 48–57.

Kropf, P., Freudenberg, M.A., Modolell, M., Price, H.P., Herath, S., Antoniazi, S., Galanos, C., Smith, D.F., and Müller, I. (2004b). Toll-like receptor 4 contributes to efficient control of infection with the protozoan parasite *Leishmania major*. Infect. Immun. 72, 1920–1928.

Koga, R., Hamano, S., Kuwata, H., Atarashi, K., Ogawa, M., Hisaeda, H., Yamamoto, M., Akira, S., Himeno, K., Matsumoto, M., and Takeda, K. (2006). TLR-dependent induction of IFN-beta mediates host defense against *Trypanosoma cruzi*. J. Immunol. *177*, 7059–7066.

Kumar, S., and Tarleton, R.L. (1998). The relative contribution of antibody production and CD8$^+$ T-cell function to immune control of *Trypanosoma cruzi*. Parasite Immunol. *20*, 207–216.

Lang, T., de Chastellier, C., Frehel, C., Hellio, R., Metezeau, P., Leao, S.d.e.S., and Antoine, J.C. (1994). Distribution of MHC class I and of MHC class II molecules in macrophages infected with *Leishmania amazonensis*. J. Cell. Sci. *107*, 69–82.

Latz, E., Schoenemeyer, A., Visintin, A., Fitzgerald, K.A., Monks, B.G., Knetter, C.F., Lien, E., Nilsen, N.J., Espevik, T., and Golenbock, D.T. (2004). TLR9 signals after translocating from the ER to CpG DNA in the lysosome. Nat. Immunol. *5*, 190–198.

Leach, T.M., and Roberts, C.J. (1981). Present status of chemotherapy and chemoprophylaxis of animal trypanosomiasis in the Eastern hemisphere. Pharmacol Ther. *13*, 91–147.

Liniger, M., Bodenmüller, K., Pays, E., Gallati, S., and Roditi, I. (2001). Overlapping sense and antisense transcription units in *Trypanosoma brucei*. Mol. Microbiol. *40*, 869–878.

Lipsitch, M. (1999). Bacterial vaccines and serotype replacement: lessons from *Haemophilus influenzae* and prospects for *Streptococcus pneumoniae*. Emerging Infect. Dis. *5*, 336–345.

Lo, S.K., Bovis, L., Matura, R., Zhu, B., He, S., Lum, H., Turco, S.J., and Ho, J.L. (1998). *Leishmania* lipophosphoglycan reduces monocyte transendothelial migration: modulation of cell adhesion molecules, intercellular junctional proteins, and chemoattractants. J. Immunol. *160*, 1857–1865.

Low, H.P., Paulin, J.J., and Keith, C.H. (1992).*Trypanosoma Cruzi* infection of BSC-1 fibroblast cells causes cytoskeletal disruption and changes in intracellular calcium levels. J Protozool *39*: 463–470.

Lozano, J., Berra, E., Municio, M.M., Diaz-Meco, M.T., Dominguez, I., Sanz, L., and Moscat, J. (1994). Protein kinase C zeta isoform is critical for kappa B-dependent promoter activation by sphingomyelinase. J. Biol. Chem. *269*, 19200–19202.

Luo, J.H., and Weinstein, I.B. (1993). Calcium-dependent activation of protein kinase C. The role of the C2 domain in divalent cation selectivity. J. Biol. Chem. *268*, 23580–23584.

Luhrs, K.A., Fouts, D.L., and Manning, J.E. (2003). Immunization with recombinant paraflagellar rod protein induces protective immunity against *Trypanosoma cruzi* infection. Vaccine. *21*, 3058–3069.

Maarouf, M., Lawrence, F., Croft, S.L., and Robert-Gero, M. (1995). Ribosomes of *Leishmania* are a target for the aminoglycosides. Parasitol. Res. *81*, 421–425.

MacMicking, J., Xie, Q.W., and Nathan, C. (1997). Nitric oxide and macrophage function. Annu. Rev. Immunol. *15*, 323–350.

Maleckar, J.R., and Kierszenbaum, F. (1983). Inhibition of mitogen-induced proliferation of mouse T and B lymphocytes by bloodstream forms of *Trypanosoma cruzi*. J. Immunol. *130*, 908–911.

Marques-da-Silva, E.A., Coelho, E.A., Gomes, D.C., Vilela, M.C., Masioli, C.Z., Tavares, C.A., Fernandes, A.P., Afonso, L.C., and Rezende, S.A. (2005). Intramuscular immunization with p36(LACK) DNA vaccine induces IFN-gamma production but does not protect BALB/c mice against *Leishmania chagasi* intravenous challenge. Parasitol Res.*98*, 67–74.

Marroquin-Quelopana, M., Oyama, S., Aguiar Pertinhez, T., Spisni, A., Aparecida Juliano, M., Juliano, L., Colli, W., and Alves, M.J. (2004). Modeling the *Trypanosoma cruzi* Tc85-11 protein and mapping the laminin-binding site. Biochem. Biophys. Res. Commun. *325*, 612–618.

Martin, D.L., Weatherly, D.B., Laucella, S.A., Cabinian, M.A., Crim, M.T., Sullivan, S., Heiges, M., Craven, S.H., Rosenberg, C.S., Collins, M.H., Sette, A., Postan, M., and Tarleton, R.L. (2006). CD8$^+$ T-cell responses to *Trypanosoma cruzi* are highly focused on strain-variant trans-sialidase epitopes. PLOS Pathog. 2, e77.

Martiny, A., Meyer-Fernandes, J.R., De Souza, W., and Vannier-Santos, M.A. (1999). Altered tyrosine phosphorylation of ERK1 MAP kinase and other macrophage molecules caused by *Leishmania* amastigotes. Mol. Biochem. Parasitol.*102*, 1–12.

Mathur, R.K., Awasthi, A., Wadhone, P., Ramanamurthy, B., and Saha, B. (2004). Reciprocal CD40 signals through p38MAPK and ERK-1/2 induce counteracting immune responses. Nat. Med. *10*, 540–544.

Mattner, J., Donhauser, N., Werner-Felmayer, G., and Bogdan, C. (2006). NKT-cells mediate organ-specific resistance against *Leishmania major* infection. Microbes Infect.*8*(2):354–62.

Mazumder, S., Maji, M., and Ali, N. (2011). Potentiating effects of MPL on DSPC bearing cationic liposomes promote recombinant GP63 vaccine efficacy: high immunogenicity and protection. PLOS Negl. Trop. Dis *5*, e1429.

McCahill, A., Warwicker, J., Bolger, G.B., Houslay, M.D., and Yarwood, S.J. (2002). The RACK1 scaffold protein: a dynamic cog in cell response mechanisms. Mol. Pharmacol. *62*, 1261–1273.

McGonigle, S., Dalton, J.P., and James, E.R. (1998). Peroxidoxins: a new antioxidant family. Parasitol. Today *14*, 139–145.

McConville, M.J., Turco, S.J., Ferguson, M.A.J., and Sacks, D.L. (1992). Developmental modification of lipophosphoglycan during the differentiation of *Leishmania-major* promastigotes to an infectious stage. EMBO J. *11*, 3593–3600.

McConville, M.J., Schnur, L.F., Jaffe, C., and Schneider, P. (1995). Structure of *Leishmania* lipophosphoglycan: inter- and intra-specific polymorphism in Old World species. Biochem. J. *310*, 807–818.

McCulloch, R., and Horn, D. (2009). What has DNA sequencing revealed about the VSG expression sites of African trypanosomes? Trends Parasitol. *25*, 359–363.

Melby, P.C., Yang, J., Zhao, W., Perez, L.E., and Cheng, J. (2001). *Leishmania donovani* p36(LACK) DNA vaccine is highly immunogenic but not protective against experimental visceral leishmaniasis. Infect Immun. *69*, 4719–4725.

Menezes, H. (1968). Protective effect of an avirulent (cultivated) strain of *Trypanosoma cruzi* against experimental infection in mice. Rev. Inst. Med. Trop. Sao Paulo. *10*,1–4.

Mikami, A., Imoto, K., Tanabe, T., Niidome, T., Mori, Y., Takeshima, H., Narumiya, S., and Numa, S. (1989). Primary structure and functional expression of the cardiac dihydropyridine-sensitive calcium channel. Nature *340*,230–233.

Miller, M.J., Wrightsman, R.A., and Manning, J.E. (1996). *Trypanosoma cruzi*: protective immunity in mice immunized with paraflagellar rod proteins is associated with a T-helper type 1 response. Exp. Parasitol. *84*, 156–167.

Miret, J., Nascimento, E., Sampaio, W., França, J.C., Fujiwara, R.T., Vale, A., Dias, E.S., Vieira, E., da Costa, R.T., Mayrink, W., Campos Neto, A., and Reed, S. (2008). Evaluation of an immunochemotherapeutic protocol constituted of N-methyl meglumine antimoniate (Glucantime) and the recombinant Leish-110f + MPL-SE vaccine to treat canine visceral leishmaniasis. Vaccine. *26*, 1585–1594.

Morales, M.A., Watanabe, R., Dacher, M., Chafey, P., Osorio y Fortea, J., Scott, D.A., Beverley, S.M., Ommen, G., Clos, J., Hem, Lenormand, S.P., Rousselle, J.C., Namane, A., and Späth, G.F. (2010). Phosphoproteome dynamics reveal heat-shock protein complexes specific to the *Leishmania donovani* infectious stage. Proc. Natl. Acad. Sci. U.S.A. *107*, 8381–8386.

Moreno, S.N., Docampo, R., Mason, R.P., Leon, W., and Stoppani A.O. (1982). Different behaviors of benznidazole as free radical generator with mammalian and *Trypanosoma cruzi* microsomal preparations. Arch. Biochem. Biophys. *218*, 585–591.

Moreno, S.N., Silva, J., Vercesi, A.E., and Docampo, R. (1994). Cytosolic-free calcium elevation in *Trypanosoma cruzi* is required for cell invasion. J. Exp. Med. *180*, 1535–1540.

Mosser, D.M., and Edelson, P.J. (1987). The third component of complement (C3) is responsible for the intracellular survival of *Leishmania major*. Nature *327*, 329–331.

Mottram, J.C., and Coombs, G.H. (1998). *Leishmania* cysteine proteinases: virulence factors in quest of a function-reply. Parasitol. Today *14*, 251–252.

Mottram, J.C., Coombs, G.H., and Alexander, J. (2004). Cysteine peptidases as virulence factors of *Leishmania*. Curr. Opin. Microbiol. *7*, 375–381.

Mucci, J., Risso, M.G., Leguizamón, M.S., Frasch, A.C., and Campetella, O. (2006). The *trans*-sialidase from *Trypanosoma cruzi* triggers apoptosis by target cell sialylation. Cell. Microbiol. *8*, 1086–1095.

Mukkada, A.J., Meade, J.C., Glaser, T.A., and Bonventre, P.F. (1985). Enhanced metabolism of *Leishmania donovani* amastigotes at acid pH: an adaptation for intracellular growth. Science *229*, 1099–1101.

Mukherjee, S., Bandyapadhyay, R., and Basu, M.K. (1988). *Leishmania donovani*: superoxide dismutase level in infected macrophages. Biosci. Rep. *8*, 131–137.

Mukherjee, A., Roy, S., Saha, B., and Mukherjee, D. (2016). Spatio-Temporal Regulation of PKC Isoforms Imparts Signaling Specificity. Front Immunol 7, 45.

Munder, M., Eichmann, K., Morán, J.M., Centeno, F., Soler, G., and Modolell, M. (1999). Th1/Th2-regulated expression of arginase isoforms in murine macrophages and dendritic cells. J. Immunol. *163*, 3771–3777.

Muñoz-Fernández, M.A., Fernández, M.A., and Fresno, M. (1992). Activation of human macrophages for the killing of intracellular *Trypanosoma cruzi* by TNF-alpha and IFN-gamma through a nitric oxide-dependent mechanism. Immunol. Lett. *33*, 35–40.

Muraille, E., De Trez, C., Brait, M., De Baetselier, P., Leo, O., and Carlier, Y. (2003). Genetically resistant mice lacking MyD88-adapter protein display a high susceptibility to *Leishmania major* infection associated with a polarized Th2 response. J. Immunol. *170*, 4237–4241.

Murray, H.W., and Nathan, C. F.(1999). Macrophage Microbicidal Mechanisms *In vivo*: Reactive Nitrogen versus Oxygen Intermediates in the Killing of Intracellular Visceral *Leishmania donovani*. J. Exp. Med. *189*, 741–746.

Murray, H.W. (2000). Treatment of visceral leishmaniasis (kala-azar): a decade of progress and future approaches. Int. J. Infect. Dis. *4*, 158–177.

Murray, H.W., Xiang, Z., and Ma, X. (2006). Responses to *Leishmania donovani* in mice deficient in both phagocyte oxidase and inducible nitric oxide synthase. Am. J. Trop. Med. Hyg. *74*, 1013–1015.

Musa, M.M., Abdoon, A.M., Nasir, B.T., Salim, Y.I., Abdel-Rahman, A.Y., and Shommein, A.M. (1994). Efficacy of Cymelarsan in the treatment of natural chronic *Trypanosoma evansi* infection in camels in the Sudan. Rev. Elev. Med. Vet. Pays Trop. *47*, 397–400.

Myler, P.J., Audleman, L., deVos, T., Hixson, G., Kiser, P., Lemley, C., Magness, C., Rickel, E., Sisk, E., Sunkin, S., Swartzell, S., Westlake, T., Bastien, P., Fu, G., Ivens, A., and Stuart, K. (1999). *Leishmania major* Friedlin chromosome 1 has an unusual distribution of protein-coding genes. Proc. Natl. Acad. Sci. USA. *96*, 2902–2906.

Nagill, R., and Kaur, S. (2011). Vaccine candidates for leishmaniasis: a review. Int. Immunopharmacol. *11*, 1464–1488.

Nandan, D., and Reiner, N.E. (1995). Attenuation of gamma interferon-induced tyrosine phosphorylation in mononuclear phagocytes infected with *Leishmania donovani*: selective inhibition of signaling through Janus kinases and Stat1. Infect Immun. *63*, 4495–4500.

Nandan, D., Lo, R., and Reiner, N.E. (1999). Activation of phospho-tyrosine phosphatase activity attenuates mitogen-activated protein kinase signaling and inhibits c-FOS and nitric oxide synthase expression in macrophages infected with *Leishmania donovani*. Infection and Immunity. *67*, 4055–4063.

Nandan, D., Yi, T., Lopez, M., Lai, C., and Reiner, N.E. (2002). *Leishmania* EF-1alpha activates the Src homology 2 domain containing tyrosine phosphatase SHP-1 leading to macrophage deactivation. J. Biol. Chem. *277*, 50190–50197.

Nogueira, R.T., Nogueira, A.R., Pereira, M.C., Rodrigues, M.M., Neves, P.C., Galler, R., and Bonaldo, M.C. (2013). Recombinant yellow fever viruses elicit $CD8^+$ T-cell responses and protective immunity against *Trypanosoma cruzi*. PLOS ONE *8*, e59347.

Norris, K.A., Bradt, B., Cooper, N.R., and So, M. (1991). Characterization of a *Trypanosoma cruzi* C3 binding protein with functional and genetic similarities to the human complement regulatory protein, decay-accelerating factor. J Immunol. *147*, 2240–2247.

Nunes, M.P., Andrade, R.M., Lopes, M.F., and DosReis, G.A. (1998). Activation-induced T-cell death exacerbates *Trypanosoma cruzi* replication in macrophages cocultured with $CD4^+$ T-lymphocytes from infected hosts. J. Immunol. *160*, 1313–1319.

Oliveira, A.C., Peixoto, J.R., De Arrada, Campos, L.B., M.A., Gazzinelli, R.T., Golenbock, D.T., Akira, S., Previato, J.O., Mendonça-Previato, L., Nobrega, A., and Bellio, M. (2004). Expression of functional TLR4 confers proinflammatory responsiveness to *Trypanosoma cruzi* glycoinositolphospholipids and higher resistance to infection with *T. cruzi*. J. Immunol. *173*, 5688–5696.

Olivier, M., Baimbridge, K.G., and Reiner, N.E. (1992). Stimulus–response coupling in monocytes infected with *Leishmania*. Attenuation of calcium transients is related to defective agonist-induced accumulation of inositol phosphates. J. Immunol. *148*, 1188–1196.

Oliveira, A.C., de Alencar, B.C., Tzelepis, F., Klezewsky, W., da Silva, R.N., Neves, F.S., Cavalcanti, G.S., Boscardin, S., Nunes, M.P., Santiago, M.F., Nóbrega, A., Rodrigues, M.M., and Bellio, M. (2010). Impaired innate immunity in Tlr4(-/-) mice but preserved $CD8^+$ T-cell responses against *Trypanosoma cruzi* in Tlr4-, Tlr2-, Tlr9- or Myd88-deficient mice. PLOS Pathog. *6*, e1000870.

Ouaissi, M.A., Cornette, J., Afchain, D., Capron, A., Gras-Masse, H., and Tartar, A. (1986). *Trypanosoma cruzi* infection inhibited by peptides modeled from a fibronectin cell attachment domain. Science *234*, 603–607.

Ouaissi, A., da Silva, A.C., Guevara, A.G., Borges, M., and Guilvard, E. (2001). *Trypanosoma cruzi*-induced host immune system dysfunction: a rationale for parasite immunosuppressive factor(s) encoding gene targeting. J. Biomed. Biotechnol. *1*, 11–17.

Panaro, M.A., Spinelli, R., Lisi, S., Sisto, M., Acquafredda, A., Fumarola, L., Mitolo, V., and Brandonisio, O. (2004). Reduced expression of the chemokine receptor CCR1 in human macrophages and U-937 cells *in vitro* infected with *Leishmania infantum*. Clin. Exp. Med. 3, 225–230.

Peluffo, G., Piacenza, L., Irigoín, F., Alvarez, M.N., and Radi, R. (2004). L-arginine metabolism during interaction of *Trypanosoma cruzi* with host cells.Trends Parasitol. *20*, 363–369.

Pépin, J., Milord, F., Khonde, A., Niyonsenga, T., Loko, L., and Mpia, B. (1994). Gambiense trypanosomiasis: frequency of, and risk factors for, failure of melarsoprol therapy. Trans. R. Soc. Trop. Med. Hyg. *88*, 447–452.

Pereira-Chioccola, V.L., Acosta-Serrano, A., De Almeida I.C., Ferguson, M.A., Souto-Padron, T., Rodrigues, M.M., Travassos, L.R., and Schenkman, S. (2000). Mucin-like molecules form a negatively charged coat that protects *Trypanosoma cruzi* trypomastigotes from killing by human anti-alpha-galactosyl antibodies. J. Cell Sci. *113*, 1299–1307.

Pereira, V.R., Lorena, V.M., Verçosa, A.F., Silva, E.D., Ferreira, A.G., Montarroyos, U.R., Silva, A.P., and Gomes, Y.M. (2003). Antibody isotype responses in BALB/c mice immunized with the cytoplasmic repetitive antigen and flagellar repetitive antigen of *Trypanosoma cruzi*. Mem. Ins.t Oswaldo Cruz. *98*, 823–825.

Perez, Brandan, C., Padilla, A.M., Xu, D., Tarleton, R.L., and Basombrio, M.A. (2011). Knockout of the dhfr-ts gene in *Trypanosoma cruzi* generates attenuated parasites able to confer protection against a virulent challenge. PLOS Negl. Trop. Dis. *5*, e1418.

Piacenza, L., Peluffo, G., Alvarez, M.N., Martinez, A., and Radi, R. (2013). *Trypanosoma cruzi* antioxidant enzymes as virulence factors in Chagas disease. Antioxid Redox Signal *19*, 723–734.

Piacenza, L., Zago, M.P., Peluffo, G., Alvarez, M.N., Basombrio, M.A., and Radi, R. (2009). Enzymes of the antioxidant network as novel determiners of *Trypanosoma cruzi* virulence. Int. J. Parasitol. *39*, 1455–1464.

Pirmez, C., Yamamura, M., Uyemura, K., Paes-Oliveira, M., Conceição-Silva, F., and Modlin, R.L. (1993). Cytokine patterns in the pathogenesis of human leishmaniasis. J. Clin. Invest. *91*, 1390–1395.

Pizzi, T., and Prager, R. (1952). Immunity to infection induced by culture of *Trypanosoma cruzi* of atenuated virulence; preliminary communication. Bol. Inf. Parasit. Chil. 7, 20–21.

Prajeeth, C.K., Haeberlein, S., Sebald, H., Schleicher, U., and Bogdan, C. (2011). *Leishmania*-infected macrophages are targets of NK cell-derived cytokines but not of NK cell cytotoxicity. Infect. Immun. 79, 2699–2708.

Procópio, D.O., Barros, H.C., and Mortara, R.A. (1999). Actin-rich structures formed during the invasion of cultured cells by infective forms of *Trypanosoma cruzi*. Eur. J. Cell Biol. 78, 911–924.

Proudfoot, L., O'Donnell, C.A., and Liew, F.Y. (1995). Glycoinositolphos-pholipids of *Leishmania major* inhibit nitric oxide synthesis and reduce leishmanicidal activity in murine macrophages. Eur. J. Immunol. 25, 745–750.

Proudfoot, L., Nikolaev, A.V., Feng, G.J., Wei, W.Q., Ferguson, M.A., Brimacombe, J.S., and Liew, F.Y. (1996). Regulation of the expression of nitric oxide synthase and leishmanicidal activity by glycoconjugates of *Leishmania* lipophosphoglycan in murine macrophages. Proc. Natl. Acad. Sci. USA. *93,* 10984–10989.

Privé, C., and Descoteaux, A. (2000). *Leishmania donovani* promastigotes evade the activation of mitogen-activated protein kinases p38, c-Jun N-terminal kinase, and extracellular signal-regulated kinase-1/2 during infection of naive ma-crophages,' Eur. J. Immunol. *30,* 2235–2244.

Puentes, S.M., Dwyer, D.M., Bates, P.A., and Joiner, K.A. (1989). Binding and release of C3 from *Leishmania donovani* promastigotes during incubation in normal human serum. J. Immunol. *143,* 3743–3749.

Qvit, N., Schechtman, D., Pena, D.A., Berti, D.A., Soares, C.O., Miao, Q., Liang, L.A., Baron, L.A., Teh-Poot, C., Martínez-Vega, P., Ramirez-Sierra, M.J., Churchill, E., Cunningham, A.D., Malkovskiy, A.V., Federspiel, N.A., Gozzo, F.C., Torrecilhas, A.C., Manso Alves, M.J., Jardim, A., Momar, N., Dumonteil, E., and Mochly-Rosen, D. (2016). Scaffold proteins LACK and TRACK as potential drug targets in kinetoplastid parasites: Development of inhibitors, Int. J. Parasitol. Drugs Drug Resist. *6,* 74–84.

Radi, R. (2013). Peroxynitrite, a stealthy biological oxidant. J. Biol. Chem. *288,* 26464–26472.

Ramos, H., Valdivieso, E., Gamargo, M., Dagger, F., and Cohen, B.E. (1996). Amphotericin B kills unicellular leishmanias by forming aqueous pores permeable to small cations and anions. J. Membr. Biol. *152,* 65–75.

Ravindran, R., Bhowmick, S., Das, A., and Ali, N. (2010). Comparison of BCG, MPL and cationic liposome adjuvant systems in leishmanial antigen vaccine formulations against murine visceral leishmaniasis. BMC Microbiol. *10,* 181.

Raynaud, J.P., Sones, R.K. and Friedheim, E.A.H. (1989). A review of cymelarsan R -a new treatment proposed for animal trypanosomiasis due to *T. evansi* and other trypanosomes of the *T. brucei* group. International Scientific Council for Trypanosomiasis Research and Control. *115,* 334–338.

Rawlings, J.S., Rosler, K.M., and Harrison, D.A. (2004). The JAK/STAT signaling pathway. J Cell Sci. *117,* 1281–1283.

Ribeiro-Gomes, F.L., Moniz-de-Souza, M.C., Alexandre-Moreira, M.S., Dias, W.B., Lopes, M.F., Nunes, M.P., and DosReis, G.A. (2007). Neutrophils activate macrophages for intracellular killing of *Leishmania major* through recruitment of TLR4 by neutrophil elastase. J. Immunol. *179,* 3988–3994.

Ritter, U., and Körner, H. (2002). Divergent expression of inflammatory dermal chemokines in cutaneous leishmaniasis. Parasite Immunol. *24,* 295–301.

Riou, G., and Benard, J. (1980). Berenil induces the complete loss of kinetoblast DNA sequences in *Trypanosoma* equiperdum. Biochem. Biophys. Res. Commun. *96,* 350–354.

Reiner, N.E., Ng, W., and McMaster, W.R. (1987). Parasite-accessory cell interactions in murine leishmaniasis. II. *Leishmania donovani* suppresses macrophage expression of class I and class II major histocompatibility complex gene products. J Immunol. *138,* 1926–1932.

Reiner, N.E., Ng, W., Ma, T., and McMaster, W.R. (1988). Kinetics of gamma interferon binding and induction of major histocompatibility complex class II mRNA in *Leishmania*-infected macrophages. Proc. Natl. Acad. Sci. U.S.A. *85,* 4330–4334.

Rezvan, H. (2013). Immunogenicity of HLA-DR1 Restricted Peptides Derived from *Leishmania major* gp63 Using FVB/N-DR1 Transgenic Mouse Model. Iran J Parasitol. *8,* 273–279.

Ropert, C., and Gazzinelli, R.T. (2004). Regulatory role of Toll-like receptor 2 during infection with *Trypanosoma cruzi*. J. Endotoxin Res. *10,* 425–430.

Roditi, I., and Lehane, M.J. (2008). Interactions between trypanosomes and tsetse flies. Curr. Opin. Microbiol. *11,* 345–351.

Rodríguez, A., Rioult, M.G., Ora, A., and Andrews, N.W. (1995). A trypanosome-soluble factor induces IP3 formation, intracellular Ca2+mobilization and microfilament rearrangement in host cells. J Cell Biol. *129*: 1263–1273.

Rodrigues, M.M., Oliveira, A.C., and Bellio, M. (2012). The Immune Response to *Trypanosoma cruzi*: Role of Toll-Like Receptors and Perspectives for Vaccine Development. J Parasitol Res. *2012,* 507874.

Ropert C., Ferreira L.R., Campos M.A., Procópio D.O., Travassos L.R., Ferguson M.A., Reis L.F., Teixeira M.M., Almeida I.C., Gazzinelli R.T. (2002). Macrophage signaling by glycosylphosphatidylinositol-anchored mucin-like glycoproteins derived from *Trypanosoma cruzi* trypomastigotes. Microbes Infect. *4,* 1015–1025.

Rosas, L.E., Keiser, T., Barbi, J., Satoskar, A.A., Septer, A., Kaczmarek, J., Lezama-Davila C.M., and Satoskar, A.R. (2005). Genetic background influences immune responses and disease outcome of cutaneous *L. mexicana* infection in mice. Int Immunol. *17,* 1347–1357.

Rosenberg, C.S., Martin, D.L., and Tarleton, R.L. (2010). $CD8^+$ T-cells specific for immuno-dominant trans-sialidase epitopes contribute to control of *Trypanosoma cruzi* infection but are not required for resistance. J Immunol. *185,* 560–568.

Rub, A., Dey, R., Jadhav, M., Kamat, R., Chakkaramakkil, S., Majumdar, S., Mukhopadhyaya, R., and Saha, B. (2009). Cholesterol depletion associated with *Leishmania major* infection alters macrophage CD40 signalosome composition and effector function. Nat. Immunol. *10,* 273–280.

Ruiz, A.M., Esteva, M., Cabeza Meckert, P., Laguens, R.P., and Segura, E.L. (1985). Protective immunity and pathology induced by inoculation of mice with different subcellular fractions of *Trypanosoma cruzi*. Acta Tropica. *42,* 299–309.

Russell, D.G., Xu, S., and Chakraborty, P. (1992). Intracellular trafficking and the parasitophorous vacuole of *Leishmania mexicana*-infected macrophages. J. Cell. Sci. *103,*1193–1210.

Ruta, S., Plasman, N., Zaffran, Y., Capo, C., Mège, J.L., and Vray, B. (1996). *Trypanosoma cruzi*-induced tyrosine phosphorylation in murine peritoneal macrophages. Parasitol Res. *82,* 481–484.

Saha, B., Das, G., Vohra, H., Ganguly, N.K., and Mishra, G.C. (1995). Macrophage-T-cell interaction in experimental visceral leishmaniasis: failure to express costimulatory molecules on *Leishmania*-infected macrophages and its implication in the suppression of cell-mediated immunity. Eur J Immunol. *25,* 2492–2498.

Saha, B., Chattopadhyay, S., Germond, R., Harlan, D.M., and Perrin, P.J. (1998). CTLA4 (CD152) modulates the Th subset response and alters the course of experimental *Leishmania major* infection. Eur. J. Immunol. *28*, 4213–4220.

Salgado-Jiménez, B., Arce-Fonseca, M., Baylón-Pacheco, L., Talamás-Rohana, P., and Rosales-Encina, J.L. (2013). Differential immune response in mice immunized with the A, R or C domain from TcSP protein of *Trypanosoma cruzi* or with the coding DNAs. Parasite Immunol. *35*, 32–41.

Scharton, T.M., and Scott, P. (1993). Natural killer cells are a source of interferon gamma that drives differentiation of $CD4^+$ T-cell subsets and induces early resistance to *Leishmania major* in mice. J. Exp. Med. *178*, 567–577.

Scharfstein, J., Schmitz, V., Morandi, V., Capella, M.M., Lima, A.P., Morrot, A., Juliano, L., and Müller-Esterl, W. (2000). Host cell invasion by *Trypanosoma cruzi* is potentiated by activation of bradykinin B(2) receptors. J. Exp. Med. *192*, 1289–1300.

Schneider, P., Rosat, J.P., Bouvier, J., Louis, J., and Bordier. C. (1992). *Leishmania major*: differential regulation of the surface metalloprotease in amastigote and promastigote stages. Exp. Parasitol. *75*, 196–206.

Seco-Hidalgo, V., De Pablos, L.M., and Osuna, A. (2015). Transcriptional and phenotypical heterogeneity of *Trypanosoma cruzi* cell populations. Open Biol. *5*, 150190.

Severn, A., Wakelam, M.J., and Liew, F.Y. (1992). The role of protein kinase C in the induction of nitric oxide synthesis by murine macrophages. Biochem. Biophys. Res. Commun. *188*, 997–1002.

Shapiro, T.A., and Englund, P.T. (1990). Selective cleavage of kinetoplast DNA minicircles promoted by antitrypanosomal drugs. Proc. Natl. Acad. Sci. U.S.A. *87*, 950–954.

Sjölander, A., Baldwin, T.M., Curtis, J.M., Bengtsson, K.L., and Handman, E. (1998). Vaccination with recombinant Parasite Surface Antigen 2 from *Leishmania major* induces a Th1 type of immune response but does not protect against infection. Vaccine. *16*, 2077–2084.

Skeiky, Y.A., Coler, R.N., Brannon, M., Stromberg, E., Greeson, K., Crane, R.T., Webb, J.R., Campos-Neto, A., and Reed, S.G. (2002). Protective efficacy of a tandemly linked, multi-subunit recombinant leishmanial vaccine (Leish-111f) formulated in MPL adjuvant. Vaccine. *20*, 3292–3303.

Silva, J.S., Twardzik, D.R., and Reed, S.G. (1991). Regulation of *Trypanosoma cruzi* infections *in vitro* and *in vivo* by transforming growth factor beta (TGF-beta). J. Exp. Med. *174*, 539–545.

Silva, G.K., Gutierrez, F.R., Guedes, P.M., Horta, C.V., Cunha, L.D., Mineo, T.W., Santiago-Silva, J., Kobayashi, K.S., Flavell, R.A., Silva, J.S., and Zamboni, D.S. (2010). Cutting edge: nucleotide-binding oligomerization domain 1-dependent responses account for murine resistance against *Trypanosoma cruzi* infection. J. Immunol. *184*, 1148–1152.

Snary, D. (1983). Cell surface glycoproteins of *Trypanosoma cruzi*: protective immunity in mice and antibody levels in human *chagasic* sera. Trans. R. Soc. Trop. Med. Hyg. *77*, 126–129.

Soong, L., Xu, J.C., Grewal, I.S., Kima, P., Sun, J., Longley, B.J., Jr., Ruddle, N.H., McMahon-Pratt D., and Flavell, R.A. (1996). Disruption of CD40–CD40 ligand interactions results in an enhanced susceptibility to *Leishmania amazoniensis* infection. Immunity. *4*, 263–273.

Soudi, S., Hosseini, A.Z., and Hashemi, S.M. (2011). Co-administration of rectal BCG and autoclaved *Leishmania major* induce protection in susceptible BALB/c mice. Parasite Immunol. *33*, 561–571.

Stempin, C., Giordanengo, L., Gea, S., and Cerban, F. (2002). Alternative activation and increase of *Trypanosoma cruzi* survival in murine macrophages stimulated by cruzipain, a parasite antigen. J Leukoc Biol. *72*, 727–734.

Stempin, C.C., Tanos, T.B., Coso, O.A., and Cerbán, F.M. (2004). Arginase induction promotes *Trypanosoma cruzi* intracellular replication in cruzipain-treated J774 cells through the activation of multiple signaling pathways. Eur J Immunol. *34*:200–209.

Storz, P., and Toker, A. (2002). 3′-phosphoinositide-dependent kinase-1 (PDK-1) in PI 3-kinase signaling. Front. Biosci. *1*, 886–902.

Sudan, R., Srivastava, N., Pandey, S.P., Majumdar, S., and Saha, B. (2012). Reciprocal regulation of protein kinase C isoforms results in differential cellular responsiveness. J Immunol. *188*, 2328–2337.

Sundar, S., Jha, T.K., Thakur, C.P., Bhattacharya, S.K., and Rai, M. (2006). Oral miltefosine for the treatment of Indian visceral leishmaniasis. Trans. R. Soc. Trop. Med. Hyg. 100(Suppl. 1), S26–33.

Taibi, A., Plumas-Marty, B., Guevara-Espinoza, A., Schöneck, R., Pessoa, H., Loyens, M., Piras, R., Aguirre, T., Gras-Masse, H., and Bossus, M. (1993). *Trypanosoma cruzi*: immunity-induced in mice and rats by trypomastigote excretory-secretory antigens and identification of a peptide sequence containing a T-cell epitope with protective activity. J Immunol. *151*, 2676–2689.

Tardieux, I., Nathanson, M.H., and Andrews, N.W. (1994). Role in host cell invasion of *Trypanosoma cruzi*-induced cytosolic-free Ca2+ transients. J. Exp. Med. *179*, 1017–1022.

Tarleton, R.L., Grusby, M.J., and Zhang, L. (2000). Increased susceptibility of Stat4-deficient and enhanced resistance in Stat6-deficient mice to infection with *Trypanosoma cruzi*. J. Immunol. *165*,1520–1525.

Taylor, J.E., and Rudenko, G. (2006). Switching trypanosome coats: what's in the wardrobe? Trends Genet. *22*, 614–620.

Thakur, C.P., and Kumar, K. (1992). Post kala-azar dermal leishmaniasis: a neglected aspect of kala-azar control programmes. Ann Trop Med Parasito. *l86*, 355–359.

Thakur, C.P., Kanyok, T.P., Pandey, A.K., Sinha, G.P., Messick, C., and Olliaro, P. (2000). Treatment of visceral leishmaniasis with injectable paromomycin (aminosidine). An open-label randomized phase-II clinical study. Trans R Soc Trop Med Hyg. *94*, 432–443.

Thomas, P.D., Campbell, M.J., Kejariwal, A., Mi, H., Karlak, B., Daverman, R., Diemer, K., Muruganujan, A., and Narechania, A. (2003). PANTHER: a library of protein families and subfamilies indexed by function. Genome Res. *13*, 2129–2141.

Todeschini, A.R., Nunes, M.P., Pires, R.S., Lopes, M.F., Previato, J.O., Mendonça-Previato, L., and DosReis,

G.A. (2002). Costimulation of host T-lymphocytes by a trypanosomal trans-sialidase: involvement of CD43 signaling. J. Immunol. *168*, 5192–5198.

Todorov, A.G., Einicker-Lamas, M., de Castro, S.L., Oliveira, M.M., and Guilherme, A. (2000). Activation of host cell phosphatidylinositol 3-kinases by *Trypanosoma cruzi* infection. J. Biol. Chem. *275*, 32182–32186.

Turco, S.J., and Descoteaux, A. (1992). The lipophosphoglycan of *Leishmania* parasites. Annu. Rev. Microbiol. *46*, 65–94.

van Zandbergen, G., Hermann, N., Laufs, H., Solbach, W., and Laskay, T. (2002). *Leishmania* promastigotes release a granulocyte chemotactic factor and induce interleukin-8 release but inhibit gamma interferon-inducible protein 10 production by neutrophil granulocytes. Inf. Immun. *70*, 4177–4184.

van Zandbergen, G., Klinger, M., Mueller, A., Dannenberg, S., Gebert, A., Solbach, W., and Laskay, T. (2004). Cutting edge: neutrophil granulocyte serves as a vector for *Leishmania* entry into macrophages. J. Immunol. *173*, 6521–6525.

Vasconcelos, J.R., Boscardin, S.B., Hiyane, M.I., Kinoshita, S.S., Fujimura, A.E., and Rodrigues, M.M. (2003). A DNA-priming protein-boosting regimen significantly improves type 1 immune response but not protective immunity to *Trypanosoma cruzi* infection in a highly susceptible mouse strain. Immunol. Cell Biol. *81*, 121–129.

Vassella, E., Reuner, B., Yutzy, B., and Boshart, M. (1997). Differentiation of African trypanosomes is controlled by a density sensing mechanism which signals cell cycle arrest via the cAMP pathway. J. Cell. Sci. *110*, 2661–2671.

Vázquez-López, R., Argueta-Donohué, J., Wilkins-Rodríguez, A., Escalona-Montaño, A., Aguirre-García, M., and Gutiérrez-Kobeh, L. (2015). *Leishmania mexicana* amastigotes inhibit p38 and JNK and activate PI3K/AKT: role in the inhibition of apoptosis of dendritic cells. Parasite Immunol. *37*, 579–589.

Venuprasad, K., Banerjee, P.P., Chattopadhyay, S., Sharma, S., Pal, S., Parab, P.B., Mitra, D., and Saha, B. (2002). Human neutrophil-expressed CD28 interacts with macrophage B7 to induce phosphatidylinositol 3-kinase-dependent IFN-gamma secretion and restriction of *Leishmania* growth. J. Immunol. *169*, 920–928.

Vergnes, B., Gourbal, B., Girard, I., Sundar, S., Drummelsmith, J., and Ouellette, M. (2007). A proteomics screen implicates HSP83 and a small kinetoplastid calpain-related protein in drug resistance in *Leishmania donovani* clinical field isolates by modulating drug-induced programmed cell death. Molecular and Cellular Proteomics *6*, 88–101.

Venturini, G., Salvati, L., Muolo, M., Colasanti, M., Gradoni, L., and Ascenzi, P. (2000). Nitric oxide inhibits cruzipain, the major papain-like cysteine proteinase from *Trypanosoma cruzi*. Biochem. Biophys. Res. Commun. *270*, 437–441.

Vespa, G.N., Cunha, F.Q., and Silva, J.S. (1994). Nitric oxide is involved in control of *Trypanosoma cruzi*-induced parasitemia and directly kills the parasite *in vitro*. Infect. Immun. *62*, 5177–5182.

Villalta, F., Smith, C.M., Ruiz-Ruano, A., and Lima, M.F. (2001). A ligand that *Trypanosoma cruzi* uses to bind to mammalian cells to initiate infection. FEBS Lett. *505*, 383–388.

Warzocha, K., Bienvenu, J., Coiffier, B., and Salles, G. (1995). Mechanisms of action of the tumor necrosis factor and lymphotoxin ligand-receptor system. Eur. Cytokine Netw. *6*, 83–96.

Wang, C.C. (1995). Molecular mechanisms and therapeutic approaches to the treatment of African trypanosomiasis. Ann. Rev. Pharmacol. Toxicol. *35*, 93–127.

Webster, J.P., and Woolhouse, M.E.J. (1999). Cost of resistance: relationship between reduced fertility and increased resistance in a snail-schistosome host–parasite system. Proc. R. Soc. *266*, 391–396.

Wilkowsky, S.E., Barbieri, M.A., Stahl, P., Isola, E.L. (2001). *Trypanosoma cruzi*: phosphatidylinositol 3-kinase and protein kinase B activation is associated with parasite invasion. Exp. Cell Res. 264, 211–218.

Woldman, I., Varinou, L., Ramsauer, K., Rapp, B., and Decker, T. (2001). The Stat1 binding motif of the interferon-gamma receptor is sufficient to mediate Stat5 activation and its repression by SOCS3. J. Biol. Chem. *276*, 45722–45728.

Woolsey, A.M., Sunwoo, L., Petersen, C.A., Brachmann, S.M., Cantley, L.C., and Burleigh, B.A. (2003). Novel PI 3-kinase-dependent mechanisms of trypanosome invasion and vacuole maturation. J Cell Sci. *116*, 3611–3622.

Wrightsman, R.A., Miller, M.J., Saborio, J.L., and Manning, J.E. (1995). Pure paraflagellar rod protein protects mice against *Trypanosoma cruzi* infection. Infect. Immun. *63*, 122–125.

Wyllie, S., Cunningham, M.L., and Fairlamb, A.H. (2004). Dual action of antimonial drugs on thiol redox metabolism in the human pathogen *Leishmania donovani*. J. Biol. Chem. *279*, 39925–39932.

Yang, J., Clarke, J.F., Ester, C.J., Young, P.W., Kasuga, M., and Holman, G.D. (1996). Phosphatidylinositol 3-kinase acts at an intracellular membrane site to enhance GLUT4 exocytosis in 3T3-L1 cells. Biochem J. *313*, 125–131.

Zer, R., Yaroslavski, I., Rosen, L., and Warburg, A. (2001). Effect of sand fly saliva on *Leishmania* uptake by murine macrophages. Int. J. Parasitol. *31*, 810–814.

Zilberstein, D., and Shapira, M. (1994). The role of pH and temperature in the development of *Leishmania* parasites. Annu. Rev. Microbiol. *48*, 449–470.

Extracellular Matrix Interacting Proteins of Trypanosomatids: Adhesion and Invasion of Host Tissues

12

Shreyasi Palit and Pijush K. Das*

Infectious Diseases and Immunology Division, CSIR-Indian Institute of Chemical Biology, Kolkata, India.

*Correspondence: pijush52@gmail.com

https://doi.org/10.21775/9781910190715.12

Abstract

Extracellular matrix (ECM), a highly dynamic, complex three-dimensional macromolecular structure is ubiquitously present in tissues forming the basal lamina. Comprising of laminin, collagen, fibronectin, elastin, proteoglycan and several other glycoproteins, ECM acts as a major mechanical barrier to almost all known invasive pathogens. Therefore, pathogens are constantly evolving measures to breach this barrier for their survival within the host. Most pathogens therefore remodel ECM either by binding to its component glycoproteins by their ECM-binding proteins or by degrading the matrix by secreting several matrix-degrading enzymes in order to adhere, invade and survive within the host tissue. Recent advances have suggested that Trypanosomatids which include *Leishmania* spp., *Trypanosoma* spp., *Phytomonas* spp., etc. also utilize a repertoire of ECM interacting proteins in order to bind and mediate downstream signalling events in order to establish their niche. This article therefore emphasizes on understanding the array of proteins of Trypanosomatids that interact with the ECM components of the host during the early stages of infection so that they can be exploited in future to develop new drugs.

Introduction

The extracellular matrix (ECM) which forms the basal lamina is a highly dynamic, complex three-dimensional macromolecular structure that provides scaffolding for the cellular components of the tissues (Nelson and Bissell, 2006; Xu *et al.*, 2009; Frantz *et al.*, 2010). It is involved in initiation and regulation of crucial biochemical and biomechanical events related to various physiological processes including tissue morphogenesis, differentiation, homeostasis, cellular signalling and transport of solutes across cellular barriers (Frantz *et al.*, 2010; Clause and Barker, 2013; Hynes, 2009). Being ubiquitously distributed, ECM provides a physical barrier between different tissues providing the host with an important defence strategy (Lemichez *et al.*, 2010). The large varieties of macromolecules that constitute the ECM have precise tissue-specific structural and chemical properties especially suited to their specific biological functions (Mouw *et al.*, 2014; Theocharis *et al.*, 2016). The highly dynamic structure of ECM comprises of two major classes of macromolecules: proteoglycans (PG) and fibrous proteins (Frantz *et al.*, 2010; Jarvelainen *et al.*, 2009; Schaefer and Schaefer, 2010). The major fibrous proteins of the ECM are collagen, fibronectin, laminin and elastin (Alberts and National Center for Biotechnology, 2004) whereas PG comprise of chains of glycosaminoglycans (GAG) covalently linked to specific core proteins (Jarvelainen *et al.*, 2009). Although ECM is essential for maintaining normal homeostasis, but abnormal composition of ECM components may result in pathological conditions (Jarvelainen *et al.*, 2009). ECM can undergo enzymatic or non-enzymatic remodelling under the influence of various stimuli, such as injury or various physiological challenges including tumorigenesis or pathogenesis (Lu *et al.*, 2011; Jakeman *et al.*, 2014; Bonnans *et al.*, 2014; Lu *et al.*, 2012). Therefore, ECM components are targeted by various intracellular pathogenic micro-organisms for adherence and invasion (Singh *et al.*, 2012). To date, several microbes belonging to diverse niches have been reported to target ECM and remodel it to establish infection either by adhering to host ECM components by various surface-bound adhesive proteins/receptors or by degrading ECM components with their surface-bound/secretory proteolytic enzymes or hijacking host proteases (plasminogen) resulting in pathogenesis (Vanlaere and Libert 2009). This chapter provides a general overview of the ECM structure and emphasizes on ECM interacting proteins of trypanosomatids (*Trypanosoma cruzi*, *Leishmania* spp., African trypanosomes and plant and insect trypanosomatids) during the early stages of infection so that they can be exploited to develop new drugs in future.

General overview of extracellular matrix components

ECM comprises of two major classes of macromolecules: fibrous proteins and proteoglycans (Järveläinen *et al.*, 2009; Frantz *et al.*, 2010; Schaefer and Schaefer, 2010) (Table 12.1 and Fig. 12.1). The proteoglycan molecules form the 'ground substance' in which the fibrous proteins remain embedded (Alberts and National Center for Biotechnology, 2004). The gel-like 'ground substance' helps in resisting compressive forces and functions to permit rapid diffusion of hormones, nutrients, and various metabolites between the blood and tissue cells while the fibrous proteins provides strength to the ECM (Alberts and National Center for Biotechnology, 2004).

Fibrous proteins

The main fibrous proteins of the ECM are collagen, elastin, fibronectin and laminin (Alberts and National Center for Biotechnology, 2004).

Collagen

The most abundant fibrous protein found in all multicellular animals is the collagen constituting 30% of their total protein mass (Alberts and National Center for Biotechnology, 2004). It constitutes the main structural element of the connective tissues and functions in directing tissue development, providing tensile strength, regulating adhesion of cells, and supporting chemotaxis and migration (Nelson and Bissell, 2006). Decades of research have revealed the presence of 28 different types of collagen and each type is comprised of homotrimers and heterotrimers of three polypeptide chains (known as α-chains) (Mouw *et al.*, 2014). The major collagen types found in connective tissues are types I, II, III, V and XI; type I being the most common and predominant collagen of skin and bone (Alberts and National Center for Biotechnology, 2004). While types IV and VII are network-forming collagens, types IX and XII are called fibril-associated collagens (Alberts and National Center for Biotechnology, 2004).

Table 12.1 Major ECM components, their distribution and function

ECM components	Distribution	Function
A. Fibrous proteins		
(i) Collagen	Basement membranes in all organs and tissues	Structural scaffold; cellular and tissue repair and development
(ii) Elastin	Connective tissues, blood vessels, heart, lungs, skin, bladder and elastic cartilage	Renders elasticity to tissues
(iii) Fibronectin (FN)	Connective tissue	Cellular adhesion, migration, growth and differentiation
(iv) Laminin	Basement membranes, connective tissues,	Structural scaffold; cellular migration; signalling
B. Proteoglycans (PG)	All tissues, cartilage and bone joints	Tissue integrity; embryonic development; cellular migration; blood coagulation; healing; inflammation; angiogenesis

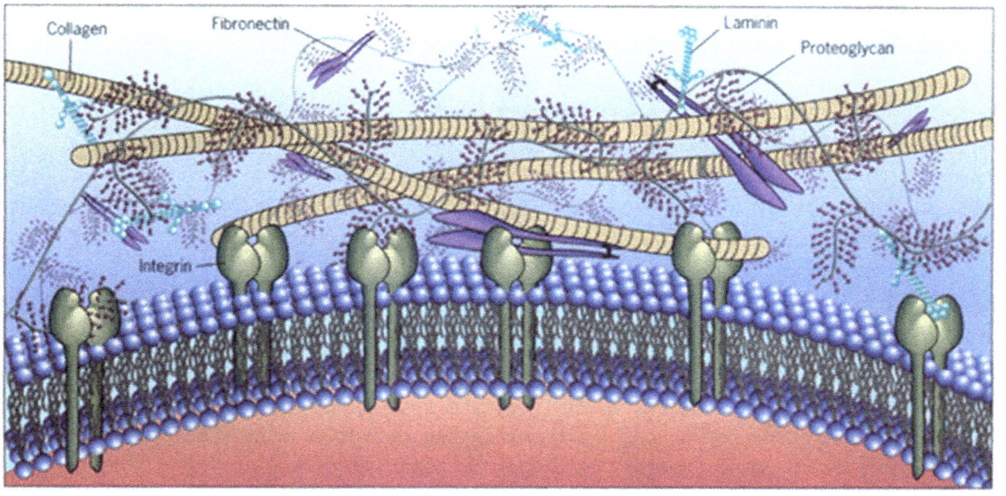

Figure 12.1 Representative figure of extracellular matrix. Reproduced with permission from Karp, G. (2009). Chapter 7: Interaction between cells and their environment – The extra cellular space. In Cell and Molecular Biology: Concepts and Experiments, Karp G, ed (John Wiley & Sons), pp. 236–243.

Elastin

Collagen associates with another major ECM fibrous protein elastin that provides recoil to tissues after transient stretch such as blood vessels, heart, lungs, skin, bladder and elastic cartilage (Theocharis *et al.*, 2016). These fibres are usually assembled during the early stages of development and are very stable due to low turnover rate (Theocharis *et al.*, 2016). They remain covered with a sheath of glycoprotein microfibrils, mainly fibrilins, which are crucial for the integrity of the elastin fibre (Alberts and National Center for Biotechnology, 2004).

Fibronectin

Fibronectin is secreted as large ECM glycoprotein and is ubiquitously expressed in a variety of cell types with critical functions in the development of vertebrates (Singh *et al.*, 2010; George *et al.*, 1993).

Laminin

Laminins are a family of large, mosaic glycoproteins that are assembled along with type IV collagen (Mouw *et al.*, 2014; Durbeej, 2010; Iorio *et al.*, 2015). It is a heterotrimer consisting of one α, one β, and one γ chain (Aumailley *et al.*, 2005). They primarily interact with other ECM components helping in organization of the ECM and play an important role in embryonic development and organogenesis (Mercurio and Shaw, 1991; Durbeej, 2010).

Proteoglycans

Proteoglycans (PG), which are abundantly found in cartilage and neural ECM, are primarily composed of glycosaminoglycan (GAG) covalently linked to a specific core protein (Iozzo and Murdoch, 1996; Schaefer and Schaefer, 2010; Bandtlow and Zimmermann, 2000; Knudson and Knudson, 2001). PGs have been classified on the basis of their core protein, localization and composition of GAG. The three major families are small leucine-rich proteoglycans (SLRPs), modular proteoglycans, and cell surface proteoglycans (Schaefer and Schaefer, 2010). The GAG chains on the protein core are unbranched polysaccharide chains composed of repeating disaccharide units that can be divided further into sulfated (chondroitin sulfate, heparansulfate and keratansulfate) and non-sulfated (hyaluronic acid) GAGs (Schaefer and Schaefer, 2010). PGs are implicated in a wide variety of cellular functions. SLRPs are known to mediate multiple signalling pathways including activation of epidermal growth factor receptor pathway (EGFR), insulin-like growth factor 1 receptor pathway (IGFIR), TGFβ pathway and regulation of inflammatory response (Schaefer and Schaefer, 2010; Schaefer and Iozzo, 2008). While modular PGs functions to regulate cell adhesion, migration and proliferation (Schaefer and Schaefer, 2010), cell-surface PGs functions as co-receptor facilitating ligand encounters with signalling receptors (Schaefer and Schaefer, 2010).

Trypanosomatids

Trypanosomatids are a very diverse group of flagellate protozoa characterized by the presence of kinetoplast DNA (kDNA), an extracellular circular network of DNA that corresponds to the mitochondrial genome of the parasite (Araújo *et al.*, 2009). Although the trypanosomatids infect a wide range of hosts, only two genera, *Leishmania* and *Trypanosoma*, are found to infect humans and manifest diseases such as leishmaniasis, Chagas' disease and African trypanosomiasis (Wallace, 1966; McGhee and Cosgrove, 1980) and *Phytomonas* infect plants (Camargo, 1999). Trypanosomiases are referred to as neglected tropical diseases, as they are largely ignored by first-world public opinion, medical science as well as pharmaceutical companies (Croft, 2008).

Leishmaniasis

Leishmaniasis that has a wide spectrum of diseased manifestations is caused by the dimorphic protozoan parasite *Leishmania* spp. that lives as a promastigote in the digestive tract of sandflies and as an amastigote in the phagolysosomes of mammalian macrophages. According to global estimates 1.5 to 2 million symptomatic cases of the disease are registered each year, resulting in 70,000 deaths (primarily from visceral leishmaniasis) (Murray *et al.*, 2005; Davies *et al.*, 2003). At least 20 species of *Leishmania* are known to infect humans, ranging from the self-healing cutaneous forms caused by *L. tropica*, *L. major*, and *L. mexicana* to the more severe mucocutaneous disease caused by *L. braziliensis* to the most severe the visceral disease caused by *L. donovani* (Sibley 2011). This protozoan therefore, requires a repertoire of adaptations to remodel host tissues to migrate to sites that are specific for their growth. This interplay between host ECM components and *Leishmania* is brought about by an array of ECM interacting proteins on the surface of the parasite (Kulkarni *et al.*, 2008; McGwire *et al.*, 2003; Bandyopadhyay *et al.*, 2002; Brittingham *et al.*, 1999; Ghosh *et al.*, 1999). A repertoire of ECM interacting proteins of *Leishmania* has been identified that facilitate parasite survival (Table 12.2).

Table 12.2 Extracellular matrix interacting proteins of glycoproteins of *Leishmania* spp.

Protein	Localization	Host ECM target or ligand	Function
gp63 or leishmanolysin or major surface protease (MSP)	Cell surface bound as well as secreted	Fibronectin, type IV collagen	Fragmentation of fibronectin resulting in decreased ROI production; inhibition of NK cell function; promotion of interphagosomal survival of parasites
Laminin-binding protein (LBP)	Cell surface bound	Laminin	Behaves as adhesin and helps in invasion of host cell
Fibronectin-binding protein	Cell surface bound	Fibronectin	Helps in parasite internalization; Increases the proximity of gp63 with fibronectin aiding in its degradation
Collagen-binding protein	Cell surface bound	Type I and type IV Collagen	Remodelling of the ECM facilitating parasite migration and internalization
Heparin-binding protein (HBP)	Cell surface bound	Heparin, Heparin sulphate and hyaluron	Promotes parasite invasion

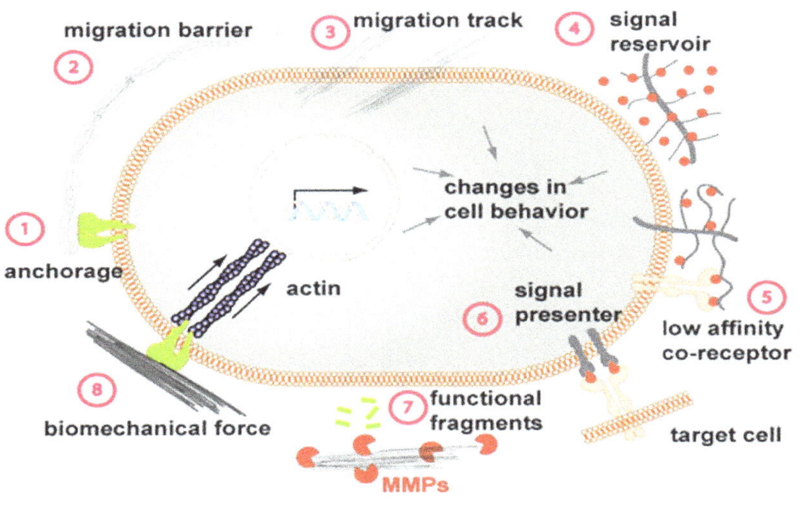

Figure 12.2 Mechanisms of ECM function. ECM performs a diverse range of functions from maintaining tissue polarity to blocking or facilitating cell migration. In addition, it also participates in cellular signalling helping in cellular communication. Reproduced with permission. © 2012 Lu, P. *et al*. Journal of Cell Biology. 196: 395–406. http://doi.org/10.1083/jcb.201102147.

Metalloproteases

Several researchers have reported a highly conserved 63 kDa protein gp63, a zinc-dependent metalloprotease to be a major surface protein on promastigotes of *Leishmania* spp. (Bouvier *et al.*, 1985; Medina-Acosta *et al.*, 1993). This metalloprotease also referred to as major surface protease(MSP)or leishmanolysin has been demonstrated to promote their interaction with murine macrophages by aiding in its migration through the ECM (McGwire *et al.*, 2003; Chakrabarty *et al.*, 1996; Liu and Chang, 1992; McGwire and Chang, 1994). Studies with gp63 knock out promastigotes of *L. major* revealed a role of the metalloprotease in direct binding to Natural Killer cells (NK cells) and inhibiting its proliferation (Lieke *et al.*, 2008). Pre-incubation of the parasite with bathophenanthroline, a zinc chelator although had no deleterious effect on the parasite restricted the migration of the parasite (McGwire *et al.*, 2003). Leishmanolysin from stationary phase promastigote is capable of digesting type IV collagen and fibronectin (McGwire *et al.*, 2003). Interestingly, the digested fragments of fibronectin as observed by SDS-PAGE differed depending on the source of GP63 used. Cell surface bound gp63 digested fibronectin into somewhat larger subunits when compared to purified gp63 (McGwire *et al.*, 2003). The degradation of fibronectin by the parasite has been reported to be associated with decreased generation of reactive oxygen intermediates (ROI) by the host macrophages facilitating the survival of amastigotes in host (McGwire *et al.*, 2003). Regardless of the condition used for digestion, laminin has been reported to be resistant to gp63 digestion (McGwire *et al.*, 2003). Importantly, leishmanolysin was found to facilitate complement inactivation in serum (Brittingham *et al.*, 1995) aiding in intraphagolysomal survival of the parasite (Brittingham *et al.*, 1999; Chaudhuri *et al.*, 1989).

Laminin-binding protein

The 67 kDa integral membrane glycoprotein of both promastigotes as well as amastigotes of *L. donovani* is responsible for the attachment of the parasite to laminin, the major glycoprotein of ECM and basement membrane (Ghosh *et al.*, 1996). Characterization of laminin-binding protein (LBP) has revealed that it has a high affinity for laminin and behaves as an adhesin, helping in invasion of host cells (Bandyopadhyay *et al.*, 2001; Ghosh *et al.*, 1999). Administration of anti-LBP antibodies have shown to reduce infectivity of the parasite in mouse model suggesting that the binding between laminin and LBP to be an important event in *Leishmania* pathogenesis (Bandyopadhyay *et al.*, 2001). Studies have revealed that brief treatment of *L. donovani* promastigotes with heat or proteolytic enzymes markedly inhibited the laminin-binding activity of LBP (Bandyopadhyay *et al.*, 2001). The binding affinity of LBP being affected by change in temperature as well as pH has been reported to be dependent on the structural conformation of both laminin as well as LBP (Bandyopadhyay *et al.*, 2002). A number of compounds such as urea, sodium chloride and N-ethylmaleimide are also known to affect the interaction between laminin and LBP (Bandyopadhyay *et al.*, 2002). Laminin-binding activity of LBP has been reported to be enhanced in presence of the cofactor zinc; the effect being saturable at physiological concentration of zinc (15 μM) (Medina-Acosta *et al.*, 1993). Zinc finger consensus sequence of B1 chain containing YIGSR sequence provides the necessary scaffolding for the binding of LBP (Bandyopadhyay *et al.*, 2002). Incubation of promastigotes with the peptide C(YIGSR)3-NH2 has been reported to phosphorylate tyrosine residues of proteins of molecular mass 115–130 kDa, although their identity remains to be elucidated (Bandyopadhyay *et al.*, 2002).

Fibronectin-binding protein

Both promastigotes and amastigotes of *Leishmania* species (*L. amazonensis, L. major, L. donovani*) express a distinct cell surface receptor for fibronectin-binding (Kulkarni *et al.*, 2008; Wyler *et al.*, 1985). Apart from this receptor, gp63, the metalloprotease can also specifically bind to human fibronectin and plays an important role in internalization of the parasite by the macrophage cells (Chakrabarty *et al.*, 1996). The binding of the surface receptor for fibronectin has been reported to increase the proximity of leishmanolysin (gp63) with fibronectin facilitating its degradation essential for inhibition of generation of reactive oxygen species (ROS) by macrophage and help in parasite survival within the host.

Collagen-binding protein

The surface of promastigotes of *L. amazonensis* and *L. mexicana* have receptors for collagen I and has been reported to move through its scaffolds (Petropolis *et al.*, 2014; Lira *et al.*, 1997). The interaction between such type I collagen receptor of the parasite with collagen I has been reported to be dose dependent and saturable (Lira *et al.*, 1997). Parasite invasion studies on collagen I 3D scaffolds have revealed remodelling of the collagen (20% degradation) possibly by the metalloproteases to facilitate their migration and internalization within host macrophage (de Menezes *et al.*, 2016). Studies reveal that a shift in collagen type from type I collagen of early stages to type III collagen during chronic phase occurs during experimental leishmaniasis (Abreu-Silva *et al.*, 2004). A higher percentage of type III collagen in the ECM decreases the stiffness of the matrix, thereby facilitating migration and invasion of the parasite (Kock *et al.*, 1997). Surface plasmon resonance imaging has also revealed that promastigotes of *Leishmania* spp. can also bind to type VI collagen of macrophages (Fatoux-Ardore *et al.*, 2014).

Heparin-binding protein

Heparin-binding protein (HBP) has been identified as adhesion molecules on the surface of several *Leishmania* spp. including promastigotes and amastigotes of *L. donovani* (Butcher *et al.*, 1992; Mukhopadhyay *et al.*, 1989; Kock *et al.*, 1997), *L. amazonensis* (Love *et al.*, 1993), *L. major* (Volf *et al.*, 2001), *L. braziliensis* (Azevedo-Pereira *et al.*, 2007; de Castro Cortes *et al.*, 2012) and *L. chagasi* (Martins *et al.*, 2015). Studies have revealed that this molecule binds not only to heparin but also to heparin sulphate and hyaluran (Fatoux-Ardore *et al.*, 2014). HBP has been reported to be one of the important factors necessary for binding and invasion of the macrophages by the parasite since pretreatment of parasites with heparin rendered them unfit for binding to the host macrophage (Butcher *et al.*, 1992). Although, HBPs have not been characterized from any species of *Leishmania*, it has been hypothesized that they resemble lectin receptors (Volf *et al.*, 2001).Studies suggest that the interaction of heparin with HBP of *L. donovani* can play a crucial role in pathogenesis as the interaction modulates various signalling pathways including inhibition of histone protein kinase activity as well as phosphorylation of 12–13 parasite proteins (Mukhopadhyay *et al.*, 1989).

Trypanosoma cruzi

Trypanosoma cruzi causes human Chagas' disease affecting 10 million people from Mexico to Argentina and Chile. The disease is transmitted either by an insect vector or during blood transfusion, organ transplantation, ingestion of contaminated food or fluids, and congenital or sexual transmission (Pina-Vazquez *et al.*, 2012). Several evidence show that *T. cruzi* interacts with host ECM components facilitating its internalization into different host tissues (Ouaissi *et al.*, 1986b; Marino *et al.*, 2003). Several ECM interacting proteins of *T. cruzi* have been identified and explored (Table 12.3).

gp63 or penetrin

Selective adhesion to ECM components – heparin, heparin sulfate as well as collagen in trypomastigotes – is mediated by a 60-kDa surface protein known as penetrin (Ortega-Barria and Pereira 1991). Such an interaction helps the parasite to adhere and migrate through the complex ECM network and invade host cells (Herrera *et al.*, 1994). Enzymatic digestion of the host cell with heparitinase and heparinase that depolarizes heparin sulphate and heparin affects the adhesion and growth of the parasite *T. cruzi* suggesting binding of heparin to be an important event in pathogenesis (Herrera *et al.*, 1994).

Propyl oligopeptidase Tc80 serine protease

A 80 kDa (Tc 80 proteinase) identified in cell-free extracts of amastigote, trypomastigote and epimastigote forms of *T. cruzi* has been reported to possess collagenase activity at neutral pH (Santana *et al.*, 1997). This proteinase activity was further reported to be secretory and its activity is strongly inhibited by $HgCl_2$, *p*-chloromercuribenzoate, tosyl-lysylchloromethane ('TLCK') and benzyloxycarbonyl-Phe-Ala-diazomethane (Santana *et al.*, 1997). This protease facilitates degradation of the Type I and Type IV collagen and disassembling of the ECM thereby help in infection establishment (Santana *et al.*, 1997).

Matrix metalloprotease-9-like proteins

An extracellular metalloprotease having matrix metalloprotease-9 (MMP-9)-like activity released by *T. cruzi* also facilitates ECM degradation and regulates parasitic infection and pathogenesis (Nogueira de Melo *et al.*, 2010). These zinc-dependent proteases are inhibited by doxycycline (Nogueira de Melo *et al.*, 2010).

GP57/51, cruzain or cruzipain

GP57/51, cruzain or cruzipain, the cysteine protease is synthesized in a regulated manner during all developmental stages of *T. cruzi* (Campetella *et al.*, 1990). Cruzipain is capable of degrading collagen, fibronectin and certain other highly antigenic proteases have been reported to be associated with the plasma membrane as well as secreted into the medium (Cazzulo *et al.*, 1990; Murta *et al.*, 1990; Cazzulo 2002). Studies have revealed that organomercurial reagents such as E-64, tosyl-L-lysine chloromethyl ketone (TLCK), cystatins and peptidyl diazomethane to be potent inhibitors of this enzyme (Lalmanach *et al.*, 1996). Specific inhibition of the enzyme can totally eliminate *T. cruzi* parasites from the host (Brak *et al.*, 2008; Barr *et al.*, 2005; Engel *et al.*, 1998).

Table 12.3 Extracellular matrix interacting proteins of glycoproteins of *Trypanosoma cruzi*

Protein	Localization	Host ECM target or ligand	Function
Penetrin (gp63)	Cell surface bound as well as secreted	Heparin, heparin sulphate and collagen	Helps in adhering and migrating through the ECM complex
POP Tc80	Secreted	Type I and IV collagen, fibronectin	Degrades collagens and disassembles the ECM of the host
Matrix metalloprotease-9-like (MMP-9-like) proteins	Cell surface bound	Not known	Facilitates ECM remodelling and pathogenesis
GP57/51, cruzain or cruzipain	Cell surface bound as well as secreted	Collagen, fibronectin and certain other proteases	Helps in infection establishment
Tc-85/gp85/TS	Cell surface	Laminin, fibronectin	Helps in adherence to the ECM, activates signalling cascade facilitating internalization of the parasite into the host

Tc85

Trypomastigote stage of *T. cruzi* expresses an array of surface glycoproteins collectively known as Tc85 as part of the gp85/trans-sialidase family (de Souza *et al.*, 2010). The concentration of Tc85 family of proteins is parasite stage-specific (de Souza *et al.*, 2010; Alves and Colli, 2008). The members of Tc85 family are known to bind to an array of ECM components such as laminin (Giordano *et al.*, 1994, 1999) and fibronectin (Ouaissi *et al.*, 1986a). A member of the polymorphic Tc-85 protein family, Tc85–11 has been reported to participate in adherence of the parasite to laminin of the ECM and help in its internalization into the host (Magdesian *et al.*, 2001). The binding of laminin and fibronectin of host ECM by *T. cruzi* dephosphorylates the paraflagellar rod proteins and α-tubulin, major structural elements of the parasite cytoskeleton (Mattos *et al.*, 2012). The cytoskeletal reorganization promotes the activation of the ERK1/2 signalling cascade, thereby, facilitating internalization of the parasite into the host cell (Dias *et al.*, 2008).

Trypanosoma brucei

The protozoan parasite *T. brucei* is responsible for causing nagana in animals and sleeping sickness in humans known as human African trypanosomiasis (HAT). It utilizes proteases to invade host tissue and establish its own niche.

T. brucei secretes a 40-kDa neutral metalloprotease that permits the parasite to degrade collagen, fibronectin and laminin and migrate through the extracellular matrix (Huet *et al.*, 1992) (Table 12.4). Several experiments revealed that the activity of this metalloprotease is inhibited by ethylene diamine tetraacetic acid (EDTA), ethylene glycol tetraacetic acid (EGTA), tetracycline and phenanthroline (de Sousa *et al.*, 2010). The highly conserved cell surface-bound matrix metalloprotease (MMP) gp63 performs several functions during the different stages of its life cycle (Pina-Vazquez *et al.*, 2012). In the later stages of the disease, extracellular release of the metalloprotease by the parasite facilitates the modification of ECM components helping it cross the blood–brain barrier (BBB) (Pina-Vazquez *et al.*, 2012).

A prolyl oligopeptidase POPTb with similar structural composition of POP from *T. cruzi* has been identified in *T. brucei* having collagenase activity (Bastos *et al.*, 2010). Associated proteases such as *T. brucei* cruzipain, brucipain and cathepsin B aid in the process of traversal across the BBB. Brucipain activates calcium-mediated signalling pathway and cruzipain activates a class of G protein-coupled receptors (GPCRs) helping the parasite to cross the BBB (McKerrow *et al.*, 1993; Mackey *et al.*, 2004; Pina-Vazquez *et al.*, 2012). All these proteases contribute an important role in degradation of the ECM by *T. brucei*, thereby helping in invasion of different organs and colonization by the parasite (Abdulla *et al.*, 2008).

Phytomonas spp.

Phytomonas spp., a digenetic parasite, infects fruits and seeds of wide variety of plants of economic importance by phytophagus insect vector (Sibley 2011). The pan-eukaryotic zinc metalloprotease gp63 is one of the major surface protease (MSP) of *Phytomonas* (d'Avila-Levy *et al.*, 2006). Although its exact role in plant host is yet to be elucidated, some evidence points that it plays an important role in

Table 12.4 Extracellular matrix interacting proteins of glycoproteins of *Trypanosoma brucei*.

Protein	Localization	Host ECM target or ligand	Function
gp63	Cell surface bound as well as secreted	ECM components	Modification of ECM components helping the parasite to cross the blood–brain barrier (BBB)
POPTb cruzipain, brucipain, cathepsin B	Secreted	ECM components	Collagenase activity; helps the parasite to cross the BBB and establish colonies

Source: Pina-Vazquez, C., Reyes-Lopez, M., Ortiz-Estrada, G., de la Garza, M., and Serrano-Luna, J. 2012. Host–parasite interaction: parasite-derived and -induced proteases that degrade human extracellular matrix. J. Parasitol. Res., 748206.

establishment of phytomonadsin, the salivary gland of the insect (d'Avila-Levy *et al.*, 2006). Another cell surface bound 67-kDa metalloprotease isolated from *P. françai* has been reported to play a role in binding to its insect host (d'Avila-Levy *et al.*, 2006).

Mechanism of extracellular matrix interaction by the extracellular matrix interacting proteins

Trypanosomatids are known to adhere to host cell ECM components and invade host tissue by using the repertoire of surface proteins that have already been discussed. These glycoproteins have been known to specifically recognize host tissue and modulate host cell signalling events in their favour and establish their niche (Masini d'Avila-Levy *et al.*, 2006).

Surface glycoproteins of *T. cruzi* have been reported to activate a protein tyrosine kinase and a parasite phospholipase C and induce calcium-mediated signalling (Ueno *et al.*, 2009). Studies also suggest stimulation of adenylate cyclase activity that facilitates Ca^{2+} release from acidocalcisomes (Burleigh and Woolsey 2002; Burleigh 2005; Rodriguez *et al.*, 1995; Low *et al.*, 1992). Quantitative analysis revealed trypomastigotes of *T. cruzi* penetrate into lysosomes of the host cell via Ca^2-dependent pathway (Low *et al.*, 1992). Transient increase of Ca^{2+}concentration in the host cell has been reported to cause a reorganization of the cytoskeleton and facilitate parasite invasion (Woolsey and Burleigh 2004; Rodriguez *et al.*, 1995; Low *et al.*, 1992). Besides surface molecules of *T. cruzi*, trypomastigotes are capable of interfering with the complement activation pathway (Araujo-Jorge *et al.*, 2008). A protease secreted by the parasite is also capable of activating the TGF-β associated with extracellular matrix (ECM) components resulting in activation of Smad 2/3 pathway facilitating infections of the heart tissue by *T. cruzi* (Araujo-Jorge *et al.*, 2008). Another mode of host cell invasion is by cytoskeletal reorganization of the paraflagellar rod proteins and α-tubulin of *T. cruzi* upon binding with laminin and fibronectin of host ECM (Mattos *et al.*, 2012). The reorganization promotes activation of the ERK1/2 signalling cascade, thereby, facilitating parasite internalization (Dias *et al.*, 2008).

Binding of ECM components such as fibronectin and type IV collagen by the ECM interacting proteins of the *Leishmania* promastigotes facilitates their degradation by proteases. Such degradation of the ECM components has been reported to be associated with enhanced migration of *Leishmania* promastigotes through ECM (McGwire *et al.*, 2003). The binding of the surface receptor for fibronectin of *Leishmania* spp. has been reported to increase the proximity of gp63 with fibronectin facilitating its degradation (Petropolis *et al.*, 2014). Degradation of ECM fibronectin of the host facilitates parasite survival by reducing the production of ROI intermediates by the host macrophage (Kulkarni *et al.*, 2008). Gp63, as already mentioned also facilitates complement inactivation in serum (Brittingham *et al.*, 1995) aiding in intra-phagolysosomal survival of the parasite (Brittingham *et al.*, 1999; Chaudhuri *et al.*, 1989). A large number of evidence suggest direct binding of metalloprotease to human natural killer (NK) cells, the crucial components of innate immunity and suppress their proliferation and gamma interferon response facilitating survival and successful infection establishment by the promastigotes of *Leishmania* spp. (Lieke *et al.*, 2008). Another mechanism of parasite survival within the host is by metalloprotease-mediated degradation of antimicrobial peptides (AMPs), also known as host defence peptides, that are capable of disrupting the membrane potential of wide range of micro-organisms leading to their death (Haines *et al.*, 2009; Kulkarni *et al.*, 2006).

Synopsis

The host–parasite relationship is a complex phenomenon and numerous micro-organisms bind to and remodel host ECM components by various ECM component binding proteins or proteases in order to establish its niche within the host. Decades of research have revealed that trypanosomatid's surface proteome confers tissue specificity for cellular invasion. The precise role of these ECM interacting proteins of trypanosomatids and the genes regulating their synthesis still remains to be elucidated. Upstream and downstream signalling pathways mediating such ECM interaction resulting in host cell invasion also remains partially understood. The identification of the genes responsible for regulation of ECM interaction as well as the regulation of signalling pathways will benefit us with better understanding of the relevance of degradation of ECM to facilitation of parasite migration *in vivo*. Elucidation of the binding region of these ECM interacting proteins of trypanosomatids and their mechanism of adherence and interaction might be potentially important for better understanding of their pathogenesis as well as for developing novel approaches of intervention in trypanosomiasis.

References

Abdulla, M.H., O'Brien, T., Mackey, Z.B., Sajid, M., Grab, D.J., and McKerrow, J.H. (2008). RNA interference of *Trypanosoma brucei* cathepsin B and L affects disease progression in a mouse model. PLoS Negl. Trop. Dis. 2, e298. https://doi.org/10.1371/journal.pntd.0000298.

Abreu-Silva, A.L., Calabrese, K.S., Mortara, R.A., Tedesco, R.C., Cardoso, F.O., Carvalho, L.O., and Goncalves da Costa, S.C. (2004). Extracellular matrix alterations in experimental murine *Leishmania (L.) amazonensis* infection. Parasitology *128*, 385-390.

Alberts, B., and National Center for Biotechnology, I. (2004). Molecular Biology of the Cell (New York, Garland).

Alves, M.J., and Colli, W. (2008). Role of the gp85/trans-sialidase superfamily of glycoproteins in the interaction of *Trypanosoma cruzi* with host structures. Subcel.l Biochem. *47*, 58-69.

Araujo-Jorge, T.C., Waghabi, M.C., Soeiro Mde, N., Keramidas, M., Bailly, S., and Feige, J.J. (2008). Pivotal role for TGF-beta in infectious heart disease: The case of *Trypanosoma cruzi* infection and consequent *Chagasic* myocardiopathy. Cytokine Growth Factor Rev. *19*, 405-413. https://doi.org/10.1016/j.cytogfr.2008.08.002.

Araujo, A., Jansen, A.M., Reinhard, K., and Ferreira, L.F. (2009). Paleoparasitology of Chagas disease – a review. Mem. Inst. Oswaldo Cruz *104* Suppl. 1, 9–16. https://dx.doi.org/10.1590/S0074-02762009000900004.

Aumailley, M., Bruckner-Tuderman, L., Carter, W.G., Deutzmann, R., Edgar, D., Ekblom, P., Engel, J., Engvall, E., Hohenester, E., Jones, J.C., *et al.* (2005). A simplified laminin nomenclature. Matrix Biol. *24*, 326-332. https://doi.org/10.1016/j.matbio.2005.05.006.

Azevedo-Pereira, R.L., Pereira, M.C., Oliveria-Junior, F.O., Brazil, R.P., Cortes, L.M., Madeira, M.F., Santos, A.L., Toma, L., and Alves, C.R. (2007). Heparin binding proteins from *Leishmania (Viannia) braziliensis* promastigotes. Vet Parasitol. *145*, 234-239.

Bandtlow, C.E., and Zimmermann, D.R. (2000). Proteoglycans in the developing brain: new conceptual insights for old proteins. Physiol. Rev. *80*, 1267-1290.

Bandyopadhyay, K., Karmakar, S., Ghosh, A., and Das, P.K. (2001). Role of 67 kDa cell surface laminin binding protein of *Leishmania donovani* in pathogenesis. J. Biochem. *130*, 141-148.

Bandyopadhyay, K., Karmakar, S., Ghosh, A., and Das, P.K. (2002). High affinity binding between laminin and laminin binding protein of *Leishmania* is stimulated by zinc and may involve laminin zinc-finger like sequences. Eur. J. Biochem. *269*, 1622–1629. https://doi.org/10.1046/j.1432-1327.2002.02793.x.

Barr, S.C., Warner, K.L., Kornreic, B.G., Piscitelli, J., Wolfe, A., Benet, L., and McKerrow, J.H. (2005). A cysteine protease inhibitor protects dogs from cardiac damage during infection by *Trypanosoma cruzi*. Antimicrob. Agents Chemother. *49*, 5160–5161. https://doi.org/10.1128/AAC.49.12.5160-5161.2005.

Bastos, I.M., Motta, F.N., Charneau, S., Santana, J.M., Dubost, L., Augustyns, K., and Grellier, P. (2010). Prolyl oligopeptidase of *Trypanosoma brucei* hydrolyzes native collagen, peptide hormones and is active in the plasma of infected mice. Microbes Infect. *12*, 457-466. https://doi.org/10.1016/j.micinf.2010.02.007.

Bonnans, C., Chou, J., and Werb, Z. (2014). Remodelling the extracellular matrix in development and disease. Nat. Rev. Mol. Cell Biol *15*, 786–801. https://doi.org/10.1038/nrm3904.

Bouvier, J., Etges, R.J., and Bordier, C. (1985). Identification and purification of membrane and soluble forms of the major surface protein of *Leishmania* promastigotes. J. Bio.l Chem. *260*, 15504–15509.

Brak, K., Doyle, P.S., McKerrow, J.H., and Ellman, J.A. (2008). Identification of a new class of nonpeptidic inhibitors of cruzain. J. Am. Chem. Soc. *130*, 6404–6410. https://doi.org/10.1021/ja710254m.

Brittingham, A., Chen, G., McGwire, B.S., Chang, K.P., and Mosser, D.M. (1999). Interaction of *Leishmania* gp63 with cellular receptors for fibronectin. Infect. Immun. *67*, 4477–4484.

Brittingham, A., Morrison, C.J., McMaster, W.R., McGwire, B.S., Chang, K.P., and Mosser, D.M. (1995). Role of the *Leishmania* surface protease gp63 in complement fixation, cell adhesion, and resistance to complement-mediated lysis. J. Immunol. *155*, 3102–3111.

Burleigh, B.A. (2005). Host cell signaling and *Trypanosoma cruzi* invasion: do all roads lead to lysosomes? Sci STKE *2005*, pe36. https://doi.org/10.1126/stke.2932005pe36.

Burleigh, B.A., and Woolsey, A.M. (2002). Cell signalling and *Trypanosoma cruzi* invasion. Cell. Microbiol. *4*, 701–711. https://doi.org/10.1046/j.1462-5822.2002.00226.x.

Butcher, B.A., Sklar, L.A., Seamer, L.C., and Glew, R.H. (1992). Heparin enhances the interaction of infective *Leishmania donovani* promastigotes with mouse peritoneal macrophages. A fluorescence flow cytometric analysis. J. Immunol. *148*, 2879–2886.

Camargo, E.P. (1999). Phytomonas and other trypanosomatid parasites of plants and fruit. Adv. Parasitol. *42*, 29–112.

Campetella, O., Martinez, J., and Cazzulo, J.J. (1990). A major cysteine proteinase is developmentally regulated in *Trypanosoma cruzi*. FEMS Microbiol Lett *55*, 145-149.

Cazzulo, J.J. (2002). Proteinases of *Trypanosoma cruzi*: patential targets for the chemotherapy of Changas desease. Curr Top Med Chem 2, 1261-1271.

Cazzulo, J.J., Cazzulo Franke, M.C., Martinez, J., and Franke de Cazzulo, B.M. (1990). Some kinetic properties of a cysteine proteinase (cruzipain) from *Trypanosoma cruzi*.

Biochim. Biophys. Acta *1037*, 186–191. https://doi.org/10.1016/0167-4838(90)90166-D.

Chakrabarty, R., Mukherjee, S., Lu, H.G., McGwire, B.S., Chang, K.P., and Basu, M.K. (1996). Kinetics of entry of virulent and avirulent strains of *Leishmania donovani* into macrophages: a possible role of virulence molecules (gp63 and LPG). J. Parasitol. *82*, 632-635.

Chaudhuri, G., Chaudhuri, M., Pan, A., and Chang, K.P. (1989). Surface acid proteinase (gp63) of *Leishmania mexicana*. A metalloenzyme capable of protecting liposome-encapsulated proteins from phagolysosomal degradation by macrophages. J. Biol. Chem. *264*, 7483–7489.

Clause, K.C., and Barker, T.H. (2013). Extracellular matrix signaling in morphogenesis and repair. Curr. Opin. Biotechnol. *24*, 830–833. https://doi.org/10.1016/j.copbio.2013.04.011.

Croft, S.L. (2008). Kinetoplastida: new therapeutic strategies. Parasite *15*, 522–527. https://doi.org/10.1051/parasite/2008153522.

d'Avila-Levy, C.M., Santos, L.O., Marinho, F.A., Dias, F.A., Lopes, A.H., Santos, A.L., and Branquinha, M.H. (2006). Gp63-like molecules in Phytomonas serpens: possible role in the insect interaction. Curr. Microbiol. *52*, 439–444. https://doi.org/10.1007/s00284-005-0222-8.

Davies, C.R., Kaye, P., Croft, S.L., and Sundar, S. (2003). Leishmaniasis: new approaches to disease control. BMJ *326*, 377–382. https://doi.org/10.1136/bmj.326.7385.377.

de Castro Cortes, L.M., de Souza Pereira, M.C., da Silva, F.S., Pereira, B.A., de Oliveira Junior, F.O., de Araujo Soares, R.O., Brazil, R.P., Toma, L., Vicente, C.M., Nader, H.B., *et al.* (2012). Participation of heparin binding proteins from the surface of *Leishmania (Viannia) braziliensis* promastigotes in the adhesion of parasites to Lutzomyia longipalpis cells (Lulo) *in vitro*. Parasit. Vectors *5*, 142. https://doi.org/10.1186/1756-3305-5-142.

de Menezes, J.P., Saraiva, E.M., and da Rocha-Azevedo, B. (2016). The site of the bite: *Leishmania* interaction with macrophages, neutrophils and the extracellular matrix in the dermis. Parasit. Vectors *9*, 264. https://doi.org/10.1186/s13071-016-1540-3.

de Sousa, K.P., Atouguia, J., and Silva, M.S. (2010). Partial biochemical characterization of a metalloproteinase from the bloodstream forms of *Trypanosoma brucei brucei* parasites. Protein J. *29*, 283–289. https://doi.org/10.1007/s10930-010-9250-8.

de Souza, W., de Carvalho, T.M., and Barrias, E.S. (2010). Review on *Trypanosoma cruzi*: host cell interaction. Int. J. Cell Biol. *2010*. https://doi.org/10.1007/10.1155/2010/295394.

Dias, W.B., Fajardo, F.D., Graca-Souza, A.V., Freire-de-Lima, L., Vieira, F., Girard, M.F., Bouteille, B., Previato, J.O., Mendonca-Previato, L., and Todeschini, A.R. (2008). Endothelial cell signalling induced by trans-sialidase from *Trypanosoma cruzi*. Cell. Microbiol. *10*, 88-99. https://doi.org/10.1111/j.1462-5822.2007.01017.x.

Durbeej, M. (2010). Laminins. Cell Tissue Res *339*, 259-268. https://doi.org/10.1007/s00441-009-0838-2.

Engel, J.C., Doyle, P.S., Hsieh, I., and McKerrow, J.H. (1998). Cysteine protease inhibitors cure an experimental *Trypanosoma cruzi* infection. J. Exp. Med. *188*, 725-734. https://doi.org/10.1084/jem.188.4.725.

Fatoux-Ardore, M., Peysselon, F., Weiss, A., Bastien, P., Pratlong, F., and Ricard-Blum, S. (2014). Large-scale investigation of *Leishmania* interaction networks with host extracellular matrix by surface plasmon resonance imaging. Infect. Immun. *82*, 594-606. https://doi.org/10.1128/IAI.01146-13.

Frantz, C., Stewart, K.M., and Weaver, V.M. (2010). The extracellular matrix at a glance. J Cell Sci *123*, 4195-4200. https://doi.org/10.1242/jcs.023820.

George, E.L., Georges-Labouesse, E.N., Patel-King, R.S., Rayburn, H., and Hynes, R.O. (1993). Defects in mesoderm, neural tube and vascular development in mouse embryos lacking fibronectin. Development *119*, 1079–1091.

Ghosh, A., Bandyopadhyay, K., Kole, L., and Das, P.K. (1999). Isolation of a laminin-binding protein from the protozoan parasite *Leishmania donovani* that may mediate cell adhesion. Biochem. J. 337, 551–558.

Ghosh, A., Kole, L., Bandyopadhyay, K., Sarkar, K., and Das, P.K. (1996). Evidence of a laminin binding protein on the surface of *Leishmania donovani*. Biochem Biophys Res Commun *226*, 101–106. https://doi.org/10.1006/bbrc.1996.1317.

Giordano, R., Chammas, R., Veiga, S.S., Colli, W., and Alves, M.J. (1994). An acidic component of the heterogeneous Tc-85 protein family from the surface of *Trypanosoma cruzi* is a laminin binding glycoprotein. Mol Biochem Parasitol *65*, 85–94.

Giordano, R., Fouts, D.L., Tewari, D., Colli, W., Manning, J.E., and Alves, M.J. (1999). Cloning of a surface membrane glycoprotein specific for the infective form of *Trypanosoma cruzi* having adhesive properties to laminin. J Biol Chem *274*, 3461-3468.

Haines, L.R., Thomas, J.M., Jackson, A.M., Eyford, B.A., Razavi, M., Watson, C.N., Gowen, B., Hancock, R.E., and Pearson, T.W. (2009). Killing of trypanosomatid parasites by a modified bovine host defense peptide, BMAP-18. PLoS Negl. Trop. Dis *3*, e373. https://doi.org/10.1371/journal.pntd.0000373.

Herrera, E.M., Ming, M., Ortega-Barria, E., and Pereira, M.E. (1994). Mediation of *Trypanosoma cruzi* invasion by heparan sulfate receptors on host cells and penetrin counter-receptors on the trypanosomes. Mol Biochem Parasitol *65*, 73–83.

Huet, G., Richet, C., Demeyer, D., Bisiau, H., Soudan, B., Tetaert, D., Han, K.K., and Degand, P. (1992). Characterization of different proteolytic activities in *Trypanosoma brucei brucei*. Biochim Biophys Acta *1138*, 213–221.

Hynes, R.O. (2009). The extracellular matrix: not just pretty fibrils. Science *326*, 1216-1219. https://doi.org/10.1126/science.1176009.

Iorio, V., Troughton, L.D., and Hamill, K.J. (2015). Laminins: roles and utility in wound repair. Adv. Wound Care (New Rochelle) *4*, 250–263. https://doi.org/10.1089/wound.2014.0533.

Iozzo, R.V., and Murdoch, A.D. (1996). Proteoglycans of the extracellular environment: clues from the gene and protein side offer novel perspectives in molecular diversity and function. FASEB J. *10*, 598–614.

Jakeman, L.B., Williams, K.E., and Brautigam, B. (2014). In the presence of danger: The extracellular matrix defensive response to central nervous system injury. Neural Regen Res 9, 377–384. https://doi.org/10.4103/1673-5374.128238.

Jarvelainen, H., Sainio, A., Koulu, M., Wight, T.N., and Penttinen, R. (2009). Extracellular matrix molecules: potential targets in pharmacotherapy. Pharmacol Rev *61*, 198-223. https://doi.org/10.1124/pr.109.001289.

Knudson, C.B., and Knudson, W. (2001). Cartilage proteoglycans. Semin Cell Dev Biol *12*, 69–78. https://doi.org/10.1006/scdb.2000.0243.

Kock, N.P., Gabius, H.J., Schmitz, J., and Schottelius, J. (1997). Receptors for carbohydrate ligands including heparin on the cell surface of *Leishmania* and other trypanosomatids. Trop Med Int Health *2*, 863–874. https://doi.org/10.1046/j.1365-3156.1997.d01-405.x.

Kulkarni, M.M., Jones, E.A., McMaster, W.R., and McGwire, B.S. (2008). Fibronectin binding and proteolytic degradation by *Leishmania* and effects on macrophage activation. Infect Immun *76*, 1738–1747. https://doi.org/10.1128/IAI.01274-07.

Kulkarni, M.M., McMaster, W.R., Kamysz, E., Kamysz, W., Engman, D.M., and McGwire, B.S. (2006). The major surface-metalloprotease of the parasitic protozoan, *Leishmania*, protects against antimicrobial peptide-induced apoptotic killing. Mol Microbiol *62*, 1484–1497. https://doi.org/ 10.1111/j.1365-2958.2006.05459.x.

Lalmanach, G., Mayer, R., Serveau, C., Scharfstein, J., and Gauthier, F. (1996). Biotin-labelled peptidyl diazomethane inhibitors derived from the substrate-like sequence of cystatin: targeting of the active site of cruzipain, the major cysteine proteinase of *Trypanosoma cruzi*. Biochem. J. *318*, 395–399. https://doi.org/ 10.1042/bj3180395.

Lemichez, E., Lecuit, M., Nassif, X., and Bourdoulous, S. (2010). Breaking the wall: targeting of the endothelium by pathogenic bacteria. Nat Rev Microbiol *8*, 93-104. https://doi.org/ 10.1038/nrmicro2269.

Lieke, T., Nylen, S., Eidsmo, L., McMaster, W.R., Mohammadi, A.M., Khamesipour, A., Berg, L., and Akuffo, H. (2008). *Leishmania* surface protein gp63 binds directly to human natural killer cells and inhibits proliferation. Clin. Exp. Immunol. *153*, 221–230. https://doi.org/10.1111/j.1365-2249.2008.03687.x.

Lira, R., Rosales-Encina, J.L., and Arguello, C. (1997). *Leishmania mexicana*: binding of promastigotes to type I collagen. Exp. Parasitol. *85*, 149–157. https://doi.org/10.1006/expr.1996.4127.

Liu, X., and Chang, K.P. (1992). Extrachromosomal genetic complementation of surface metalloproteinase (gp63)-deficient *Leishmania* increases their binding to macrophages. Proc. Natl. Acad. Sci. U.S.A. *89*, 4991–4995.

Love, D.C., Esko, J.D., and Mosser, D.M. (1993). A heparin-binding activity on *Leishmania* amastigotes which mediates adhesion to cellular proteoglycans. J. Cell Biol. *123*, 759–766.

Low, H.P., Paulin, J.J., and Keith, C.H. (1992). *Trypanosoma cruzi* infection of BSC-1 fibroblast cells causes cytoskeletal disruption and changes in intracellular calcium levels. J. Protozool. *39*, 463–470.

Lu, P., Takai, K., Weaver, V.M., and Werb, Z. (2011). Extracellular matrix degradation and remodeling in development and disease. Cold Spring Harb. Perspect. Biol. 3. https://doi.org/ 10.1101/cshperspect.a005058.

Lu, P., Weaver, V.M., and Werb, Z. (2012). The extracellular matrix: a dynamic niche in cancer progression. J Cell Biol *196*, 395-406. https://doi.org/10.1083/jcb.201102147.

Mackey, Z.B., O'Brien, T.C., Greenbaum, D.C., Blank, R.B., and McKerrow, J.H. (2004). A cathepsin B-like protease is required for host protein degradation in *Trypanosoma brucei*. J. Biol. Chem. *279*, 48426–48433. https://doi.org/10.1074/jbc.M402470200.

Magdesian, M.H., Giordano, R., Ulrich, H., Juliano, M.A., Juliano, L., Schumacher, R.I., Colli, W., and Alves, M.J. (2001). Infection by *Trypanosoma cruzi*. Identification of a parasite ligand and its host cell receptor. J. Biol. Chem. *276*, 19382–19389. https://doi.org/10.1074/jbc.M011474200.

Marino, A.P., Silva, A.A., Pinho, R.T., and Lannes-Vieira, J. (2003). *Trypanosoma cruzi* infection: a continuous invader-host cell cross talk with participation of extracellular matrix and adhesion and chemoattractant molecules. Braz. J. Med. Biol. Res. *36*, 1121–1133.

Martins, T.V., de Carvalho, T.V., de Oliveira, C.V., de Paula, S.O., Cardoso, S.A., de Oliveira, L.L., and Marques-da-Silva Ede, A. (2015). *Leishmania chagasi* heparin-binding protein: Cell localization and participation in *L. chagasi* infection. Mol. Biochem. Parasitol. *204*, 34–43. https://doi.org/10.1016/j.molbiopara.2015.12.005.

Masini, d'Avila-Levy, C., de Almeida Dias, F., Nogueira de Melo, A.C., Martins, J.L., De Carvalho Santos Lopes, A.H., Souza Dos Santos, A.L., Vermelho, A.B., and Branquinha, M.H. (2006). Insights into the role of gp63-like proteins in lower trypanosomatids. FEMS Microbiol. Lett. *254*, 149–156. https://doi.org/10.1111/j.1574-6968.2005.00022.x.

Mattos, E.C., Schumacher, R.I., Colli, W., and Alves, M.J. (2012). Adhesion of *Trypanosoma cruzi* trypomastigotes to fibronectin or laminin modifies tubulin and paraflagellar rod protein phosphorylation. PLoS One *7*, e46767. https://doi.org/10.1371/journal.pone.0046767.

McGhee, R.B., and Cosgrove, W.B. (1980). Biology and physiology of the lower Trypanosomatidae. Microbiol Rev *44*, 140-173.

McGwire, B., and Chang, K.P. (1994). Genetic rescue of surface metalloproteinase (gp63)-deficiency in *Leishmania amazonensis* variants increases their infection of macrophages at the early phase. Mol. Biochem. Parasitol. *66*, 345–347.

McGwire, B.S., Chang, K.P., and Engman, D.M. (2003). Migration through the extracellular matrix by the parasitic protozoan *Leishmania* is enhanced by surface metalloprotease gp63. Infect. Immun. *71*, 1008–1010. https://doi.org/10.1128/IAI.71.2.1008-1010.2003.

McKerrow, J.H., Sun, E., Rosenthal, P.J., and Bouvier, J. (1993). The proteases and pathogenicity of parasitic protozoa. Annu. Re.v Microbiol. *47*, 821-853. https://doi.org/10.1146/annurev.mi.47.100193.004133.

Medina-Acosta, E., Beverley, S.M., and Russell, D.G. (1993). Evolution and expression of the *Leishmania* surface proteinase (gp63) gene locus. Infect Agents Dis *2*, 25–34.

Mercurio, A.M., and Shaw, L.M. (1991). Laminin binding proteins. Bioessays *13*, 469-473. https://doi.org/10.1002/bies.950130907.

Mouw, J.K., Ou, G., and Weaver, V.M. (2014). Extracellular matrix assembly: a multiscale deconstruction. Nat Rev Mol Cell Biol *15*, 771–785. https://doi.org/10.1038/nrm3902.

Mukhopadhyay, N.K., Shome, K., Saha, A.K., Hassell, J.R., and Glew, R.H. (1989). Heparin binds to *Leishmania donovani* promastigotes and inhibits protein phosphorylation. Biochem J *264*, 517–525. https://doi.org/10.1042/bj2640517.

Murray, H.W., Berman, J.D., Davies, C.R., and Saravia, N.G. (2005). Advances in leishmaniasis. Lancet *366*, 1561–1577. https://doi.org/10.1016/S0140-6736(05)67629-5.

Murta, A.C., Persechini, P.M., Padron Tde, S., de Souza, W., Guimaraes, J.A., and Scharfstein, J. (1990). Structural and functional identification of GP57/51 antigen of *Trypanosoma cruzi* as a cysteine proteinase. Mol. Biochem. Parasitol. *43*, 27–38.

Nelson, C.M., and Bissell, M.J. (2006). Of extracellular matrix, scaffolds, and signaling: tissue architecture regulates development, homeostasis, and cancer. Annu Rev Cell Dev Biol *22*, 287-309. https://doi.org/10.1146/annurev.cellbio.22.010305.104315.

Nogueira, de Melo, A.C., de Souza, E.P., Elias, C.G., dos Santos, A.L., Branquinha, M.H., d'Avila-Levy, C.M., dos Reis, F.C., Costa, T.F., Lima, A.P., de Souza Pereira, M.C., *et al.* (2010). Detection of matrix metallopeptidase-9-like proteins in *Trypanosoma cruzi*. Exp. Parasitol. *125*, 256–263. https://doi.org/10.1016/j.exppara.2010.01.023.

Ortega-Barria, E., and Pereira, M.E. (1991). A novel *T. cruzi* heparin-binding protein promotes fibroblast adhesion and penetration of engineered bacteria and trypanosomes into mammalian cells. Cell *67*, 411–421. https://doi.org/10.1016/0092-8674(91)90192-2.

Ouaissi, M.A., Cornette, J., Afchain, D., Capron, A., Gras-Masse, H., and Tartar, A. (1986a). *Trypanosoma cruzi* infection inhibited by peptides modeled from a fibronectin cell attachment domain. Science *234*, 603–607. https://doi.org/10.1126/science.3094145.

Ouaissi, M.A., Cornette, J., and Capron, A. (1986b). Identification and isolation of *Trypanosoma cruzi* trypomastigote cell surface protein with properties expected of a fibronectin receptor. Mol. Biochem. Parasitol. *19*, 201–211.

Petropolis, D.B., Rodrigues, J.C., Viana, N.B., Pontes, B., Pereira, C.F., and Silva-Filho, F.C. (2014). *Leishmania amazonensis* promastigotes in 3D Collagen I culture: an *in vitro* physiological environment for the study of extracellular matrix and host cell interactions. Peer J. 2, e317. https://doi.org/10.7717/peerj.317.

Pina-Vazquez, C., Reyes-Lopez, M., Ortiz-Estrada, G., de la Garza, M., and Serrano-Luna, J. (2012). Host-parasite interaction: parasite-derived and -induced proteases that degrade human extracellular matrix. J. Parasitol. Re.s *2012*, 748206. https://doi.org/10.1155/2012/748206.

Rodriguez, A., Rioult, M.G., Ora, A., and Andrews, N.W. (1995). A trypanosome-soluble factor induces IP3 formation, intracellular Ca^{2+} mobilization and microfilament rearrangement in host cells. J Cell Biol *129*, 1263-1273.

Santana, J.M., Grellier, P., Schrevel, J., and Teixeira, A.R. (1997). A *Trypanosoma cruzi*-secreted 80 kDa proteinase with specificity for human collagen types I and IV. Biochem J *325 (Pt 1)*, 129-137. https://doi.org/10.1042/bj3250129.

Schaefer, L., and Iozzo, R.V. (2008). Biological functions of the small leucine-rich proteoglycans: from genetics to signal transduction. J Biol Chem *283*, 21305-21309. https://doi.org/10.1074/jbc.R800020200.

Schaefer, L., and Schaefer, R.M. (2010). Proteoglycans: from structural compounds to signaling molecules. Cell Tissue Res *339*, 237–246. https://doi.org/10.1007/s00441-009-0821-y.

Sibley, L.D. (2011). Invasion and intracellular survival by protozoan parasites. Immunol Rev *240*, 72-91. https://doi.org/10.1111/j.1600-065X.2010.00990.x.

Singh, B., Fleury, C., Jalalvand, F., and Riesbeck, K. (2012). Human pathogens utilize host extracellular matrix proteins laminin and collagen for adhesion and invasion of the host. FEMS Microbiol Rev *36*, 1122-1180. https://doi.org/10.1111/j.1574-6976.2012.00340.x.

Singh, P., Carraher, C., and Schwarzbauer, J.E. (2010). Assembly of fibronectin extracellular matrix. Annu. Rev. Cell Dev. Biol. *26*, 397–419. https://doi.org/10.1146/annurev-cellbio-100109-104020.

Theocharis, A.D., Skandalis, S.S., Gialeli, C., and Karamanos, N.K. (2016). Extracellular matrix structure. Adv. Drug Deliv. Rev. *97*, 4–27. https://doi.org/10.1016/j.addr.2015.11.001.

Ueno, N., Bratt, C.L., Rodriguez, N.E., and Wilson, M.E. (2009). Differences in human macrophage receptor usage, lysosomal fusion kinetics and survival between logarithmic and metacyclic *Leishmania infantum chagasi* promastigotes. Cell. Microbiol. *11*, 1827–1841. https://doi.org/ 10.1111/j.1462-5822.2009.01374.x.

Vanlaere, I., and Libert, C. (2009). Matrix metalloproteinases as drug targets in infections caused by gram-negative bacteria and in septic shock. Clin. Microbio.l Rev. *22*, 224–239, Table of Contents. https://doi.org/10.1128/CMR.00047-08.

Volf, P., Svobodova, M., and Dvorakova, E. (2001). Bloodmeal digestion and *Leishmania major* infections in Phlebotomus duboscqi: effect of carbohydrates inhibiting midgut lectin activity. Med. Vet. Entomol. *15*, 281–286. https://doi.org/10.1046/j.0269-283x.2001.00308.x.

Wallace, F.G. (1966). The trypanosomatid parasites of insects and arachnids. Exp. Parasitol. *18*, 124–193. https://doi.org/10.1016/0014-4894(66)90015-4.

Woolsey, A.M., and Burleigh, B.A. (2004). Host cell actin polymerization is required for cellular retention of *Trypanosoma cruzi* and early association with endosomal/lysosomal compartments. Cell. Microbiol. *6*, 829–838. https://doi.org/10.1111/j.1462-5822.2004.00405.x.

Wyler, D.J., Sypek, J.P., and McDonald, J.A. (1985). *In vitro* parasite-monocyte interactions in human leishmaniasis: possible role of fibronectin in parasite attachment. Infect.Immun. *49*, 305–311.

Xu, R., Boudreau, A., and Bissell, M.J. (2009). Tissue architecture and function: dynamic reciprocity via extra- and intra-cellular matrices. Cancer Metastasis Rev. *28*, 167.176. https://doi.org/ 10.1007/s10555-008-9178-z

Effects of Phospholipid Analogues on Trypanosomatids

13

Wanderley de Souza[1,2*], Joseane Godinho[1,2], Emile Barrias[1,2,3], Marina Roussaki[4], Juliany Cola Fernandes Rodrigues[1,2,5] and Theodora Calogeropoulou[4]

[1]Laboratório de Ultraestrutura Celular Hertha Meyer, Instituto de Biofísica Carlos Chagas Filho, Universidade Federal do Rio de Janeiro, Rio de Janeiro, Brazil.
[2]Instituto Nacional de Ciência e Tecnologia de Biologia Estrutural e Bioimagem, Rio de Janeiro, Brazil.
[3]Laboratório de Microscopia Aplicada a Ciências da Vida, Instituto Nacional de Metrologia, Normalização, Tecnologia e Qualidade-INMETRO, Duque de Caxias, Rio de Janeiro, Brazil.
[4]National Hellenic Research Foundation, Institute of Biology, Medicinal Chemistry and Biotechnology, Athens, Greece.
[5]Núcleo Multidisciplinar de Pesquisa em Biologia (NUMPEX-BIO), Polo Avançado de Xerém, Universidade Federal do Rio de Janeiro, Duque de Caxias, Brazil.

*Correspondence: wsouza@biof.ufrj.br

https://doi.org/10.21775/9781910190715.13

Abstract

Diseases caused by pathogenic protozoa that belong to the Trypanosomatidae include Chagas' disease and sleeping sickness, which are caused by *Trypanosoma cruzi* and *Trypanosoma brucei* (rhodesiense and gambiense subspecies), respectively, and the various forms of leishmaniasis, caused by several species of the *Leishmania* genus. Although significant efforts have been directed at developing new chemotherapeutic strategies against trypanosomatids, the compounds routinely used in clinical practice are those that were identified many years ago (pentavalent antimonials, pentamidine, or liposomal amphotericin B for leishmaniasis; nifurtimox or benzinidazole for Chagas' disease). Several groups are working on the identification of new and more specific drug targets in trypanosomatids. One important recognized target is the phospholipids (PLs) biosynthesis in view of the important role played by these molecules in cell structure and functions. In this chapter we review basic aspects of the PLs biosynthetic pathways both in mammalian cells and in trypanosomatids. In addition special emphasis is given to the phospholipid analogues (PAs) that constitute a new class of pharmaceutical compounds that exhibit excellent biocompatibility and an especial amphiphilicity making them appropriate to be employed as drugs with broad biological activity. The various possibilities to synthesize these compounds either by organic and/or enzymatic synthetic steps are discussed. Mention is made to the process of molecular hybridization which allows the synthesis of new hybrid molecules by combining two pharmacophores such as PAs with trifluralin. Several PAs have been tested against parasitic protozoa and some of them have shown excellent anti-parasite activity. One of them, miltefosine, has been introduced in clinics to treat leishmaniasis. We also review the effect of several PAs on the growth, morphology, cell cycle, lipid accumulation, and mitochondrion function in the parasites. These compounds are able to kill the parasites through mechanisms such as autophagy and/or a apoptotic-like cell death process.

Introduction

Diseases caused by pathogenic protozoa, including those whose biological agents belong to the Trypanosomatidae family, represent important public health problems in several regions of the world. As examples, we can mention Chagas' disease and sleeping sickness, which are caused by *Trypanosoma cruzi* and *Trypanosoma brucei* (*rhodesiense* and *gambiense* subspecies), respectively, and the various forms of leishmaniasis, caused by several species of the *Leishmania* genus. In addition, some species of the genus *Phytomonas* are responsible for important plant diseases and still others belonging to the *T. brucei* complex are pathogenic for both wild and animals of economic interest. Although significant efforts have been directed at developing new chemotherapeutic strategies against trypanosomatids, the compounds routinely used in clinical practice are those that were identified many years ago. For instance, the current treatment for clinical leishmaniasis manifestations present problems such as low efficacy, high toxicity, parasite resistance, difficulty of acquisition by patients, and high cost (Barrett and Croft, 2012, Chawla and Madhubala, 2010; Singh *et al.*, 2012). The pentavalent antimonials, used in Brazil and most parts of the world as the first line of treatment, have been available since the beginning of the twentieth century. When this first-choice therapy fails, another drug such as pentamidine, liposomal amphotericin B, or possibly paramomicin can be substituted. The use and efficacy of these second-line drugs may also vary between different clinical manifestations and regions of the world in which they are used. In the case of *T. cruzi*, the treatment is based on two drugs: nifurtimox and benznidazole. Independently of the compounds utilized, the treatment is not effective for the chronic phase of Chagas' disease (reviewed by De Souza *et al.*, 2013). It is clear that investment in research and development of new compounds to treat these diseases is very low when compared to other diseases mainly found in more affluent countries. For instance, between 2000 and 2011, only 3.8% of newly developed drugs were destined to treat diseases now recognized as neglected diseases, which may affect approximately one sixth of the world's population (DNDi Latin America, 2012). Several groups are working on the identification of new and specific drug targets in trypanosomatids in addition to testing thousands of already synthesized compounds for reasons other than treating trypanosomatids. One of the main targets is lipid biosynthesis. Indeed, significant advances have been reported with the use of inhibitors of the sterol biosynthesis pathway (Review in Macedo-Silva *et al.*, 2015). On the other hand, it is well recognized that PLs are essential components in cell structure and function and play a key role in several essential cellular functions. Herein, we will review some basic aspects of the PL biosynthetic pathway, comparing the differences in composition and metabolic processes between mammalian cells and trypanosomatids. In addition, we will describe the efforts by several groups to develop new compounds that interfere with PLs biosynthesis.

Phospholipids and their importance for the cell

Phospholipids (PLs) represent the major lipids found in biological membranes, and they play important roles in several biological processes. Among the most abundant phospholipids are the phosphoglycerides, which are characterized by the presence of three-carbon glycerol backbones. The most abundant phospholipid in mammalian cells is phosphatidylcholine (PC). However, a variety of other phospholipids are also present and have important functions as membrane constituents: (1) phosphatidylethanolamine (PE); (2) phosphatidylserine (PS); (3) sphingomyelin (SM); (4) cardiolipin (CL); and (5) phosphatidylinositol (PI) and its phosphorylated derivatives. The PLs contain a polar head group and two hydrocarbon tails, which consist of different length fatty acids. Over the past 25 years, phospholipids have been characterized as key regulatory molecules involved in different cell signalling pathways, sometimes acting as signalling molecules or second messengers (Berridge and Irvine, 1984; Nishizuka, 1986). Another pivotal example of the importance of phospholipids for cellular metabolism is the participation of phosphatidylserine in apoptotic cell death (Fadok *et al.*, 1992, 2000). One characteristic feature of the phospholipids is spontaneous bilayer formation in an aqueous environment. This property allows the assembly of membranes, which delineate each cell and each organelle and has a fundamental role in the establishment of several intracellular compartments.

Phospholipids biosynthetic pathways in mammalian cells and in trypanosomatids

The first key molecule involved in mammalian phospholipid biosynthesis is phosphatidic acid (PA), which is used to produce diacylglycerol *per se* or CDP-diacylglycerol, two phospholipid precursors essential for the synthesis of all mammalian phospholipids. PA can be produced by two different reactions from either glycerol-3-phosphate or dihydroxyacetone phosphate (Fig. 13.1). These reactions are catalysed by different enzymes such as glycerol-3-phosphate acyltransferase (GPAT), phosphatidic acid phosphatase-1 (PAP-1), and CDP-diacylglycerol synthase (CDP-GDS) that have been found in the membranes of endoplasmic reticulum and mitochondrion (reviewed in Vance, 2014). In 1956, using rat liver as an enzyme source, Kennedy and Weiss (1956) elucidated a pathway for the *de novo* biosynthesis of PE and PC. The PE and PC branches of this Kennedy pathway are based upon the formation of characteristic high-energy intermediates: (1) CDP-ethanolamine used for the synthesis of PE and (2) CDP-choline used for the synthesis of PC. As a consequence, the two branches of the Kennedy pathway are often referred to as the CDP-ethanolamine and CDP-choline pathways, respectively. However, nowadays in the genomic era, several genes coding for proteins involved in the PL biosynthesis have been identified resulting in a significant increase in the basic knowledge of the PL biosynthetic pathway and PL sorting to different organelles (Vance and Vance, 2004). All mammalian cells make PC via the CDP-choline pathway (Kennedy pathway) (Fig. 13.1) (Kennedy and Weiss, 1956; Vance, 2014). First, extracellular choline is imported into the cell and rapidly phosphorylated to phosphocholine by the cytosolic enzyme choline kinase. In the next reaction, the CTP: phosphocholine cytidylyltransferase generates CDP-choline, an enzyme that exists as

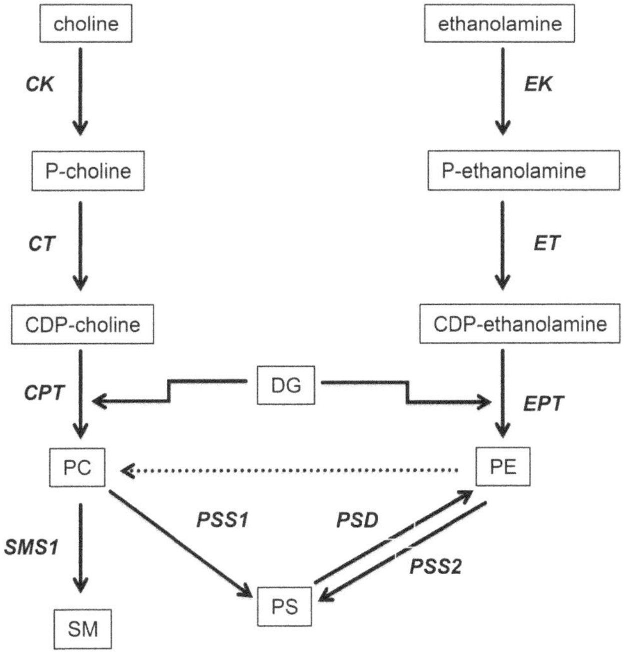

Figure 13.1 Summary diagram showing the basic steps of the biosynthetic pathway for PC, PE, and PS in mammalian cells. CK, choline kinase; CPT, CDP-choline:1,2-diacylglycerol cholinephosphotransferase; CT, CTP:phosphocholine cytidylyltransferase; EK, ethanolamine kinase; ET, CTP: phosphoethanolamine cytidyl-yltransferase; EPT, CDP-ethanolamine:1,2-diacylglycerol ethanol- aminephosphotransferase; PC, phosphatidylcholine; PE, phosphatidylethanolamine; PEMT, phosphatidylethanolamine *N*-methyltransferase; PS, phosphatidylserine; PSD, phosphatidyl-serine decarboxylase; PSS, phosphatidylserine synthase; SM, sphingomyelin; SMS1, sphingomyelin synthase-1.

two isoforms (α and β), which are activated upon binding to membranes. Interestingly, the α isoform contains a nuclear localization sequence that is absent in the β; thus, the first isoform appears initially to be located in the nucleus and is then translocated to the cytosol, while the second is only extranuclear (Pelech and Vance, 1984; Cornell and Northwood, 2000; Fagone and Jackowski, 2013). Another important enzyme in this pathway is CDP-choline: 1,2-diacylglycerol cholinephosphotransferase, an integral membrane protein primarily localized to the ER, which catalyses the transfer of diacylglycerol to CDP choline to synthesize PC. In the alternative pathway for PC synthesis, PE is converted to PC by three sequential methylation reactions that are all catalysed by PE *N*-methyltransferase, an enzyme bound to ER membranes (Fig. 13.1) (Vance, 2013, 2015). This group of reactions is known as the Greenberg pathway and predominates in lower eukaryotes such as fungi and some protozoan parasites. In mammalian cells, the exception occurs in hepatocytes, in which around 30% of total PC is produced by PE methylation (Walkey *et al.*, 1997). With regard to the synthesis of PS, two distinct enzymes, PS-synthase 1 and 2, are enriched in a specific domain of the ER called mitochondria-associated membranes (MAM) (Stone and Vance, 2000). These synthases catalyse two reactions in which serine replaces either the choline head group of PC (PS synthase-1) or ethanolamine head group of PE (PS synthase 2) (Vance and Tasseva, 2013). Finally, there are two distinct biosynthetic routes to synthesize PE in mammalian cells that occur in the mitochondrion and endoplasmic reticulum (ER) (Vance and Vance, 2008; Vance and Tavessa, 2013). In the CDP-ethanolamine pathway, the first enzymes are cytosolic, whereas the last enzyme is bound to ER membranes. As shown in Fig. 13.1, there are two initial steps that convert ethanolamine to CDP-ethanolamine, which finally will be converted to PE by the integral ER membrane protein CDP-ethanolamine:1,2-diacylglycerol ethanolaminephosphotransferase. The second method to synthesize PE in mammalian cells requires PS decarboxylase, an enzyme that is restricted to the inner mitochondrial membrane (Borkenhagen *et al.*, 1961; Vance and Tavessa, 2013); in this pathway, PS is transported to mitochondrion where it should be decarboxylated and generate PE that can be exported to other organelles by unknown mechanisms (Shiao *et al.*, 1995; Kainu *et al.*, 2013).

After describing how PLs are synthesized in mammalian cells, it is important to look at what happens in trypanosomatids. Several studies have analysed the phospholipid composition of the three main pathogenic trypanosomatids (*Trypanosoma brucei*, *Leishmania* spp. and *Trypanosoma cruzi*) (Ramakrishnan *et al.*, 2013). African trypanosomes have a membrane glycerophospholipid composition similar to higher eukaryotes, with PC (45-60% of total lipid phosphorous) and PE (10-20% of total lipid phosphorus) representing the two most abundant PLs followed by phosphatidylinositol (PI) (6-12%), PS (4%), CL (2%-4%) and phosphatidylglycerol (PG) (2%) as minor membrane phospholipids (Dixon and Williamson, 1970; Patnaik *et al.*, 1993; Carroll and McCrorie, 1986; Gonzalez *et al.*, 2012). Similarly, in *Leishmania*, PC is the most abundant (30-50%), followed by PE (10%) and PI (10%); for PA, CL, and PG, it has been found in only small traces and is almost undetectable (Wassef *et al.*, 1985). There are some controversies in relation to the presence of PS in *Leishmania*. It was reported in several studies (Wassef *et al.*, 1985; Zheng *et al.*, 2010; Ramos *et al.*, 2008; Van Zandbergern *et al.*, 2006). However, mass spectrometry analysis and serine labelling failed to detect this lipid in *L. major* promastigotes (Zhang *et al.*, 2007; Zufferey *et al.*, 2003; Weingärtner *et al.*, 2012). Another relevant piece of information is that the glycerophospholipids of *Leishmania* and *T. brucei* consist of high amounts of ether-type molecular species (Richmond *et al.*, 2010; Patnaik *et al.*, 1993; Signorell *et al.*, 2008); both parasites contain saturated and unsaturated C18 chains with some small differences in the double bound position (Beach *et al.*, 1979). In *T. cruzi* epimastigotes, the phospholipid analysis showed that they are composed mainly by PC (44%), followed by PE (28%), PI (12%), SM (4%), and small traces of CL, PA, and lyso-PC and PS (Oliveira *et al.*, 1977). The sphingophospholipid classes, sphingomyelin (SM), inositol phosphorylceramide (IPC), and ethanolamine phosphorylceramide (EPC), constitute 10-15% of the total lipid phospholipids in *T. brucei* (Dixon and Williamson, 1970; Patnaik *et al.*, 1993; Sutterwala *et al.*, 2007, 2008). The presence of SM and IPC has also been demonstrated in *T. cruzi* epimastigotes (Oliveira *et al.*, 1997; Quiñones *et al.*, 2004; Bertello *et al.*, 1995; Uhrig *et al.*, 1996), whereas in *Leishmania*, IPC represents the major sphingophospholipid class, which does not synthesize SM, in contrast to that in trypanosome parasites (Zhang *et al.*, 2003; Kaneshiro *et al.*, 1986). In *T. cruzi* and *T. brucei*, the Kennedy pathway represents the only route for PC synthesis. Genomic analysis showed that they lack genes encoding for PE *N*-methyltransferase and phosphoethanolamine methyltransferase (Ramakrishnan *et al.*, 2013). Several studies confirmed these observations as determined by labelling experiments using ethanolamine (Gibellini *et al.*, 2009; Rifkin *et al.*, 1995) or serine (Signorell *et al.*, 2008). These data confirmed that trypanosomes are auxotrophic for choline and have to up take choline from the extracellular environment. Different from trypanosomes, *Leishmania* spp. genomes revealed genes encoding candidate PE *N*-methyltransferase (Lykidis, 2007) suggesting that the *Leishmania* genus possess another route for PC synthesis. However, the enzymes and pathways have to be experimentally characterized. Several works have shown that in *T. brucei* PE synthesis is dependent of CDP-ethanolamine pathway. RNAi-mediated down-regulation of ethanolamine kinase, ethanolamine-phosphate cytidyltransfarase, and EPT in *T. brucei* procyclics resulted in mitochondrial morphology changes, defects in cytokinesis, and growth arrest (Signorell *et al.*, 2008; Farine and Bütikofer, 2013; Signorell *et al.*, 2009), demonstrating that the Kennedy pathway is essential for parasite survival; they also demonstrated that ethanolamine-phosphate cytidyltransfarase is essential for bloodstream forms (Gibellini *et al.*, 2009). For *Leishmania* and *T. cruzi*, genomic analysis allowed the identification of putative genes for all three enzymes of the CDP-ethanolamine pathway; these findings indicated that this PE synthetic route is present in the Kinetoplastida order (Lykidis, 2007; Zhang and Beverley, 2010). However, PE may also be synthesized by two other alternative routes: (1) production of PE from PS decarboxylation (Signorell *et al.*, 2008; Gibellini *et al.*, 2009) or (2) formation of ethanolamine-phosphate from degradation of sphingolipids, which can then be used in the CDP-ethanolamine pathway (Zhang *et al.*, 2007; Saba *et al.*, 1997).

Studies carried out with *T. brucei* showed that inositol-3-phosphate synthase is essential for the viability of bloodstream forms, indicating that *de novo* synthesis of *myo*-inositol is required for parasite survival (Martin and Smith, 2005). Thus, some studies have confirmed that *T. brucei* has two different enzymes (Gonzalez-Salgado *et al.*, 2012; Smith and Butikofer, 2010; Martin and Smith, 2005,2006): (1) *myo*-inositol synthase, which is required for GPI synthesis and is located in the ER and (2) PI synthase, which, in bloodstream form, has dual localization (ER and Golgi complex). These observations were supported by another procyclic study showing that exogenous *myo*-inositol is essential for bulk PI and IPC production but not for GPI formation (Gonzales-Salgado *et al.*, 2012). This group also identified a *myo*-inositol transporter located in the plasma membrane and Golgi complex. Gene orthologues for *myo*-inositol-3-phosphate synthase and PI synthase have been identified in *T. congolense*, *T. vivax*, *T. cruzi*, and *Leishmania* spp., suggesting that *de novo* synthesis of *myo*-inositol occurs in all kinetoplastids (Drew *et al.*, 1995; Seyfang and Landfear, 1999). Gene orthologues for *myo*-inositol transporter were also found for other kinetoplastids including *Leishmania* spp. (Ramakrishnan *et al.*, 2013). Available data indicate that amastigotes from several *Leishmania* species contain host-derived SLs and PLs, which could be degraded to generate intermediates for lipid synthesis or serve as precursors for headgroup remodelling. Uptake of PLs may involve membrane-associated ATP-binding cassette (ABC) transporters. Genomes of *L. major* and *L. infantum* contain a family (42 in total) of these transporters. Several of these ABC transporters have been characterized and they possess activity to inwardly translocate PC, PE, PS, alkylglycerophosphocholine, and hexadecylphosphocholine (miltefosine). Other members of the family (such as the ABCG-like transporters) confer resistance to alkyl-phospholipids when overexpressed (Castanys-Muñoz *et al.*, 2007).

Table 13.1 summarizes some of the basic knowledge trying to compare the differences between mammalian cells and kinetoplastids about phospholipid composition, biosynthetic pathways and organelles and compartments that are involved in the routes of the phospholipid synthesis.

Table 13.1 Summary of phospholipid composition, biosynthetics pathways and organelles and compartments involved in the phospholipid synthesis in mammalian cells and kinetoplastids

	Mammalian cells	Kinetoplastids
Pathway	Kennedy[a]	Kennedy (*T. brucei* and *T. cruzi*) Greenberg or PE methylation (activity of PMTs) (*Leishmania spp.*)
Phospholipid composition	PC (most abundant), PE, PS, SM, CL, and PI (varies according to cell type or membrane composition)	Varies according to genus or species
		Trypansoma brucei: PC (45–60%); PE (10–20%); PI (6–12%); PS (4%; CL (2–4%); PG (2%); SM, IPC, EPC (10–15% of the total phospholipids)
		Leishmania spp.: PC (30–50%); PE (10%); PI (10%); PA, CL, PG (small traces almost undetectable); PS (controversy); IPC is the major sphingophospholipid class. It does not synthesize SM
		Trypanosoma cruzi: PC (44%); PE (28%); PI (12%); SM (4%); CL, PA lyso-PC, PS (small traces); presence of SM and IPC
Organelles and compartments involved	Endoplasmic reticulum, mitochondrion, cytosol	Endoplasmic reticulum, mitochondrion, cytosol

[a]Exception for hepatocytes where around 30% of PC is produced by PE methylation.
CL, cardiolipin; EPC, ethanolamine phosphorylcermide; IPC, inositol phosphorylceramide; PC, phosphatidylcholine; PE, phosphatidylethanolamine; PS, phosphatidylserine; PI, phosphatidylinositol; PG, phosphatidylgrlycerol; SM, sphingomyelin.

Design and synthesis of phospholipid analogues

Phospholipid analogues (PAs) are a new class of pharmaceutical compounds that exhibit excellent biocompatibility and an especial amphiphilicity making them appropriate to be employed as drugs with broad biological activity. Phospholipids are phosphorus-containing lipids with a polar and a non-polar portion in their structures, while the variation in head groups and aliphatic chains leads to the existence of a wide variety of these types of molecules.

According to several sources, phospholipids can be divided into natural and synthetic analogues (van Hoogevest and Wendel, 2014). During the last decades, the general demand for synthetic phospholipids mainly for drug development processes has increased; therefore new synthetic routes to produce phospholipids had to be developed.

Phospholipids can be manufactured either by organic and/or enzymatic synthetic steps. Chemical synthesis of phospholipids can be divided into semi-synthesis and total synthesis. In the simplest approach, a phospholipid can be converted to an analogue by modifications of its functional groups. Thus, semi-synthesis of phospholipids refers to changing the head or the tail groups or both; when compared with total synthesis, this process requires less reaction steps. The choice of synthetic route is dependent on the (commercial) availability of key intermediates or starting materials (Li *et al.*, 2015).

Phospholipids can be categorized into different classes according to the alcohols contained in the lipid portion. Thus, they can be divided into glycerophospholipids and sphingomyelins (Li *et al.*, 2015). In particular, glycerophospholipids can be further classified as ether or ester glycerophospholipids or lysophospholipids (D'Arrigo and Servi, 2010) by the type of bonding and position of the aliphatic moieties with respect to the glycerol backbone (Li *et al.*, 2015). The preparation of acyl- or ether-substituted glyceryl phospholipid derivatives involves stereospecific synthesis of the appropriately substituted glycerol backbone, which is then further phosphorylated using a variety of P(V) or P(III) reagents as shown in Fig. 13.2. These reagents are also employed for the phosphorylation of a variety of alcohols other than glycerol to generate ether phospholipids, the synthesis of which will be the main focus of this section.

Synthesis of ether phospholipids using P(V) reagents

The most widely used phosphorus (V) reagent is phosphorus oxychloride ($POCl_3$). A typical synthesis of ether phospholipid analogues (EP) is depicted in Fig. 13.3 (Lukác *et al.* 2009). The strategy involves phosphorylation of the appropriate alcohol by $POCl_3$ and subsequent attachment of the desired headgroup. This can be achieved in a one pot procedure by reaction of the resulting alkyl dichloridophosphate with the headgroup precursor and subsequent hydrolysis to afford the EP analogue (Fig. 13.3). Alternatively, the alkyl dichloridophosphate is hydrolysed to phosphoric acid, which is then converted to the pyridinium salt. This salt can be activated by a variety of reagents such as TPS-Cl (2,4,6-tri-isopropylbenzenesulfonylcholride) or MSNT (1-(mesitylene-2-sulfonyl)-3-nitro-1H-1,2,4-triazole), after which it reacts with the head group precursor (Fig. 13.3) (Lukác *et al.*, 2009; Calogeropoulou *et al.*, 2008, Papanastasiou *et al.*, 2010).

A modified procedure (reported by Merck) can also be applied for acid sensitive alcohols and utilizes phosphorus tri-1,2,4-triazolide as the phosphorylating reagent, which is generated *in situ* from $POCl_3$ and 1,2,4-triazole (Gluesenkamp, 2012).

An alternative approach was developed by Zumbuehl's lab. This approach enabled functional group modifications of the lipid portion after phosphorylation and involved transformation of the resulting phosphorochloridate to the corresponding phosphoramidate, a simple protecting group that is readily removable under the same conditions used for BOC deprotection (Fig. 13.4) (Fedotenko *et al.*, 2010).

In particular, a number of methods have been developed for the preparation of choline-substituted EPs. Thus, primary alcohol phosphorylation using bromoethylphosphoric acid dichloride followed by

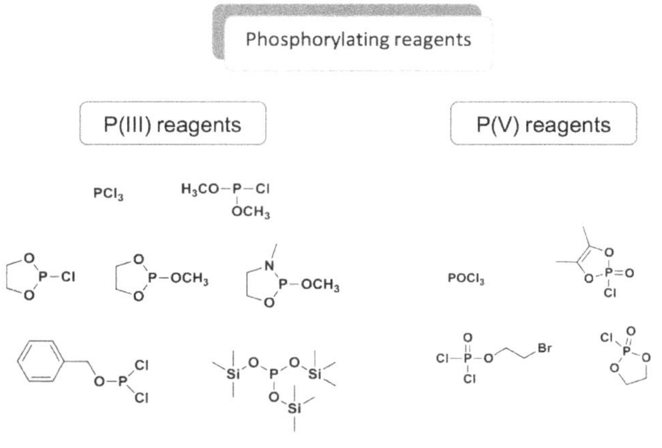

Figure 13.2 Phosphorylating reagents.

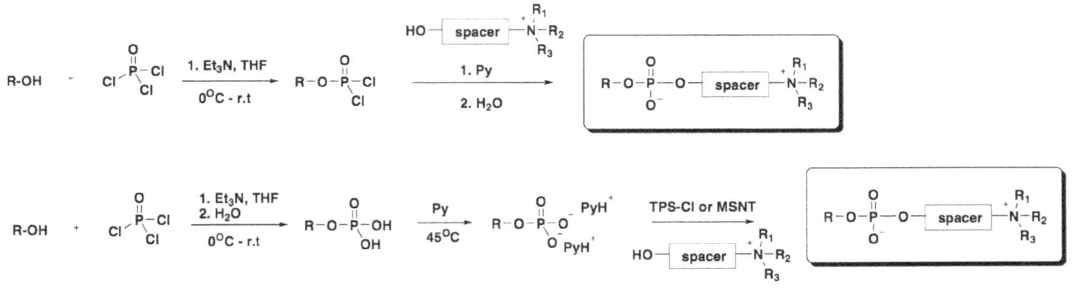

Figure 13.3 General synthetic methodology of phospholipid analogues.

hydrolysis and reaction with trimethylamine yields alkylphosphocholines (Fig. 13.5) (Hirt *et al.*, 1958; Eibl *et al.*, 1978).

In a similar approach, the appropriate alcohol is phosphorylated with $POCl_3$ followed by 2-bromoethanol. Hydrolysis yields the bromoethyl-substituted phosphodiester, which in turn reacts with trimethylamine to produce the final EPs (Fig. 13.6) (North *et al.*, 2009).

Dioxophospholane derivatives have been used as phosphorylating agents, which can be opened upon treatment with trimethylamine to yield the choline-substituted EPs. This approach was employed by Magolda and colleagues to generate alkyl and substituted alkylphosphocholines (Fig. 13.7) (Magolda *et al.*, 1985).

An improvement of the method depicted in Fig. 13.7 involved the development of 2-chloro-1,3,2-dioxaphospholane-2-oxide as a phosphorylating reagent, which results in production of phospholane intermediates that are then transformed to EPs upon ring opening with trimethylamine (Fig. 13.8) (Pinchuk *et al.*, 2006; Alam *et al.*, 2012).

Furthermore, development of the cyclic enediol phosphorchloridate reagent allows for ring opening of the intermediate cyclic phosphate with choline chloride to afford the phosphocholine EP after hydrolysis (Fig. 13.9) (Ramirez *et al.*, 1978).

The synthesis of phosphonamidate choline derivatives involves sequential reaction of $POCl_3$ with an alcohol, phenol, and the hydrogen bromide salt of 2-bromoethylamine. The bromoethylphosphonoamidate derivative reacts with trimethylamine in EtOH/MeOH (1:2) to yield the phosphonamidate derivative (Fig. 13.10) (North *et al.*, 2009).

Finally, alkylphosphonocholines can be synthesized using 3-chloropropylphosphonic acid as the phosphorylating reagent according to the procedure of Kley and co-workers (Fig. 13.11) (Kley *et al.*, 1998).

Synthesis of ether phospholipids using P(III) reagents

Phosphorus(V)-based reagents are generally less reactive than phosphorus(III)-reagents which can be oxidized after the addition of the first alcohol. The most common P(III) reagents are phosphoramidites, which are mainly used for nucleotide synthesis.

Zaffalon and colleagues developed benzyloxydichlorophosphine (BODP), which has lower cost, higher stability, and greater ease of handling than phosphoramidite reagents and was successfully used for the synthesis of various artificial choline phospholipids as shown in Fig. 13.12 (Zaffalon *et al.*, 2011).

In the search for an efficient method for the synthesis of EPs, Hendrickson and Hendrickson (2001) developed a phosphorylation approach using a dialkylchlorophosphite and I_2. The active intermediate is a phosphoryl iodide formed by oxidation of the phosphite with I_2. Bromoethanol (2-bromoethanol), dimethyl chlorophosphite, and an alcohol are converted to a phosphate triester with high yield in a one-pot reaction. Subsequently,

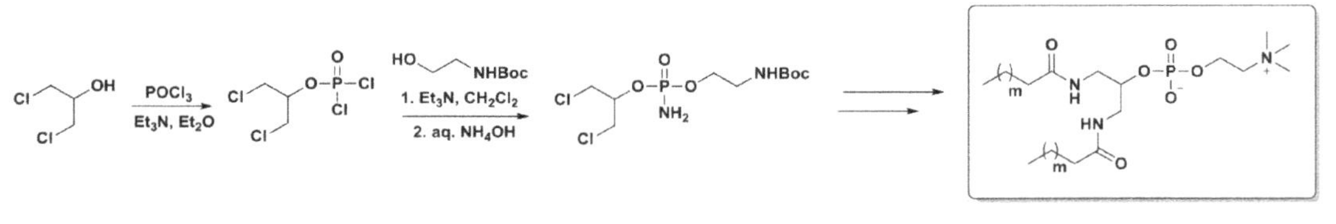

Figure 13.4 Synthesis of EPs using the phosphoramidate approach.

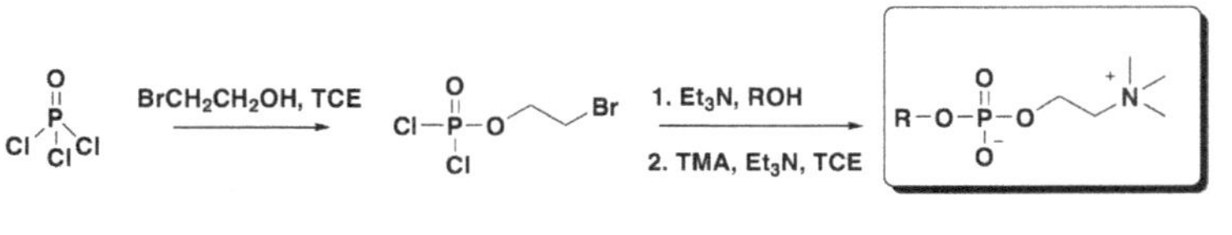

Figure 13.5 Synthesis of EPs by amination of phosphatidic acid bromoethylesters.

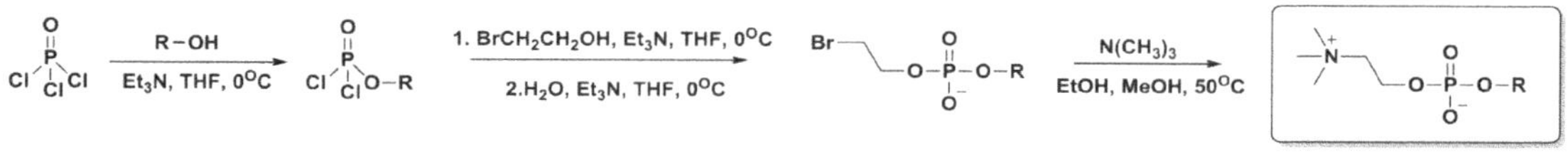

Figure 13.6 Alkylphosphocholine synthesis via bromoethyl-substituted phosphodiesters.

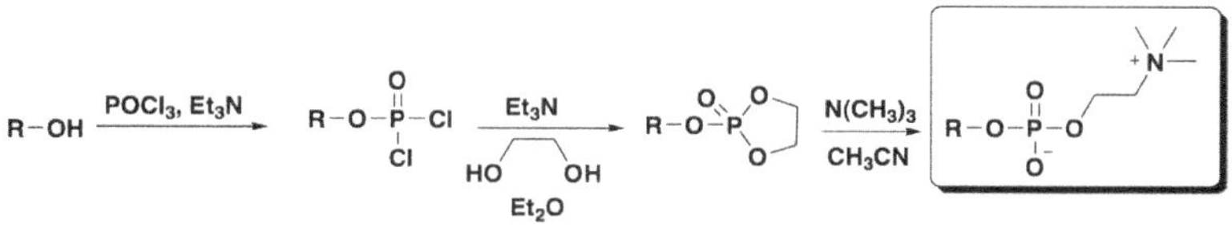

Figure 13.7 Alkylphosphocholine synthesis through cyclic phosphates.

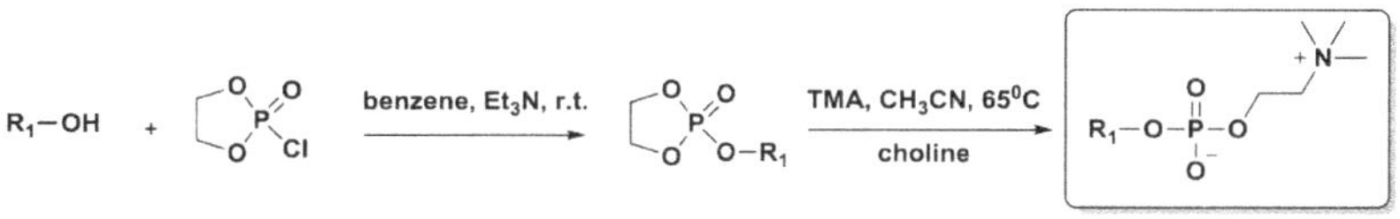

Figure 13.8 Alkylphosphocholine synthesis using 2-chloro-1,3,2-dioxaphospholane-2-oxide.

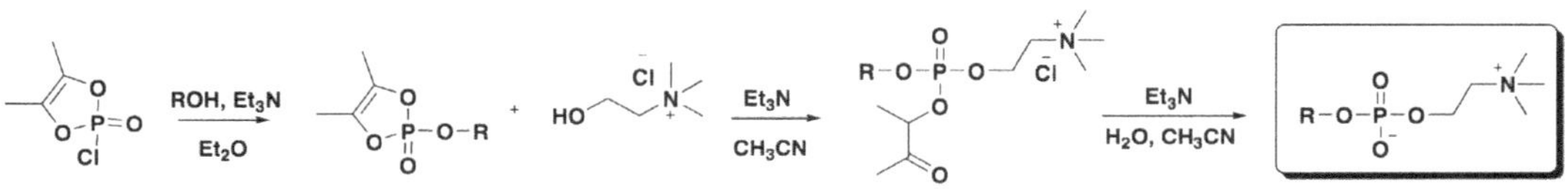

Figure 13.9 Synthesis of EPs using cyclic enediol phosphorchloridate.

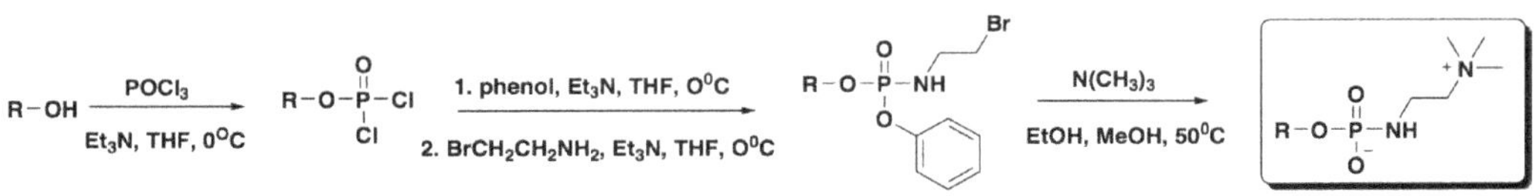

Figure 13.10 Synthesis of phosphonamidate choline derivatives.

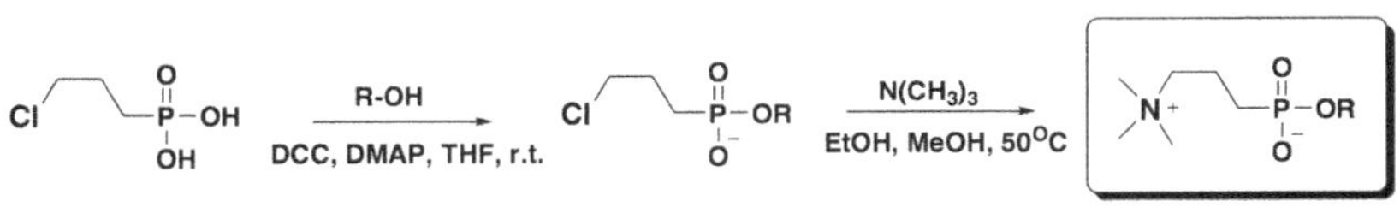

Figure 13.11 Synthesis of alkylphosphonocholines.

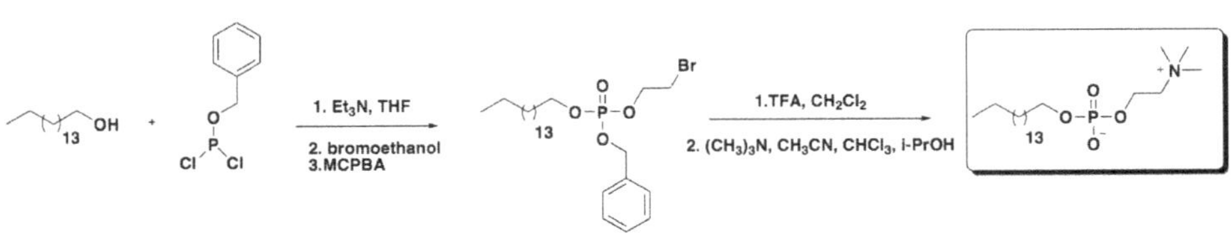

Figure 13.12 Synthesis of alkylphosphocholines using the BODP reagent.

treatment with aqueous trimethylamine yields the desired EP (Fig. 13.13).

Another high-yielding synthetic route for alkylphosphocholines was reported by Erukulla *et al.* (1994). Reaction of the appropriate alcohol with commercially available ethylene chlorophosphite gives the corresponding cyclic phosphite, which undergoes oxidation and ring opening with bromine in CH_2Cl_2 to yield (2-bromoethyl)phosphate ester; hydrolysis of the P–Br bond and quaternization with aqueous trimethylamine generates the final phospholipid (Fig. 13.14).

In conclusion, taking into consideration all of the available synthetic options, various tailored phospholipid derivatives can be developed in order to obtain information about their bioactivity.

Effect of phospholipid analogues on the growth and morphology of trypanosomatids

The first metabolically stable analogues of lysophosphatidylcholine were synthesized in the late 1960s with the primary goal of identifying novel immunomodulatory compounds. For the production of metabolically resistant analogues to acetyltransferases and lysophospholipases, the C1 and C2 carbons of the glycerol chain in the phosphatidylcholine molecule were modified by exchanging the ester linkages for ethers. Among these new synthetic compounds, edelfosine (1-*O*-octadecyl-2-methoxy-sn-glycero-3-phosphocholine) was found to have potent anti-tumour activity, both *in vitro* and *in vivo*. However, clinical trials have shown reduced efficacy for this compound. Currently, edelfosine is used only for the treatment of acute leukaemia (Van Blitterswijk and Verheij, 2008). At the end of 1980, Eibl and Unger synthesized hexadecylphosphocholine (HePC), known as miltefosine, which does not contain the glycerol chain. This process resulted in the production of an alkylphospholipid (APL). Early clinical trials using miltefosine as a systemic anti-tumour agent have failed because it has haemolytic properties when administered intravenously, although it can be administered orally and topically. The first reports of APL activity against trypanosomatid growth in *Leishmania donovani* were published in the 1980s (Achterberg and Gercken, 1987a; Achterberg and Gercken, 1987b). *L. donovani* was treated with 1-acyl lysophospholipids, 1-O-alkenyl-sn-glycero-3-phosphoethanolamine, 1-O-hexadecyl-sn-glycero-3-phosphocholine, 1-O-hexadecyl-sn-glycerol, and with the unnatural analogue, perifosine. These natural compounds at concentrations ranging between 2.1 and 10.9 μM were capable of reducing cell growth to 50% of the controls, while the unnatural alkyl lysophospholipid analogue was 10-fold more toxic. Miltefosine was tested *in vitro* against promastigotes and amastigotes of *L. donovani, L. major, L. tropica, L. aethiopica, L. mexicana,* and *L. panamensis.* In these tests, *L. donovani* was shown to be the most sensitive of all species, and *L. major* was the least sensitive (Escobar *et al.*, 2001, Achterberg and Gercken, 1987a; Croft *et al.*, 1987). For *L. major*, miltefosine demonstrated *in vitro* effects against promastigotes in the low micromolar range. HePC acts by inhibiting choline transport through the plasma membrane (Zufferey and Mamoun, 2002). In the case of *L.*

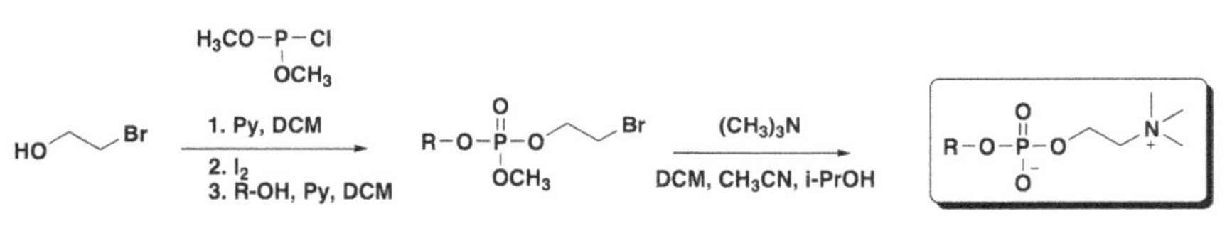

Figure 13.13 Synthesis of choline-substituted EPs using dialkylchlorophosphite.

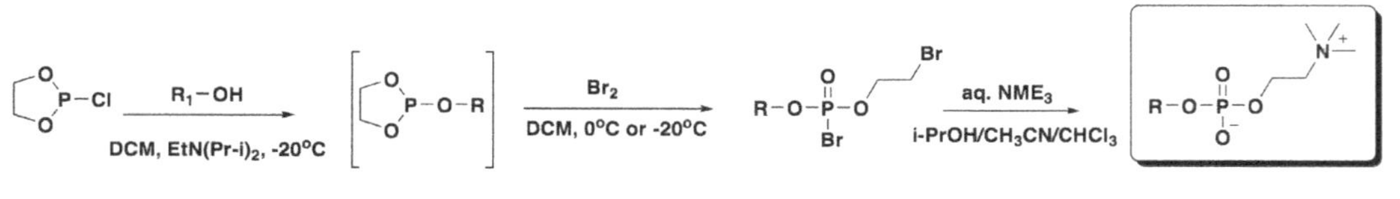

Figure 13.14 Synthesis of alkylphosphocholines using ethylene chlorophosphite.

mexicana, miltefosine inhibited the activity of coenzyme A acyltransferase (glycosomal enzyme involved in alkyl-PC metabolism) in a dose-dependent manner with an inhibitory concentration of 50 μM (Lux *et al.*, 2000). A Phase III trial using an oral dose of 50–100 mg/day over 28 days resulted in a long-term cure rate of 94%, and in this context, miltefosine was registered in India in 2002 for the first time for use in clinical therapy (Barratt, 2009; Sundar *et al.*, 2002). This class of drug, which consists of the first phospholipid analogues (edelfosine and ilmofosine), was initially used and designed for treatment of various types of cancer. Miltefosine was also considered to be a potential drug against visceral leishmaniasis treatment (Croft *el al.*, 1987, Kuhlencord *et al.*, 1992). Despite the variable response to treatments, the use of miltefosine was approved as the first drug of choice in India and Germany for treating cases of visceral leishmaniasis and in Colombia to treat cutaneous leishmaniasis (Soto *et al.*, 2004). However, data from the literature show that this species has peculiarities that often make it more resistant to treatments, both due to intrinsic properties and the type of infection caused by the protozoan (Rangel *et al.*, 1996). The first data related to the treatment of *T. cruzi* with APLs were obtained by Croft *et al.* (1996) and Santa-Rita *et al.* (2000). The Croft group (1996) tested the *in vitro* activities of four anti-cancer alkyl-lysophospholipids [ET-18-OCH3 (edelfosine), hexadecylphosphocholine (miltefosine), ilmofosine, and SRI 62–834] against intracellular amastigotes and the activity range was between 0.2 and 5.0 μM. The same compounds were tested in *T. brucei* trypomastigotes and the activity ranged from 7.0 to 50.8 μM. Santa-Rita *et al.* (2000) also tested some APLs (ET-18-OCH_3, hexadecylphosphocholine and ilmofosine) against the three developmental stages of *T. cruzi*. Under different experimental conditions, incubation with bloodstream trypomastigotes resulted in higher activity of the compounds in comparison with crystal violet indicating that they may be promising for use in parasite extermination in blood bags. The ED_{50} values were 13.4 ± 2.8 μM and 11.7 ± 0.6 μM for amastigotes and epimastigotes after 24 h of treatment, respectively. ET-18-OCH_3 (0.3 and 0.6 μM) inhibited epimastigotes differentiation to trypomastigotes (Dm28c clone) in the range 40%-57%. This drug (3.75-15 μM) also caused a time- and concentration-dependent inhibition of amastigotes intracellular proliferation in heart muscle cells with ED_{50} values of 14.3 ± 4.2, 8.9 ± 1.9 and 6. 8 ± 0.4 μM, after 1, 2, and 3 days of treatment, respectively. Pre-treatment of the parasite with this drug inhibited its internalization into the host cell. Interestingly, the intracellular differentiation of amastigotes to trypomastigotes was not hampered by the drug. Miltefosine's effects were also tested on *T. cruzi* resistant strains (Colombian) and strains that are partially resistant (Y) to treatment with clinical compounds, and the results obtained suggest that miltefosine induces the production of tumour necrosis factor α (TNF-α) and radicals of nitric oxide in macrophages *in vitro*, although the cytotoxic effects and the levels of resistance for both strains is still very high. *In vivo* studies with BALB/c mice infected with the Y strain indicated that orally administered miltefosine was able to promote survival and reduce parasitaemia to levels similar to those observed with benznidazole, the drug currently used to treat Chagas' disease. After four months of treatment, the levels of parasitaemia were undetectable in blood and tissue samples, thus suggesting that miltefosine has anti-protozoal effect in *T. cruzi* resistant strains (Saraiva *et al.*, 2002). Edelfosine was able to inhibit the incorporation of L-[methyl-14C] methionine into phosphatidylcholine with an IC_{50} of approximately 2 μM, thus indicating this step was the primary site of edelfosine on *de novo* PC synthesis from the Greenberg pathway in *T. cruzi* (Lira *et al.*, 2001). Evidence of the excellent antileishmanial activity was demonstrated *in vitro* and *in vivo* experiments. Many studies have shown that there are inter- and intraspecific differences in *Leishmania* susceptibility to different drugs, including miltefosine and other already tested phospholipid analogues. Studies *in vitro* with *L. amazonenis* showed an IC_{50} of 7.5 μM for miltefosine, 4.2 μM for edelfosine, and 2.2 μM for ilmofosine (Santa-Rita *et al.*, 2004), and 3.6 μM for perifosin (Cabrera-Serra *et al.*, 2007). The mechanism of APL action is different from classical drugs, since their molecular targets are primarily located in cell membranes. This selectivity for the plasma membrane is due to its chemical structure consisting of a polar head group and a long non-polar hydrocarbon chain. This structure allows the compound to easily fit within the lipid bilayer and present detergent properties that lead to cellular lysis in high concentrations (Van Blitterswijk and Verheij, 2008). Intracellular accumulation of miltefosine occurs first by binding to plasma membrane followed by internalization in the parasite cell body and other intracellular targets such as organelles and enzymes involved in metabolic pathways. However, the exact mechanism of miltefosine's action has not yet been deciphered. In mammalian cell cultures, the effects are mainly on signal transduction, lipid metabolism, and calcium homeostasis (Rakotomanga *et al.*, 2007). Miltefosine's actions in parasites also include interference in several mechanisms: (1) signal transduction pathways; (2) biosynthesis of glycosylphosphatidylinositol anchors; and 3) sterol biosynthesis (Rakotomanga *et al.*, 2007). *In L. donovani* and *T. cruzi*, one of the main effects of miltefosine is on the phosphatidylcholine biosynthetic pathway (Greenberg's pathway), and its action is more selective for the protozoan synthetic versus the mammalian synthetic pathway (Kennedy's route) (Lira *et al.*, 2001, Saraiva et *al.*, 2002). Miltefosine's immunomodulatory properties have been proposed as an explanation for its contribution to the anti-leishmanial activity (Wadhone *et al.*, 2009; Zeisig *et al.*, 1995).

Phospholipid derivatives such as miltefosine are believed to penetrate the membranes by means of inward translocation (flip-flop mechanism), which is a protein-mediated and energy-dependent process. The protein transporter that participates in this action is a P-type ATPase, called the miltefosine transporter in *L. donovani* (LdMT). It belongs to the aminophospholipid subfamily of the translocators and is expressed in the plasma membrane where it controls the translocation of phospholipids (Perez-Victoria, *et al.*, 2001). For drug binding to the membrane, its phospholipid composition is also relevant. *L. donovani* strains resistant to miltefosine present differences in fatty acid and sterol composition (Rakotomanga *et al.*, 2007). Recently, it was shown that miltefosine significantly interferes with the plasma membrane structure of *L. amazonensis* promastigotes. As a result of this type of interference, the molecular dynamics of the membranes are increased, causing cell lysis and inhibition of cell growth. When inserted into the membrane, miltefosine interacts with the lipid component and exerts detergent activity at the lipid–protein interface causing nuclear micellar aggregates at the hydrophobic sites of the proteins and also destabilizing the proteins by increasing the dynamics and conformational expansions. This interaction leads to changes in membrane permeability due to the insertion of the hydrophilic portion of miltefosine within the membrane (Moreira *et al.*, 2014).

In vivo tests in mice models of visceral infection by *L. donovani* and *L. infantum* after 4 weeks of oral treatment with miltefosine at doses of 10 mg/kg/day and 20 mg/kg/day, resulted in parasite load reduction when compared with those treated with pentostam (Kuhlencord, 1992). Based on data from the literature, the effects of miltefosine vary according to the *Leishmania* spp. and disease; thus, the doses used in monotherapy stay between 20, 30, 40 and 50 mg/kg/day. At the end of 21 days of treatment with these doses, the lesions were significantly reduced when compared to the controls and glucantime-treated mice, being able to affirm that clinically they were cured. When comparing these data from the literature it was possible to observe that the effective dose of miltefosine varied according to the *Leishmania* species, since animals infected with *L. donovani* and treated for four weeks with miltefosine were cured after receiving a dose of 20 mg/kg/day (Kuhlencord, 1992); in addition, in *L. major* cutaneous infections the effective dose of miltefosine after ten days of treatment was 25 mg/kg/day (Aguiar *et al.*, 2009). In order to validate the effective dose of the drug, however, it was necessary to monitor the animals 6 months after treatment. At the end of this period it was possible to conclude that the animals that remained cured were those treated with doses of 40 mg/kg/day and 50 mg/kg/day. Although the animals presented a debilitated profile when they received the dose of 50 mg/kg/day, they regained their weight throughout the observation period (Godinho *et al.*, 2012). The use of miltefosine is an important improvement in leishmaniasis treatment since it has the great advantage of oral administration, efficient action, and moderate toxicity in patients ranging from newborns to the elderly (Sinderman *et al.*, 2004). Since the main side-effects occur in the gastrointestinal system at high doses (> 150 mg/day), it could be teratogenic, which is of major concern clinically (Sindermann and Engel, 2006). Miltefosine was used to treat patients in Colombia with cutaneous leishmaniasis caused by *L. panamensis*. A cure of 91% was reported, whereas in

Guatemalan regions in which infections occurred mainly as a result of infection with *L. braziliensis* and *L. guyanensis*, a cure of 53% was reported. A study carried out in Brazil showed that there was a 75% cure in patients treated with miltefosine when compared with 53% cure obtained with pentavalent antimonials. However, there was no difference between the efficacy of these two drugs in the treatment of children with 70% and 68% cure for miltefosine and pentavalent antimonials, respectively (Soto *et al.*, 2004; Machado *et al.*, 2010).

In Fig. 13.15, panels A–B show a summary of the main features and known mechanisms of action of miltefosine against *Leishmania spp.*, *T. cruzi* and *T. brucei*, showing some of the published information in the last decades and the importance of this class of compounds in the search for new chemotherapeutic agents against the neglected diseases caused by these important protozoan parasites.

The results described above stimulated the development of new more potent and less toxic phospholipid analogues, using miltefosine as a prototype compound to synthesize several new alkylphospholipid analogues with potential activity anti-*Leishmania* and *Trypanosoma*. Extensive structure activity relationship studies by the Calogeropoulou laboratory showed that the phosphocholine group appears to be preferred in relation to other quaternary amine containing head groups, while introduction of rings in the lipid portion decreased the toxicity but maintained the

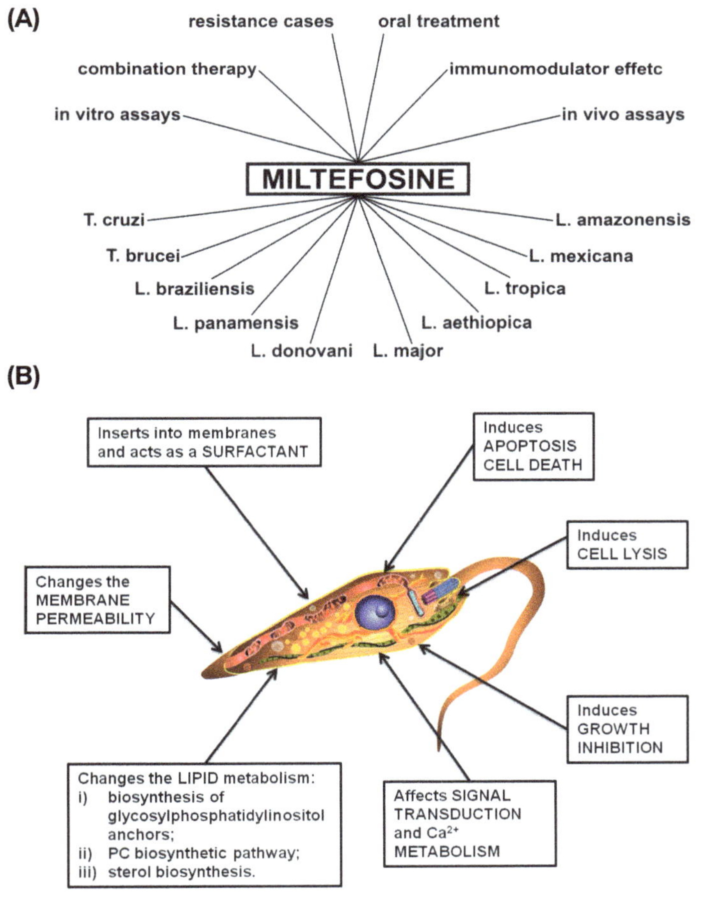

Figure 13.15 Some of the discoveries of the last decades about the miltefosine and its action against protozoan parasites *Leishmania spp.*, *T. cruzi* and *T. brucei*. (A) Some and special features about miltefosine. Genus and species susceptible. (B) Summary of the main features and known mechanisma of action of miltefosine.

potency of the resulting EPs (some of the reactions were explained above) (Avlonitis *et al.*, 2003, Kapou *et al.*, 2007, Calogeropoulou *et al.*, 2008, Papanastasiou *et al.*, 2010).

Molecular hybridization is a modern tool of the medicinal chemistry to synthesize compounds with special biological activities. In this context, TC95 is a new hybrid molecule synthesized by combining two pharmacophores with known biological activity against trypanosomatids. It was synthesized incorporating the dinitroaniline moiety in the lipid portion of miltefosine, thus producing a new alkylphospholipid analogue (Godinho *et al.*, 2013). Initially, this compound was chosen from 13 alkylphospholipid analogues as the most promising because it displayed IC_{50} values of 2.6 and 1.2 µM for the promastigotes and intracellular amastigotes of *L. amazonensis* after 48 h of treatment, respectively (Godinho *et al.*, 2013). Previous studies in *Leishmania* species showed that miltefosine treatment resulted in mitochondrial alterations, appearance of multinucleated cells, changes in plasma membrane, and formation of autophagic structures and cellular alterations suggestive of death by apoptosis (Santa-Rita *et al.*, 2004). TC95 significantly altered the shape of *L. amazonensis* promastigotes after 24, 48, and 72 hours of treatment with different concentrations. When analysing the promastigotes by scanning electron microscopy (SEM – Figs. 13.16 and 13.17), the shape alterations became evident; several cells were observed presenting changes in the flagellar membrane and flagellar pocket, and more than two flagella, indicating that promastigotes treated with TC95 failed to finalize the cell cycle probably arresting in cytokinesis. The changes observed in the cell body morphology of the treated promastigotes may be related to significant alterations in phospholipid and sterol composition of the plasma membrane as previously described for miltefosine (Rakotomanga *et al.*, 2007; Lira *et al.*, 2001). These changes in lipid composition may consequently result in remodelling of cytoskeleton and lipid accumulation in lipid droplets. In *T. cruzi*, miltefosine also

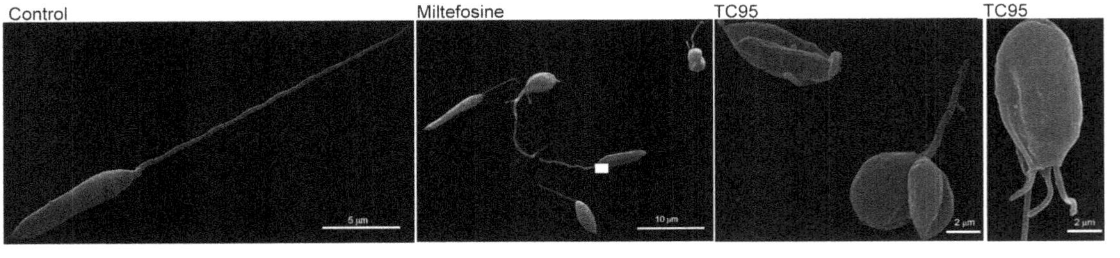

Figure 13.16 Scanning electron microscopy of *Leishmania amazonensis* promastigotes control and treated with different concentrations of miltefosine and TC95. Control – General overview of a parasite without any alteration in its shape and surface. Miltefosine – promastigotes treated with 25 µM miltefosine for 48 h. TC95 – promastigotes treated with 2 µM TC95 for 24 h.

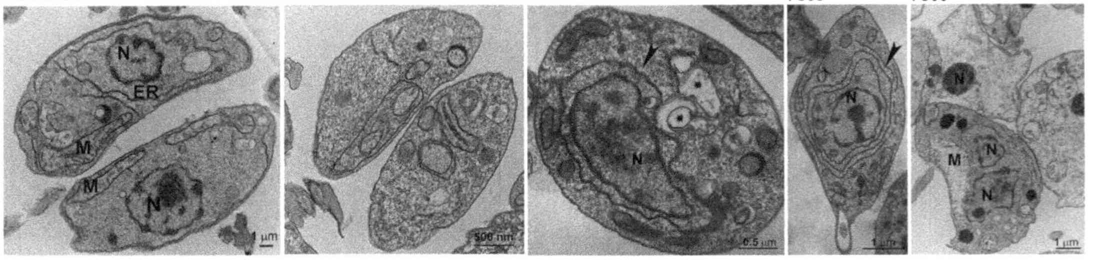

Figure 13.17 Ultrathin sections of *Leishmania amazonensis* promastigotes control and treated with miltefosine and TC95. Control: general overview of a control parasite without any alteration. Miltefosine: promastigotes treated with 50 µM miltefosine for 48 h. TC95: promastigotes treated with 4 µM TC95 for 48 h.

induced inhibition of the C-22 desaturase (one important enzyme of sterol biosynthesis) probably as a consequence of phospholipid biosynthesis inhibition in the Greenberg pathway (Lira *et al.*, 2001). Using transmission electron microscopy, the main changes observed after treatment of promastigotes and amastigotes with TC95 for only 24h were concentrated in the mitochondrion. The mitochondrion was swollen and presented with severe alterations such as disorganization of its structure, rupture of its membrane, and damage in the cristae, which also showed vesiculation (Fig. 13.17). The mitochondrion is a central organelle that regulates important and essential cellular processes such as oxidative phosphorylation, energy balance, cellular signalling, growth, and ROS production. Thus, mitochondrion is an important organelle in the regulation of cell death process in eukaryotic cells, also including the trypanosomatids (Rodrigues and De Souza, 2008; De Souza *et al.*, 2009). In addition to mitochondrial alterations, intense vacuolization of the cytosol, alterations in the flagellum, and presence of ER involving organelles (typical structures found in autophagy) were also observed (Rodrigues *et al.*, 2007). These alterations have already been described in the literature after treatment with APLs (miltefosine, edelfosine, and ilmofosine), however with TC95 they appeared more frequently and were already intense in the first 24h of treatment. All these alterations in the ultrastructure may also be related to stress situations induced by several different inhibitors, since some of them were also observed after treatment of *T. cruzi* with taxol, for example, a drug that interferes with the dynamic of microtubules (reviewed in Rodrigues *et al.*, 2007).

Cell death processes and immunomodulatory effects induced by phospholipid analogues

At present, there are several reports confirming the initial observation that miltefosine and lysophospholipid analogues can induce programmed cell death by apoptosis in susceptible mammalian cells using a mechanism that is connected to the PC biosynthesis inhibition (Engelmann *et al.*, 1996; Urbina, 2006; Wieder *et al.*, 1999). The experimental evidence for this conclusion is that miltefosine induced an increase of diacylglycerol and triacylglycerol biosynthesis in KB-cells with concomitant reduction of PC prior to the DNA fragmentation (Engelmann *et al.*, 1996). This effect completely correlated with the inhibition of PC biosynthesis, which also resulted in inhibition of sphingomyelin biosynthesis leading to a significant increase in ceramide cellular levels that are an important trigger for apoptotic cell death (Wieder *et al.*, 1998, 1999). These data are supported by the prevention of cell death when the miltefosine-treated mammalian cells are incubated with fumonisin-B (an inhibitor of ceramide biosynthesis) (Wieder *et al.*, 1998). Thus, in summary there is evidence that supports the correlation between the inhibition of PC biosynthesis and induction of apoptotic cell death by lysophospholipid analogues. Evidence includes several reported findings: (1) arrest of cell cycle in different cell lineages when PC biosynthesis is inhibited (Cui and Houweling, 2002; Wieder 1998; Wright *et al.*, 2004); (2) mutations in CTP: phosphocholine cytidylyltransferase (CCT) result in cells that are sensitive to temperature variation and lose cell viability when incubated in a non-permissive growth temperature (Cui *et al.*, 1996; Wright *et al.*, 2004); (3) apoptosis can be induced by short-chain ceramides that can inhibit the CCT or cholinephosphotransferase (CPT), an effect that can be reversed by incubation of cells with PC (Cui and Houweling, 2002; Ramos *et al.*, 2000); and (4) programmed cell death that can also be induced by other known inhibitors of PC biosynthesis such as farnesol, geranylgeranol, and chelerythrine (Cui and Houweling, 2002).

Apoptotic cell death in unicellular eukaryotes is a very controversial issue because there is a lack of information about the molecular machinery and cell signalling pathways involved in the programming of death (Proto *et al.*, 2013). However, there are several publications showing that treatment of protozoan parasites with different classes of inhibitors induced a type of cell death that shares many features with apoptotic cell death observed in higher eukaryotes (Gannavaram and Debrabant, 2012). In 2004, Paris *et al.* demonstrated for the first time that miltefosine induced a cell death process in *L. donovani* promastigotes, which presented phenotypes typically found in mammalian cells such as cell shrinkage, DNA fragmentation, and phosphatidylserine exposure, without alterations in plasma membrane integrity (Paris *et al.*, 2004). The same phenotype was also observed by another group working independently with *L. donovani* (Verma and Dey, 2004). Other studies also showed the induction of apoptosis-like death in *L. amazonensis, L. major, L. tropica*, and arsenite-resistant *L. donovani* (Verma *et al.*, 2007; Marinho *et al.*, 2011; Khademvatan *et al.*, 2011). In some of these, they also revealed that miltefosine is able to induce mitochondrial membrane depolarization and release of cytochrome c with consequent activation of cytoplasm proteases that have a caspase-like activity (Verma *et al.*, 2007).

Miltefosine has been also proposed as an immunomodulatory molecule interfering efficiently with the immune system of the different models of studies (Dorlo *et al.*, 2012; Zeisig *et al.*, 1995; Wadhone *et al.*, 2009; Escobar *et al.*, 2001; Verhaar *et al.*, 2014). In the last several years in different preclinical models, there are several studies that have confirmed the efficiency of miltefosine in inducing cytokine release (Hochhuth *et al.*, 1992; Eue, 2002; Hilgard *et al.*, 1997). In macrophages infected with *L. donovani* and deficient for interferon-γ, miltefosine enhanced interferon-γ receptors and restored both the production of interferon-γ and Th1/Th2 balance by promoting interleukin-12-dependent Th1 response (Wadhone *et al.*, 2009). However, these data still remain controversial because a recent study demonstrated that miltefosine did not alter the release of the cytokines interleukin 10 and 12 or tumour necrosis factor-α (Griewank *et al.*, 2010). In addition, miltefosine has been tested for the treatment of skin lesions in mastocytosis patients, because it was able to modulate mast cells resulting in their activation and release of immune mediators (Weller *et al.*, 2008; Hartmann *et al.*, 2010).

A critical view of the use of miltefosine for treatment of leishmaniasis

As described above, miltefosine is being used to treat patients with visceral leishmaniasis in some countries, especially in India, Nepal, and Bangladesh (Pandley *et al.*, 2016; Sundar *et al.*, 2015; Osyn *et al.*, 2014; Ramesh *et al.*, 2015). Several attempts at using it to treat cutaneous leishmaniasis in Latin America have been made. Clinical studies clearly show the differences in susceptibility of miltefosine for treating *L. donovani* infection in which some groups reported a rapid parasitic clinical cure of 94%, while the efficacy in cases caused by *L. brasiliensis* was only 33% in Guatemala and varied from 58% to 88% in Bolivia (Soto *et al.*, 2007; Soto *et al.*, 2008). At a certain point, miltefosine was the drug of choice for the visceral leishmaniasis elimination programme in India, Nepal, and Bangladesh, probably in part due to its ease of oral use (Sundar *et al.*, 2008a). The absence of medical control in the treatment program caused the patients to prematurely discontinue treatment, and suboptimal doses ultimately led to the development of parasitic resistance (Sundar and Murray, 2005). After one decade of use of the drug on the Indian subcontinent, its efficacy declined to 90.3%, and the relapse rate doubled (Sundar *et al.*, 2012); 10.8% and 20.0% of patients from Nepal also relapsed within 6 and 12 months after miltefosine treatment, respectively (Rijal *et al.*, 2013). Other cases of resistance have been reported in patients treated with miltefosine, mainly due to its long half-life (100-200 h). In India, where visceral leishmaniasis is an anthroponosis, cases are more frequent in relation to the places in which this disease is a zoonosis. Recently, Dr Das (personal communication) who coordinates an extensive leishmaniasis treatment programme in India and Nepal, communicated that his group is no longer using miltefosine to treat visceral leishmaniasis. The option for treating these patients is now liposomal amphotericin B. Thus, new therapeutic strategies need to be developed. The choices of drug combination therapy between miltefosine, liposomal amphotericin B, and paromomycin are options that require more extensive studies (Sundar, 2008; Sundar and Charkravarty, 2015; Sundar and Singh, 2016).

References

Alam, Md.M., Joh, E.H., Kim, Y., Oh, Y.I., Hong, J., Kim, B., Kim, D.H., and Lee, Y.S. (2012). Synthesis and biological evaluation of cyclopentane-linked alkyl phosphocholines as potential anticancer agents that act by inhibiting Akt phosphorylation. Eur. J. Med. Chem. 47, 485–492.

Avlonitis, N., Lekka, E., Detsi, A., Koufaki, M., Calogeropoulou, T., Scoulica, E., Siapi, E., Kyrikou, I., Mavromoustakos, T., Tsotinis, A., *et al.* (2003). Antileishmanial ring-substituted ether phospholipids. J. Med. Chem. *46*, 755–767. https://doi.org/10.1021/jm020972c.

Beach, D.H., Holz, G.G., and Anekwe, G.E. (1979). Lipids of *Leishmania* promastigotes. J. Parasitol. 65, 201–216.

Berridge, M.J., and Irvine, R.F. (1984). Inositol trisphosphate, a novel second messenger in cellular signal transduction. Nature *312*, 315–321.

Bertello, L.E., Gonçalvez, M.F., Colli, W., and de Lederkremer, R.M. (1995). Structural analysis of inositol phospholipids from *Trypanosoma cruzi* epimastigote forms. Biochem. J. *310*, 255–261.

Borkenhagen, L.F., Kennedy, E.P., and Fielding, L. (1961). Enzymatic formation and decarboxylation of phosphatidylserine. J Biol Chem. *236*, PC28-PC32.

Calogeropoulou, T., Angelou, P., Detsi, A., Fragiadaki, I., and Scoulica, E. (2008). Design and synthesis of potent antileishmanial cycloalkylidene-substituted ether phospholipid derivatives. J. Med. Chem. *51*, 897–908.

Carroll, M., and McCrorie, P. (1986). Lipid composition of bloodstream forms of *Trypanosoma brucei brucei*. Comp. Biochem. Physiol., B *83*, 647–651.

Cornell, R.B., and Northwood, I.C. (2000). Regulation of CTP:phosphocholine cytidylyltransferase by amphitropism and relocalization. Trends Biochem. Sci. *25*, 441–447.

Cui, Z., and Houweling, M. (2002). Phosphatidylcholine and cell death. Biochim. Biophys. Acta *1585*, 87–96.

Cui, Z., Houweling, M., Chen, M.H., Record, M., Chap, H., Vance, D.E., Terce, F.(1996). A genetic defect in phosphatidylcholine biosynthesis triggers apoptosis in Chinese hamster ovary cells. J. Biol. Chem. *271*, 14668—14671.

D'Arrigo, P., and Servi, S. (2010). Synthesis of Lysophospholipids. Molecules. *15*, 1354–1377.

Dixon, H., and Williamson, J. (1970). The lipid composition of blood and culture forms of *Trypanosoma lewisi* and *Trypanosoma rhodesiense* compared with that of their environment. Comp Biochem Physiol. *33*, 111–128.

Dorlo, T.P., Balasegaram, M., Beijnen, J.H., and de Vries, P.J. (2012). Miltefosine: a review of its pharmacology and therapeutic efficacy in the treatment of leishmaniasis. J. Antimicrob. Chemother. *67*, 2576–2597. https://doi.org/10.1093/jac/dks275.

Drew, M.E., Langford, C.K., Klamo, E.M., Russell, D.G., Kavanaugh, M.P., and Landfear, S.M. (1995). Functional expression of a myo-inositol/H+ symporter from *Leishmania donovani*. Mol. Cell. Biol. *15*, 5508–5515.

Eibl, H., and Nicksch, A. (1978). The synthesis of phospholipids by direct amination, Chem. Phys. Lipids. *22*, 1–8.

Engelmann, J., Henke, J., Willker, W., Kutscher, B., Nössner, G., Engel, J., and Leibfritz, D. (1996). Early stage monitoring of miltefosine induced apoptosis in KB-cells by multinuclear NMR spectroscopy. Anticancer Res. *16*, 1429–1439.

Erukulla, R.K., Byun, H.S., and Bittman, R. (1994). Antitumor Phospholipids: A One-pot Introduction of a Phosphocboline Moiety into Lipid Hydroxy Acceptors. Tetrahedron Lett. *35*, 5783–5784.

Escobar, P., Yardley, V., and Croft, S.L. (2001). Activities of hexadecylphosphocholine (miltefosine), AmBisome, and sodium stibogluconate (Pentostam) against *Leishmania donovani* in immunodeficient scid mice. Antimicrob. Agents Chemother. *45*, 1872–5.

Eue, I. (2002). Hexadecylphosphocholine selectively upregulates expression of intracellular adhesion molecule-1 and class I major histocom-patibility complex antigen in human monocytes. J. Exp. Ther. Oncol. 2:333–6.

Fadok, V.A., Bratton, D.L., Rose, D.M., Pearson, A., Ezekewitz, R.A., and Henson, P.M. (2000). A receptor for phosphatidylserine-specific clearance of apoptotic cells. Nature *405*, 85–90. https://doi.org/10.1038/35011084.

Fadok, V.A., Voelker, D.R., Campbell, P.A., Cohen, J.J., Bratton, D.L., and Henson, P.M. (1992). Exposure of phosphatidylserine on the surface of apoptotic lymphocytes triggers specific recognition and removal by macrophages. J. Immunol. *148*, 2207–2216.

Fagone, P., and Jackowski, S. (2013). Phosphatidylcholine and the CDP-choline cycle. Biochim. Biophys. Acta *1831*, 523–532. https://doi.org/10.1016/j.bbalip.2012.09.009.

Farine, L., and Bütikofer, P. (2013). The ins and outs of phosphatidylethanolamine synthesis in *Trypanosoma brucei*. Biochim. Biophys. Acta *1831*, 533–542. https://doi.org/10.1016/j.bbalip.2012.09.008.

Fedotenko, I.A., Zaffalon, P.L., Favarger, F., and Zumbuehl, A. (2010). The synthesis of 1,3-diamidophospholipids. Tetrahedron Lett. *51*, 5382–5384.

Gannavaram, S., and Debrabant, A. (2012). Programmed cell death in *Leishmania*: biochemical evidence and role in parasite infectivity. Front Cell Infect Microbiol *2*, 95. https://doi.org/10.3389/fcimb.2012.00095.

Gibellini, F., Hunter, W.N., and Smith, T.K. (2009). The ethanolamine branch of the Kennedy pathway is essential in the bloodstream form of *Trypanosoma brucei*. Mol. Microbiol. *73*, 826–843. https://doi.org/10.1111/j.1365-2958.2009.06764.x.

Gluesenkamp, K.H., Synthesis of phosphoric esters. US Patent No 8,940,907 (25 January, 2015).

Gonzalez-Salgado, A., Steinmann, M.E., Greganova, E., Rauch, M., Mäser, P., Sigel, E., *et al.* (2012). Myo-inositol uptake is essential for bulk inositol phospholipid but not glycosylphosphatidylinositol synthesis in *Trypanosoma brucei*. J Biol Chem. *287*, 13313–23.

Griewank, K., Gazeau, C., Eichhorn, A., *et al.* (2010). Miltefosine efficiently eliminates *Leishmania major* amastigotes from infected murine dendritic cells without altering their immune functions. Antimicrob Agents Chemother. *54*, 652–9.

Hartmann, K., Siebenhaar, F., Belloni, B., *et al.* (2010). Effects of topical treatment with the raft modulator miltefosine and clobetasol in cutaneous mastocytosis: a randomized, double-blind, placebo-controlled trial. Br. J. Dermatol. *162*, 185–90.

Hendrickson, E.K., and Hendrickson, H.S. (2001). Efficient synthesis of the cholinephosphate phospholipid headgroup. Chem. Phys. Lipids *109*, 203–207.

Hilgard, P., Klenner, T., Stekar, J., Nössner, G., Kutscher, B., and Engel, J. (1997). D-21266, a new heterocyclic alkylphospholipid with antitumour activity. Eur. J. Cancer 33, 442–446.

Hirt, R., and Berchtold, R. (1958). Zur Synthese der Phosphatide. 2. Eine neue Synthese der Lecithine. Pharm Acta Helv 33, 349–356.

Hochhuth, C.H., Vehmeyer, K., Eibl, H., and Unger, C. (1992). Hexadecylphosphocholine induces interferon-gamma secretion and expression of GM-CSF mRNA in human mononuclear cells. Cell. Immunol. *141*, 161–168.

Kainu, V., Hermansson, M., Hänninen, S., Hokynar, K., and Somerharju, P. (2013). Import of phosphatidylserine to and export of phosphatidylethanolamine molecular species from mitochondria. Biochim. Biophys. Acta *1831*, 429–437. https://doi.org/10.1016/j.bbalip.2012.11.003.

Kaneshiro, E.S., Jayasimhulu, K., and Lester, R.L. (1986). Characterization of inositol lipids from *Leishmania donovani* promastigotes: identification of an inositol sphingophospholipid. J Lipid Res. *27*, 1294–303.

Kapou, A., Benetis, N.P., Avlonitis, N.,, Calogeropoulou, T., Koufaki, M., Scoulica, E., Nikolaropoulos, S.S., and Mavromoustakos, T. (2007). 3D-Quantitative structure–activity relationships of synthetic antileishmanial ring-substituted ether phospholipids. Bioorg Med Chem. *15*, 1252–1265.

Kennedy, E.P., and Weiss, S.B. (1956). The function of cytidine coenzymes in the biosynthesis of phospholipides. J. Biol. Chem. *222*, 193–214.

Kent, C. (1995). Eukaryotic phospholipid biosynthesis. Annu. Rev. Biochem. *64*, 315–343. https://doi.org/10.1146/annurev.bi.64.070195.001531.

Khademvatan, S., Gharavi, M.J., Rahim, F., and Saki, J. (2011). Miltefosine-induced apoptotic cell death on *Leishmania major* and *L. tropica* strains. Korean J. Parasitol. *49*, 17–23. https://doi.org/10.3347/kjp.2011.49.1.17.

Kley, J., Unger, C., and Massing, U. (1998). Synthesis of Isosteric Phosphono Analogs of Biologically Active Alkylphosphocholines. Monatsh Chem. *129*, 173–185.

Letsinger, R.L., and Lunsford, W.B. (1976). Synthesis of thymidine oligonucleotides by phosphite triester intermediates. J. Am. Chem. Soc. *98*, 3655–3661.

Li, J., Wang, X., Zhang, T., Wang, C., Huang, Z., Luo, X., Deng, Y. (2015). A review on phospholipids and their main applications in drug delivery systems. Asian J. Pharm. Sci. *10*, 81–98.

Liu, H., Zhu, Z., Kang, H., Wu, Y., Sefan, K., and Tan, W. (2010). DNA-based micelles: synthesis, micellar properties and size-dependent cell permeability. Chemistry *16*, 3791–3797. https://doi.org/10.1002/chem.200901546.

Liu, X., Stocker, B.L., and Seeberger, P.H. (2006). Total synthesis of phosphatidylinositol mannosides of Mycobacterium tuberculosis. J. Am. Chem. Soc. *128*, 3638–3648. https://doi.org/10.1021/ja0565368.

Lukác, M., Mrva, M., Fischer-Fodor, E., Lacko, I., Bukovský, M., Miklásová, N., Ondriska, F., and Devínsky, F. (2009). Synthesis and biological activity of dialkylphosphocholines. Bioorg. Med. Chem. Lett. *19*, 6346–6349. https://doi.org/10.1016/j.bmcl.2009.09.079.

Lukác, M., Mojzis, J., Mojzisová, G., Mrva, M., Ondriska, F., Valentová, J., Lacko, I., Bukovský, M., Devínsky, F., and Karlovská, J. (2009). Dialkylamino and nitrogen heterocyclic analogues of hexadecylphosphocholine and cetyltrimetylammonium bromide: Effect of phosphate group and environment of the ammonium cation on their biological activity. Eur J Med Chem. *44*, 4970–4977.

Lykidis, A. (2007). Comparative genomics and evolution of eukaryotic phospholipid biosynthesis. Prog. Lipid Res. *46*, 171–199.

Magolda, R.L., and Johson, P.R. (1985). A new efficient and versatile synthesis of alkyl phosphocholines, Tetrahedron Lett. *26*, 1167–1170.

Marinho, F.d.e.A., Gonçalves, K.C., Oliveira, S.S., Oliveira, A.C., Bellio, M., d'Avila-Levy, C.M., Santos, A.L., and Branquinha, M.H. (2011). Miltefosine induces programmed cell death in *Leishmania amazonensis* promastigotes. Mem. Inst. Oswaldo Cruz *106*, 507–509.

Martin, K.L., and Smith, T.K. (2006). Phosphatidylinositol synthesis is essential in bloodstream form *Trypanosoma brucei*. Biochem. J. *396*, 287–295.

Martin, K.L., and Smith, T.K. (2005). The myo-inositol-1-phosphate synthase gene is essential in *Trypanosoma brucei*. Biochem Soc Trans. 33, 983–5.

Martin, S.F., Josey, J.A., Wong, Y.L., and Dean, D.W. (1994). General Method for the Synthesis of Phospholipid Derivatives of 1,2-O-Diacyl-sn-Glycerols. J Org Chem. *59*, 4805–4820.

Nishizuka, Y. (1986). Studies and perspectives of protein kinase C. Science *233*, 305–312.

North, E.J., Osborne, D.A., Bridson, P.K., Baker, D.L., Parrill, A.L. (2009). Autotaxin structure–activity relationships revealed through lysophosphatidylcholine analogs Bioorg Med Chem. *17*, 3433–3442.

Oliveira, M.M., Timm, S.L., and Costa, S.C. (1977). Lipid composition of *Trypanosoma cruzi*. Comp. Biochem. Physiol., B *58*, 195–199.

Ostyn, B., Hasker, E., Dorlo, T.P., Rijal, S., Sundar, S., Dujardin, J.C., and Boelaert, M. (2014). Failure of miltefosine treatment for visceral leishmaniasis in children and men in South-East Asia. PLOS ONE *9*, e100220. https://doi.org/10.1371/journal.pone.0100220.

Pandey, K., Ravidas, V., Siddiqui, N.A., Sinha, S.K., Verma, R.B., Singh, T.P., Dhariwal, A.C., Das Gupta, R.K., and Das, P. (2016). Pharmacovigilance of Miltefosine in Treatment of Visceral Leishmaniasis in Endemic Areas of Bihar, India. Am. J. Trop. Med. Hyg. *95*, 1100–1105.

Papanastasiou, I., Prousis, K.C., Georgikopoulou, K., Pavlidis, T., Scoulica, E., Kolocouris, N., and Calogeropoulou, T. (2010). Design and synthesis of new adamantyl-substituted antileishmanial ether phospholipids. Bioorg. Med. Chem. Lett. *20*, 5484–5487.

Paris, C., Loiseau, P.M., Bories, C., Breard, J.(2004). Miltefosine induces apoptosis-like death in *Leishmania donovani* promastigotes. Antimicrob. Agents Chemother. *48*, 852–859.

Patnaik, P.K., Field, M.C., Menon, A.K., Cross, G.A., Yee, M.C., and Bütikofer, P. (1993). Molecular species analysis of phospholipids from *Trypanosoma brucei* bloodstream and procyclic forms. Mol Biochem Parasitol. *58*, 97–105.

Pelech, S.L., and Vance, D.E. (1984). Regulation of phosphatidylcholine biosynthesis. Biochim. Biophys. Acta *779*, 217–251.

Pinchuk, A.N., Rampy, M.A., Longino, M.C., Scott Skinner, R.W., Gross, M.D., Weichert, J.P., and Counsell, R.E. (2006). Synthesis and Structure–Activity Relationship Effects on the Tumor Avidity of Radioiodinated Phospholipid Ether Analogues. J Med Chem. *49*, 2155–2165.

Proto, W.R., Coombs, G.H., and Mottram, J.C. (2013). Cell death in parasitic protozoa: regulated or incidental? Nat. Rev. Microbiol. *11*, 58–66. https://doi.org/10.1038/nrmicro2929.

Quiñones, W., Urbina, J.A., Dubourdieu, M., and Luis Concepción, J. (2004). The glycosome membrane of *Trypanosoma cruzi* epimastigotes: protein and lipid composition. Exp. Parasitol. *106*, 135–149. https://doi.org/10.1016/j.exppara.2004.03.006.

Ramakrishnan, S., Serricchio, M., Striepen, B., and Bütikofer, P. (2013). Lipid synthesis in protozoan parasites: a comparison between kinetoplastids and apicomplexans. Prog. Lipid Res. *52*, 488–512. https://doi.org/10.1016/j.plipres.2013.06.003.

Ramesh, V., Singh, R., Avishek, K., Verma, A., Deep, D.K., Verma, S., and Salotra, P. (2015). Decline in clinical efficacy of oral miltefosine in treatment of post-kalazar dermal leishmaniasis (PKLD) in India. PLOS Neg. Trop. Dis. https://doi.org/10.1371/journal.pntd.0004093.

Ramirez, F., Okazaki, H., and Marecek, J.F. (1978). Synthesis of Lecithin Analogues by Means of Cyclic Enediol Phosphates. Derivatives of 1-Octadecanol and of Cholesterol. J Org Chem. *43*, 2331–2334.

Ramos, B., Salido, G.M., Campo, M.L., and Claro, E. (2000). Inhibition of phosphatidylcholine synthesis precedes apoptosis induced by C2-ceramide: protection by exogenous phosphatidylcholine. Neuroreport. *11*, 3103–3108.

Ramos, R.G., Libong, D., Rakotomanga, M., Gaudin, K., Loiseau, P.M., and Chaminade, P. (2008). Comparison between charged aerosol detection and light scattering detection for the analysis of *Leishmania* membrane phospholipids. J Chromatogr A. *1209*, 88–94.

Richmond, G.S., Gibellini, F., Young, S.A., Major, L., Denton, H., Lilley, A., and Smith, T.K. (2010). Lipidomic analysis of bloodstream and procyclic form *Trypanosoma brucei*. Parasitology *137*, 1357–1392. https://doi.org/10.1017/S0031182010000715.

Rifkin, M.R., Strobos, C.A., and Fairlamb, A.H. (1995). Specificity of ethanolamine transport and its further metabolism in *Trypanosoma brucei*. J. Biol. Chem. *270*, 16160–16166.

Rodrigues, J.C., Bernardes, C.F., Visbal, G., Urbina, J.A., Vercesi, A.E., and de Souza, W. (2007). Sterol methenyl transferase inhibitors alter the ultrastructure and function of the *Leishmania amazonensis* mitochondrion leading to potent growth inhibition. Protist *158*, 447–456.

Saba, J.D., Nara, F., Bielawska, A., Garrett, S., and Hannun, Y.A. (1997). The BST1 gene of *Saccharomyces cerevisiae* is the sphingosine-1-phosphate lyase. J. Biol. Chem. *272*, 26087–26090.

Serricchio, M., and Bütikofer, P. (2012). An essential bacterial-type cardiolipin synthase mediates cardiolipin formation in a eukaryote. Proc. Natl. Acad. Sci. U.S.A. *109*, E954–61. https://doi.org/10.1073/pnas.1121528109.

Seyfang, A., and Landfear, S.M. (1999). Substrate depletion upregulates uptake of myo-inositol, glucose and adenosine in *Leishmania*. Mol. Biochem. Parasitol. *104*, 121–130.

Shiao, Y.-J., Lupo, G., and Vance, J.E. (1995). Evidence that phosphatidylserine is imported into mitochondria via a mitochondria-associated membrane and that the majority of mitochondrial phosphatidylethanolamine is derived from decarboxylation of phosphatidylserine. J Biol Chem *270*, 11190–11198.

Signorell, A., Rauch, M., Jelk, J., Ferguson, M.A., and Bütikofer, P. (2008). Phosphatidylethanolamine in *Trypanosoma brucei* is organized in two separate pools and is synthesized exclusively by the Kennedy pathway. J. Biol. Chem. *283*, 23636–23644. https://doi.org/10.1074/jbc.M803600200.

Signorell, A., Gluenz, E., Rettig, J., Schneider, A., Shaw, M.K., Gull, K., *et al.* (2009). Perturbation of phosphatidylethanolamine synthesis affects mitochondrial morphology and cell cycle progression in procyclic form *Trypanosoma brucei*. Mol Microbiol. 72:1068–79.

Smith, T.K., and Bütikofer, P. (2010). Lipid metabolism in *Trypanosoma brucei*. Mol. Biochem. Parasitol. *172*, 66–79. https://doi.org/10.1016/j.molbiopara.2010.04.001.

Stone, S.J., and Vance, J.E. (2000). Phosphatidylserine synthase-1 and -2 are localized to mitochondria-associated membranes. J. Biol. Chem. *275*, 34534–34540. https://doi.org/10.1074/jbc.M002865200.

Sundar, S., and Chakravarty, J. (2015). An update on pharmacotherapy for leishmaniasis. Expert Opin Pharmacother *16*, 237–252. https://doi.org/10.1517/14656566.2015.973850.

Sundar, S., and Singh, A. (2016). Recent developments and future prospects in the treatment of visceral leishmaniasis. Ther Adv Infect Dis *3*, 98–109. https://doi.org/10.1177/2049936116646063.

Sutterwala, S.S., Hsu, F.F., Sevova, E.S., Schwartz, K.J., Zhang, K., Key, P., Turk, J., Beverley, S.M., and Bangs, J.D. (2008). Developmentally regulated sphingolipid synthesis in African trypanosomes. Mol. Microbiol. *70*, 281–296. https://doi.org/10.1111/j.1365-2958.2008.06393.x.

Sutterwala, S.S., Creswell, C.H., Sanyal, S., Menon, A.K., and Bangs, J.D. (2007). De novo sphingolipid synthesis is essential for viability, but not for transport of glycosylphosphatidylinositol-anchored proteins, in African trypanosomes. Eukaryot Cell 6, 454–64.

Thuong, N., and Charbier, P. (1974). N'130-Nouvelle Methode de Preparation de la Phosphorylcholine, de la Phosphorylhomocholine et de Leurs Derivatives. *Bull. Soc. Chim. Fr. 14*, 667–671.

Uhrig, M.L., Couto, A.S., Colli, W., and de Lederkremer, R.M. (1996). Characterization of inositolphospholipids in *Trypanosoma cruzi* trypomastigote forms. Biochim. Biophys. Acta *1300*, 233–239.

Urbina, J.A. (2006). Mechanisms of action of lysophospholipid analogues against trypanosomatid parasites. Trans. R. Soc. Trop. Med. Hyg. 100 (Suppl. 1), S9–S16.

van Hoogevest, P., and Wendel, A. (2014). The use of natural and synthetic phospholipids as pharmaceutical excipients. Eur J Lipid Sci Technol *116*, 1088–1107. https://doi.org/10.1002/ejlt.201400219.

van Zandbergen, G., Bollinger, A., Wenzel, A., Kamhawi, S., Voll, R., Klinger, M., Müller, A., Hölscher, C., Herrmann, M., Sacks, D., *et al.* (2006). *Leishmania* disease development depends on the presence of apoptotic promastigotes in the virulent inoculum. Proc. Natl. Acad. Sci. U.S.A. *103*, 13837–13842.

Vance, D.E. (2013). Physiological roles of phosphatidylethanolamine N-methyltransferase. Biochim. Biophys. Acta *1831*, 626–632. https://doi.org/10.1016/j.bbalip.2012.07.017.

Vance, D.E. (2014). Phospholipid methylation in mammals: from biochemistry to physiological function. Biochim. Biophys. Acta *1838*, 1477–1487. https://doi.org/10.1016/j.bbamem.2013.10.018.

Vance, D.E., and Ridgway, N.D. (1988). The methylation of phosphatidylethanolamine. Prog. Lipid Res. *27*, 61–79.

Vance, D.E., and Vance, J.E. (2008). Phospholipid biosynthesis in eukaryotes. In: Vance DE, Vance JE, editors. Biochemistry of Lipids, Lipoproteins and Membranes. Amsterdam: Elsevier; 5th Edition, pp. 213–244.

Vance, J.E. (2015). Phospholipid synthesis and transport in mammalian cells. Traffic *16*, 1–18. https://doi.org/10.1111/tra.12230.

Vance, J.E., and Tasseva, G. (2013). Formation and function of phosphatidylserine and phosphatidylethanolamine in mammalian cells. Biochim. Biophys. Acta *1831*, 543–554. https://doi.org/10.1016/j.bbalip.2012.08.016.

Vance, J.E., and Vance, D.E. (2004). Phospholipid biosynthesis in mammalian cells. Biochem. Cell Biol. *82*, 113–128. https://doi.org/10.1139/o03-073.

Verhaar, A.P., Wildenberg, M.E., Peppelenbosch, M.P., Hommes, D.W., and van den Brink, G.R. (2014). Repurposing miltefosine for the treatment of immune-mediated disease? J. Pharmacol. Exp. Ther. *350*, 189–195. https://doi.org/10.1124/jpet.113.212654.

Verma, N.K., and Dey, C.S. (2004). Possible mechanism of miltefosine-mediated death of *Leishmania donovani*. Antimicrob. Agents Chemother. *48*, 3010–3015. https://doi.org/10.1128/AAC.48.8.3010-3015.2004.

Verma, N.K., Singh, G., and Dey, C.S. (2007). Miltefosine induces apoptosis in arsenite-resistant *Leishmania donovani* promastigotes through mitochondrial dysfunction. Exp. Parasitol. *116*, 1–13.

Wadhone, P., Maiti, M., Agarwal, R., Kamat, V., Martin, S., and Saha, B. (2009). Miltefosine promotes IFN-gamma-dominated anti-leishmanial immune response. J. Immunol. *182*, 7146–7154. https://doi.org/10.4049/jimmunol.0803859.

Walkey, C.J., Donohue, L.R., Bronson, R., Agellon, L.B., and Vance, D.E. (1997). Disruption of the murine gene encoding phosphatidylethanolamine N-methyltransferase. Proc. Natl. Acad. Sci. U.S.A. *94*, 12880–12885.

Wanderley, J.L., and Barcinski, M.A. (2010). Apoptosis and apoptotic mimicry: the *Leishmania* connection. Cell. Mol. Life Sci. *67*, 1653–1659. https://doi.org/10.1007/s00018-010-0291-0.

Wanderley, J.L., Pinto da Silva, L.H., Deolindo, P., Soong, L., Borges, V.M., Prates, D.B., de Souza, A.P., Barral, A., Balanco, J.M., do Nascimento, M.T., *et al.* (2009). Cooperation between apoptotic and viable metacyclics enhances the pathogenesis of Leishmaniasis. PLOS ONE *4*, e5733. https://doi.org/10.1371/journal.pone.0005733.

Wassef, M.K., Fioretti, T.B., and Dwyer, D.M. (1985). Lipid analyses of isolated surface membranes of *Leishmania donovani* promastigotes. Lipids *20*, 108–115.

Weingärtner, A., Kemmer, G., Müller, F.D., Zampieri, R.A., Gonzaga dos Santos, M., Schiller, J., and Pomorski, T.G. (2012). *Leishmania* promastigotes lack phosphatidylserine but bind annexin V upon permeabilization or miltefosine treatment. PLOS ONE *7*, e42070. https://doi.org/10.1371/journal.pone.0042070.

Weller, K., Artuc, M., Jennings, G., Friedrichson, T., Guhl, S., dos Santos, R.V., Sünder, C., Zuberbier, T., and Maurer, M. (2009). Miltefosine inhibits human mast cell activation and mediator release both *in vitro* and *in vivo*. J. Invest. Dermatol. *129*, 496–498. https://doi.org/10.1038/jid.2008.248.

Wieder, T., Orfanos, C.E., and Geilen, C.C. (1998). Induction of ceramide-mediated apoptosis by the anticancer phospholipid analog, hexadecylphosphocholine. J. Biol. Chem. *273*, 11025–11031.

Wieder, T., Reutter, W., Orfanos, C.E., and Geilen, C.C. (1999). Mechanisms of action of phospholipid analogs as anticancer compounds. Prog. Lipid Res. *38*, 249–259.

Wright, M.M., Howe, A.G., and Zaremberg, V. (2004). Cell membranes and apoptosis: role of cardiolipin, phosphatidylcholine, and anticancer lipid analogues. Biochem. Cell Biol. *82*, 18–26. https://doi.org/10.1139/o03-092.

Zaffalon, P.L., and Zumbuehl, A. (2011). BODP – A Versatile Reagent for Phospholipid Synthesis. Synthesis *5*, 778–782.

Zeisig, R., Rudolf, M., Eue, I., and Arndt, D. (1995). Influence of hexadecylphosphocholine on the release of tumor necrosis factor and nitroxide from peritoneal macrophages *in vitro*. J. Cancer Res. Clin. Oncol. *121*, 69–75.

Zhang, K., and Beverley, S.M. (2010). Phospholipid and sphingolipid metabolism in *Leishmania*. Mol. Biochem. Parasitol. *170*, 55–64. https://doi.org/10.1016/j.molbiopara.2009.12.004.

Zhang, K., Pompey, J.M., Hsu, F.F., Key, P., Bandhuvula, P., Saba, J.D., Turk, J., and Beverley, S.M. (2007). Redirection of sphingolipid metabolism toward de novo synthesis of ethanolamine in *Leishmania*. EMBO J. *26*, 1094–1104.

Zhang, K., Showalter, M., Revollo, J., Hsu, F.F., Turk, J., and Beverley, S.M. (2003). Sphingolipids are essential for differentiation but not growth in *Leishmania*. EMBO J. *22*, 6016–6026. https://doi.org/10.1093/emboj/cdg584.

Zheng, L., T'Kind, R., Decuypere, S., von Freyend, S.J., Coombs, G.H., and Watson, D.G. (2010). Profiling of lipids in *Leishmania donovani* using hydrophilic interaction chromatography in combination with Fourier transform mass spectrometry. Rapid Commun Mass Spectrom *24*, 2074–82.

Zufferey, R., Allen, S., Barron, T., Sullivan, D.R., Denny, P.W., Almeida, I.C., Smith, D.F., Turco, S.J., Ferguson, M.A., and Beverley, S.M. (2003). Ether phospholipids and glycosylinositolphospholipids are not required for amastigote virulence or for inhibition of macrophage activation by *Leishmania major*. J. Biol. Chem. *278*, 44708–44718. https://doi.org/10.1074/jbc.M308063200.

Index

www.ingramcontent.com/pod-product-compliance
Ingram Content Group UK Ltd.
Pitfield, Milton Keynes, MK11 3LW, UK
UKHW061959041026
15591UKWH00016B/1091

9 781910 190715